AF356157

Creation of a Low-Carbon Healthy Building Environment with Intelligent Technologies

Creation of a Low-Carbon Healthy Building Environment with Intelligent Technologies

Editors

Shi-Jie Cao
Dahai Qi
Junqi Wang

Basel • Beijing • Wuhan • Barcelona • Belgrade • Novi Sad • Cluj • Manchester

Editors
Shi-Jie Cao
School of Architecture
Southeast University
Nanjing
China

Dahai Qi
Department of Civil and
Building Engineering
Université de Sherbrooke
Sherbrooke
Canada

Junqi Wang
School of Architecture
Southeast University
Nanjing
China

Editorial Office
MDPI AG
Grosspeteranlage 5
4052 Basel, Switzerland

This is a reprint of articles from the Special Issue published online in the open access journal *Buildings* (ISSN 2075-5309) (available at: https://www.mdpi.com/journal/buildings/topical_collections/ Creat_Carb_Healt_Build_Envir_Intel_Tech).

For citation purposes, cite each article independently as indicated on the article page online and as indicated below:

Lastname, A.A.; Lastname, B.B. Article Title. *Journal Name* **Year**, *Volume Number*, Page Range.

ISBN 978-3-7258-2385-7 (Hbk)
ISBN 978-3-7258-2386-4 (PDF)
doi.org/10.3390/books978-3-7258-2386-4

Contents

About the Editors

Shi-Jie Cao

Shi-Jie Cao is a full Professor and the director of 'Center for Sustainable Built Environment, CSBE' at School of Architecture, Southeast University; the founding Director for Jiangsu Province Engineering Research Center of Urban Heat and Pollution Control; a Visiting Professor of Institute of Sustainability, Global Centre for Clean Air Research (GCARE), University of Surrey, UK. He firstly initiates the muti-disciplinary direction of Sustainable Built Environment System and awarded the National Science Fund for Distinguished Young Scholars in the area. He serves as Deputy Editor for Indoor and Built Environment, Associate Editor of Sustainable Cities and Society (Special Issue), Journal of Cleaner Production, Editorial Board members for 10 international journals. He is now the President of International Society of Built Environment (ISBE), Committee Members of Chinese Society for Urban Agglomeration, etc. He has published more than 100 SCI papers (H index:35; H10 index:74); 10 papers have been awarded as ESI articles; the BEST PAPER AWARD of "Energy and Buildings" in 2021; awarded as World's Top 2% Scientists in from 2020 and 2023 in the area of built environment design. Due to his leadership and contribution of industry application for sustainable built environment design and control, he won the First Prize of China Award for Science and Technology in Construction in 2021 (Ministry of Housing and Urban-Rural Development in China). He has been always interested and active in international academic collaborations, such as Harvard (Health Initiatives), Concordia (Sustainable Cities) etc. He edited 'Handbook of Ventilation Design in Built Environment' invited by IET publisher in UK. He has been served as committee member and given speech for national and international conferences more than 50 times. He serves as reviewers for international journals and assessor/panel members for International Funding Agencies, such as NSFC, RGC (Hongkong), UKRI, EPSRC, etc.

Dahai Qi

Dahai Qi received his PhD in Building Engineering from Concordia University in 2016. He was then a postdoctoral fellow at the Centre for Zero Energy Building Studies at Concordia University. He graduated from Nanjing Normal University (China) with a Bachelor's degree in Building Environment and Equipment Engineering in 2006 before proceeding to a Master's degree in Heating, Ventilation and Air Conditioning. In 2018, he joined the Department of Civil and Building Engineering at the Université de Sherbrooke. Prof. Qi has done extensive research on the building's thermal air flow and its effects on building safety, human health, building energy, and thermal comfort. He has been a major member of many national and international projects on energy management and building efficiency, and he has also done substantial work in the field of heat and the mass transfer of building physics, supported by the NSERC/HydroQuébec Industrial Research Chair, the National Institute of Standards and Technology (NIST, USA), and the Air Management and Control Association (AMCA, USA). He received the ASHRAE grant in 2013 (ASHRAE) and the "Chinese Government Award for Self-financed Students Abroad" in 2016.

Junqi Wang

Junqi Wang is an Associate Professor in the School of Architecture, Southeast University. He graduated from the City University of Hong Kong with a PhD degree (during which he visited and exchanged with the Department of Automation of Tsinghua University). He is engaged in researching intelligent sensing and the control of built environments and big data in urban environmental systems. He was a research associate at the Department of Building Science and Technology, City University of Hong Kong. Currently, he is the executive deputy director of the Jiangsu Urban Built Environment Heat and Pollution Co-control Engineering Research Center, an Editorial Board Member (Early Career) of the international journal *Sustainable Cities and Society* (JCR Q1), and a member of the Editorial Board for *Frontiers in Sustainable Cities*. He is also a member of the Intelligent Building Committee of CAHVAC and a member of the Intelligent Building Division of the IEEE Robotics and Automation Society.

Preface

The building sector is responsible for approximately one-third of global carbon emissions. As worldwide efforts to reduce carbon emissions are being increasingly implemented, there is a growing urgency to reduce carbon emissions from buildings. However, designing and operating building systems that provide a healthy and comfortable indoor environment while minimizing carbon emissions can be challenging. Therefore, creating a low-carbon yet comfortable building environment presents a significant research challenge that requires an interdisciplinary approach, incorporating expertise in regard to building environments, automatic control, architecture, and artificial intelligence. This reprint aims to highlight recent innovative research and developments in the building environment, energy, and intelligent technology environments.

Shi-Jie Cao, Dahai Qi, and Junqi Wang
Editors

Article

Quantifying the Influence of Different Block Types on the Urban Heat Risk in High-Density Cities

Binwei Zou [1], Chengliang Fan [1,2,*] and Jianjun Li [1,*]

[1] School of Architecture and Urban Planning, Guangzhou University, Guangzhou 510006, China; zoubinwei@e.gzhu.edu.cn
[2] State Key Laboratory of Subtropical Building and Urban Science, Guangzhou 510640, China
* Correspondence: chengliang.fan@gzhu.edu.cn (C.F.); lijianjun@gzhu.edu.cn (J.L.)

Abstract: Urbanization and climate change have led to rising urban temperatures, increasing heat-related health risks. Assessing urban heat risk is crucial for understanding and mitigating these risks. Many studies often overlook the impact of block types on heat risk, which limits the development of mitigation strategies during urban planning. This study aims to investigate the influence of various spatial factors on the heat risk at the block scale. Firstly, a GIS approach was used to generate a Local Climate Zones (LCZ) map, which represents different block types. Secondly, a heat risk assessment model was developed using hazard, exposure, and vulnerability indicators. Thirdly, the risk model was demonstrated in Guangzhou, a high-density city in China, to investigate the distribution of heat risk among different block types. An XGBoost model was used to analyze the impact of various urban spatial factors on heat risk. Results revealed significant variations in heat risk susceptibility among different block types. Specifically, 33.9% of LCZ 1–4 areas were classified as being at a high-risk level, while only 23.8% of LCZ 6–9 areas fell into this level. In addition, the pervious surface fraction (PSF) had the strongest influence on heat risk level, followed by the height of roughness elements (HRE), building surface fraction (BSF), and sky view factor (SVF). SVF and PSF had a negative impact on heat risk, while HRE and BSF had a positive effect. The heat risk assessment model provides valuable insights into the spatial characteristics of heat risk influenced by different urban morphologies. This study will assist in formulating reasonable risk mitigation measures at the planning level in the future.

Keywords: heat risk; spatial factors; local climate zone; XGBoost; block scale

Citation: Zou, B.; Fan, C.; Li, J. Quantifying the Influence of Different Block Types on the Urban Heat Risk in High-Density Cities. *Buildings* **2024**, *14*, 2131. https://doi.org/10.3390/buildings14072131

Academic Editor: Rafik Belarbi

Received: 7 June 2024
Revised: 6 July 2024
Accepted: 10 July 2024
Published: 11 July 2024

1. Introduction

1.1. Background

Climate change and urbanization have the potential to exacerbate the urban heat island (UHI) and extreme heat events, posing significant heat-related health risks in urban built environments [1]. Heat events can elevate the risk of various health issues, including heart, respiratory, and kidney diseases, as well as increase morbidity and mortality rates [2–4]. For example, the mortality rate in China has surged to 110–140% of the 1995–2014 level due to the rising heat risk [5]. It is crucial to assess the heat-related health risks in urban development [6,7]. In the summer of 2022, it was estimated that 9100 deaths in Germany were attributed to heat-related causes [8]. In India, an estimated 1116 people die from heat waves every year [9]. Many studies focus on exploring urban heat risk assessment, aiming to understand the risk pattern within communities and across cities. Urban heat risk assessment is crucial for revealing risk variation patterns and proposing heat risk mitigation strategies.

1.2. Literature Review

The interplay between urbanization, climate change, population growth, and aging can exacerbate health risks associated with high temperatures [1]. In general, heat risk is regarded as the level of health risk posed to humans by extremely high temperatures. This risk is particularly acute among socioeconomically and physically vulnerable groups, including the elderly, disabled individuals, infants, and the impoverished, as well as those who are frequently exposed to outdoor environments, such as outdoor activity participants and workers [10]. Empirical assessments of heat-related health risk have primarily relied on two methods. One approach involves identifying the spatial distribution of the risk by establishing a regression model. For example, a distributed lagged nonlinear model was used to evaluate the heat risk of 51 regions in Seoul [11]. Similarly, a weighted regression model can be used to determine areas of high heat risk in different regions of Singapore, taking into account UHI intensity and the proportion of the elderly population [12]. Another popular method for assessing heat risk is the "Crichton Risk Triangle" framework, which integrates societal, population, economic, and urban morphology aspects of a city. This framework considers the following three key dimensions: hazard, exposure, and vulnerability. Hazard encompasses elements that have the potential to trigger risks, with extremely high temperatures and heatwaves significantly increasing the likelihood of heat-related hazards [13]. Exposure primarily concerns the populations that are subjected to high temperatures, particularly the individuals engaging in outdoor activities [14]. Vulnerability serves as an indicator of an individual's resilience to high temperatures, which is influenced by diverse factors such as encompassing income, education, and age level [15]. Many studies on heat risk assessments have been conducted at the administrative units. For example, urban heat island intensity, population density, and elderly population ratio were used to assess heat risk of Singapore [16]. Results are generally reported at the administrative unit level, limiting their ability to provide block-scale information on the spatial distribution of risk levels. Similarly, some studies have conducted risk assessments at the administrative level [11,17], but few of them have elucidated the relationship between urban morphology and heat risk at block. Block-scale assessments are crucial as they represent specific urban block planning and are the directly perceived areas by human inhabitants.

Urban morphology significantly influences the microclimate by altering local wind patterns and solar radiation, subsequently affecting temperature distributions and heat risk within various block scales [18,19]. Local Climate Zones (LCZ) is a powerful tool to represent different block types [20], which can be used to quantify the association between urban morphology and the thermal environment [21–23]. For example, the heat characteristics and their spatiotemporal patterns can be thoroughly analyzed based on the categorization of a LCZ [24]. Remote sensing data revealed UHI effect intensity across different LCZs, with built-up areas intensifying UHI, while land-cover areas (e.g., LCZ D and LCZ G) mitigate it. This establishes a linkage between the thermal environment and LCZs in urban areas, thereby facilitating the application of an LCZ in heat risk assessments [25]. Similarly, an LCZ map has been used to quantify heat risk level in different LCZ types. Results indicated that at least 60% of LCZs 1-5 were designated as high-risk areas, while LCZ 6 was deemed to be more suitable for implementing measures to mitigate heat hazards [26]. Additionally, an LCZ map can be integrated with urban population mortality rates to assess heat-related health risk. The findings revealed that densely built-up areas exhibit higher risk levels compared to land-cover areas [27]. In summary, the LCZ, which aggregates various urban morphology indicators, can facilitate a comprehensive understanding of the relationship between different block types and heat risk.

Revealing the correlation between urban morphology factors and heat risks is essential for developing effective mitigation strategies to reduce these risks. The complex spatial morphology significantly impacts the urban thermal environment, rendering a singular linear indicator insufficient to capture its complex relationship with heat risks. Various machine learning algorithms, such as neural networks [28] and random forests [29], provide invaluable tools for exploring the nonlinear relationships among multiple morphology factors. These algorithms have been widely used to investigate the correlation between urban areas, the urban thermal environment, and the health of residents [30,31]. For example, the XGBoost algorithm was employed to explore the correlation between the morphology factors of European cities and the urban thermal environment. The findings indicate that approximately two-thirds of temperature variations within cities can be explained by urban morphological features [32]. Additionally, the random forest algorithm was used to investigate the influence of urban morphology on Land Surface Temperature (LST). Results showed that high building density positively affects LST, while the floor area ratio exhibited a negative impact [33]. Previous studies have successfully elucidated the relationship between urban structure and its thermal environment, there remains a gap in methodology for interpreting how different urban spatial factors influence heat risk.

1.3. Research Objectives and Structure

To address the gap, this study proposes a heat risk assessment model to investigate heat risk at various block types. Firstly, urban morphology factors were obtained using multi-source data to map LCZs. The proposed model for assessing heat risk was demonstrated in Guangzhou, a high-density city in China. Furthermore, the correlation between the heat risk level and different LCZs was quantified through spatial correlation. Finally, the XGBoost model was applied to interpret the sensitivity of urban morphology on block-level heat risk. The primary novelties of this study can be summarized as follows: (1) proposing a novel heat risk assessment model by considering urban morphology factors; (2) revealing the correlation between urban morphology factors and heat risk, which facilitates the development of mitigation strategies to reduce high risk areas; and (3) quantifying the spatial autocorrelation of different heat risk levels and identifying their specific reasons these variations at a block scale. This study provides support for explaining the relationship between heat risk and urban morphology factors at block scale, thereby facilitating the development of healthy and sustainable building designs.

This paper introduces the study area and data sources in Section 2. Section 3 explains the calculation methods for risk indicators and the heat risk assessment methodology. Section 4 presents the results. Sections 5 and 6 provide the conclusions and discussion, respectively.

2. Study Area and Data

2.1. Study Area

Guangzhou is located in the core of the Greater Bay Area in China (Figure 1), spanning between 112°57′~114°3′ E longitude and 22°26′~23°56′ N latitude. It lies within a subtropical humid climate zone, characterized by a hot summer and a warm winter, with a monthly average temperature ranging from 14 °C to 28 °C. With a population of approximately 18.81 million inhabitants and an urbanization rate of 86.48% [34], Guangzhou serves as a prototypical high-density city that has undergone rapid urbanization, resulting in the significant expansion of its built-up area. This expansion has given rise to considerable heat-related health hazards. Therefore, examining the heat risk of various block units is paramount to developing effective mitigation strategies and fostering a more sustainable urban environment.

2.2. Data Collection and Pre-Processing

2.2.1. Dataset for Urban Morphology Factors

This study integrated various data sources for Guangzhou, including urban morphology, vegetation coverage, water body coverage, and land cover information (Table 1). First, key urban morphology indicators such as Sky View Factor (SVF), Height of Roughness Elements (HRE), and Building Surface Fraction (BSF) were calculated using the building footprints and heights obtained from the architectural data of Guangzhou City. Then, the Impervious Surface Fraction (ISF) and Pervious Surface Fraction (PSF) were derived using vegetation coverage data and water body coverage data. Lastly, the land cover was classified using the land cover data.

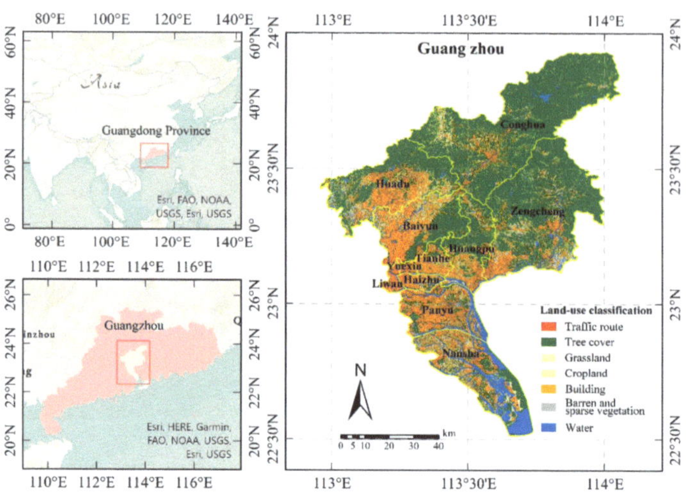

Figure 1. The location of Guangzhou, China.

Table 1. Data acquisition and their application in this study.

Theme	Source	Period	Resolution	Application
Building footprint	https://www.resdc.cn/Default.aspx (accessed on 4 December 2023)	2019	-	LCZ mapping
Water cover	https://www.openstreetmap.org/ (accessed on 4 December 2023)	2020	-	LCZ mapping
Green cover	https://www.openstreetmap.org/ (accessed on 4 December 2023)	2020	-	LCZ mapping
Land use [35]	https://zenodo.org/ (accessed on 5 December 2023)	2022	1 m	LCZ mapping
Landsat-8	https://www.usgs.gov/ (accessed on 5 December 2023)	2015–2020	30 m	Hazard/Exposure calculation
Population density	https://www.worldpop.org/ (accessed on 8 December 2023)	2020	100 m	Hazard/Exposure calculation
Population density (>65)	https://www.worldpop.org/ (accessed on 8 December 2023)	2020	100 m	Hazard/Exposure calculation
Night-time Light	http://59.175.109.173:8888/app/login.html (accessed on 10 December 2023)	2019	130 m	Hazard/Exposure calculation
Anthropogenic heat flux	https://dataverse.harvard.edu/ (accessed on 7 February 2024)	2019	500 m	Hazard calculation
Mobile signaling data	China Unicom mobile phone	July 2022	-	Residents' activity preference

2.2.2. Dataset for Heat Risk Assessment

In this study, some risk indicators were calculated using Landsat-8 data. Specifically, the TIRS10 band (thermal infrared band) of Landsat-8, spanning from July to August between 2015 and 2020, with the atmospheric correction algorithm, provided the summer average LST, which represents the hazard dimension. In addition, the Near-infrared (NIR) and Red-band of Landsat-8 were utilized for calculating the Normalized Difference Vegetation Index (NDVI). Meanwhile, the green band (Green), NIR, and Short-Wave Infrared band one (SWIR1) of Landsat-8 were used to calculate the Enhanced Water Index (EWI). Population density (PD) represents the exposure dimension. Both the density of population over 65 years old (OPD) and Nighttime Light data (NTL) were used to calculate the vulnerability dimension. The selection of these heat risk indicators were based on the previous research [26]. Due to differences in resolution, all data were resampled to achieve a 30 m resolution for calculating risk values.

3. Methodology

3.1. Framework for Heat Risk Assessment

Heat risk assessment is a comprehensive outcome that integrates urban multi-source data preprocessing, LCZs mapping, heat risk mapping, and spatial patterns analysis. Figure 2 presents the workflow of the heat risk assessment. Firstly, urban morphology factors, building data, and land cover data are obtained to generate an LCZ map, which is subsequently mapped using a GIS-based approach. Then, three heat risk assessment indicators were selected, including heat hazard, heat vulnerability, and heat exposure. Heat hazard reflects environmental temperature severity, influenced by the urban environment, which is crucial for heat risk assessments. When daily maximum temperatures exceed 35 °C, it is considered to be high heat, posing potential negative effects on human health. [36]. Heat vulnerability reflects an individual's physical state and capacity to cope with heat risks, influencing resistance, response, and recovery [37–39]. Meanwhile, heat exposure is usually described as the presence of people, environmental functions, infrastructure, and cultural assets in areas and contexts susceptible to adverse impacts [40]. These three dimensions are then multiplied to generate the overall heat risk map. Finally, spatial autocorrelation and the XGBoost-SHAP methods are employed to investigate the relationship between urban spatial morphology and heat risk levels.

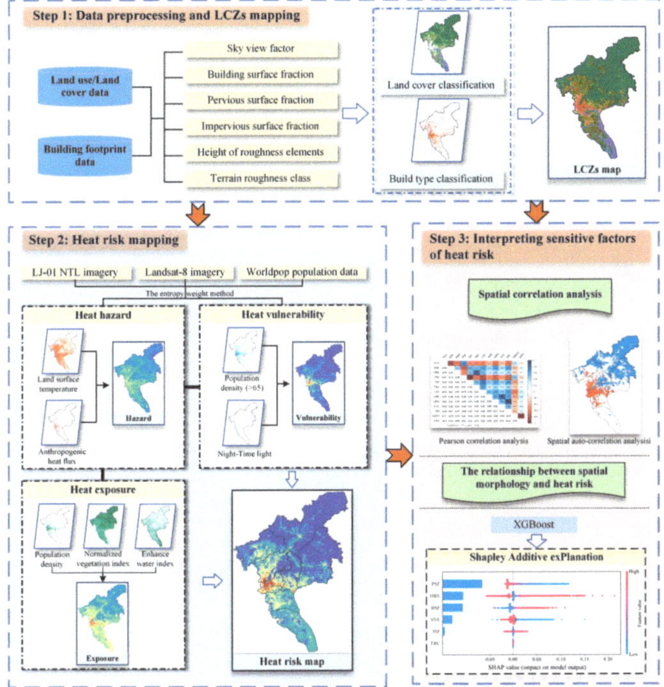

Figure 2. Workflow of the heat risk assessment.

3.2. Improved Heat Risk Assessment Model

The proposed GIS-based risk assessment model is based on Crichton's risk triangle [41], including hazard, vulnerability, and exposure indicators. Risk values in different blocks were calculated using Equation (1). The natural breakpoint method is used to categorize heat risks into seven levels.

$$\text{HRK} = \text{hazard} \times \text{vulnerability} \times \text{exposure} \tag{1}$$

The entropy weight method is used to determine the weight of each indicator. Range normalization transforms multi-source data into a unified range by using Equations (2) and (3) [42]. Information entropy is defined in (Equations (4) and (5)), and the weight is calculated in (Equation (6)). These weights are used to compile indicators within layers, as shown in (Equation (7)).

$$X = \begin{bmatrix} x_{11} & \cdots & x_{1n} \\ \vdots & \ddots & \vdots \\ x_{m1} & \cdots & x_{mn} \end{bmatrix} \tag{2}$$

$$Y_{ij} = 0.1 + \frac{X_{ij} - min(X_i)}{max(X_i) - min(X_i)} \times (0.9 - 0.1) \tag{3}$$

where X is the original indicator matrix composed of m research units and n indicators; X_{ij} is the original value of the i-th research unit and the j-th indicator; and Y_{ij} are the standardized values.

$$P_{ij} = \frac{Y_{ij}}{\sum_{i=1}^{n} Y_{ij}}, \, i = 1, \ldots, n, j = 1, \ldots, m \tag{4}$$

$$E_j = -\ln(n)^{-1} \sum_{i=1}^{n} P_{ij} \ln P_{ij} \tag{5}$$

where P_{ij} is the variation of the indicator size; E_j is information entropy; and n represents n research units.

$$w_i = \frac{1 - E_j}{k - \sum E_j} (j = 1, 2, \ldots, m) \tag{6}$$

$$S_i = \sum_{j=1}^{m} w_j * x_{ij} \tag{7}$$

where w_i represents indicator weight and S_i represents the comprehensive indicator value.

3.2.1. Heat Hazard

Land Surface Temperature (LST) was chosen as the primary indicator of heat hazards. In addition, anthropogenic heat, resulting from human activities like industrial labor, transportation, and metabolism raises environmental temperature, contributing to microclimate differences at the block scale [43]. Therefore, anthropogenic heat was considered in the hazard indicators and data from a previous publication [44]. This study calculated the summer average LST values using the radiative transfer method, known for accurate temperature estimation [45,46]. Landsat-8 imagery were available from Google Earth Engine of summer days from July to August between 2015 and 2020.

The two indicators were standardized within the range of 0.1–0.9. The standardized indicators were then multiplied by their corresponding weights to obtain the heat hazard value (Equation (8)).

$$Hazard = LST \times w_L + AHF \times w_A \tag{8}$$

3.2.2. Heat Vulnerability

The elderly is less resilient to high temperatures and more susceptible to related illnesses. Consequently, population density (>65 years old) (OPD) was chosen as a vulnerability indicator. Furthermore, Night-Time Light (NTL) positively correlates with economic development [47,48], which can indirectly reflect income levels [49,50] and individual resilience to heat risks. Both variables were standardized to a range of 0.1–0.9. The standardized indicators were then multiplied by their respective weights to calculate the vulnerability value (Equation (9)).

$$Vulnerability = OPD \times w_N - NTL \times w_o \tag{9}$$

3.2.3. Heat Exposure

Population density (PD) data were included as one of indicators of heat exposure, which can be obtained from the WorldPop website. Vegetation coverage, as reflected by the Normalized Difference Vegetation Index (NDVI), provides shade and helps reduce environmental temperatures, while water bodies absorb heat, also mitigating heat risks. These three indicators were standardized to a range of 0.1–0.9. Subsequently, the standardized indicators were multiplied by their corresponding weights to calculate the exposure value (Equation (10)).

$$Exposure = PD \times w_P - NDVI \times w_N - EWI \times w_E \tag{10}$$

NDVI serves as a vegetation layer indicator (Equation (11)) [51], and EWI clarifies water body boundaries using remote sensing (Equation (12)) [52]. Therefore, PD, NDVI, and EWI were chosen as heat exposure indicators.

$$NDVI = \frac{NIR - R}{NIR + R} \tag{11}$$

$$EWI = \frac{Green - (NIR + SWIR1)}{Green + (NIR + SWIR1)} \tag{12}$$

where NIR is the near-infrared band; R is the red band; $Green$ is the green band; and $SWIR1$ is the short-wave infrared band one.

3.3. Block Types Classification

LCZ types were classified into 17 categories that encapsulate morphology characteristics and microclimate changes, consisting of ten built-up areas (LCZ 1~LCZ 10) and seven land cover areas (LCZ A~LCZ G). Each category possesses distinct surface characteristics that impact the urban microclimate. Previous studies have established a unified framework for evaluating urban zoning based on diverse surface characteristics [53,54]. The GIS-based classification method has proven to be effective in urban climate zone classification, accurately reflecting the 3D morphology of different block types, thereby justifying its application in generating the LCZ map [55]. The key processes for LCZ mapping are as follows: (1) determining the LCZ grid resolution; (2) assessing urban spatial morphology; and (3) utilizing fuzzy classification and majority voting methods to categorize LCZs.

3.3.1. Grid Resolution

Establishing an appropriate boundary scale significantly enhances LCZ classification accuracy. Although urban morphology can be segmented, the thermal climate remains stable within a specific area, influenced by factors such as surface roughness, architectural geometric properties, and weather conditions. In general, an LCZ's diameter ranges from 400 to 1000 m grid resolution [56]. Based on previous research, a 240 m × 240 m grid resolution has been successfully applied for LCZ classification in Guangzhou [57,58]. Consequently, this study adopts this resolution (240 m × 240 m).

3.3.2. Assessing Urban Morphology Factors

The key parameters that define urban canyon geometry include H (mean building height on both street canyon sides), W (horizontal canyon extent), and L (canyon length). These parameters are used to calculate BSF [33], HRE, and SVF. In addition, land cover interacts with the atmosphere, altering thermal conditions within the canyon. Surface characteristics play a crucial role in influencing latent heat flux, which in turn affects temperature dynamics [59]. Land cover can be categorized into impermeable surfaces (e.g., asphalt and concrete) and permeable surfaces (e.g., soil, water, and vegetation). These can be quantitatively represented by the fraction of impermeable surface (ISF) and the fraction of permeable surface (PSF), as shown in Table 2.

Table 2. Calculation method for urban morphology factors.

Property	Methods	Formulas	Description
SVF	SAGA GIS	$SVF = \frac{S_{sky}}{S_{total}}$ [60]	where S_{sky} indicates the visible sky area in the model space, m^2; and S_{total} indicates the total sky in the model space, m^2.
BSF	Building footprints, ArcGIS pro	$BSF = \frac{S_b}{S_{total}}$ [60]	where S_b indicates the total building footprint area, m^2; and S_{total} indicates the total block area, m^2.
HRE	Building height, ArcGIS	$HRE = \frac{\sum_i^n S_i * H_i}{S_{total}}$ [60]	where S_i indicates the building footprint area, m^2; H_i indicates the building height, m; and n indicates the count of typical buildings within a block.
PSF	Green cover, water cover, ArcGIS pro	$PSF = \frac{S_p}{S_{total}}$ [60]	where S_p indicates the total pervious area, m^2; and S_{total} indicates the total block area, m^2.
ISF	ArcGIS pro	$ISF = 1 - (BSF + PSF)$ [61]	where S_i indicates the total impervious area, m^2; and S_{total} indicates the total block area, m^2.
TRC	Davenport classification of terrain roughness [62]	$Z_0 = f_0 \overline{Z_H}$	where Z_0 represents the surface roughness length; f_0 represents the empirical coefficient; and $\overline{Z_H}$ represents the height of the surface elements [63].

3.3.3. LCZ Mapping and Validation

Remote sensing- and GIS-based methods are widely employed for LCZ classification, leveraging urban spatial data [64]. Furthermore, GIS-based methods offer valuable function to further investigate the relationship between heat risk and urban morphology factors [65]. This study adopts a GIS-based method to classify different LCZ types, distinguishing between built-up areas and land cover areas based on building footprints [56]. Specifically, land use data were utilized to identify land cover areas, while urban morphology data were used to map built-up areas. In addition, the fuzzy classification and majority voting methods were employed to classify built-up areas, as demonstrated in Figure 3. Figure 3a shows an example of a spatial morphology element, showcasing the application of a trapezoidal linear function to determine the fuzzy membership of each LCZ type. Figure 3b demonstrates a map of the spatial morphology element, while Figure 3c shows each neighborhood block has a dominant LCZ type. Moreover, this study utilized an area-based assessment method to determine the proportion of classified categories within the total area. This information is then leveraged to construct a confusion matrix, enabling the calculation of the overall accuracy for the LCZ classification.

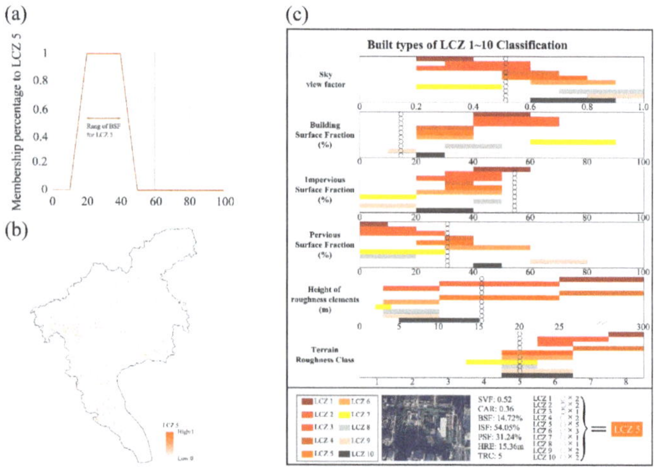

Figure 3. The process of LCZ mapping. (**a**) Linear fuzzy membership percentage. (**b**) Example of BSF in LCZ 5. (**c**) Built-up area classification.

3.4. Spatial Correlation Analysis

3.4.1. Pearson's Correlation

Pearson's correlation is used to quantify the linear association between urban morphology and heat risk factors [66]. Pearson's correlation coefficient varies from -1 to 1 (Equation (13)), and r = 1 indicates a perfect positive correlation. r = -1 indicates a perfect negative correlation.

$$r = \frac{\sum_{i=1}^{n} \left(X_i - \overline{X}\right)\left(Y_i - \overline{Y}\right)}{\sqrt{\sum_{i=1}^{n}\left(X_i - \overline{X}\right)^2}\sqrt{\sum_{i=1}^{n}\left(Y_i - \overline{Y}\right)^2}} \tag{13}$$

where X_i and Y_i represent the i-th observation value of the two variables, respectively; and $\overline{X}$ and $\overline{Y}$ represent the mean of the two variables, respectively.

3.4.2. Spatial Autocorrelation Method

Spatial autocorrelation is the similarity between adjacent data values resulting from spatial interaction and diffusion. This dependence weakens or disappears when the distance between the data values increases. Moran's I is an indicator used to measure spatial autocorrelation by comparing variable values in neighboring regions [67]. In this study, the Moran's I-based univariate local indicator of spatial autocorrelation (LISA) detects spatial clustering of heat risk, as shown in Equation (14):

$$I = \frac{N\sum_{i=1}^{n}\sum_{j=1}^{n} w_{ij}(x_i - \overline{x})(x_j - \overline{x})}{\left(\sum_{i=1}^{n}\sum_{j=1}^{n} w_{ij}\right)\sum_{i=1}^{n}(x_i - \overline{x})^2} \tag{14}$$

where N represents the number of observations (points or polygons); w_i and w_j represent the variable values at the locations i and j, respectively; w_{ij} represents the weight indicating the relationship between location i and location j; and $\overline{x}$ represents the mean of all observation values. Moran's I value ranges from +1 to -1. An amount of +1 indicates strong positive spatial autocorrelation, 0 indicates perfect randomness, and -1 suggests dispersion.

Bivariate Moran's I detects the spatial autocorrelation between urban morphology and heat risk, interpreting their spatial interrelationship. It consists of the following two patterns: clustering (high–high, low–low), and dispersion (high–low, low–high). The calculation method is shown in Equation (15) [68].

$$I = \frac{x_i - \overline{x}}{\sum_{i}(x_i - \overline{x})^2}\sum_{j} w_{ij}\left(x_j - \overline{x}\right) \tag{15}$$

where x_i and x_j are the value of the attributes x at location i and j; $\overline{x}$ is the average value of the census tract; and w_{ij} is the spatial weight matrix.

3.5. Interpretable Machine Learning Model

The XGBoost machine learning model, renowned for its resistance to nonlinearity, inherent feature selection, and interpretability, was chosen for this study to analyze urban morphology indicators and heat risk levels [69]. In this study, the XGBoost model was employed, with indicators of urban morphology serving as the independent variables and urban heat risk values serving as the dependent variables. The XGBoost model was configured with a learning rate of 0.1 and a maximum tree depth of 3. To train the XGBoost model, 70% of the data were used as the training set, and 30% as the test set. The model's performance was assessed through metrics including coefficient of determination (R^2), Mean Squared Error (MSE), and Root Mean Squared Error (RMSE).

The interpretability of the "black box model" is crucial for this study. Shapley Additive Explanations (SHAP), based on game theory, provide post hoc interpretation, elucidating the outputs of any machine learning model. The core principle of SHAP is to compute the marginal contribution of features to the model's output, allowing for the interpretation of the "black box model" on both global and local levels. Previous research has demonstrated that SHAP's interpretation of the XGBoost model yields spatial effect results comparable to those of the Spatial Lag Model and Multi-scale Geographically Weighted Regression [70]. Therefore, this study employs SHAP for interpretation. For each predicted sample, the SHAP model generates a "Shapley value", which represents the sum of the values assigned to each feature.

4. Results

4.1. Classification Results of LCZ Types

In this study, the overall accuracy of LCZ mapping using the GIS-based method was 85.5%, which is fulfill the high-precision requirement for assessing heat risk [26]. The spatial pattern of LCZ results is shown in Figure 4a. Land cover areas accounted for 84.4% of all LCZs. LCZ A-B and LCZ C-D were the most common land covers, accounting for 66.6% and 20.5%, respectively (Figure 4b). They were mainly concentrated in northern hilly areas. Built-up areas accounted for 15.6% of all LCZs. LCZ 5 and LCZ 2 were the dominant type, accounting for 24.7% and 20.8%, respectively. They were concentrated in the central areas of Guangzhou city. The rest of the LCZ types and their proportion are ranked as follows (Figure 4b): LCZ 10, LCZ 9, LCZ 4, LCZ 6, LCZ 8, LCZ 1, and LCZ 3.

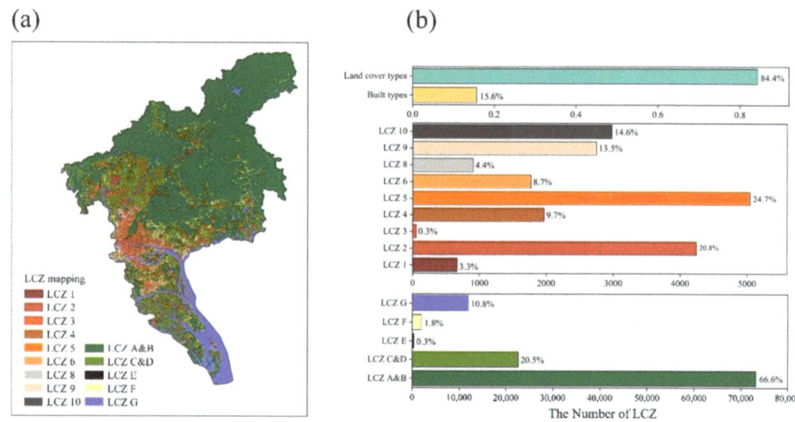

Figure 4. GIS-based block type results: (**a**) LCZ mapping; (**b**) proportion of each block type.

4.2. Analysis of Heat Risk Distribution under Different Block Type

4.2.1. Spatial Distribution of Hazard, Vulnerability, and Exposure Indicators

Three risk indicators were categorized into the following seven levels using the natural breaks method in ArcGIS Pro: very low, lower, low, medium, high, higher, and very high. The spatial distribution of the three indicators is shown in Figure 5. Figure 5a shows that hazard values range from 0.13 to 0.79. High-risk areas (i.e., high and higher levels) were mostly clustered in the west and scattered in the east. These high-risk areas share features like impermeable and exposed surfaces, receiving abundant solar radiation, resulting in rapid surface warming. Figure 5b shows that vulnerability values range from 0.10 to 0.90. High-risk areas appeared in the area around Yuexiu district, the historic center in Guangzhou. This area boasts a dense elderly population that is particularly susceptible to the adverse effects of high temperatures. Other districts exhibit clusters of medium-to low-risk areas, while the northern hills and southern riverine regions are marked by low-risk areas. Moreover, Figure 5c indicates that exposure values span from 0.10 to 0.75. High-risk levels are also found in Yuexiu, since it is also composed of numerous built-up areas with an insufficient number of shaded green spaces. Similarly, the Pearl River area exhibited similar risks. The rest of the area shows a mix of low-risk and lower-risk levels.

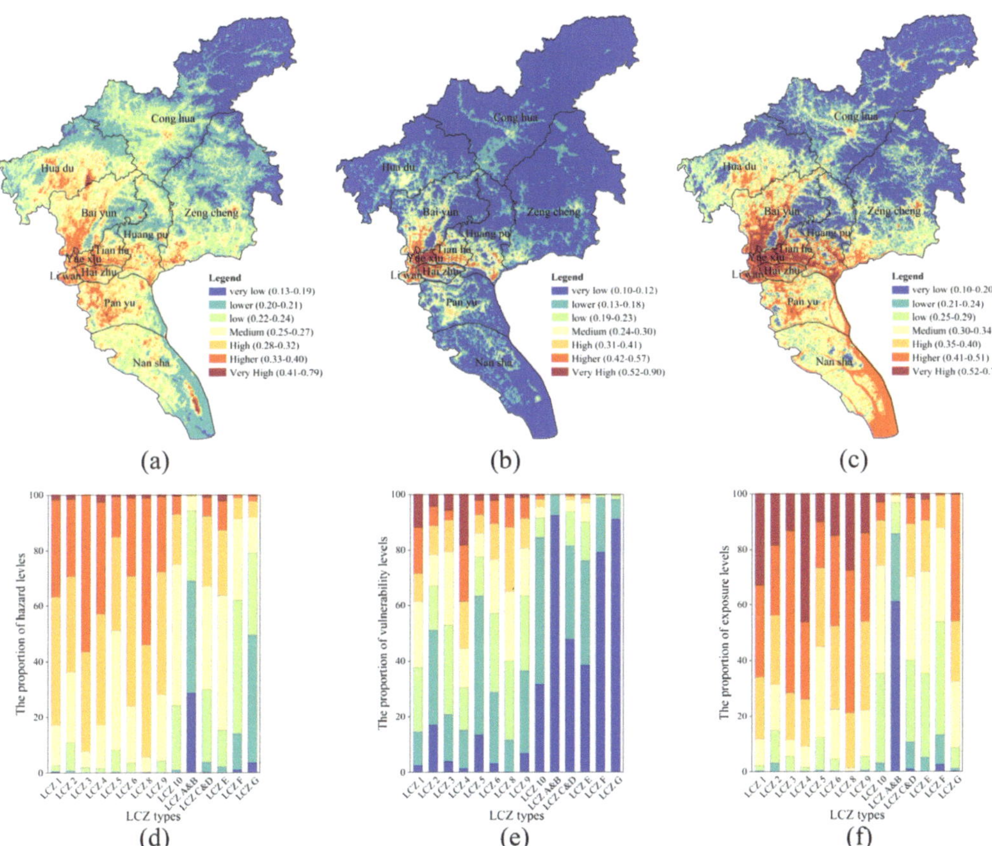

Figure 5. The spatial distribution of (**a**) hazard, (**b**) vulnerability, (**c**) exposure, and the proportion of (**d**) hazard, (**e**) vulnerability, and (**f**) exposure with different LCZ types.

4.2.2. Spatial Distribution of Heat Risk Levels

The distribution of each LCZ corresponding to the heat risk level is shown in Figure 6. It indicates that heat risk levels were higher in the city center compared to the suburbs. Very high–high-risk areas and very low–low-risk areas were primarily distributed in the central and northern areas. Very high-risk areas were concentrated in central of city (Figure 6a), which can be attributed to the dense population and a significant elderly population residing in this area. Compared to the heat hazard in Figure 5a, the risk may be significantly reduced if social factors are incorporated using the proposed assessment model. The traditional heat risk assessment method may overestimate high heat risk since it only considers natural factors. However, the residents in these areas had a strong resilience to high temperatures when considering social factors.

In addition, built-up areas posed a higher heat risk than land covers, as shown in Figure 6b. For example, LCZ 4 had the most very high- and high-risk areas, accounting for 15.48% and 21.07%, respectively. Conversely, low-risk areas were the most in LCZ A and B, accounting for 93.81%, followed by LCZ F at 57.98%. This is evident in the higher heat risk values for built-up types in comparison to those for land covers, as shown in Figure 7. LCZ 4 had the widest risk distribution, with mean heat risk values of 0.21, followed by LCZ 1 and LCZ 8. In general, LCZs 1–4, which have high building and population density, are more likely to be affected by heat risks.

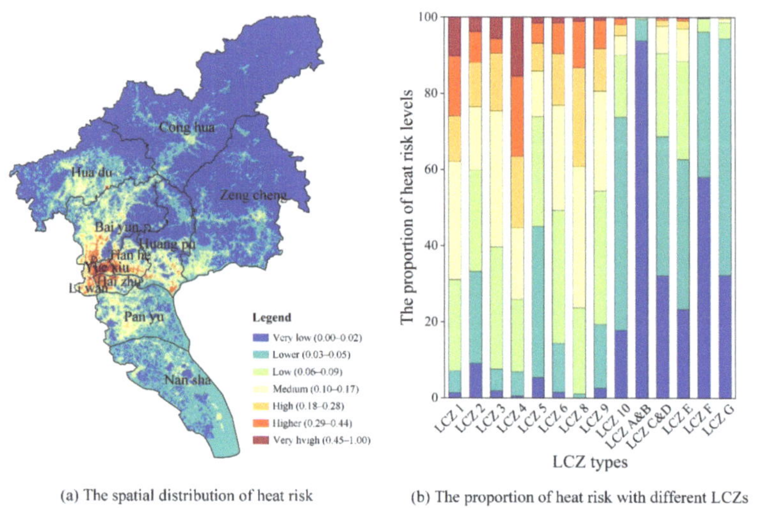

(a) The spatial distribution of heat risk   (b) The proportion of heat risk with different LCZs

Figure 6. The spatial distribution and the proportion of heat risk within different LCZs.

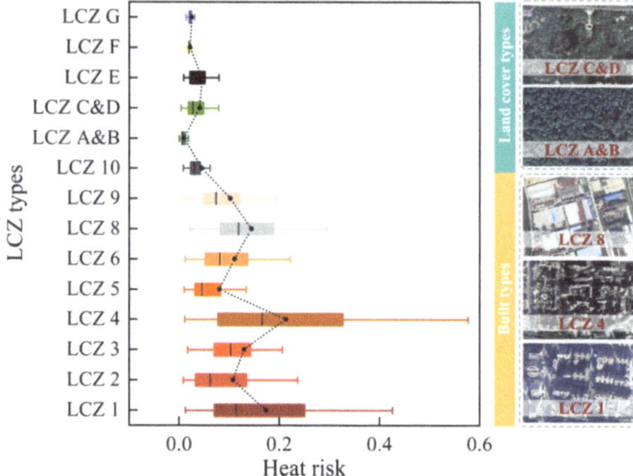

Figure 7. Boxplot diagrams of the range of heat risk within different LCZs.

4.2.3. Spatial Autocorrelation between Different Heat Risk Levels

The local indicator of spatial autocorrelation (LISA) was used to understand the spatial clustering of heat risk in Guangzhou. It explains the risk distribution in different built-up areas (Figure 8). Moran's I was 0.933, the z-score was 667.934, and the p-value was less than 0.01, indicating a significant positive correlation in the clustering of heat risk. High-risk areas (high–high clusters) were concentrated in the central area of Guangzhou. Low-risk areas (low–low clusters) were concentrated in the northern mountainous areas and southern farmland areas. Compared to the suburbs, urban built-up areas were more likely to exhibit heat risks related to high temperatures. This is true especially for the built-up areas in LCZs 1–5, which produce a clustering effect of heat risk. Parks and adjacent urban built-up areas have a variety of clusters, such as the low–high clusters in the city center. This proves that in these areas, the risk value is lower due to the abundant vegetation coverage.

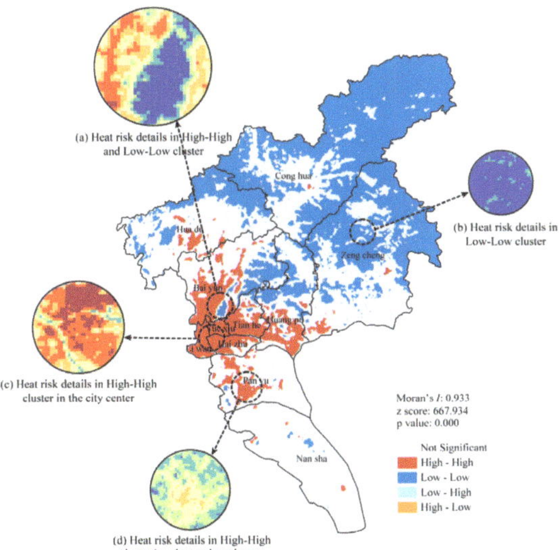

Figure 8. Spatial relationship of heat risk using LISA method.

4.2.4. Relationship between Heat Risk and People's Activity Preferences

This study used the signaling data of China Unicom mobile phones in July 2022 to explore people's activity preferences in the following three spaces: parks, shopping malls, and pedestrian streets, as shown in Figure 9. Figure 10a indicates that the hourly number of visitors in the pedestrian street (LCZs 2) are far greater than that in the park (LCZs A and B) and shopping mall (LCZs 4), with a higher heat risk level. Compared to park and shopping center visitors, the frequency of pedestrian street visitors shows greater variability and periodicity. The number of visitors to pedestrian streets decreases on weekdays and significantly increases on weekends, while parks and shopping centers are almost unaffected. Similarly, the average hourly number of elderly visitors in the street was greater than that in the park and the shopping mall (Figure 10b). Overall, open activity spaces classified as "built types", such as pedestrian streets, attract larger crowds, including the elderly. Secondly, parks and green spaces categorized as "land cover types" provide shaded areas for cooling and recreation. In addition, a male-to-female ratio was above 1 (Figure 10c). Men favored parks, preferring low-risk areas like LCZ A and B. Conversely, women preferred shopping malls and streets despite there being a high heat risk. To safeguard women's health, measures addressing their severe heat risk are needed. Future considerations should include strategies to mitigate heat risk in outdoor spaces, particularly in pedestrian streets [71].

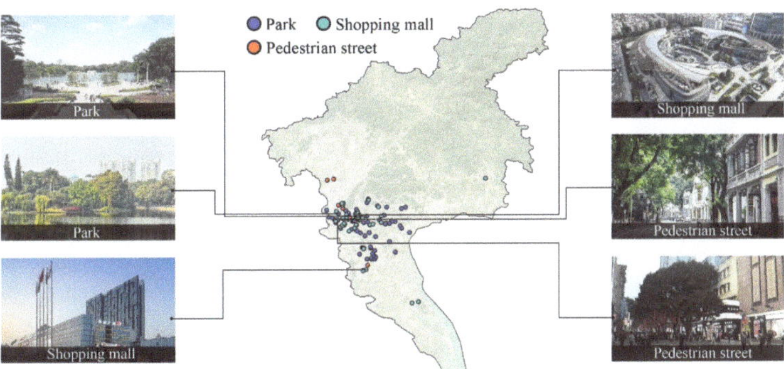

Figure 9. Three types of space in study area.

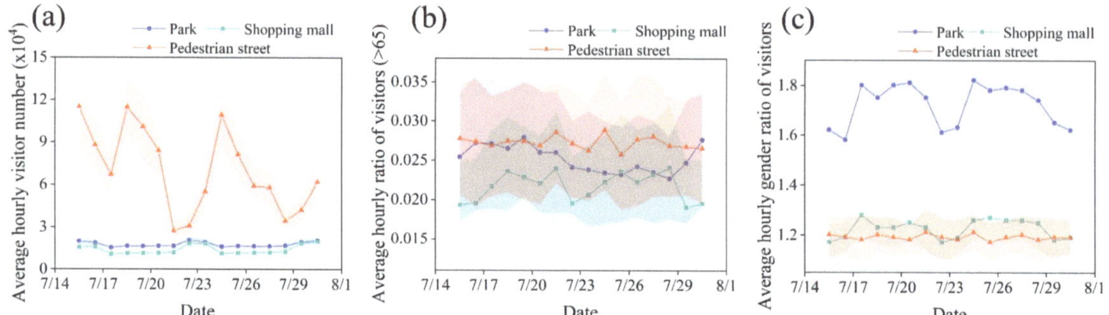

Figure 10. Different outdoor spaces that people visit depending on their activity preferences and (**a**) the number of visitors, (**b**) the density of visitors (>65), and (**c**) the gender of visitors.

4.3. Relationship between Urban Morphology and Heat Risk

4.3.1. Sensitivity Analysis of Different Urban Morphology Factors

The sensitivity of urban morphology factors to urban heat risk indicators was quantified using Pearson's correlation heatmap, as shown in Figure 11. The map can also be used to analyze the appropriateness of exploring urban heat risk through the analysis of urban morphology. The results show that the *p*-value between all indicators was <0.001, and the confidence interval was 95%, indicating that urban morphology and heat risk factors are statistically significant. In terms of heat risk indicators, LST, AHF, OPD, NTL, PD, and EWI exhibited consistency. There was a negative correlation between SVF and PSF and a positive correlation between BSF, HRE, ISF, and TRC, which is consistent with the heat risk spatial clustering in Section 4.3.2. In contrast, NDVI exhibited an opposite situation. The relationship between LST and urban morphology was the highest overall, with an absolute average Pearson's relationship coefficient of 0.64. The highest coefficient appeared between LST and the urban morphology indicator PSF, reaching −0.79. The relationship between EWI and urban morphology was the lowest overall, with an absolute average Pearson's correlation coefficient of 0.15. The lowest coefficient appeared between EWI and morphology indicator HRE, with a coefficient of only 0.092. While urban morphology may not directly influence heat exposure and vulnerability, it can exert an indirect impact by influencing heat risk factors. For instance, higher BSF can accommodate a larger population, while higher PSF can mitigate the surrounding temperatures.

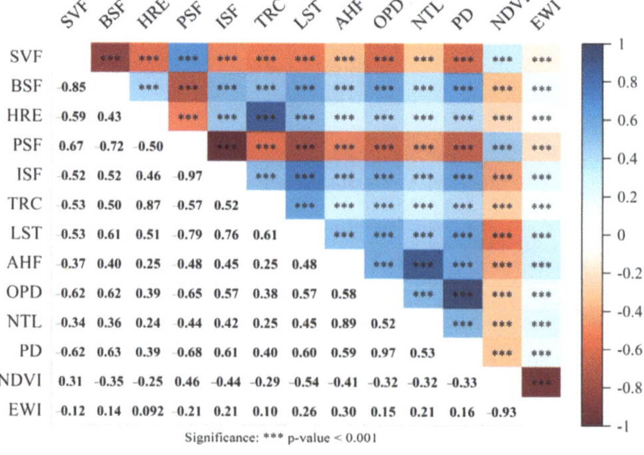

	SVF	BSF	HRE	PSF	ISF	TRC	LST	AHF	OPD	NTL	PD	NDVI	EWI
SVF		***	***	***	***	***	***	***	***	***	***	***	***
BSF	-0.85		***	***	***	***	***	***	***	***	***	***	***
HRE	-0.59	0.43		***	***	***	***	***	***	***	***	***	***
PSF	0.67	-0.72	-0.50		***	***	***	***	***	***	***	***	***
ISF	-0.52	0.52	0.46	-0.97		***	***	***	***	***	***	***	***
TRC	-0.53	0.50	0.87	-0.57	0.52		***	***	***	***	***	***	***
LST	-0.53	0.61	0.51	-0.79	0.76	0.61		***	***	***	***	***	***
AHF	-0.37	0.40	0.25	-0.48	0.45	0.25	0.48		***	***	***	***	***
OPD	-0.62	0.62	0.39	-0.65	0.57	0.38	0.57	0.58		***	***	***	***
NTL	-0.34	0.36	0.24	-0.44	0.42	0.25	0.45	0.89	0.52		***	***	***
PD	-0.62	0.63	0.39	-0.68	0.61	0.40	0.60	0.59	0.97	0.53		***	***
NDVI	0.31	-0.35	-0.25	0.46	-0.44	-0.29	-0.54	-0.41	-0.32	-0.32	-0.33		***
EWI	-0.12	0.14	0.092	-0.21	0.21	0.10	0.26	0.30	0.15	0.21	0.16	-0.93	

Figure 11. Heatmap of Pearson's correlation between urban morphology and heat risk indicators.

4.3.2. Spatial Relationship between Heat Risk and Urban Morphology

The spatial dependence of heat risk and lagged urban morphology is understood through bivariate LISA maps and scatter plots, which explain the spatial relationship between heat risk and urban morphology factors (Figure 12). Moran's I of HRE (MI: 0.345), BSF (MI: 0.548), ISF (MI: 0.497), and TRC (MI: 0.329) were significantly positive, exerting a positive impact on heat risk. Conversely, SVF (MI: −0.561) and PSF (MI: −0.565) had a negative effect on heat risk, with significantly strong correlation. The corresponding cluster maps indicate the significant local correlation between the heat risk and urban morphology. Specifically, the indicators observed in the central area of Guangzhou, which has high–high clusters and dominant built-up areas, underscore that dense and compact urban morphology with fewer green spaces are more prone to heat risk in the summer. In contrast, the same indicators in the northern hilly and riverine areas, which have low–low clusters and dominant land cover areas, are less susceptible to heat risk. SVF and PSF in Guangzhou have resulted in high–low and low—high clusters, showing how vegetation coverage and sky openness can have a negative effect on heat risk. For instance, downtown Guangzhou exhibits low risk values due to high sky openness and high vegetation coverage in the north. This can be attributed to the heat absorption capacity of the vegetation. The smaller the SVF, the more the solar radiation can be blocked with shade.

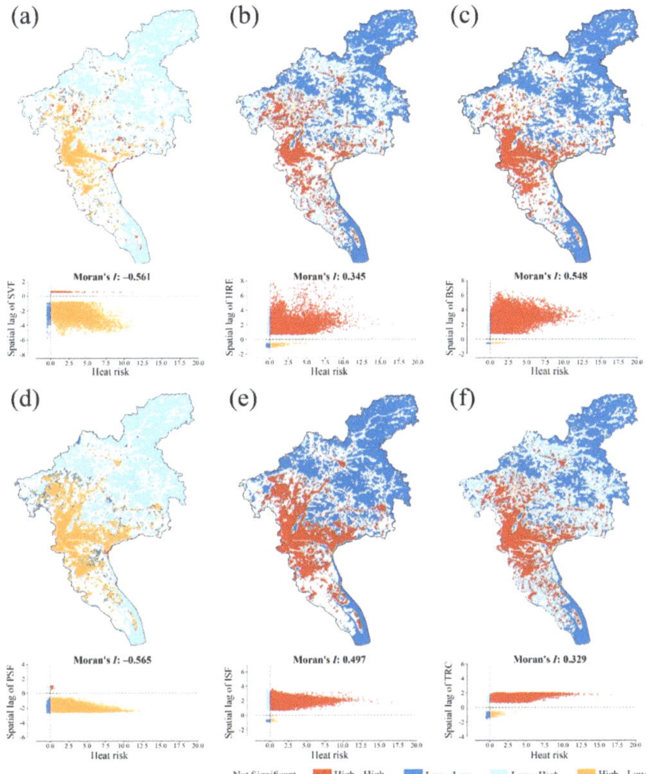

Figure 12. Spatial relationship between urban morphology and heat risk using bivariate LISA: (**a**) SVF, (**b**) HRE, (**c**) BSF, (**d**) PSF, (**e**) ISF, and (**f**) TRC.

4.3.3. Effect of Urban Morphology Factors on Heat Risk

Urban morphology influences regional temperature, population, and the economy. Examining the influence of urban morphology indicators on heat risk can provide valuable insights for future urban planning and development. The XGBoost-SHAP model helps in understanding the effect of urban morphology on heat risk (Figure 13). This study identified PSF as the predominant factor influencing heat risk, followed by HRE, BSF, SVF, ISF, and TRC. It is speculated that because 84.4% of the study area comprises land covers with low heat risk values, it has increased the effect of PSF in model training. The effects of HRE and BSF were almost consistent, and their correlation with LST was closer. This result can be attributed to higher buildings and dense layouts that block solar radiation and reduce the absorption of heat by the surface.

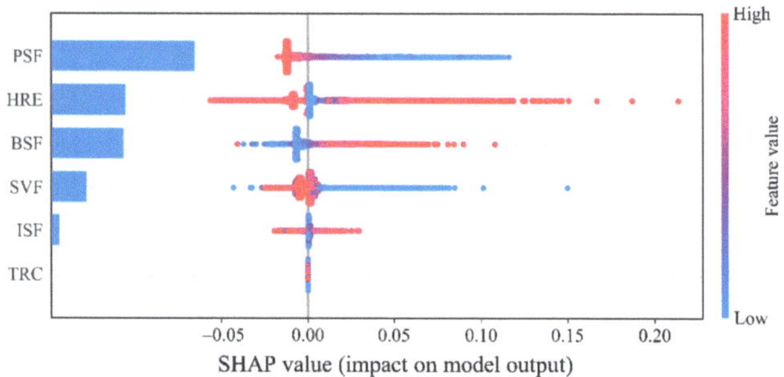

Figure 13. SHAP values of different urban morphology indicators.

To further clarify the impact of urban morphology indicators on heat risk, the dependence plot shows the nonlinear relationship between individual urban morphology indicators and heat risk indicators (Figure 14). Overall, SVF and PSF showed a negative correlation, and the dependence between the two was the strongest (Figure 14a,c). The negative impact of SVF became increasingly evident when the positive impact exceeded 0.75. As the SVF increases, the sky is more visible and, thus, the population density is lower, meaning that the heat risk is also lower. Conversely, a higher SVF can lead to a greater gain in solar radiation. The surface temperature then rises, increasing the heat risk. Moreover, PSF had a negative impact on heat risk after it exceeded 0.6. The impact of BSF and HRE on heat risk roughly presented an inverted U-shape (Figure 14b,e), which indicates that the impact on heat risk gradually increased as BSF and HRE increased. However, the impact declined after reaching a certain level. Nevertheless, the overall impact remained positive. It is worth noting that too large or too small a BSF can have a negative impact on heat risk. Similarly, HRE can have a significant negative impact at 15 m because the self-built houses in the suburbs are mostly 15 m tall. In this case, the regional heat risk was high. ISF had a positive impact on heat risk when the SHAP value was higher than 0.7, and vice versa (Figure 14d).

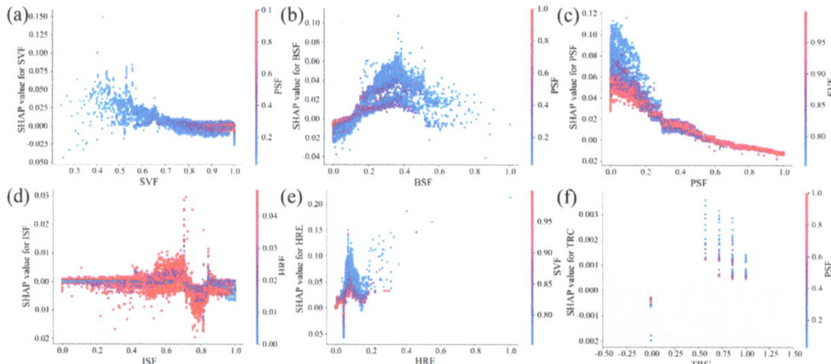

Figure 14. Dependence of the feature based on SHAP values: (**a**) SVF, (**b**) BSF, (**c**) PSF, (**d**) ISF, (**e**) HRE, and (**f**) TRC.

5. Discussion

This study developed a GIS-based risk assessment model to analyze the distribution of heat risk distribution across different block types. The results indicate that LCZs 1–4, characterized by high building and population density, were more prone to high heat risk. Conversely, LCZ 10 and LCZ E exhibited low heat risk levels, despite being in high hazard zones in certain regions, primarily due to their lower population density. It was found that low-risk areas are present in these areas, such as urban parks (LCZ A~LCZ E). In addition, this study observed a tendency for spatial clustering of heat risk, with the potential to transform entire districts into high-risk areas, as shown in Figure 10. Moreover, the low-high clustering of heat risk was observed at the junction of high-value areas dominated by built-up areas, and low-value areas dominated by green spaces. This indicates that green spaces have a significant role in reducing heat risk. Urban planners can effectively reduce heat risk through the cross-construction of green spaces and built-up areas based on the findings of this study.

Urban morphology correlates with surface temperature, the economy, population, etc. (Figure 12), which confirms the appropriateness of exploring the correlation between urban heat risk and urban morphology. This study found that PSF has the greatest influence on heat risk. Higher PSF means being able to absorb more solar radiation, raising LST, while population and economy are also closely related to PSF [72]. The reason may be that the proportion of green spaces in the study area is too large, increasing the influence of PSF on heat risk. Further research needs to be conducted in areas with dominant built-up types. SVF and PSF showed a decreasing impact on urban heat risk. This is because as SVF increases, it blocks solar radiation and reduces surface temperature. PSF has the same effect due to lower thermal conductivity and greater heat capacity of PSF than that of the ISF (Figure 14a,c). HRE and BSF showed an increase and then a decrease, particularly due to population density. The above results indicate that the GIS-based assessment model can explain the relationship between urban spatial morphology and heat risk.

This study has several limitations that could potentially introduce inaccuracies in the heat risk results. Firstly, the GIS-based method for mapping LCZs is limited to areas with a comprehensive database of building information. This approach may not be suitable for smaller cities where data acquisition is challenging or incomplete. Secondly, this study considered only six risk indicators due to the difficulty of obtaining precise block-scale data. However, numerous factors that influence individuals' adaptability to heat risks, including demographic, social, and economical factors, deserve further consideration in future research. Future studies should prioritize urban built-up areas, characterized by intense human activities and a higher likelihood of heat risks. Thirdly, a broader range of socioeconomic factors should be considered, as they may serve as potential indicators of vulnerability or exposure in future urban heat risk assessments, encompassing aspects such as the accessibility of parks and hospitals, education levels, and outdoor activity patterns of residents. Due to data limitations, this study has provided limited exploration between outdoor heat risk and human activity preferences. Human activity preferences are also influenced by temporal dimensions and differences in indoor versus outdoor heat risk within the same area. Finally, the psychological impact on the perception of heat risk is an essential consideration. Residents' perception of heat risk may be influenced by mood or cognitive levels, potentially affecting their resilience to heat risk. This could serve as a risk indicator to be included in future research.

6. Conclusions

This study proposes an urban heat risk assessment model to explore the influence of urban morphology and block types on heat risk. Firstly, urban morphology factors, building data, and land cover data were obtained to generate an LCZ map using a GIS-based approach. Then, three heat risk assessment indicators were selected, including heat hazard, heat vulnerability, and heat exposure. These risk indicators are then multiplied to generate an overall heat risk map. The proposed model was demonstrated in Guangzhou, a densely populated city in China. Finally, spatial autocorrelation and the XGBoost-SHAP methods were employed to investigate the relationship between urban spatial morphology and heat risk levels.

Results indicated that heat risk levels in built-up areas surpassed those in land covers, with LCZ 4 exhibiting the highest heat risk, boasting a hazard ratio of 55.23%. Conversely, LCZ 10 and LCZ 5 were identified as low-risk areas, accounting for 90.10% and 73.94%, respectively. In residential and commercial areas, such as LCZ 4 and LCZ 1, it is recommended to adopt mitigation strategies for heat risk. However, in LCZ 9 and LCZ 10, where population density is low, the adoption of such strategies may not be necessary. In addition, different block types not only reflect temperature differences, but also lead to spatial heterogeneity in social activities. For example, LCZ 1 and LCZ 2, which have higher population densities, exhibit higher heat risks. These results can be used for developing block-scale urban planning strategies, optimizing the overall spatial morphology of the city, and reducing the health risks due to high temperatures. Furthermore, when considering urban morphology factors, such as SVF and PSF, they had a negative effect on heat risk, while BSF, HRE, and TRC had a more positive effect. This can be attributed to the effect of solar radiation on the surface temperature and the tendency of people to congregate in areas with diverse building types. By blocking solar radiation, urban spaces absorb less heat, resulting in lower temperatures. Additionally, areas with sparser, lower buildings tend to have lower population densities, which in turn contributes to a more effective reduction in heat risk at the block level. Therefore, potential strategies for mitigating urban heat risk could involve increasing tree planting for shading and evenly distributing the population to create more conducive living environments.

Urban planning shapes the development of diverse block types, characterized by variations in building height, density, and pervious surface fraction across various urban areas. These disparities in spatial factors directly influence the absorption and reflection of solar radiation, contributing to temperature fluctuations within different built environments. Additionally, the functions and population capacities of different blocks influence residents' living behaviors, resulting in a variability in population density. By considering the combined impacts of both natural and social environments, these factors ultimately result in spatial disparities in heat risk. The proposed heat risk assessment model can assist urban planners in evaluating and implementing mitigating strategies for heat risk at the community and neighborhood level, thereby enhancing the safety of the built environment for future residents.

Author Contributions: Conceptualization: B.Z. and C.F.; Data curation: C.F. and B.Z.; Formal analysis: B.Z., C.F. and J.L.; Methodology: B.Z., C.F. and J.L.; Writing—original draft preparation: B.Z.; Writing—review and editing: C.F., J.L.; Funding acquisition: C.F.; Resources: J.L.; Supervision: C.F. and J.L. All authors have read and agreed to the published version of the manuscript.

Funding: This study was supported by Guangdong Basic and Applied Basic Research (No. 2023A1515012138), Guangzhou Science and Technology Project (No. 2024A04J3355), Guangdong Philosophy and Social Science Planning Project (No. GD24YGL28), Open Foundation of the State Key Laboratory of Subtropical Building and Urban Science (No. 2023KA01), Science and Technology Program of Guangzhou University (No. PT252022006). This paper is also supported by Guangzhou University Graduate Innovation Ability Development Program.

Data Availability Statement: The datasets generated and analyzed during the current study are available from the corresponding author on reasonable request.

Conflicts of Interest: The authors declare that they have no known competing financial interests or personal relationships that could have appeared to influence the work reported in this paper.

Nomenclature

AHF	Anthropogenic heat flux
BSF	Building surface fraction
EWI	Enhanced water index
Green	Green band
HRE	Height of roughness elements
ISF	Impervious surface fraction
LISA	Local indicator of spatial autocorrelation
LST	Land surface temperature
MSE	Mean squared error
NDVI	Normalized difference vegetation index
NIR	Near-infrared
NTL	Night-time light
OPD	Density of population over 65
PD	Population density
PSF	Pervious surface fraction
R^2	Coefficient of determination
RMSE	Root mean squared error
SHAP	Shapley additive explanations
SVF	Sky view factor
SWIR1	Short-wave infrared band
TRC	Terrain roughness class

References

1. Ebi, K.L.; Capon, A.; Berry, P.; Broderick, C.; de Dear, R.; Havenith, G.; Honda, Y.; Kovats, R.S.; Ma, W.; Malik, A. Hot Weather and Heat Extremes: Health Risks. *Lancet* **2021**, *398*, 698–708. [CrossRef] [PubMed]
2. Chaseling, G.K.; Iglesies-Grau, J.; Juneau, M.; Nigam, A.; Kaiser, D.; Gagnon, D. Extreme Heat and Cardiovascular Health: What a Cardiovascular Health Professional Should Know. *Can. J. Cardiol.* **2021**, *37*, 1828–1836. [CrossRef] [PubMed]
3. Lien, T.-C.; Tabata, T. Regional Incidence Risk of Heat Stroke in Elderly Individuals Considering Population, Household Structure, and Local Industrial Sector. *Sci. Total Environ.* **2022**, *853*, 158548. [CrossRef] [PubMed]
4. Sorensen, C.; Hess, J. Treatment and Prevention of Heat-Related Illness. *N. Engl. J. Med.* **2022**, *387*, 1404–1413. [CrossRef] [PubMed]
5. Zhang, G.; Sun, Z.; Han, L.; Iyakaremye, V.; Xu, Z.; Miao, S.; Tong, S. Avoidable Heat-Related Mortality in China during the 21st Century. *NPJ Clim. Atmos. Sci.* **2023**, *6*, 81. [CrossRef]
6. Wang, S.; Sun, Q.C.; Huang, X.; Tao, Y.; Dong, C.; Das, S.; Liu, Y. Health-Integrated Heat Risk Assessment in Australian Cities. *Environ. Impact Assess. Rev.* **2023**, *102*, 107176. [CrossRef]
7. He, B.-J.; Zhao, D.; Dong, X.; Zhao, Z.; Li, L.; Duo, L.; Li, J. Will Individuals Visit Hospitals When Suffering Heat-Related Illnesses? Yes, But…. *Build. Environ.* **2022**, *208*, 108587. [CrossRef]
8. Huber, V.; Breitner-Busch, S.; He, C.; Matthies-Wiesler, F.; Peters, A.; Schneider, A. Heat-Related Mortality in the Extreme Summer of 2022: An Analysis Based on Daily Data. *Dtsch. Arztebl. Int.* **2024**, *121*, 79. [CrossRef]
9. de Bont, J.; Nori-Sarma, A.; Stafoggia, M.; Banerjee, T.; Ingole, V.; Jaganathan, S.; Mandal, S.; Rajiva, A.; Krishna, B.; Kloog, I. Impact of Heatwaves on All-Cause Mortality in India: A Comprehensive Multi-City Study. *Environ. Int.* **2024**, *184*, 108461. [CrossRef] [PubMed]
10. Fatima, S.H.; Rothmore, P.; Giles, L.C.; Bi, P. Outdoor Ambient Temperatures and Occupational Injuries and Illnesses: Are There Risk Differences in Various Regions within a City? *Sci. Total Environ.* **2022**, *826*, 153945. [CrossRef]
11. Jang, J.; Lee, W.; Choi, M.; Kang, C.; Kim, H. Roles of Urban Heat Anomaly and Land-Use/Land-Cover on the Heat-Related Mortality in the National Capital Region of South Korea: A Multi-Districts Time-Series Study. *Environ. Int.* **2020**, *145*, 106127. [CrossRef] [PubMed]
12. Zhu, W.; Yuan, C. Urban Heat Health Risk Assessment in Singapore to Support Resilient Urban Design—By Integrating Urban Heat and the Distribution of the Elderly Population. *Cities* **2023**, *132*, 104103. [CrossRef]
13. Dai, X.; Liu, Q.; Huang, C.; Li, H. Spatiotemporal Variation Analysis of the Fine-Scale Heat Wave Risk along the Jakarta-Bandung High-Speed Railway in Indonesia. *Int. J. Environ. Res. Public Health* **2021**, *18*, 12153. [CrossRef] [PubMed]
14. Park, C.Y.; Thorne, J.H.; Hashimoto, S.; Lee, D.K.; Takahashi, K. Differing Spatial Patterns of the Urban Heat Exposure of Elderly Populations in Two Megacities Identifies Alternate Adaptation Strategies. *Sci. Total Environ.* **2021**, *781*, 146455. [CrossRef]

15. Freychet, N.; Hegerl, G.C.; Lord, N.S.; Lo, Y.E.; Mitchell, D.; Collins, M. Robust Increase in Population Exposure to Heat Stress with Increasing Global Warming. *Environ. Res. Lett.* **2022**, *17*, 064049. [CrossRef]

16. Chen, B.; Xie, M.; Feng, Q.; Li, Z.; Chu, L.; Liu, Q. Heat Risk of Residents in Different Types of Communities from Urban Heat-Exposed Areas. *Sci. Total Environ.* **2021**, *768*, 145052. [CrossRef] [PubMed]

17. Huang, X.; Li, Y.; Guo, Y.; Zheng, D.; Qi, M. Assessing Urban Risk to Extreme Heat in China. *Sustainability* **2020**, *12*, 2750. [CrossRef]

18. Liu, Y.; Li, Q.; Yang, L.; Mu, K.; Zhang, M.; Liu, J. Urban Heat Island Effects of Various Urban Morphologies under Regional Climate Conditions. *Sci. Total Environ.* **2020**, *743*, 140589. [CrossRef] [PubMed]

19. Wang, B.; Sun, S.; Duan, M. Wind Potential Evaluation with Urban Morphology-A Case Study in Beijing. *Energy Procedia* **2018**, *153*, 62–67. [CrossRef]

20. Oke, T.R. The Energetic Basis of the Urban Heat Island. *Q. J. R. Meteorol. Soc.* **1982**, *108*, 1–24. [CrossRef]

21. Jin, L.; Pan, X.; Liu, L.; Liu, L.; Liu, J.; Gao, Y. Block-Based Local Climate Zone Approach to Urban Climate Maps Using the UDC Model. *Build. Environ.* **2020**, *186*, 107334. [CrossRef]

22. Liu, L.; Liu, J.; Jin, L.; Liu, L.; Gao, Y.; Pan, X. Climate-Conscious Spatial Morphology Optimization Strategy Using a Method Combining Local Climate Zone Parameterization Concept and Urban Canopy Layer Model. *Build. Environ.* **2020**, *185*, 107301. [CrossRef]

23. Maharoof, N.; Emmanuel, R.; Thomson, C. Compatibility of Local Climate Zone Parameters for Climate Sensitive Street Design: Influence of Openness and Surface Properties on Local Climate. *Urban. Clim.* **2020**, *33*, 100642. [CrossRef]

24. Wu, J.; Liu, C.; Wang, H. Analysis of Spatio-Temporal Patterns and Related Factors of Thermal Comfort in Subtropical Coastal Cities Based on Local Climate Zones. *Build. Environ.* **2022**, *207*, 108568. [CrossRef]

25. Zhang, L.; Nikolopoulou, M.; Guo, S.; Song, D. Impact of LCZs Spatial Pattern on Urban Heat Island: A Case Study in Wuhan, China. *Build. Environ.* **2022**, *226*, 109785. [CrossRef]

26. Ma, L.; Huang, G.; Johnson, B.A.; Chen, Z.; Li, M.; Yan, Z.; Zhan, W.; Lu, H.; He, W.; Lian, D. Investigating Urban Heat-Related Health Risks Based on Local Climate Zones: A Case Study of Changzhou in China. *Sustain. Cities Soc.* **2023**, *91*, 104402. [CrossRef]

27. Savić, S.; Marković, V.; Šećerov, I.; Pavić, D.; Arsenović, D.; Milošević, D.; Dolinaj, D.; Nagy, I.; Pantelić, M. Heat Wave Risk Assessment and Mapping in Urban Areas: Case Study for a Midsized Central European City, Novi Sad (Serbia). *Nat. Hazard.* **2018**, *91*, 891–911. [CrossRef]

28. Wu, Y.; Feng, J. Development and Application of Artificial Neural Network. *Wirel. Pers. Commun.* **2018**, *102*, 1645–1656. [CrossRef]

29. Liu, H.; Liang, J.; Liu, Y.; Wu, H. A Review of Data-Driven Building Energy Prediction. *Buildings* **2023**, *13*, 532. [CrossRef]

30. Chen, Y.; Zheng, L.; Song, J.; Huang, L.; Zheng, J. Revealing the Impact of Urban Form on COVID-19 Based on Machine Learning: Taking Macau as an Example. *Sustainability* **2022**, *14*, 14341. [CrossRef]

31. Chung, J.; Lee, Y.; Jang, W.; Lee, S.; Kim, S. Correlation Analysis between Air Temperature and MODIS Land Surface Temperature and Prediction of Air Temperature Using TensorFlow Long Short-Term Memory for the Period of Occurrence of Cold and Heat Waves. *Remote Sens.* **2020**, *12*, 3231. [CrossRef]

32. Zekar, A.; Milojevic-Dupont, N.; Zumwald, M.; Wagner, F.; Creutzig, F. Urban Form Features Determine Spatio-Temporal Variation of Ambient Temperature: A Comparative Study of Three European Cities. *Urban. Clim.* **2023**, *49*, 101467. [CrossRef]

33. Sun, Y.; Gao, C.; Li, J.; Wang, R.; Liu, J. Quantifying the Effects of Urban Form on Land Surface Temperature in Subtropical High-Density Urban Areas Using Machine Learning. *Remote Sens.* **2019**, *11*, 959. [CrossRef]

34. Guangzhou Statistics Bureau. Available online: https://lwzb.gzstats.gov.cn:20001/datav/admin/home/www_nj/ (accessed on 29 December 2023).

35. Li, Z.; He, W.; Cheng, M.; Hu, J.; Yang, G.; Zhang, H. SinoLC-1: The First 1-Meter Resolution National-Scale Land-Cover Map of China Created with the Deep Learning Framework and Open-Access Data. *Earth Syst. Sci. Data* **2023**, *2023*, 1–38.

36. Kinney, P.L.; Ge, B.; Sampath, V.; Nadeau, K. Health-Based Strategies for Overcoming Barriers to Climate Change Adaptation and Mitigation. *J. Allergy Clin. Immunol.* **2023**, *152*, 1053–1059. [CrossRef] [PubMed]

37. Ho, H.C.; Knudby, A.; Chi, G.; Aminipouri, M.; Lai, D.Y.-F. Spatiotemporal Analysis of Regional Socio-Economic Vulnerability Change Associated with Heat Risks in Canada. *Appl. Geogr.* **2018**, *95*, 61–70. [CrossRef] [PubMed]

38. Xiang, Z.; Qin, H.; He, B.-J.; Han, G.; Chen, M. Heat Vulnerability Caused by Physical and Social Conditions in a Mountainous Megacity of Chongqing, China. *Sustain. Cities Soc.* **2022**, *80*, 103792. [CrossRef]

39. Zhang, M.; Wang, H.; Jin, W.; Dijk, M.P.V. Assessing Heat Wave Vulnerability in Beijing and Its Districts, Using a Three Dimensional Model. *Int. J. Glob. Warm.* **2019**, *17*, 297–314. [CrossRef]

40. IPCC. *Climate Change 2022:Impacts, Adaptation and Vulnerability*; GIEC: Geneva, Switzerland, 2022.

41. Crichton, D. The Risk Triangle. *Nat. Disaster Manag.* **1999**, *102*, 102–103.

42. Dong, J.; Peng, J.; He, X.; Corcoran, J.; Qiu, S.; Wang, X. Heatwave-Induced Human Health Risk Assessment in Megacities Based on Heat Stress-Social Vulnerability-Human Exposure Framework. *Landsc. Urban. Plan.* **2020**, *203*, 103907. [CrossRef]

43. Li, J.; Sun, R.; Liu, T.; Xie, W.; Chen, L. Prediction Models of Urban Heat Island Based on Landscape Patterns and Anthropogenic Heat Dynamics. *Landsc. Ecol.* **2021**, *36*, 1801–1815. [CrossRef]

44. Qian, J.; Zhang, L.; Schlink, U.; Meng, Q.; Liu, X.; Janscó, T. High Spatial and Temporal Resolution Multi-Source Anthropogenic Heat Estimation for China. *Resour. Conserv. Recycl.* **2024**, *203*, 107451. [CrossRef]
45. Sajib, M.Q.U.; Wang, T. Estimation of Land Surface Temperature in an Agricultural Region of Bangladesh from Landsat 8: Intercomparison of Four Algorithms. *Sensors* **2020**, *20*, 1778. [CrossRef] [PubMed]
46. Sekertekin, A. Validation of Physical Radiative Transfer Equation-Based Land Surface Temperature Using Landsat 8 Satellite Imagery and SURFRAD in-Situ Measurements. *J. Atmos. Sol. Terr. Phys.* **2019**, *196*, 105161. [CrossRef]
47. Xiao, Q.-L.; Wang, Y.; Zhou, W.-X. Regional Economic Convergence in China: A Comparative Study of Nighttime Light and GDP. *Front. Phys.* **2021**, *9*, 525162. [CrossRef]
48. Zhao, Z.; Tang, X.; Wang, C.; Cheng, G.; Ma, C.; Wang, H.; Sun, B. Analysis of the Spatial and Temporal Evolution of the GDP in Henan Province Based on Nighttime Light Data. *Remote Sens.* **2023**, *15*, 716. [CrossRef]
49. Zhao, N.; Cao, G.; Zhang, W.; Samson, E.L. Tweets or Nighttime Lights: Comparison for Preeminence in Estimating Socioeconomic Factors. *ISPRS J. Photogramm. Remote Sens.* **2018**, *146*, 1–10. [CrossRef]
50. Liu, C.; Wang, C.; Xu, Y.; Liu, C.; Li, M.; Zhang, D.; Liu, G.; Li, W.; Zhang, Q.; Li, Q. Correlation Analysis between Nighttime Light Data and Socioeconomic Factors on Fine Scales. *IEEE Geosci. Remote. Sens. Lett.* **2022**, *19*, 1–5.
51. Varlamova, E.; Solovyev, V. Study of NDVI Vegetation Index in East Siberia under Global Warming. In Proceedings of the 22nd International Symposium on Atmospheric and Ocean Optics: Atmospheric Physics, Tomsk, Russia, 30 June–3 July 2016; SPIE: Bellingham, WA, USA, 2016; Volume 10035, pp. 1190–1195.
52. Liu, S.; Wu, Y.; Zhang, G.; Lin, N.; Liu, Z. Comparing Water Indices for Landsat Data for Automated Surface Water Body Extraction under Complex Ground Background: A Case Study in Jilin Province. *Remote Sens.* **2023**, *15*, 1678. [CrossRef]
53. Rodler, A.; Leduc, T. Local Climate Zone Approach on Local and Micro Scales: Dividing the Urban Open Space. *Urban. Clim.* **2019**, *28*, 100457. [CrossRef]
54. Wu, Y.; Sharifi, A.; Yang, P.; Borjigin, H.; Murakami, D.; Yamagata, Y. Mapping Building Carbon Emissions within Local Climate Zones in Shanghai. *Energy Procedia* **2018**, *152*, 815–822. [CrossRef]
55. Chen, Y.; Hu, Y. The Urban Morphology Classification under Local Climate Zone Scheme Based on the Improved Method-A Case Study of Changsha, China. *Urban. Clim.* **2022**, *45*, 101271. [CrossRef]
56. Stewart, I.D.; Oke, T.R. Local Climate Zones for Urban Temperature Studies. *Bull. Am. Meteorol. Soc.* **2012**, *93*, 1879–1900. [CrossRef]
57. Huang, X.; Liu, A.; Li, J. Mapping and Analyzing the Local Climate Zones in China's 32 Major Cities Using Landsat Imagery Based on a Novel Convolutional Neural Network. *Geo Spat. Inf. Sci.* **2021**, *24*, 528–557. [CrossRef]
58. Lau, K.K.-L.; Ren, C.; Shi, Y.; Zheng, V.; Yim, S.; Lai, D. Determining the Optimal Size of Local Climate Zones for Spatial Mapping in High-Density Cities. In Proceedings of the 9th International Conference on Urban Climate jointly with 12th Symposium on the Urban Environment, Toulouse, France, 20–24 July 2015; pp. 20–24.
59. Chen, T.; Sun, A.; Niu, R. Effect of Land Cover Fractions on Changes in Surface Urban Heat Islands Using Landsat Time-Series Images. *Int. J. Environ. Res. Public Health* **2019**, *16*, 971. [CrossRef]
60. Fan, C.; Zou, B.; Li, J.; Wang, M.; Liao, Y.; Zhou, X. Exploring the Relationship between Air Temperature and Urban Morphology Factors Using Machine Learning under Local Climate Zones. *Case Stud. Therm. Eng.* **2024**, *55*, 104151. [CrossRef]
61. Unger, J.; Lelovics, E.; Gál, T. Local Climate Zone Mapping Using GIS Methods in Szeged. *Hungarian Geogr. Bull.* **2014**, *63*, 29–41. [CrossRef]
62. Davenport, A.G.; Grimmond, C.S.B.; Oke, T.R.; Wieringa, J. Estimating the Roughness of Cities and Sheltered Country. 12 Th Conf. on Applied Climatology, 8–11 May 2000, Asheville, NC. *Am. Meteorol. Soc. Search* **2000**, *96*.
63. Hammond, D.; Chapman, L.; Thornes, J. Roughness Length Estimation along Road Transects Using Airborne LIDAR Data. *Meteorol. Appl.* **2012**, *19*, 420–426. [CrossRef]
64. Ching, J.; Mills, G.; Bechtel, B.; See, L.; Feddema, J.; Wang, X.; Ren, C.; Brousse, O.; Martilli, A.; Neophytou, M. WUDAPT: An Urban Weather, Climate, and Environmental Modeling Infrastructure for the Anthropocene. *Bull. Am. Meteorol. Soc.* **2018**, *99*, 1907–1924. [CrossRef]
65. Estacio, I.; Babaan, J.; Pecson, N.; Blanco, A.; Escoto, J.; Alcantara, C. GIS-Based Mapping of Local Climate Zones Using Fuzzy Logic and Cellular Automata. *Int. Arch. Photogramm. Remote Sens. Spatial Inf. Sci.* **2019**, *42*, 199–206. [CrossRef]
66. Pearson, K. *On the Theory of Contingency and Its Relation to Association and Normal Correlation*; Drapers' Co. Memoirs: London, UK, 1904.
67. Moran, P.A. Notes on Continuous Stochastic Phenomena. *Biometrika* **1950**, *37*, 17–23. [CrossRef] [PubMed]
68. Anselin, L.; Syabri, I.; Kho, Y. GeoDa: An Introduction to Spatial Data Analysis. In *Handbook of Applied Spatial Analysis: Software Tools, Methods and Applications*; Springer: Berlin/Heidelberg, Germany, 2009; pp. 73–89.
69. Chen, T.; Guestrin, C. Xgboost: A Scalable Tree Boosting System. In Proceedings of the 22nd Acm Sigkdd International Conference on Knowledge Discovery and Data Mining, San Francisco, CA, USA, 13–17 August 2016; pp. 785–794.
70. Li, Z. Extracting Spatial Effects from Machine Learning Model Using Local Interpretation Method: An Example of SHAP and XGBoost. *Comput. Environ. Urban. Syst.* **2022**, *96*, 101845. [CrossRef]

71. Chen, S.; Bao, Z.; Lou, V. Assessing the Impact of the Built Environment on Healthy Aging: A Gender-Oriented Hong Kong Study. *Environ. Impact Assess. Rev.* **2022**, *95*, 106812. [CrossRef]
72. Schug, F.; Frantz, D.; van der Linden, S.; Hostert, P. Gridded Population Mapping for Germany Based on Building Density, Height and Type from Earth Observation Data Using Census Disaggregation and Bottom-up Estimates. *PLoS ONE* **2021**, *16*, e0249044. [CrossRef]

 *buildings*

Article

An Improved Zonal Ventilation Control Method of Waiting Hall of High-Speed Railway Station Based on Real-Time Occupancy

Pei Zhou [1,*], Jintao Zhou [1], Yu Tang [2], Zicheng Ma [2], Ming Yao [2], Jian Zhu [1] and Huanyu Si [1]

[1] School of Civil Engineering, Hefei University of Technology, Hefei 230009, China; 202110593@mail.hfut.edu.cn (J.Z.); zhujian@hfut.edu.cn (J.Z.); hfutshy@163.com (H.S.)
[2] The First Company of China Eighth Engineering Bureau Ltd., Jinan 250013, China; tangzjbj@163.com (Y.T.); richen3@163.com (Z.M.); yaozjbi@163.com (M.Y.)
* Correspondence: peizhou@hfut.edu.cn

Abstract: The random movement of occupants in a high-speed railway station results in a more complex indoor environment. In this study, the indoor thermal environment and the thermal comfort in summer were investigated via field measurements and questionnaires in the waiting hall of a high-speed railway station. The results showed that there was an uneven horizontal temperature distribution in the area, and over 30% of the passengers were dissatisfied with the air conditioning system. In order to improve the control of the indoor temperature as well as reduce the energy consumption of the air conditioning system, an improved zonal control strategy and AMPC control optimization algorithm based on real-time people are proposed, and different control strategies are modeled and simulated using MATLAB/Simulink. It is concluded that the improved zonal control method proposed in this paper can save 28.04% of the fan energy consumption compared with the traditional control strategy.

Keywords: high-speed railway station; indoor thermal environment; thermal comfort; Simulink; zonal ventilation control

Citation: Zhou, P.; Zhou, J.; Tang, Y.; Ma, Z.; Yao, M.; Zhu, J.; Si, H. An Improved Zonal Ventilation Control Method of Waiting Hall of High-Speed Railway Station Based on Real-Time Occupancy. *Buildings* **2024**, *14*, 1783. https://doi.org/10.3390/buildings14061783

Academic Editor: Theodore Stathopoulos

Received: 17 May 2024
Revised: 4 June 2024
Accepted: 7 June 2024
Published: 13 June 2024

1. Introduction

According to the data released by the China National Railway Administration, the total operating mileage of the national railway reached 150,000 km in 2022, with high-speed railways accounting for a total operating mileage of 40,000 km [1]. By 2030, the total mileage of the high-speed railways in China will exceed 45,000 km, covering over 80% of the cities nationwide [2]. Modern high-speed railway station buildings, as large public buildings, have a high internal space compared to office buildings, with no internal partitions and a large flow of occupants, which results in a more complex indoor environment [3,4]. In addition, unlike other public buildings, modern high-speed railway station buildings have evolved from simple waiting areas to complex urban spaces with diverse functions and services, which has brought about issues of passengers' thermal comfort and energy consumption. Previous studies in terms of thermal comfort have shown that recommendations can be made to ensure passengers' thermal comfort in waiting areas as well as to set different thermal environment parameters to avoid unnecessary energy waste in other non-occupied areas [5,6].

Research on indoor thermal comfort has been carried out for several decades. The Predicted Mean Vote (PMV) and Percentage of People Dissatisfied (PPD) methods proposed by Fanger [7] and the adaptive model proposed by De Dear and G Brager [8] are widely used to evaluate the thermal comfort of indoor environments; high-speed railway station buildings are included as well. Chirag Deb [9] studied the indoor thermal comfort of a train station in southern India in summer, and the results showed that the passengers had a high tolerance and adaptability to the environment. Furthermore, the duration of time passengers spend in the waiting hall also affects thermal comfort [10]. In fact, passengers' adaptation to the indoor thermal environment is higher than the results predicted by the PMV–PPD model, and, with increasing time spent, passengers' real thermal neutral temperature gradually reaches the predicted thermal neutral temperature [11]. Ye Yuan [12] investigated the dynamic thermal response of passengers throughout the departure process and determined their specific thermal comfort needs in different functional zones. The range of thermo-neutral temperature variations for passengers in different functional zones was derived by means of a questionnaire survey and calculations in order to provide a reference for improving the thermal comfort of passengers and optimizing the indoor environment.

In fact, setting the design parameters of an indoor thermal environment directly affects the thermal comfort of the indoor occupants, and it plays a vital role in the energy consumption of the building's air conditioning system. Therefore, the critical aspect to ensuring indoor thermal comfort and achieving the energy-efficient operation of the air conditioning system lies in setting appropriate indoor thermal environment parameters. However, due to the complex nature of large indoor spaces, obtaining their thermal environment parameters has become a primary concern. Currently, there are two main methods for predicting the indoor airflow and temperature distribution: micro-scale models and macro-scale models. Micro-scale models use Computational Fluid Dynamics (CFDs) based on the Navier–Stokes equations to calculate detailed indoor environments and obtain values for all the relevant parameters. However, for large indoor spaces under dynamic conditions, accurate results can be obtained but at a relatively high calculation cost [13,14]. The macro-scale models are mainly divided into node models and zone models [15]. Node models assume that the indoor environment is uniformly distributed, thereby ignoring the indoor airflow and temperature distribution. Therefore, these models can quickly simulate the dynamic changes in the thermal environment but cannot predict the complex airflow and temperature changes in large indoor spaces.

The zone model is a simulation model between the node model and CFD model, proposed by Lebrun in 1970 [16]. This model can balance the accuracy and computational cost of the model while considering the potential of multi-zone coupling, and it can predict the indoor environmental conditions easily, especially in large open-space buildings. Lu [17] proposed an adaptive zone method used in an atrium building, which can achieve similar accuracy with fewer zones compared to the traditional zoning methods. Bauman F [18] proposed a zoning model for waiting rooms that provides more accurate calculations of the load both in occupant and non-occupant zones, and the ventilation effects were evaluated, thus establishing the foundation for further energy-efficient designs.

A precise mathematical model is the fundamental aspect for achieving energy-saving control in HVAC systems. The energy-saving control in HVAC systems has made great progress in the past few decades. The most common HVAC system control method is still traditional PID control. However, when complex and variable indoor environments occur in real circumstances, the traditional PID control often cannot achieve the desired control effects. Therefore, new control methods or algorithms are continuously being developed, such as fuzzy PID control strategies [19], self-disturbance control methods [20], etc. Model predictive control (MPC) is an online optimization control algorithm with good robustness, excellent control performance, and low requirements for model accuracy. Therefore, it can be applied to optimize the building automation control systems, especially in HVAC system control [21]. In recent years, many researchers have utilized model predictive control methods for intelligent control in HVAC systems. Tang [22] proposed an MPC method for optimizing the operation of cold storage integrated central air conditioning systems during rapid DR events, which effectively reduced the building energy consumption as expected by the power grid, improved the indoor environments during DR events, and significantly reduced the indoor maximum temperatures without consuming additional energy. Wang [23] designed a gray-box coupled model for office buildings and proposed a model predictive control strategy to manage building energy consumption and indoor air quality. They also evaluated the energy-saving potential of the proposed control strategies in simulations in different climates. Ren C [24] et al. proposed a zonal demand ventilation control strategy based on a fast predictive zoning model to balance the energy savings and indoor infection rate of air conditioning systems, and the results showed that the proposed ZDCV control strategy improved the energy saving efficiency by 34% and simultaneously reduced the indoor infection rate.

This paper investigates the thermal environments of high-speed railway stations, especially for the waiting areas, adopting on-site measurements and questionnaire surveys to analyze the thermal environments in waiting areas. Additionally, an improved zoning model is employed to simulate the internal thermal environment of the target indoor space. The mathematical relationship for heat transfer within the waiting space is determined based on energy conservation principles, using real-time passenger flow data as parameters in Simulink to establish a simulation model, which is different from the previous studies. To clarify, two points that distinguish this research from the previous studies include the following: 1. in the improved zonal model, the thermal coupling effect was considered in the control model, which is different from the traditional zonal model with clear physical partitions; in this study, the zonal division is defined as virtual walls, namely without real walls or partitions. However, heat and mass transfer may occur across these virtual boundaries or between the adjacent subzones. 2. We used the real occupancy as the input of the control simulation, which can handle the supply airflow rate flexibly and maintain the indoor thermal environment to meet its requirements based on the real number of passengers. The temperature response and energy consumption under different control strategies are compared and analyzed. The rest of this paper is organized as follows: Section 2 will introduce the research methodology; the thermal environment will be summarized in Section 3, such as thermal images, indoor temperature distribution, CO_2 distribution, thermal sensation vote analysis, and so on; in Section 4, the mathematical model and the control algorithm will be addressed; Sections 5 and 6 include the simulation results as well as conclusions and discussions.

2. Research Methodology

The proposed method is shown in Figure 1, which mainly contains four parts; the first part is to obtain the parameters of the indoor thermal environment and questionnaire results through on-site measurement. The second part is to analyze the thermal indoor environment, and a zonal control strategy is proposed in this part to deal with the above issues acquired from part 1; the third part is the modeling and validation process of the proposed control strategies. The advanced control algorithm as the optimized control scheme is selected to optimize the control performance. Then, in order to verify the effectiveness of the proposed control method, the traditional zonal control method and the proposed scheme are compared and evaluated; finally, the energy consumption of the air conditioning system is predicted and compared.

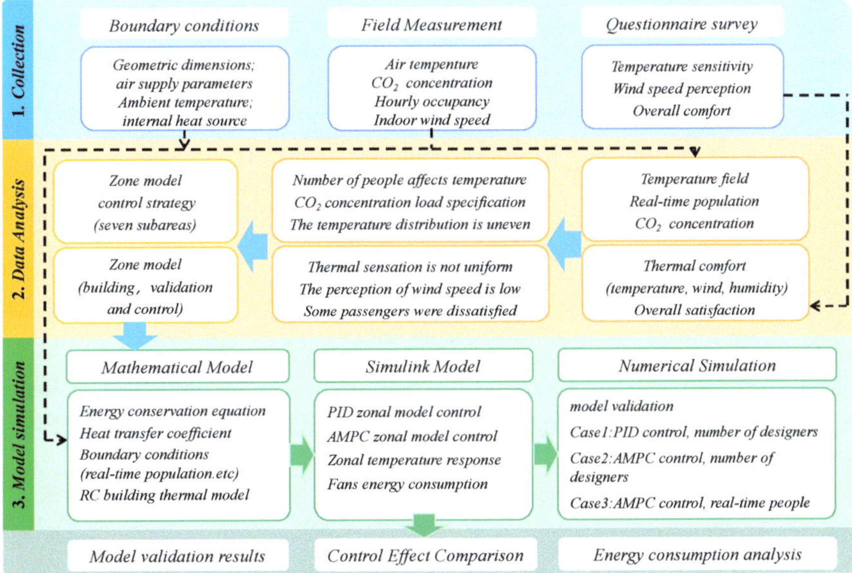

Figure 1. Flowchart of the research methodology.

2.1. Research Object and Method

The second-floor waiting hall of the high-speed railway station is 345 m long from north to south and 145 m wide from east to west, with a total area of approximately 49,536 m². It is symmetrical from east to west, with a height of 21 m, slightly higher at the central skylight area, reaching a height of 24.4 m. The total waiting hall within the station is approximately 31,398 m². The cooling air supply inlets in the station are nozzle diffusers, with multiple of them installed parallel above the waiting area and ticket checking gates 3 m above the floor, on-site photos are shown in Figure 2 which marked as a, b, and c on the bottom.

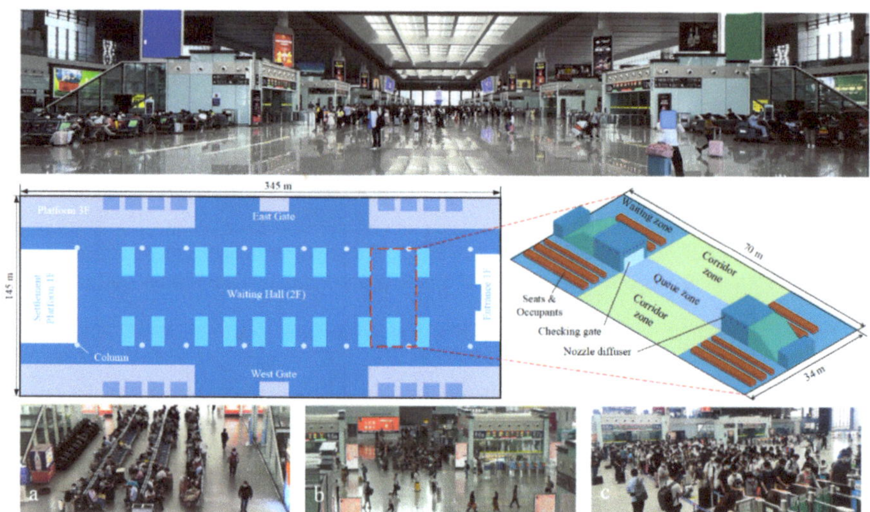

Figure 2. Schematic diagram and measurement area of the high-speed rail station.

Due to the large area of the waiting hall and high internal structural repetition, measurements are conducted only in the waiting areas on both sides of checking gates 4A, 5A, 4B, and 5B, as well as the central aisle. The measurement area is approximately 70 × 34 m, totaling about 2380 m². Indoor air temperature, CO_2 concentration, and other environmental parameters are measured. The data are collected from 7:00 to 20:30. Additionally, a questionnaire survey is conducted to gather passengers' thermal comfort perceptions and evaluations of the waiting thermal environment.

The entire measurement area is divided into three zones based on functionality: waiting zone with fixed seats, queueing zone, and aisle or corridor zone. The measured area includes two ticket checking gates and four waiting areas on both sides, with two waiting areas on each side of the gates. The aisle area is located between waiting areas on the same side, while the queueing area is between the two ticket checking gates, as illustrated in Figure 1.

2.1.1. Questionnaire

This study employs the Questionnaire Star online survey platform to create electronic questionnaires, which are randomly distributed to waiting passengers within the station via QR code sharing. The questionnaire consists of over 20 questions designed to gather comprehensive information on human responses to the thermal environment. The content of the questionnaire is primarily described as follows: gender, age, thermal sensation in the current area, overall thermal comfort, etc. The thermal sensation voting in the questionnaire utilizes the ASHRAE 7-point thermal sensation scale [21], while the wind sensations are set according to the thermal sensation voting scale.

2.1.2. Site Measurement of Thermal Environment

The measurement of thermal environment parameters mainly involves indoor parameter collection. The detailed parameters of the testing instruments are shown in Table 1. The measured parameters include indoor air temperature, indoor air velocity, and indoor CO_2 concentration. Due to the absence of obvious radiant heat sources within the measurement area, this study does not measure black globe temperature. According to ASHRAE 55 standard [25], the measurement point height is set at 1.1 m (which is breathing zone height when seated).

Table 1. Monitoring instruments for the thermal environment.

Instruments	Parameters	Range	Accuracy
Thermal Imager-TiS50	Temperature	−20~+450 °C	0.02 °C
Thermal	Air velocity	0~30 m/s	±1%
Anemometer-ST866A	Air temperature	0~45 °C	±1 °C
CO_2 detector-AR8200	CO_2 concentration	350~9999 ppm	±(30 + 5%) ppm
Laser rangefinder-DL331070L	Distance	0.05~70 m	±3 mm
Infrared thermometer-AS842A	Temperature	−50~600 °C	±1.5 °C

3. Site Measurement Analysis

3.1. Thermal Environment Measurement Rests

3.1.1. Overall Thermal Environment

Figure 3 shows the thermal images of different areas within the waiting hall, taken on 8 July 2022 at 10:00 a.m. with an outdoor temperature of approximately 34 °C. The captured areas include the waiting area, queueing area, and main entrance. As depicted in Figure 3, the maximum temperature difference within the captured areas exceeds 10 °C, with a maximum difference of 11.7 °C (Figure 3c) and a minimum difference of 10.2 °C (Figure 3b), indicating significant temperature variations.

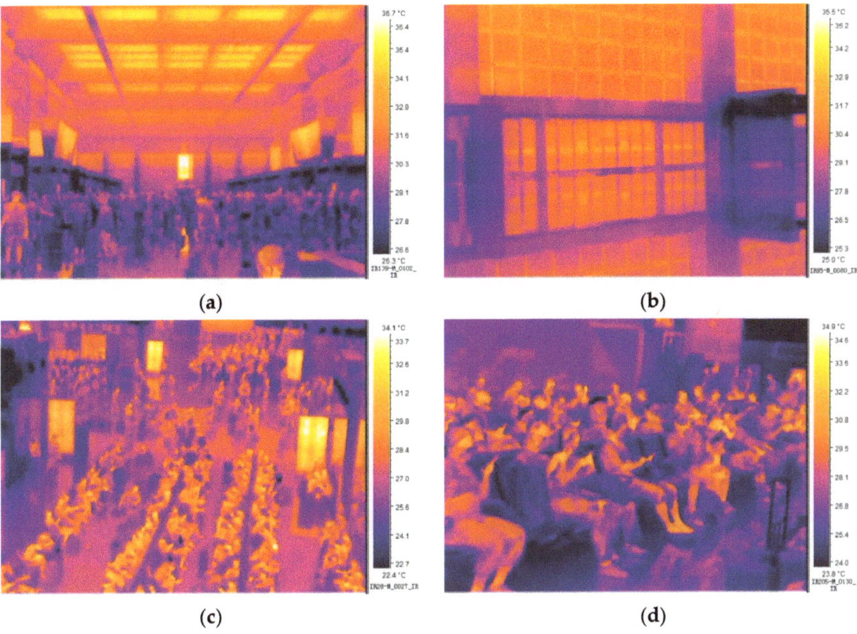

Figure 3. Thermal imaging of different areas of the waiting hall: (**a**) an overview of the waiting hall; (**b**) main entrance; (**c**) waiting area; (**d**) waiting passengers.

Figure 3a provides an overview thermal image of the waiting hall, showing a temperature difference of 10.4 °C. Temperature stratification is observed within the waiting hall, with higher temperatures in the upper space compared to the lower space, particularly noticeable from the ticket checking gate height. The station hall's ceiling is composed of glass skylights, resulting in a significant influence from outdoor conditions and overall higher temperatures in this area.

Figure 3c,d depict the temperature distribution within the waiting area. As shown in the figure, the main heat source in the waiting area is the heat generated by the flowing people, while other heat sources such as advertising boards, lighting, commercial small shops, etc., have negligible effects and can be disregarded in the following control process.

3.1.2. Horizontal Indoor Air Temperature Distribution

A total of 112 temperature measurement points are set up within the measurement area, with 16 points in each subzone. The overall temperature distribution within the measurement area is calculated using MATLAB's built-in linear interpolation method; the generated temperature contour is shown in Figure 4 (inwhich blue to red gradient indicating temperature increasing). The maximum measured temperature within the measurement area is 28.2 °C, while the minimum is 25.7 °C, and the maximum temperature difference is 2.3 °C. The temperature variations and zonal average temperatures are summarized in Table 2. From the figure, it can be observed that there is a significant uneven distribution of cold and hot spots within the measurement region.

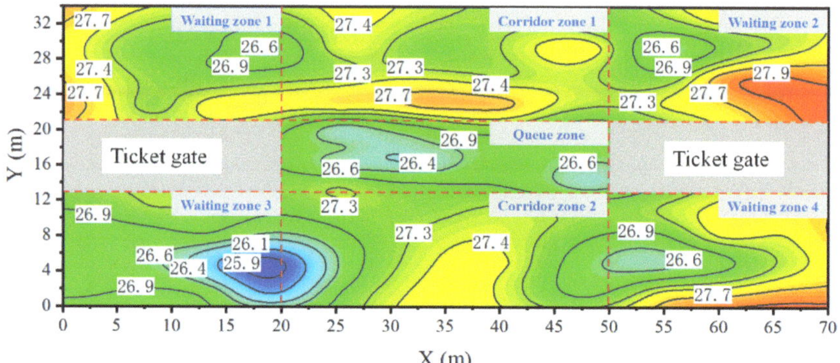

Figure 4. Temperature distribution in the measurement area.

Table 2. Measured temperature ranges and average values in each subzone.

Measurement Area	Waiting Zone				Corridor Zone		Queue Zone
	1	2	3	4	1	2	
Temperature range (°C)	26.5~27.7	26.5~28.2	26.7~27.5	26.5~27.8	27~27.8	26.7~27.7	25.7~27.3
Average temperature (°C)	27.25	27.5	26.7	27.4	27.4	27.15	26.7

To explore the reasons for the uneven temperature distribution within the measurement area, waiting zone 3 (waiting area 5A) is selected for temperature measurements under different conditions. The difference in the number of people is counted and measured, with three people density values set as follows: $\lambda = 25\%$, $\lambda = 50\%$, and $\lambda = 75\%$, corresponding to approximately 20 people, 40 people, and 60 people, respectively (here, $\lambda = 100\%$ means designed number of passengers in this area). Five measurement points are established in waiting zone 3, and the distribution of measurement points (point 1 to 5) and temperature measurement results are shown in Figure 5.

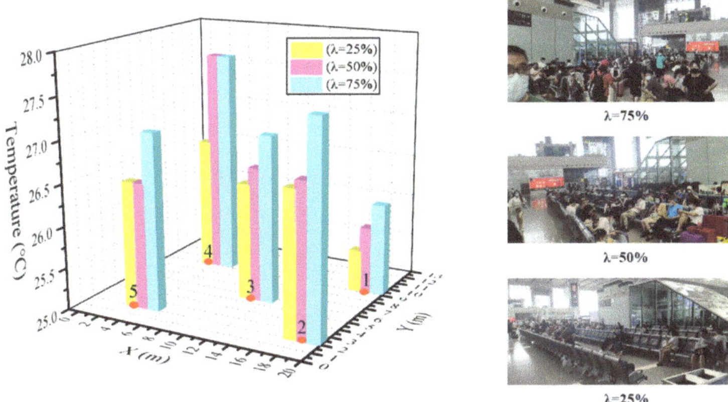

Figure 5. Comparison of the thermal environment of waiting zone 3 with different occupancy densities.

Figure 5 shows the temperature measurement results of waiting zone 3 under different occupancy density conditions. The X and Y axes represent the positions of the measurement points within the area, while the Z axis indicates the measured temperature. As shown in the figure, the highest measured temperature is 27.5 °C (measurement point 4, λ = 75%), and the lowest is 25.5 °C (measurement point 1, λ = 25%), with a maximum temperature difference of 2 °C. When occupancy accounts for 25% of the area, the highest measured temperature is 26.7 °C and the lowest is 25.5 °C, resulting in a temperature difference of 1.2 °C. With an occupancy rate of 50%, the temperature difference reaches 1.9 °C; the temperature difference is 1.4 °C with 75% of designed number of passengers. Under different occupancy density conditions, there is a temperature difference of over 1 °C between measurement points in waiting zone 3. Under varying occupancy densities, measurement point 1, which is closer to the air supply diffuser, consistently exhibits the lowest temperature, while measurement point 4, farther from the air supply diffuser, shows the highest temperature. Furthermore, at the same measurement point, the temperature measured when occupancy density is λ = 25% is consistently lower than when λ = 75%. Therefore, the temperature variation within the waiting zone correlates negatively with the number of occupants and their distance from the air supply diffuser. It is noted here that the air supply diffusers are operated with constant speed.

3.1.3. Indoor CO_2 Distribution

CO_2 concentration is one of the important factors affecting human comfort. Measurements of CO_2 concentration and the number of people in waiting zone 3 are conducted at different times of the day, with the measurement point located in the center of waiting zone 3 (see Figure 4), and the results are shown in Figure 6. The orange curve in the graph represents the number of people, while the blue bars represent CO_2 concentration. From the figure, it can be observed that, before 9:00, the number of people and CO_2 concentration show similar changing trends. From 10:00 to 17:00, the number of people remains relatively stable, with fluctuations of less than 10 people, and, during this period, the CO_2 concentration in waiting zone 3 also remains relatively stable. After 18:00, there is a significant decrease in the number of people, and, at the same time, the CO_2 concentration decreases slowly but follows the same decreasing trend. Therefore, there is a positive correlation between the change in CO_2 concentration and the change in the number of people, with a slight time lag. It is interesting that the overall CO_2 concentrations are less than 600 ppm, which represents a good indoor air quality (lower than the designed value, normally less than 1000 ppm); the reason for this can be explained by the tall space and opened indoor area since the CO_2 concentration can be easily diluted in this space.

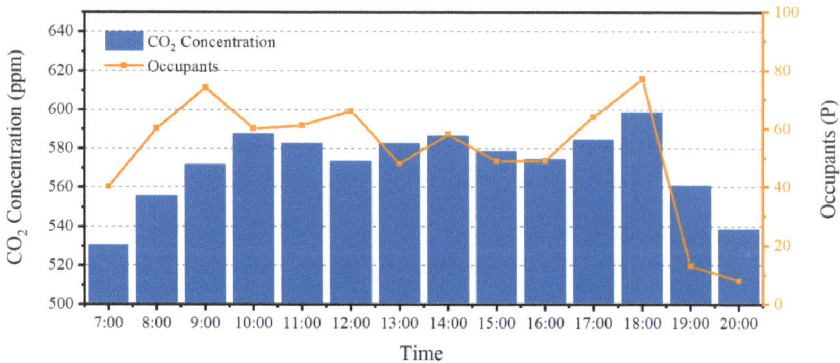

Figure 6. Variations in CO_2 concentration and number of people in waiting zone 3.

3.1.4. Questionnaire Results

This study collected a total of 542 valid questionnaires, including 335 males and 207 females, with a male-to-female ratio of approximately 6:4. The survey results are shown in Figure 7.

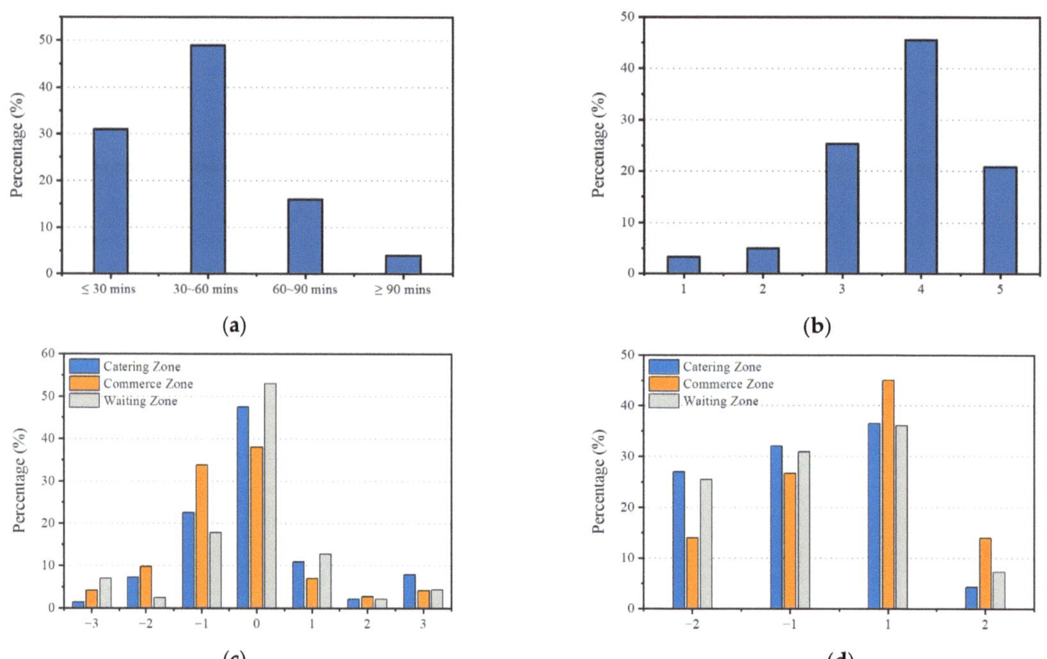

Figure 7. Results of the questionnaire: (**a**) waiting time for passengers; (**b**) passenger satisfaction with the air conditioning system; (**c**) thermal sensation vote; (**d**) wind sensation vote.

1. Passenger Waiting Time Statistics

Figure 7a shows the waiting times of passengers. As depicted, 35.7% of passengers have a waiting time of 30 min or less, 46.3% wait between 30 and 60 min, 13% wait between 60 and 90 min, and only 4% wait for over 90 min. Therefore, passengers generally have short waiting times at the high-speed railway stations, indicating high mobility among passengers.

2. Overall Satisfaction

Using a five-point scale, the overall satisfaction with the air conditioning system in summer is evaluated. As shown in Figure 7b, over 66% of passengers express satisfaction with the overall air conditioning system, with 45.57% rating 4 (satisfied) and 20.85% rating 5 (very satisfied). Meanwhile, among those who rated 1 to 3 points, the majority scored the air conditioning system 3 points (average), with only 3.32% expressing dissatisfaction (1 point). From the figure, it is evident that over 30% of passengers estimate the air conditioning system as average or below, indicating potential issues with temperature control in the high-speed railway station.

3. Thermal Sensation Statistics

Figure 7c displays the thermal sensation voting results of passengers in different areas of the waiting hall. As shown in the three different areas of the waiting hall, the proportion of passengers voting for neutral thermal sensation (0) is the highest. Among them, the waiting zone has the highest proportion of neutral thermal sensation, followed by the dining zone, while the commercial area's thermal sensation voting tends to lean towards slightly cool (−1) and neutral (0). Therefore, the overall thermal comfort is higher in the waiting zone, followed by the dining area, while the commercial area is generally cooler. Moreover, thermal sensation varies among passengers in different areas of the waiting hall.

4. Passenger Perception of Airflow

Figure 7d presents the voting results of passengers' perception of airflow in different areas of the waiting hall. As shown in the figure, the overall perception of airflow in each area is neutral. However, over 50% of passengers consider the airflow as too strong in both the dining and waiting areas. Moreover, the draft perception is achieved by over 40% of passengers in commercial area. Thus, the overall perception of airflow in the waiting hall is relatively high. It can be concluded that the mean thermal sensation in the dining and waiting areas is −1, indicating a perception of high airflow, while the perception of airflow in the commercial area is more balanced, with the overall thermal sensation close to neutral.

3.2. PMV Model

The air temperature and indoor air velocity measured on-site are introduced into the Predicted Mean Vote (*PMV*) model proposed by ASHRAE 55-2020 [25] to calculate the *PMV* and Predicted Percentage of Dissatisfied (*PPD*) values for the four waiting areas. The calculation formulas are as follows, and the results are shown in Table 3.

$$PMV = -\frac{7}{83}T_R v_R + \frac{28}{75}T_R - \frac{689}{74} \tag{1}$$

$$PPD = 100 - 95exp(-0.03353 \cdot PMV^4 - 0.2179 \cdot PMV^2) \tag{2}$$

where T_R is air temperature, °C; v_R is air velocity, m/s.

Table 3. Calculated PMV in measuring zone.

Measurement Area	Waiting Zone				Corridor Zone		Queue Zone
	1	2	3	4	1	2	
Temperature (°C)	27.25	27.5	26.7	27.4	27.4	27.15	26.7
Wind speed (m/s)	0.40	0.55	0.79	0.55	0.28	0.49	0.90
PMV	−0.05	−0.32	−1.12	−0.35	0.27	−0.29	−1.37
PPD	5.05%	7.13%	31.43%	7.55%	6.51%	6.75%	43.92%

ISO-7730-2005, which is an international standard of analytical determination and interpretation of thermal comfort using calculation of *PMV* and *PPD* indices and local thermal comfort criteria [7], when the *PMV* is between −0.5 and +0.5 and the *PPD* is less than 10%, it indicates a comfortable thermal environment. From the table, it can be observed that waiting area 4 and the queueing area do not perform well in terms of thermal comfort and tend to be overall cooler. The data from the table indicate that the main reason for this is the overall high airflow velocity, while the temperature remains within the comfortable range. During on-site measurements, the air conditioning diffusers throughout the entire waiting hall are almost at full operation, especially in main stream areas within the jet flow entrainment region, leading to a strong draft sensation and more severe energy consumption.

In summary, there is a significant uneven distribution of temperature and velocity in the waiting zones of the high-speed railway station, resulting in a discrepancy in experience for passengers in terms of thermal comfort. This is evident from the survey results and *PMV* indicators, showing poor thermal comfort perception in certain waiting areas. It can be concluded that the air conditioning and ventilation strategies employed at the high-speed railway station have several issues in controlling the regional thermal environment. Therefore, in the following section, this study will use professional software to model the target space, simulate temperature response with real-time passenger flow, and further investigate the potential energy conservation of the air conditioning system with proposed control algorithms.

4. Mathematical Control Model

4.1. Model Parameters

As shown in Figure 1, the studied area is simplified into three parts: the waiting area, the queueing area, and the aisle area. From the thermal images, it can be seen that there is a significant temperature stratification above the ticket checking gates, which are approximately 5 m high. Therefore, the height of the target area model in this study is set to 5 m. The dimensions of the waiting area (L × W × H) are 20 m × 13 m × 5 m, the queueing area is located between the two ticket checking gates, with dimensions of 30 m × 8 m × 5 m, and the aisle area has dimensions of 30 m × 13 m × 5 m.

Boundary conditions such as the wall temperature of the waiting area and the supply air temperature are measured. The air supply diffusers in the waiting area are set on both side walls, with 8 on each side. The air supply outlets in the queueing area are installed above the ticket checking gates, with a total of 14 on both sides of the gates. There are no air outlets in the aisle area. Additionally, the specifications of all air supply nozzle diffusers are consistent, with an outer geometric diameter of 0.45 m and an actual air outlet diameter of 0.25 m.

Based on the results obtained from the thermal imaging camera, the wall temperature of the waiting area is 31 °C, and the supply air temperature is 21 °C. The air supply velocity from the outlets measured by the anemometer is 4 m/s, the outdoor air temperature in the measurement area is 29 °C, and the indoor air temperature in the measurement area at 8 a.m. is 26 °C.

The determination of the number of people is particularly important as the main heat source in the study area [26]. From 8:00 a.m. to 8:00 p.m., photos are taken every 30 min to count the number of people in the waiting area, queueing area, and aisle area, resulting in 12 h of data on the changes in the number of people in the measurement area. The dynamic changes in the number of passengers in each area are shown in Figure 8. It can be observed from the figure that the number of people in the waiting area and the aisle area are relatively stable. The density of people in waiting areas 3 and 4 fluctuates around 0.33 people/m², while the density in waiting areas 1 and 2 remains stable at around 0.17 people/m². The number of people in the aisle area remains at a lower level, with an average density of only 0.04 people/m². The changes in the number of people in the queueing area are more significant. Due to the gathering of people at the ticket checking gates when the train arrives and departs, there are drastic fluctuations in the number of people, with a maximum difference exceeding 130 people. The maximum density of people in each area at different times is 0.62 people/m^2, which is lower than the design density of 0.67 people/m^2 [27]. Therefore, traditional ventilation control strategies based on design occupancy are prone to consume more energy than those based on real-time occupancy.

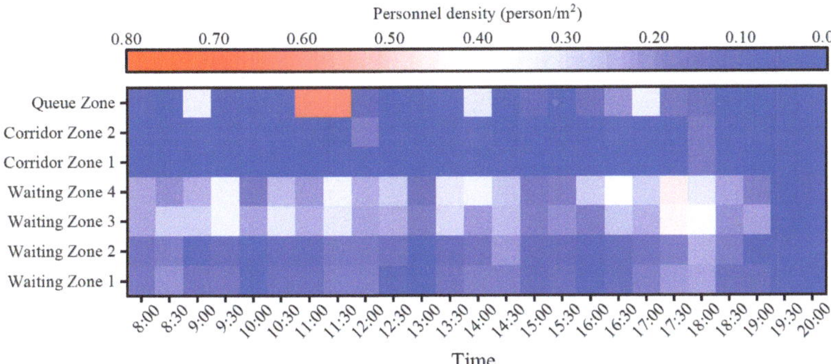

Figure 8. Calculation results of dynamic occupancy distribution in different areas.

4.2. Mathematical Model

The measurement area of this study is located indoors with few internal devices and no obvious heat sources. Therefore, the main internal heat source is occupants. To simplify the model, other heat sources (advertising board, lighting, small shops, equipment, and heat from the ceiling level) are ignored. Thus, according to the principle of energy conservation, the simplified equation for the measurement area can be written as follows:

$$\rho C_p V \frac{dt_n}{d\tau} = \rho C_p G (t_s - t_n) + Q_1 + Q_2 \tag{3}$$

where ρ is the air density in kg/m^3, C_p is the specific heat capacity of the room in J/(kg·°C), V is the volume of the room in m^3, G is the air supply volume in m^3/s, T_s and T_n are the supply air temperature and indoor air temperature, respectively, in Celsius, Q_1 and Q_2 are, respectively, the heat transferred by the maintenance structure and the heat generated by occupant, kW. However, the spatial area of the measurement zone is separated by two ticket checking points in the middle. Moreover, since occupants are not uniformly distributed in the area but vary spatially over time, it is unreasonable to assume that the temperature distribution inside the entire area is uniform. Therefore, this model is not suitable for temperature control in the measurement area.

This paper proposes a method for coordinating temperature zoning control in large open spaces by incorporating thermal coupling effects. This method divides the large space into multiple sub-regions while considering the impact of thermal coupling effects between adjacent regions on the control of the air conditioning system [28,29]. In this study, the measurement area is divided into seven sub-regions based on their functions and load characteristics: four waiting areas, two aisle areas, and one queue area, with a total of five controllable sub-regions since there are no air vents in the aisles. Since there are no physical partitions between zones in the zoning model, when actual energy exchange occurs between adjacent zones, the heat transfer between neighboring zones may be unequal due to differences in airflow on both sides. Even with the same airflow, differences in heat transfer may occur due to turbulence, a phenomenon known as thermal coupling. In order to accurately describe the energy exchange between adjacent zones during this process, some researchers have proposed the definition of a heat exchange coefficient that can be used to represent thermal coupling between adjacent zones [30–32]:

$$Q_{\text{coup}} = k_c (t_i - t_j) \tag{4}$$

where k_c is heat exchange coefficient between adjacent zones, kW/°C; t_i, t_j is temperature of adjacent intervals, °C.

Based on Equation (4), the simplified subzone equation can be obtained accordingly; the energy conservation equation in waiting area can be written as

$$\rho C_p V \frac{dt_n}{d\tau} = \rho C_p G(t_s - t_n) + h_w A_w(t_w - t_n) + h_a A_a(t_a - t_n) + k_c(t_c - t_n) + Q \tag{5}$$

the energy conservation equation in aisle area can be written as

$$\rho C_p V \frac{dt_n}{d\tau} = h_a A_a(t_a - t_n) + k_{c,1}(t_{c,1} - t_n) + k_{c,2}(t_{c,2} - t_n) + k_{c,3}(t_{c,3} - t_n) + Q \tag{6}$$

where ρ is air density, kg/m^3; C_p is room heat capacity, kJ/(kg·°C); V is room volume, m^3; G is air supply flow-rate, (m^3/s); t_w, t_a, t_c are side wall temperature, outside air temperature, and its adjacent zonal temperature, °C; h_w, h_a represents the convective heat transfer coefficient between the side wall surface and the outside air, as well as between the air inside the waiting area and the air outside the waiting area, kW/(m^2·°C); k_c stands for the heat exchange coefficient between the waiting area and adjacent areas, kW/°C; and Q is internal heat gain generated by occupants, kW.

4.3. Dynamic Thermal Model

In order to predict the thermal response of the building under given conditions (such as changes in the number of people), a dynamic thermal model of the building is developed. Detailed white-box models (the mathematical model can be directly acquired from the relationship between input and output) require high computational costs, while black-box models (the completely unknown relationship between input and output) require a large amount of data for training; therefore, this study adopts a gray-box model (the relationship between the input and output is partly known in the control system) to simulate building heat transfer. Considering that complex models increase the computational time for control simulations, the gray-box model is simplified.

The simplified model possesses both controllability and prediction accuracy; thus, it is widely used for simulating thermal changes in indoor spaces. The controlled area studied in this paper is located within the building, where solar radiation is low and there are no significant radiant heat sources; hence, radiative heat transfer can be neglected. Similar to an electrical network, the thermal changes within each sub-area can be represented by a 1R1C model. This model uses thermal resistance and thermal capacitance to represent the thermal characteristics of the building, including the building envelope, external environment, indoor air, and thermal coupling between subzones. The heat transfer between each component of the model is depicted in Figure 9. The air temperature within each sub-area is influenced by the cold air delivered by the HVAC system, external environment, internal heat gains within the sub-area, and thermal coupling between adjacent areas. Therefore, the thermal balance equations for each subzone can be revised as follows (Equations (7)–(13)):

$$C_{in,1} \frac{dT_{in,1}}{d\tau} = \frac{T_w - T_{in,1}}{R_{w,1}} + \frac{T_{out} - T_{in,1}}{R_{a,1}} + \frac{T_{in,5} - T_{in,1}}{R_{5,1}} + Q_{in,1} + Q_{HVAC,1} \tag{7}$$

$$C_{in,2} \frac{dT_{in,2}}{d\tau} = \frac{T_w - T_{in,2}}{R_{w,2}} + \frac{T_{out} - T_{in,2}}{R_{a,2}} + \frac{T_{in,5} - T_{in,2}}{R_{5,2}} + Q_{in,2} + Q_{HVAC,2} \tag{8}$$

$$C_{in,3} \frac{dT_{in,3}}{d\tau} = \frac{T_w - T_{in,3}}{R_{w,3}} + \frac{T_{out} - T_{in,3}}{R_{a,3}} + \frac{T_{in,7} - T_{in,3}}{R_{7,3}} + Q_{in,3} + Q_{HVAC,3} \tag{9}$$

$$C_{in,4} \frac{dT_{in,4}}{d\tau} = \frac{T_w - T_{in,4}}{R_{w,4}} + \frac{T_{out} - T_{in,4}}{R_{a,4}} + \frac{T_{in,7} - T_{in,4}}{R_{7,4}} + Q_{in,4} + Q_{HVAC,4} \tag{10}$$

$$C_{in,5} \frac{dT_{in,5}}{d\tau} = \frac{T_{out} - T_{in,5}}{R_{a,5}} + \frac{T_{in,1} - T_{in,5}}{R_{5,1}} + \frac{T_{in,2} - T_{in,5}}{R_{5,2}} + \frac{T_{in,6} - T_{in,5}}{R_{5,6}} + Q_{in,5} \tag{11}$$

$$C_{in,6}\frac{dT_{in,6}}{d\tau} = \frac{T_{in,5} - T_{in,6}}{R_{5,6}} + \frac{T_{in,7} - T_{in,6}}{R_{7,6}} + Q_{in,6} + Q_{HVAC,6} \tag{12}$$

$$C_{in,7}\frac{dT_{in,7}}{d\tau} = \frac{T_{out} - T_{in,7}}{R_{a,7}} + \frac{T_{in,3} - T_{in,7}}{R_{3,7}} + \frac{T_{in,4} - T_{in,7}}{R_{4,7}} + \frac{T_{in,6} - T_{in,7}}{R_{7,6}} + Q_{in,7} \tag{13}$$

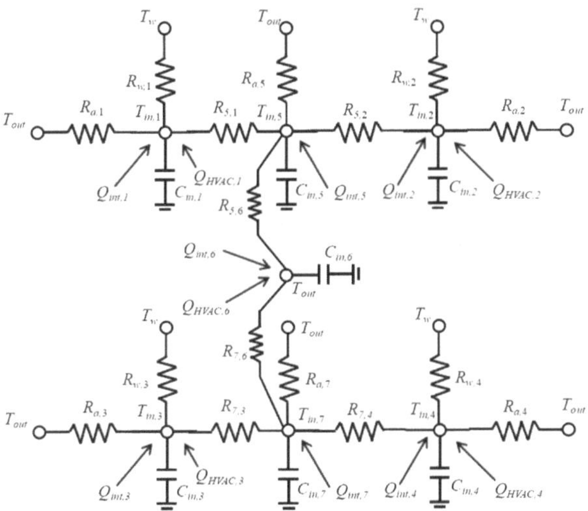

Figure 9. Schematic of RC gray-box building thermal model.

In the equation, C and R, respectively, represent the thermal capacitance and thermal resistance of the model; T represents temperature, with subscripts *in*, *out*, and *w* denoting the air inside each subzone, the external area, and the surface of the envelope structure, respectively; Q_{in} denotes the internal heat gains of each subzone; and Q_{HVAC} represents the cooling capacity provided by the HVAC system for each subzone.

The internal heat gains of the subzones are only from human body heat dissipation, and the occupants in the waiting area are in a state of light activity. According to the "Energy Efficiency Design Standard for Public Buildings [33]", the average heat dissipation per person in transportation buildings is 134 W/person. Therefore, the internal heat gain Q_{in} is expressed as

$$Q_{in} = P \times q \tag{14}$$

In the equation, P represents the number of occupants within the area; q denotes the average heat dissipation per person, which is 134 W/per person.

Additionally, for the convenience of calculating the cooling capacity provided by the HVAC system for each subzone, it is assumed that the air inside each subzone is uniformly mixed. The cooling capacity Q_{HVAC} provided by the HVAC system is calculated using the heat transfer equation:

$$Q_{HVAC,i} = \rho C G_i \left(T_{supply,i} - T_{return,i} \right) \tag{15}$$

where ρ, C are air density and heat capacity; G_i is subzone supply airflow rate, m^3/s; and $T_{supply,i}$, $T_{return,i}$ are zonal supply and return air temperature, °C.

It is worth noting that, at the same time, the above dynamic thermal model of the building has also been simplified: due to the controlled area being within the building, the wall temperature and surrounding environmental temperature remain relatively stable and have minimal impact on the air temperature inside; thus, they are set as constant temperatures. The heat transfer resistance between adjacent zones in the dynamic thermal model of the building is closely related to the heat exchange coefficient k_c between zones, which is a function of the temperature difference between adjacent zones. Therefore, during the calculation process, the heat transfer resistance between adjacent zones varies at each time step. The adaptive model predictive control (AMPC) adopted in this study can adjust the parameters of the model based on real-time data updates, self-correcting at each time step to improve control accuracy by correcting prediction results.

4.4. Air Conditioning System Control Optimization Based on Model Predictive Control

4.4.1. Control Mechanism of Adaptive Model Predictive Control

Adaptive model predictive control (AMPC) combines the principles of model predictive control (MPC) and adaptive control to handle situations where there is uncertainty, variation, or unknown parameters in the system model. The key aspect of adaptive MPC is its ability to adjust the model parameters in real time based on measurements using recursive or adaptive parameter update methods, thus continuously updating the system model to reflect changes in its dynamic behavior. Subsequently, based on the updated model and the current system state, an optimization algorithm is used to solve an optimization problem at each sampling time to compute the optimal control input sequence. The objective of the optimization problem is to minimize prediction errors while satisfying constraints to obtain the optimal control strategy. Based on the optimal control input sequence obtained, only the first control signal is applied and executed, and the process of model updating, optimization problem solving, and control signal application is repeated in the next sampling time. Through this iterative process, adaptive MPC can track real-time changes in the system's dynamic characteristics and adjust accordingly based on real-time information.

Figure 10 illustrates a schematic diagram of the AMPC for the air conditioning system in the waiting area of the high-speed railway station under real-time passenger conditions. In the structure of AMPC, the dynamic model of the system is determined based on measured boundary conditions. To obtain an adaptive model as the control model for MPC to predict the current output, the model parameters are updated and adjusted using the output results from the previous time step. Based on the prediction results, the optimal control signal for the system is determined by solving an optimization problem under the conditions of the objective function and constraints. In this study, the optimized control signal is the cooling capacity provided by the air conditioning supply fans in each area.

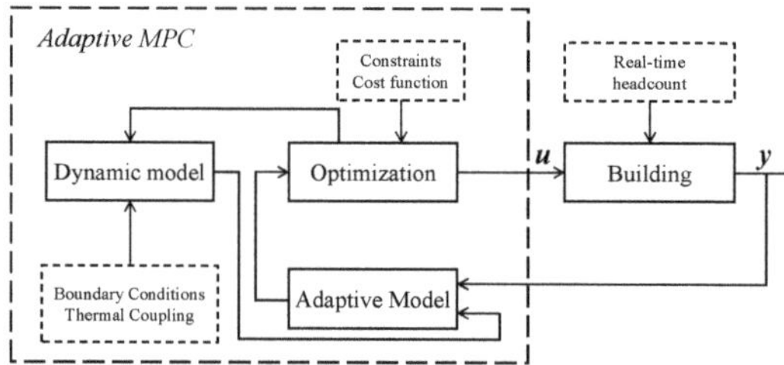

Figure 10. Schematic diagram of AMPC control principle based on real-time number of people.

4.4.2. MPC Control Model

To establish the adaptive model predictive control controller, it is necessary to develop the dynamic thermal model of the controlled area and an adaptive model for updating model parameters. Since the issue investigated in this paper is a Multiple Input Multiple Output (MIMO) problem, a state-space model is adopted to describe the system dynamics. In order to clearly express the relationship between the inputs and outputs of the control system, the dynamic thermal model of the building (i.e., Equations (7)–(13)) is linearized and transformed into a continuous-time linear state-space model, as shown in Equation (16).

$$dx/dt = Ax + Bu \tag{16}$$

Here, system state matrix $x = [T_{in,1},T_{in,2},T_{in,3},T_{in,4},T_{in,5},T_{in,6},T_{in,7}]^{\mathrm{T}}$; input matrix $u = [Q_{in,i},Q_{HVAC,i},T_w,T_{out}]^{\mathrm{T}}$ ($i = 1\sim7$). System matrix:

$$A = \begin{bmatrix} \frac{-1}{C_{in,1}\cdot R_{w,1}} + \frac{-1}{C_{in,1}\cdot R_{a,1}} + \frac{-1}{C_{in,1}\cdot R_{5,1}} & 0 & 0 & 0 & \frac{1}{C_{in,1}\cdot R_{5,1}} & 0 & 0 \\[2mm] 0 & \frac{-1}{C_{in,2}\cdot R_{w,2}} + \frac{-1}{C_{in,2}\cdot R_{a,2}} + \frac{-1}{C_{in,2}\cdot R_{5,2}} & 0 & 0 & \frac{1}{C_{in,2}\cdot R_{5,2}} & 0 & 0 \\[2mm] 0 & 0 & \frac{-1}{C_{in,3}\cdot R_{w,3}} + \frac{-1}{C_{in,3}\cdot R_{a,3}} + \frac{-1}{C_{in,3}\cdot R_{7,3}} & 0 & 0 & 0 & \frac{1}{C_{in,3}\cdot R_{7,3}} \\[2mm] 0 & 0 & 0 & \frac{-1}{C_{in,4}\cdot R_{w,4}} + \frac{-1}{C_{in,4}\cdot R_{a,4}} + \frac{-1}{C_{in,4}\cdot R_{7,4}} & 0 & 0 & \frac{1}{C_{in,4}\cdot R_{7,4}} \\[2mm] \frac{1}{C_{in,5}\cdot R_{1,5}} & \frac{1}{C_{in,5}\cdot R_{2,5}} & 0 & 0 & \frac{-1}{C_{in,5}\cdot R_{a,5}} + \frac{-1}{C_{in,5}\cdot R_{1,5}} + \frac{-1}{C_{in,5}\cdot R_{2,5}} + \frac{-1}{C_{in,5}\cdot R_{6,5}} & \frac{1}{C_{in,5}\cdot R_{6,5}} & 0 \\[2mm] 0 & 0 & 0 & 0 & \frac{1}{C_{in,6}\cdot R_{6,5}} & \frac{-1}{C_{in,6}\cdot R_{6,5}} + \frac{-1}{C_{in,6}\cdot R_{7,6}} & \frac{1}{C_{in,6}\cdot R_{7,5}} \\[2mm] 0 & 0 & \frac{1}{C_{in,7}\cdot R_{7,3}} & \frac{1}{C_{in,7}\cdot R_{7,4}} & 0 & \frac{1}{C_{in,7}\cdot R_{7,6}} & \frac{-1}{C_{in,7}\cdot R_{a,7}} + \frac{-1}{C_{in,7}\cdot R_{3,7}} + \frac{-1}{C_{in,7}\cdot R_{4,7}} + \frac{-1}{C_{in,7}\cdot R_{6,7}} \end{bmatrix}_{7\times7}$$

input matrix:

$$B = \begin{bmatrix} -\frac{1}{C_{in,1}} & 0 & 0 & 0 & 0 & 0 & 0 & \frac{1}{C_{in,1}} & 0 & 0 & 0 & 0 & 0 & 0 & \frac{1}{C_{in,1}R_{w,1}} & \frac{1}{C_{in,1}R_{a,1}} \\[1mm] 0 & -\frac{1}{C_{in,2}} & 0 & 0 & 0 & 0 & 0 & 0 & \frac{1}{C_{in,2}} & 0 & 0 & 0 & 0 & 0 & \frac{1}{C_{in,2}R_{w,2}} & \frac{1}{C_{in,2}R_{a,2}} \\[1mm] 0 & 0 & -\frac{1}{C_{in,3}} & 0 & 0 & 0 & 0 & 0 & 0 & \frac{1}{C_{in,3}} & 0 & 0 & 0 & 0 & \frac{1}{C_{in,3}R_{w,3}} & \frac{1}{C_{in,3}R_{a,3}} \\[1mm] 0 & 0 & 0 & -\frac{1}{C_{in,4}} & 0 & 0 & 0 & 0 & 0 & 0 & \frac{1}{C_{in,4}} & 0 & 0 & 0 & \frac{1}{C_{in,4}R_{w,4}} & \frac{1}{C_{in,4}R_{a,4}} \\[1mm] 0 & 0 & 0 & 0 & 0 & 0 & 0 & 0 & 0 & 0 & 0 & \frac{1}{C_{in,5}} & 0 & 0 & 0 & \frac{1}{C_{in,5}R_{a,5}} \\[1mm] 0 & 0 & 0 & 0 & 0 & -\frac{1}{C_{in,6}} & 0 & 0 & 0 & 0 & 0 & 0 & \frac{1}{C_{in,6}} & 0 & 0 & 0 \\[1mm] 0 & 0 & 0 & 0 & 0 & 0 & 0 & 0 & 0 & 0 & 0 & 0 & 0 & \frac{1}{C_{in,7}} & 0 & \frac{1}{C_{in,7}R_{a,7}} \end{bmatrix}_{7\times16}$$

To determine the values of the block parameters, R and C, for the dynamic thermal model, the supply air temperature, air temperature, and wall temperature in each area are recorded and fed into the model. The calculated R and C are shown in Table 4.

Table 4. Parameters of dynamic building thermal model.

Parameter	Value (J/K)	Parameter	Value (K/W)	Parameter	Value (K/W)	Parameter	Value (K/W)
$C_{in,1}$	1.5756×10^6	$R_{w,1}$	5.2083×10^{-4}	$R_{a,1}$	0.0013	$R_{1,5}$	0.5495×10^{-4}
$C_{in,2}$	1.5756×10^6	$R_{w,2}$	5.2083×10^{-4}	$R_{a,2}$	0.0013	$R_{2,5}$	0.5495×10^{-4}
$C_{in,3}$	1.5756×10^6	$R_{w,3}$	5.2083×10^{-4}	$R_{a,3}$	0.0013	$R_{3,7}$	0.5495×10^{-4}
$C_{in,4}$	1.5756×10^6	$R_{w,4}$	5.2083×10^{-4}	$R_{a,4}$	0.0013	$R_{4,7}$	0.5495×10^{-4}
$C_{in,5}$	2.3634×10^6	$R_{w,5}$	—	$R_{a,5}$	5.5556×10^{-4}	$R_{5,6}$	0.3571×10^{-4}
$C_{in,6}$	1.4544×10^6	$R_{w,6}$	—	$R_{a,6}$	—	$R_{6,7}$	0.3571×10^{-4}
$C_{in,7}$	2.3634×10^6	$R_{w,7}$	—	$R_{a,7}$	5.5556×10^{-4}		

4.4.3. Model Discretization

Discretization of the model involves converting the continuous state-space building thermal model into a discrete state-space model based on sampling time, which is then applied to the MPC controller. Combining the continuous-time state-space building thermal model as shown in Equation (16) with the cooling load demand model, the discrete-time state-space model can be represented as Equations (17) and (18), which is used for predicting the evolution of the system in the MPC optimal control strategy.

$$x_{k+1} = A_d x_k + B_d u_k \tag{17}$$

$$y_k = C_d x_k \tag{18}$$

where A_d and B_d represent the discrete results of the system matrix A and the input matrix B at the sampling time, respectively; y_k is the predicted result vector containing the indoor air temperatures of each subzone.

4.4.4. Control Conditions

MPC control mainly consists of four parts: the objective function, constraints, system dynamics, and current state. Among these, system dynamics and current state refer to the building dynamic model and initial conditions. The objective of the MPC controller is to minimize the power consumption of the air conditioning fan under real-time passenger flow conditions. Therefore, at each time step, the objective of the MPC controller, as shown in Equation (19), is to achieve stable temperature control and minimize the air conditioning fan power demand within the prediction horizon, denoted as "N". The prediction horizon (N) at each sampling time refers to the duration from the next time step to the end of the simulated control.

$$\min J = \sum_{0}^{N-1} \left(X_{in}^k - X_{ref} \right)^T Q \left(X_{in}^k - X_{ref} \right) + \left(U^K \right)^T R \left(U^K \right) + \left(X_{in}^N - X_{ref} \right)^T F \left(X_{in}^N - X_{ref} \right) \tag{19}$$

where X_{ref} is the setpoint for air temperature; the three terms in the equation represent the weighted sum of errors, the weighted sum of inputs, and the weighted sum of terminal errors, respectively. Q, R, and F are weights assigned to each term according to the desired control effect.

The constraints of the MPC controller are defined by Equations (20) and (21), which include limits on indoor air temperature and the cooling capacity provided by the air conditioning system. Throughout the control period, the indoor air temperature is constrained to be within 0.5 °C of the setpoint, and the cooling capacity provided by the air conditioning system should not be less than 0 W. Utilizing the aforementioned objective function, inequality constraints, and discrete-time state-space model, the MPC controller is formulated as a linear optimization problem. Due to its high computational efficiency and ease of solution, it facilitates optimal control in practical applications.

$$-0.5\,^{\circ}\mathrm{C} \leq T_{\mathrm{MPC}}^K - T_{\mathrm{ref}}^K \leq 0.5\,^{\circ}\mathrm{C} \tag{20}$$

$$0\,\mathrm{W} \leq Q_{\mathrm{HVAC}} \tag{21}$$

4.5. Test Platform

This study utilizes MATLAB/Simulink software (Version R2023a) to test the designed adaptive MPC controller to optimize the control effectiveness of the air conditioning system in the high-speed railway station waiting zone under real-time passenger conditions. In Simulink, a dynamic thermal model of the waiting area is established, and the accuracy of this model is validated using site measurement results.

The initial temperature of the area is set to 26 °C, and there are a total of 8 diffusers in each waiting area. The model considers the occupants within the area as internal heat sources, and detailed parameter settings of the model can be found in Table 5.

Table 5. Setup of the boundary condition.

	Waiting Zone (4)	Queue Zone	Corridor Zone (2)	Remarks
Room size	20 m × 13 m × 5 m	30 m × 8 m × 5 m	30 m × 13 m × 5 m	
Supply air inlet		Round nozzle; Φ = 0.25 m		
Number of supply air inlets	8	14	—	
Supply airflow rate	1.57 m^3/s	2.74 m^3/s	—	
Supply air temperature		21 °C		
Internal heat gains		Occupants; 134 W/person		[34]
Enclosure (checking gate)		31 °C		
The air temperature outside the control area		29 °C		
Heat exchange coefficient (thermal coupling)		280 W/(m^2·°C)		[30]
Convective heat transfer coefficient (wall with air)		9.6 W/(m^2·°C)		[35]
Convective heat transfer coefficient (air with air)		12 W/(m^2·°C)		[31]

The zonal control model of the waiting area established using Simulink is shown in Figure 11. The variable air volume system (VAV) employs adaptive MPC control. The simulation time is set to 12 h, with the set temperature for each zone being T_{ref} = 28 °C. The supply air temperature, t_s, remains at 21 °C constant. The heat source for the model is the real-time number of occupants within a 12 h period. The dynamic simulation time step is set to 1 s.

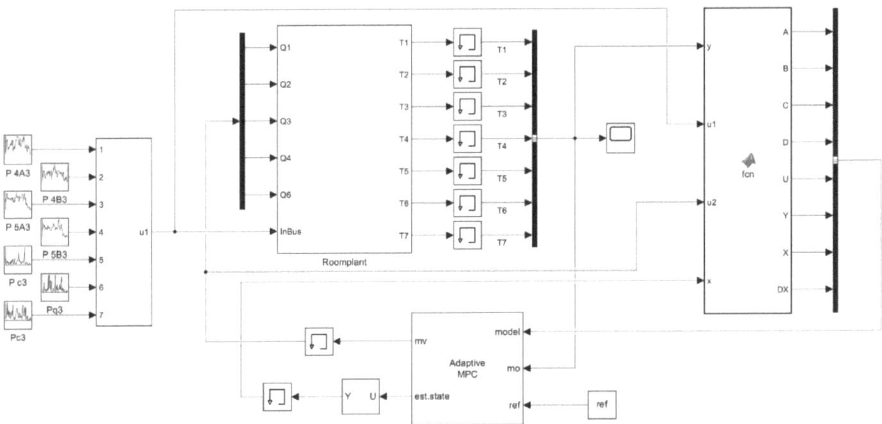

Figure 11. Simulation model of the measurement area.

In order to test the control performance of the proposed adaptive MPC control method and to validate the energy-saving control effect of the proposed control strategy on the operation of the air conditioning system under real-time passenger conditions, this study designed three scenarios. The specific details are shown in Table 6. Among these, the designed capacity for waiting areas 1 and 3, based on the seating arrangement within the waiting area, is 42 people, while, for waiting areas 2 and 4, it is 82 people. According to the Railway Passenger Station Design Code [36], the capacity for aisle areas is designed for 10 people, and, for queueing areas, it is designed for 40 people.

Table 6. Simulation scenarios.

Cases	Controller	Internal Heat Gains
Case 1	PID control	Design number of people
Case 2	AMPC control	
Case 3		Real-time number of people

5. Analysis of Simulation Results

5.1. Model Validation

The box plots show the measured values, and the yellow lines show the predicted values. Based on the actual measurement data, the dynamic thermal model of the building is validated. A comparison between the predicted and measured temperatures of waiting area 3 is shown in Figure 12. To quantify the deviation between the predicted and measured data, two metrics are used to evaluate the predictive performance: Mean Absolute Error (MAE) and Root Mean Square Error (RMSE), as shown in Table 7. Therefore, the model established in this study can accurately simulate the temperature variations within the area and serve as a benchmark model.

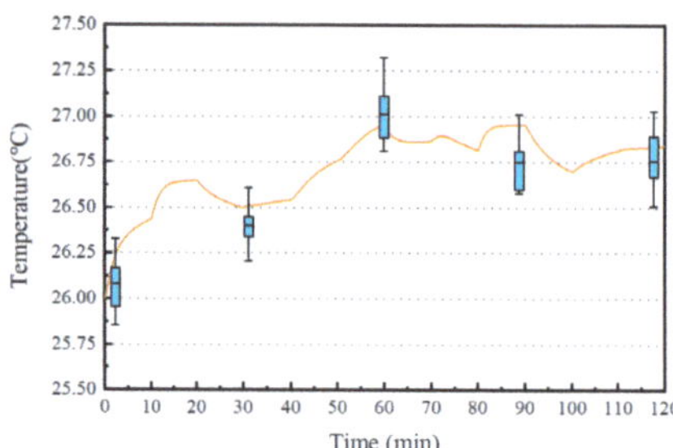

Figure 12. Comparison between the predicted and actual indoor air temperatures.

Table 7. Accuracy indices of the dynamic building thermal model.

	MAE (°C)	RMSE (°C)
Indoor Air Temperature	0.176	0.206

5.2. Simulation Results of Air Temperature under Different Conditions

Based on the design conditions in Table 6, the temperature responses of each subzone are simulated and analyzed. The model parameters are consistent for each condition. The simulated temperature results of the air conditioning operation for 2 h under different conditions are shown in Figure 13.

Figure 13a shows the temperature control results of case 1. In case 1, each sub-area is independently controlled using PID controllers while considering inter-zone thermal coupling. To compare with the proposed adaptive model predictive control (AMPC) results, the temperature is set to 28 ± 0.5 °C. As shown in the figure, the temperature trends in each sub-area are basically the same, but none of them can be controlled within the design temperature range. Moreover, the temperature variations in the queueing area and the aisle area are greater than that in the waiting area, with the temperature difference in the queueing area exceeding 2 °C with variations in the number of people. Therefore, the PID controller cannot provide accurate control effects for the proposed zoning control method.

Figure 13b shows the temperature responses of each area in case 2. Case 2 adopts the adaptive model predictive control, with the control conditions as described in Section 4.4.1, and the main heat source within the area remains the design occupancy. It can be observed from the figure that, compared to the PID control, the AMPC can significantly improve the temperature control effect. After 30 min, the air temperatures in each respective area are controlled within the design temperature range. It is noted that the air temperature in the queueing area is consistently lower than in other areas, with an average temperature difference of 0.37 °C. The control setting time is approximately 11 min, with a maximum overshoot of 1.7 °C, indicating good control performance. To be more in accordance with the actual operation of the high-speed rail station's air conditioning system, the proposed AMPC control method uses the actual passenger flow in each area as the main heat source. The temperature control response of each sub-area under this condition is shown in Figure 13c; at the beginning of the control, the temperature reaches the temperature setpoint at 12 min, with all four waiting areas reaching the designed values around 23 min. Meanwhile, under actual passenger conditions, the temperature changes when the number of people are minimal, with only the queueing area experiencing significant changes due to drastic fluctuations in the number of passengers, resulting in temperature variations exceeding the control range from 15 to 25 min and from 70 to 80 min. The aisle areas are adjacent to the queueing area, and the effect of inter-area thermal coupling causes the air temperatures in these two areas to follow the same trend as the queueing area. However, due to the lower number of people in these areas, the temperature changes exceed the set temperature limit. Therefore, under the conditions of the AMPC controller, the temperature responses of each sub-area within the waiting area are within a reasonable range, indicating an ideal temperature control effect.

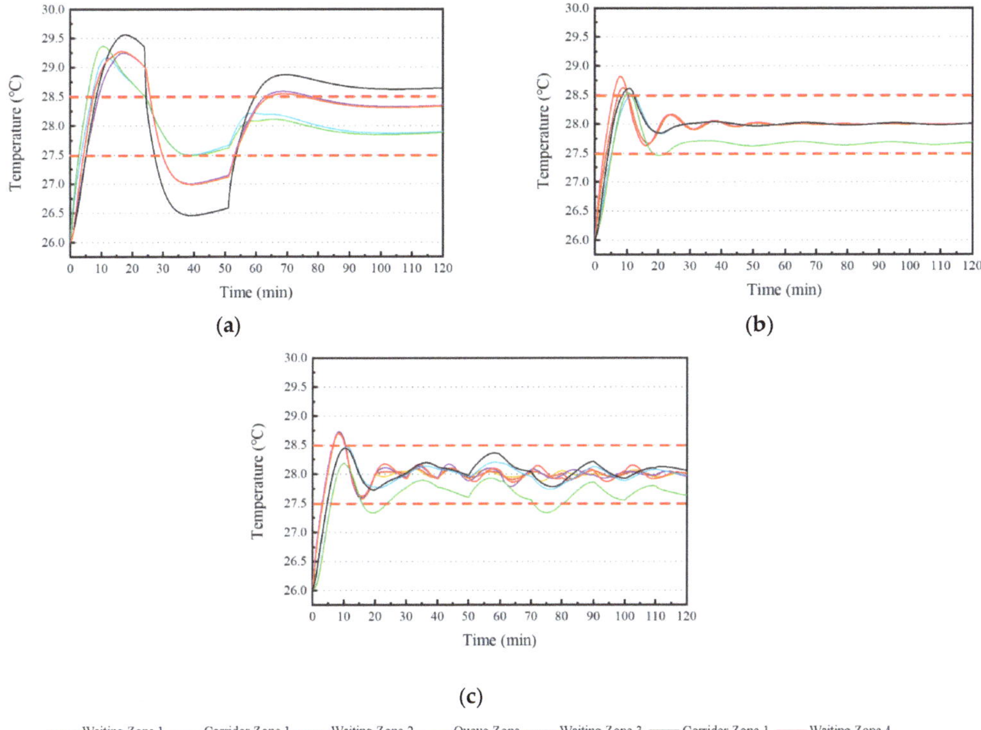

Figure 13. Simulation results of temperature in each area under different working conditions: (**a**) Case 1; (**b**) Case 2; (**c**) Case 3.

In conclusion, the control effect of the proposed AMPC control method is significantly better than that of the traditional PID control method. Moreover, under real-time passenger conditions, the AMPC controller can quickly control the temperature within the required range.

5.3. Fan Energy Consumption

The performance of the control scenarios with actual occupancy as the heat source input undoubtedly exceeds the performance of those with design occupancy. However, the resulting issue of fan energy consumption should not be overlooked. The energy consumption of ventilation systems is directly proportional to the air volume and can be calculated using a formula in Ref. [37]. The main difference between case 1 and case 2 lies in the system's response dynamic performance indicators, while their energy consumption is almost the same, so no fan energy consumption estimation is created. The total energy consumption of the fans running for 12 h in cases 2 and 3 is calculated as shown in Table 8, and the energy consumption of each area is shown in Figure 14. Using case 2 as a benchmark, the fan energy saving rate of the zoning control method based on the real-time occupancy proposed in this paper is approximately 10% to 35% (black square dot), providing an overall energy saving rate of 28.04%.

Table 8. Energy consumption of ventilation (12 h).

Cases	Energy Consumption (kWh)	Energy Saving Ratio
Case 1	309.13	—
Case 2	311.72	—
Case 3	224.31	28.04%

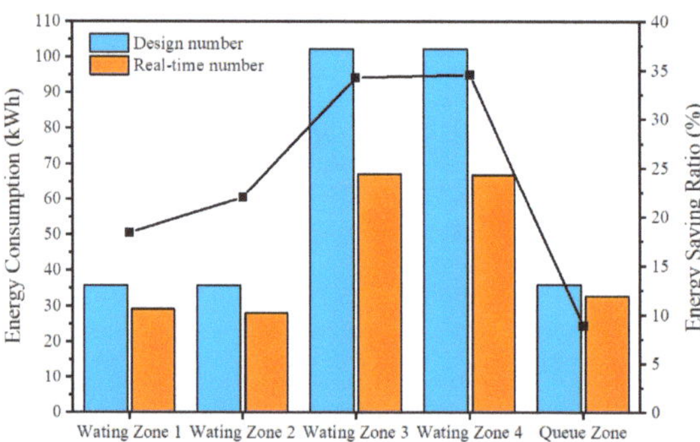

Figure 14. Fan energy consumption and energy saving rate (12 h).

6. Conclusions

This paper aims to optimize the ventilation control strategy in the high-speed rail station waiting area. The field surveys of the waiting hall revealed issues related to the thermal environment. An improved zoning model was proposed and Simulink (R2023a) software was utilized to simulate different design scenarios, and the following conclusions were drawn:

There is a significant thermal imbalance phenomenon in the high-speed rail station waiting area. The maximum temperature difference in the waiting area is 2.3 °C, mainly caused by the uneven distribution of people in the space.

This paper employs a zoning control model to simulate the dynamic temperature changes in the research area. By introducing the coefficient of heat exchange between the adjacent zones under the thermal coupling effect (which is different from the previous studies in which the zonal model has clear physical partitions), the temperature simulation results of the zoning control model are in accordance with the actual temperature measurements; thus, the proposed zoning model can accurately simulate real operating conditions.

The zoning control strategy based on real-time occupancy proposed in this paper can reduce the controller adjustment time and decrease the fan energy consumption. Compared to the traditional control methods, the comprehensive energy-saving rate is approximately 28.04%.

7. Limitations and Future Work

In reality, the thermal coupling process between adjacent zones is relatively complex. Factors such as turbulent flow near virtual boundaries and temperature differences can affect the value of the heat exchange coefficient. In practical processes, the heat exchange coefficient between adjacent zones should vary in real time. However, this paper's analysis is limited to validating the effectiveness of the optimization ventilation control strategy based on the real-time occupancy in ventilation energy savings under the condition where the heat exchange coefficient varies linearly with the temperature difference between the neighboring zones. Therefore, the next stage of research needs to verify the effectiveness of the optimization control strategy under conditions where the heat exchange coefficient varies in real time due to differences in turbulence intensity and other factors. Additionally, this paper only selected one high-speed rail station as the research subject. Different high-speed rail station building layouts and dynamic changes in passengers may vary. Future applications and validations of the proposed energy-saving control strategies should be conducted in more high-speed rail buildings to improve the universality of the proposed optimization control strategies. Furthermore, the heat exchange processes between the subzones in the zoning method are complex, and there is a strong correlation among the temperature changes between the adjacent areas. The optimization algorithm used by the AMPC controller proposed in this paper cannot achieve rapid and accurate adjustment when facing complex situations. Therefore, the future research should also adopt better optimization control methods to further improve the performance of the controller, enabling the efficient energy-saving control of the ventilation and air conditioning systems in high-speed rail station waiting areas.

Author Contributions: Conceptualization, P.Z. and J.Z. (Jintao Zhou); methodology, P.Z.; site measurement, Y.T., Z.M., H.S., and M.Y.; control simulation, P.Z., J.Z. (Jintao Zhou), and Y.T.; writing—original draft preparation, P.Z., J.Z. (Jintao Zhou), and H.S.; writing—review and editing, Y.T., Z.M., and J.Z. (Jian Zhu), supervision, M.Y. and J.Z. (Jian Zhu), funding acquisition, Z.M. and J.Z. (Jian Zhu). All authors have read and agreed to the published version of the manuscript.

Funding: This work is supported by Returned Overseas Innovation and Entrepreneurship Support Program of Anhui Province (No.2022LCX020) and The National University Student Innovation and Entrepreneurship Training Program (202310359040).

Data Availability Statement: The data presented in this study are available in article.

Conflicts of Interest: Authors Yu Tang, Zicheng Ma and Ming Yao were employed by the company The First Company of China Eighth Engineering Bureau Ltd. The remaining authors declare that the research was conducted in the absence of any commercial or financial relationships that could be construed as a potential conflict of interest.

References

1. Outline of the Fourteenth Five-Year Plan for National Economic and Social Development of the People's Republic of China and the Vision 2035. 2021. Available online: http://www.xinhuanet.com/2021-03/13/c_1127205564.htm (accessed on 7 June 2024).
2. National Development and Reform Commission. Mid-Long Term Railway Network Plan (2016–2025). Available online: https://zfxxgk.ndrc.gov.cn/web/iteminfo.jsp?id=366 (accessed on 7 June 2024). (In Chinese)
3. Lan, B.; Yu, Z.J.; Huang, G. Study on the impacts of occupant distribution on the thermal environment of tall and large public spaces. *Build. Environ.* **2022**, *218*, 109134. [CrossRef]
4. Du, X. Investigation of indoor environment comfort in large high-speed railway stations in Northern China. *Indoor Built Environ.* **2020**, *29*, 54–66. [CrossRef]
5. Su, X.; Yuan, Y.; Wang, Z.; Liu, W.; Lan, L.; Lian, Z. Human thermal comfort in non-uniform thermal environments: A review. *Energy Built Environ.* **2023**, *5*, 853–862. [CrossRef]
6. Yang, L.; Zhao, S.; Zhai, Y.; Gao, S.; Wang, F.; Lian, Z.; Duanmu, L.; Zhang, Y.; Zhou, X.; Cao, B.; et al. The Chinese thermal comfort dataset. *Sci. Data* **2023**, *10*, 662. [CrossRef]
7. ISO 7730:2005; Moderate Thermal Environments Determination of the PMV and PPD Indices and Specification of the Conditions for Thermal Comfort. International Organization for Standardization: Geneva, Switzerland, 2005.
8. De Dear, R.; Brager, G.S. Developing an adaptive model of thermal comfort and preference. *ASHRAE Trans.* **1998**, *104*, 145–167.
9. Deb, C.; Ramachandraiah, A. Evaluation of thermal comfort in a rail terminal location in India. *Build. Environ.* **2010**, *45*, 2571–2580. [CrossRef]
10. Wang, C.; Li, C.; Xie, L.; Wang, X.; Chang, L.; Wang, X.; Li, H.X.; Liu, Y. Thermal environment and thermal comfort in metro systems: A case study in severe cold region of China. *Build. Environ.* **2023**, *227*, 109758. [CrossRef]
11. Jia, X.; Cao, B.; Zhu, Y.; Huang, Y. Field studies on thermal comfort of passengers in airport terminals and high-speed railway stations in summer. *Build. Environ.* **2021**, *206*, 108319. [CrossRef]
12. Yuan, Y.; Yue, H.; Chen, H.; Song, C.; Liu, G. Passenger thermal comfort in the whole departure process of high-speed railway stations: A case study with thermal experience and metabolic rate changes in summer. *Energy Build.* **2023**, *291*, 113105. [CrossRef]

13. Gao, R.; Zhang, H.; Li, A.; Wen, S.; Du, W.; Deng, B. Research on optimization and design methods for air distribution system based on target values. *Build. Simul.* **2021**, *14*, 721–735. [CrossRef]
14. Wijesooriya, K.; Mohotti, D.; Lee, C.-K.; Mendis, P. A technical review of computational fluid dynamics (CFD) applications on wind design of tall buildings and structures: Past, present and future. *J. Build. Eng.* **2023**, *74*, 106828. [CrossRef]
15. Lu, Y.; Dong, J.; Liu, J. Zonal modelling for thermal and energy performance of large space buildings: A review. *Renew. Sustain. Energy Rev.* **2020**, *133*, 110241. [CrossRef]
16. Lebrun, J. Exigences Physiologiques et Modalités Physiques de la Climatisation par Source Statique Concentrée. Rédaction et Administration. 1971. Available online: https://bibliotheque.insa-lyon.fr/recherche/viewnotice/id_sigb/26801/id_int_bib/1 (accessed on 7 June 2024). (In French)
17. Lu, Y.; Wang, Z.; Liu, J.; Dong, J. Zoning strategy of zonal modeling for thermally stratified large spaces. *Build. Simul.* **2021**, *14*, 1395–1406. [CrossRef]
18. Bauman, F.S.; Dally, A. *Underfloor Air Distribution (UFAD) Design Guide*; American Society of Electroplated Plastics: Washington, DC, USA, 2003.
19. Ezber, S.; Akdoğan, E.; Gemici, Z. Fuzzy Logic Based Heating and Cooling Control in Buildings Using Intermittent Energy. In Proceedings of the International Symposium on Intelligent Manufacturing and Service Systems, Istanbul, Turkey, 26–28 May 2023; Springer: Berlin/Heidelberg, Germany, 2023.
20. Heng, L.F. Research on the Application of Self-Immunity Control Technology in the Temperature Control System of Variable Air Volume Air Conditioner. Master's Thesis, Qingdao University of Science and Technology, Qingdao, China, 2018.
21. O'Dwyer, E.; De Tommasi, L.; Kouramas, K.; Cychowski, M.; Lightbody, G. Prioritised objectives for model predictive control of building heating systems. *Control Eng. Pract.* **2017**, *63*, 57–68. [CrossRef]
22. Tang, R.; Wang, S. Model predictive control for thermal energy storage and thermal comfort optimization of building demand response in smart grids. *Appl. Energy* **2019**, *242*, 873–882. [CrossRef]
23. Wang, X.; Dong, B.; Zhang, J.J. Nationwide evaluation of energy and indoor air quality predictive control and impact on infection risk for cooling season. *Build. Simul.* **2023**, *16*, 205–223. [CrossRef] [PubMed]
24. Ren, C.; Yu, H.; Wang, J.; Zhu, H.-C.; Feng, Z.; Cao, S.-J. Zonal demand-controlled ventilation strategy to minimize infection probability and energy consumption: A coordinated control based on occupant detection. *Environ. Pollut.* **2024**, *345*, 123550 [CrossRef] [PubMed]
25. *ANSI/ASHRAE Standard 55*; Thermal Environmental Conditions for Human Occupancy. ASHRAE: Atlanta, GA, USA, 2020.
26. Wang, J.; Jiang, L.; Yu, H.; Feng, Z.; Castaño-Rosa, R.; Cao, S.-J. Computer vision to advance the sensing and control of built environment towards occupant-centric sustainable development: A critical review. *Renew. Sustain. Energy Rev.* **2024**, *192*, 114165 [CrossRef]
27. De Dear, R.J.; Brager, G.S. Thermal comfort in naturally ventilated buildings: Revisions to ASHRAE Standard 55. *Energy Build.* **2002**, *34*, 549–561. [CrossRef]
28. Huang, H.; Chen, L.; Hu, E. A neural network-based multi-zone modelling approach for predictive control system design in commercial buildings. *Energy Build.* **2015**, *97*, 86–97. [CrossRef]
29. Yang, X.; Wang, H.; Su, C.; Wang, X.; Wang, Y. Heat transfer between occupied and unoccupied zone in large space building with floor-level side wall air-supply system. *Build. Simul.* **2020**, *13*, 1221–1233. [CrossRef]
30. Zhou, P.; Wang, S.; Zhou, J.; Hussain, S.A.; Liu, X.; Gao, J.; Huang, G. A modelling method for large-scale open spaces orientated toward coordinated control of multiple air-terminal units. *Build. Simul.* **2023**, *16*, 225–241. [CrossRef] [PubMed]
31. Wang, H.; Zhou, P.; Guo, C.; Tang, X.; Xue, Y.; Huang, C. On the calculation of heat migration in thermally stratified environment of large space building with sidewall nozzle air-supply. *Build. Environ.* **2019**, *147*, 221–230. [CrossRef]
32. Zhang, S.; Cheng, Y.; Huan, C.; Lin, Z. Heat removal efficiency based multi-node model for both stratum ventilation and displacement ventilation. *Build. Environ.* **2018**, *143*, 24–35. [CrossRef]
33. *ASHRAE Handbook—Fundamentals*; ASHRAE: Atlanta, GA, USA, 2021; Available online: https://www.ashrae.org/technical-resources/ashrae-handbook/description-2021-ashrae-handbook-fundamentals (accessed on 7 June 2024).
34. *GB50189-2015*; Design Standard for Energy Efficiency of Public Buildings. Ministry of Housing and Urban-Rural Development of the People's Republic of China: Beijing, China, 2015.
35. Emmel, M.G.; Abadie, M.O.; Mendes, N. New external convective heat transfer coefficient correlations for isolated low-rise buildings. *Energy Build.* **2007**, *39*, 335–342. [CrossRef]
36. Design Code for Railway Passenger Stations. China Industry Standard-Railway CN-TB. 27 June 2018. Available online: https://www.nra.gov.cn/xwzx/xwxx/xwlb/202204/t20220405_280153.shtml (accessed on 7 June 2024).
37. Wang, J.; Huang, J.; Feng, Z.; Cao, S.-J.; Haghighat, F. Occupant-density-detection based energy efficient ventilation system: Prevention of infection transmission. *Energy Build.* **2021**, *240*, 110883. [CrossRef]

Article

Contemporary Evaporative Cooling System with Indirect Interaction in Construction Implementations: A Theoretical Exploration

Pinar Mert Cuce [1,2], Erdem Cuce [3,4,*] and Saffa Riffat [5]

1   Department of Architecture, Faculty of Engineering and Architecture, Recep Tayyip Erdogan University, Zihni Derin Campus, 53100 Rize, Turkey; pinar.mertcuce@erdogan.edu.tr

2   Department of Energy Systems Engineering, Faculty of Engineering and Architecture, Recep Tayyip Erdogan University, Zihni Derin Campus, 53100 Rize, Turkey

3   Department of Mechanical Engineering, Faculty of Engineering and Architecture, Recep Tayyip Erdogan University, Zihni Derin Campus, 53100 Rize, Turkey

4   Department of School of Engineering and the Built Environment, Birmingham City University, Birmingham B4 7XG, UK

5   Department of Architecture and Built Environment, University of Nottingham, Nottingham NG7 2RD, UK; saffa.riffat@nottingham.ac.uk

*   Correspondence: erdem.cuce@erdogan.edu.tr

Abstract: The construction sector, including in developed countries, plays a notable part in the overall energy consumption worldwide, being responsible for 40% of it. In addition to this, heating, ventilating and air-conditioning (HVAC) systems constitute the largest share in this sector, accounting for 40% of energy usage in construction and 16% globally. To address this, stringent rules and performance measures are essential to reduce energy consumption. This study focuses on mathematical optimisation modelling to enhance the performance of indirect-contact evaporative cooling systems (ICESs), a topic with a significant gap in the literature. This modelling is highly comprehensive, covering various aspects: (1) analysing the impact of the water-spraying unit (WSU) size, working air (WA) velocity and hydraulic diameter (D_h) on the evaporated water vapour (EWV) amount; (2) evaluating temperature and humidity distribution for a range of temperatures without considering humidity at the outlet of the WSU, (3) presenting theoretical calculations of outdoor temperature (T_{out}) and humidity with a constant WSU size and air mass flow rate (MFR), (4) examining the combined effect of the WA MFR and relative humidity (ϕ) on T_{out} and (5) investigating how T_{out} influences the indoor environment's humidity. The study incorporates an extensive optimisation analysis. The findings indicate that the model could contribute to the development of future low-carbon houses, considering factors such as the impact of T_{out} on indoor ϕ, the importance of low air velocity for achieving a low air temperature, the positive effects of D_h on outdoor air and the necessity of a WSU with a size of at least 8 m for adiabatic saturation.

Keywords: buildings; HVAC; energy consumption; energy-efficient solution; optimisation

Citation: Cuce, P.M.; Cuce, E.; Riffat, S. Contemporary Evaporative Cooling System with Indirect Interaction in Construction Implementations: A Theoretical Exploration. *Buildings* **2024**, *14*, 994. https://doi.org/10.3390/buildings14040994

Academic Editor: Alessandro Prada

Received: 29 February 2024
Revised: 23 March 2024
Accepted: 26 March 2024
Published: 3 April 2024

1. Introduction

The clean and effective use of energy production is becoming increasingly important with the increase in the importance and awareness of environmental problems that have come as a result of global warming [1–6]. The phenomenon of global warming induces the creation of greenhouse gases, impacting both nature and humans. This is fuelled by the unregulated surge in energy consumption, reflective of irregular increases in the global human population and economic advancements, coupled with the escalating combustion of fossil fuels. In spite of dedicated endeavours to reduce the interval between traditional energy sources and renewables, only approximately 30% of the overall global energy need is fulfilled by green energy technologies [7]. In this context, there is widespread agreement within the scientific community that conducting investigations into the efficient control and utilisation of energy resources and effectively decreasing the consumption of energy is imperative. This issue affects not just one sector, but the entire spectrum of industries. Hence, if each sector implemented measures to curtail the energy demand, the escalating greenhouse gas emissions globally could potentially be halted. Urgent and intensified actions must be taken to achieve stability as a result. A review of literature studies indicates that a substantial portion of global energy consumption is attributed to buildings. As an illustration, in the European Union (EU), buildings contribute 40% of the overall energy consumption [8]. Tzeiranaki et al. [9] reported that the total energy loss for the EU in 2020 was equivalent to approximately 1086 million tons of oil.

In a study conducted years later, regarding the present annual greenhouse gas emissions within the UK territory, standing at 454 $MtCO_2$, approximately 91 $MtCO_2$ was shown to be related to the functioning of structures. Notably, 90% of these emissions stemmed from domestic buildings. The emissions directly associated with constructing new buildings in the UK during 2018 were estimated to range between 17.0 and 18.5 $MtCO_2$. Specifically, domestic buildings were shown to be responsible for 9.4 and 8.9 $MtCO_2$, consistently [10]. Hence, the issue of energy consumption is not only significantly high, but also has an evident impact on the environment.

Buildings' energy consumption is allocated not only to heating, but also to cooling, making the impact of cooling significant. According to a study by Rashad et al. [11], the demand for cooling is on the rise in tandem with the increased desire for enhanced comfort in buildings. Evaporative cooling is highlighted as a strong contender to meet this demand due to its cost-effectiveness [12]. Evaporative cooling, as underscored by Chen et al. [13], is presented as a feasible substitute for mechanical vapour compression in air conditioning, demanding roughly one-fourth of the electric power compared to vapour compression refrigeration [14].

Evaporative cooling systems currently available can be categorised into direct-contact (DC) evaporative cooling as well as ICESs. In evaporative cooling through DC, the air in need of cooling is brought into DC with the LWF and the need can be met through a heat transfer (HT) between the WA and the LWF [15]. Evaporative cooling through DC is suitable for usage primarily in arid, hot environmental conditions, or in spaces that necessitate both cooling and humidification [16]. Conversely, in ICESs, the interior air decreases in temperature through auxiliary air, known as the WA, which undergoes cooling by means of evaporation [17]. From a thermodynamic perspective, the moist pathway takes in heat from the dry pathway through the process of water evaporation, effectively cooling the non-moist pathway, whilst the latent heat of vaporising water is released into the WA. The depiction in Figure 1 illustrates the setup of ICESs.

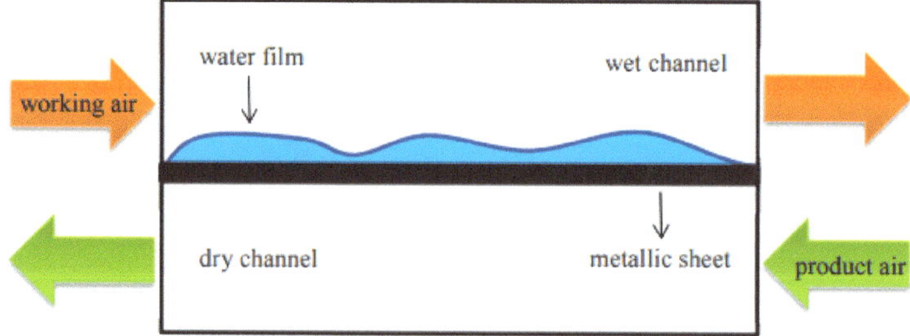

Figure 1. A comprehensive diagram illustrating a counter-flow ICES.

The metal panel situated amid the product and working air generally functions as a device that facilitates the exchange of heat between the two air streams, available in plate form [18,19], tube form [18,20] and heat pipe form [18,21]. ICESs possess the utility of cooling the product air without altering its specific humidity [22]. Moreover, ICESs were chosen for this study due to their avoidance of incorporating contaminated water droplets into the system, a concern that can arise during operation, posing significant health risks [23]. As a result, these systems have been extensively employed in residential structures as HVAC systems for over a century [17].

The theoretical examination of ICESs is relatively intricate, given that the air-cooling process entails simultaneous thermal and mass diffusion at the interface between the water film and air. Numerous efforts have been undertaken to achieve a precise comprehension of the thermal characteristics of these systems. Previous theoretical studies utilised 1D mathematical models [24–28]. Erens and Dreyer [29] assessed three analytical models, revealing that the optimal configuration for the cooling system is achieved with the proportion of airflow velocity between the primary and secondary components of approximately 1.4, presuming an equal MFR. Tsay [30] conducted a numerical investigation of a reverse-flow and moist surface heat exchanger, revealing that a significant portion of the energy transfer over the water film was soaked through the film vaporisation process. Guo and Zhao [31] provided a comprehensive examination of an ICES. The impact of diverse factors, including the speeds of the fundamental and auxiliary air flows, channel width, inlet ϕ and the plate's wettability, on thermal efficiency was explored. They asserted that a decreased inlet ϕ of secondary air, increased plate wettability and a greater velocity rate pertaining to the secondary air in relation to the primary air resulted in an enhanced efficiency of the refrigeration framework. Halasz [32] introduced a comprehensive mathematical model for evaporative cooling systems. The model was constructed by developing a set of four fractional derivative formulas that described evaporation without maintaining an isentropic condition process with a variable watercourse orientation, chilled fluid and air. Riffat and Zhu [33] introduced a mathematical thermal and mass diffusion model in their innovative ICES design, incorporating permeable ceramic and thermal conduit elements. Their findings suggested the potential for achieving substantial cooling capacity in arid and breezy weathers. They emphasised the importance of setting the indoor air velocity at an appropriate level of 0.6 m/s for optimal efficiency. Adam et al. [34] investigated heat and mass transfer in an ICES, specifically focusing on a crossflow arrangement. A numerical model was developed to predict condensation in the primary air and was validated against both numerical and experimental data. The simulations explored different condensation scenarios in the dry channel under severe operating conditions. The results showed that condensation states depended on these factors, with lower wettability factors delaying and reducing condensation. Additionally, increasing the secondary air velocity enhanced condensation and improved the cooling capacity. Belarbi et al. [35] theoretically explored water spray evaporation as a means for natural temperature diminishment in dwellings. Some researchers have engaged in analogous modelling efforts [36,37]. Maheshwari et al. [38] assessed the potential energy savings of ICESs. They noted a significantly higher energy-saving potential in interior areas compared to coastal areas. The study revealed that an ICES in interior areas was 30% further economically efficient in contrast to a conventional air-conditioning system.

A comprehensive examination of the existing literature clearly indicated numerous endeavours towards the theoretical exploration of ICESs. Nevertheless, it is apparent from previous research that achieving a satisfactory alignment between experimental and theoretical outcomes has proven challenging due to the intricate nature of the process and the inherent limitations in assumptions. In addition, there is a pressing need for further studies dedicated to innovative ICESs explicitly tailored for residential buildings. Consequently, to achieve a more thorough understanding of the techno-economic evaluation within such systems, it becomes imperative to undertake a comprehensive analysis that encompasses both theoretical research and experimental validation. Hence, in response to the existing gap in the literature regarding comprehensive modelling or validation through experimental studies, this research introduces a mathematical framework for the inventive design of ICESs, which has already undergone testing at a trial site in the southeast of the United Kingdom. It is anticipated that this study, which meticulously considers various factors, is likely to significantly contribute to future perspectives by offering a detailed insight into the novel modelling of temperature reduction systems along with its initial findings.

2. An Innovative Evaporative Cooling System

This investigation constituted the initial segment of a study focused on an innovative configuration of an ICES. The primary objective of this phase was to discern the operational system features via a theoretical examination. In this context, the study examined the influences of operational and environmental factors, including the temperatures inside and outside, the MFR, the φ levels of the supplied and the WA and the channel configuration of the heat exchanger, on the comprehensive efficiency of the system. The setup was developed and built to provide ventilation and cooling for residential structures.

To this end, a test residence was erected in southeastern UK, illustrated in Figure 2, with the system installed beneath the roof. The outer measurements of the testing structure were 7 m in size, 3 m in broadness, as well as 4.3 m in height. It featured a pair of windows, along with a roof skylight (Velux window), all equipped with double glazing. Nevertheless, it was evident from the test site that the study was conducted in conditions representative of a summer climate. Table 1 provides the size of the elements of the trial dwelling. The walls and roof were insulated with 150 mm of mineral wool, while the floor had a 150 mm insulation layer.

Figure 2. A residence designated for testing purposes located in the southeastern region of the UK.

Table 1. Measurements of the elements in the experimental residence.

	Trial Residence		Components		
	Exterior	Interior	Windows	Velux Window	Door
Size (m)	7	6.7	1.2	0.75	1.5
Width (m)	3	2.7	0.1	0.1	0.1
Height (m)	4.3	2.45–3.3	1	0.9	2.05

The suggested ICES could be divided into two primary components: the heat exchanger unit and the WSU. The system was designed to cool the stagnant indoor air through humidification in the WSU. Afterwards, the cooled air was directed to the polycarbonate heat exchanger (PHE) unit until the isentropic saturation circumstance was reached. This process enabled a reduction in the $T_{fa,in}$ at the reverse-fluid heat exchange systems. Four PHE sheets were manufactured by connecting 0.2 mm thick PHE sheets using an adhesive with high thermal conductivity, creating a cross sectional area of 10 mm^2. Figure 3 depicts the square cross-section channels of the PHE for both the inflowing and outflowing air. Two of the PHE sheets were positioned at the facade of the dwelling, whilst the remaining two were at the rear, as illustrated in Figure 4.

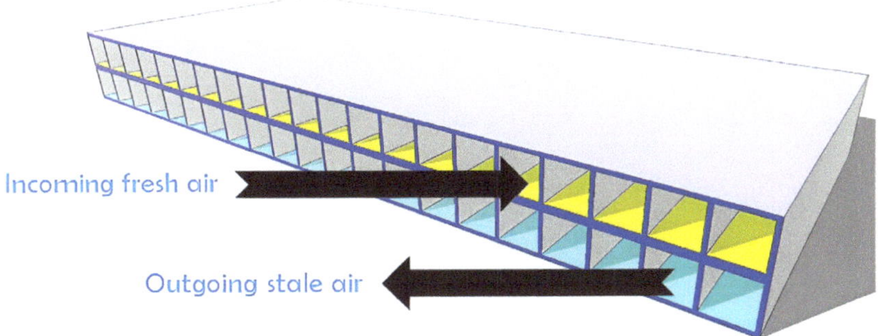

Figure 3. Square-shaped channels of the heat exchange unit for the warm and cold flows.

Figure 4. Square-shaped channels of the heat exchange model for the warm as well as cold flows.

The measurements and the wall measure of how thick the PHE are provided in Table 2.

Table 2. Specifications regarding the size of the polycarbonate heat exchanger.

Heat Exchange Plate			Wall Thickness (mm)		
Size (m)	Width (m)	Depth (m)	Upper	Central	Lower
1.7	0.425	0.01	0.8	0.2	0.8

The termination of every per PHE sheet inside the structure was accompanied by an exhaust duct linked to the top conduits and an intake duct linked to the inferior conduits. The exhaust and intake ducts had a square shape with inner sizes measuring 60×60 mm^2. Both the exhaust and intake ducts were subsequently incorporated into rectangular channels with interior measurements of 105×50 mm^2. Ultimately, the oblong conduits linked to fans with variable air velocities. The outer extremities of the PHE concluded below the roof slates, facilitating the drainage of condensed water into the gutters. Conversely, air purifiers would draw fresh air from under the overhangs. The WSU was regulated with a valve, controlling the MFR for the sprayed water. Spraying was carried out using tap water circulating within a copper pipe, having internal dimensions measuring 8 mm in diameter and 1 m in size, positioned at the central point of the exhaust channel immediately prior to the thermal interchange apparatus. A straightforward representation of the entire system is presented in Figure 5.

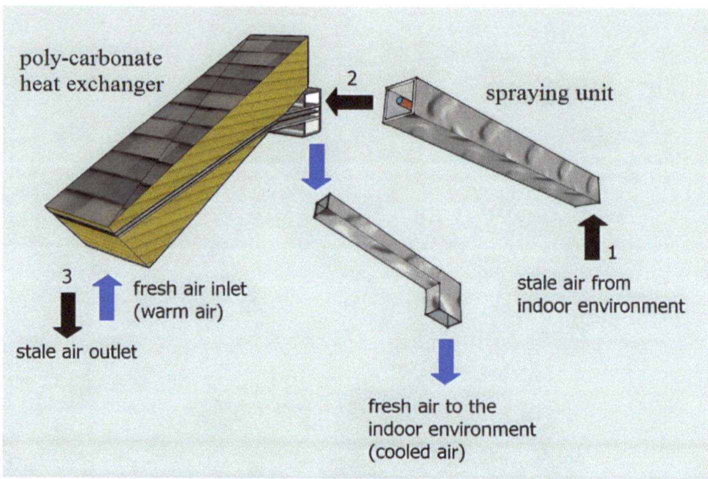

Figure 5. Illustration of the suggested ICES setup.

3. Theoretical Framework of the System

In this part, a precise mathematical description of the innovative ICES is introduced. The model was divided into two components: a theoretical examination of the WSU and a numerical analysis of the PHE unit. Initially, the rate of water film evaporation into the WA was calculated under various operating conditions, and the thermophysical characteristics of the WA were established at the exit of the WSU. A HT comparison for flow within the channels was utilised to assess the rate of evaporation. The rate of evaporation depended on the mass transfer coefficient (K), which was identified by the Sherwood number (Sh). The Sh, which expresses the relationship between convective and diffusive K, was determined using correlations involving Reynolds (Re) and Schmidt (Sc) numbers. To calculate the previously talked about unitless quantities, the thermophysical air attributes at the film temperature (T_f) were operated. T_f was expressed as stated below:

$$T_f = \frac{T_{lwf} + T_{wa,in}}{2} \tag{1}$$

The flow's characteristic was subsequently established using the Re. The Re is a dimensionless parameter indicating the relationship between the inertial and viscous forces, providing insight into the significance of these forces under specific flow conditions [39]. For noncircular tube flows, the Re was calculated based on D_h. As a reminder, below we wrote out the Re formulation:

$$Re = \frac{\rho V D_h}{\mu} \tag{2}$$

where
ρ: density (kg/m^3);
V: velocity (m/s);

μ: dynamic viscosity ($\frac{kg}{ms}$).

$$D_h = \frac{4A}{P} \tag{3}$$

In Equation (3), A represents the duct cross-sectional area, while p indicates the perimeter submerged in the liquid. In the other calculated demand formulation, the Sc can be found, and it was written out as follows:

$$Sc = \frac{v}{Dm} \tag{4}$$

where

v: kinematic viscosity ($\frac{m^2}{s}$);

Dm: mass diffusivity ($\frac{m^2}{s}$).

As apparent from the formulation, the Sc is a unitless factor expressed as the rate between the momentum and mass diffusivities. It can be employed to delineate fluid flows involving concurrent processes of momentum and mass diffusion convection [40]. Afterwards, the Sh was calculated using the subsequent equation:

$$Sh = \frac{KD_h}{Dm} \tag{5}$$

Sh is a unitless parameter that signifies the convective ratio to dispersive K. The mean K was determined by employing the mean Sh in accordance with the below formula:

$$\overline{K} = \frac{\overline{Sh}dm}{D_h} \tag{6}$$

The line above the abbreviations in the formula means the average.

To calculate the Sh, the subsequent partially theoretical equation established by Frossling [41] could be applied:

$$Sh = 2 + 0.552\sqrt{Re}\sqrt[3]{Sc} \tag{7}$$

Regarding the water vapour spread coefficient, the curve obtained through the regression adjustment derived from the dataset of Bolz and Tuve [42] could be employed:

$$Dm = 1.656 \times 10^{-10}T^2 + 4.479 \times 10^{-8}T - 2.775 \times 10^{-6} \tag{8}$$

The MFR was determined through the disparity in the humidity concentration between the WSU and the WA. The water vapour's fractional pressure at the WSU expressed vapour pressure of water at saturation in vapour of water vapour concentration (c_w) at the WSU was the water vapour density determined using the fractional pressure and temperature. The concentration at the WSU is known as the mass fraction of water vapour (mf_w).

$$mf_w = \frac{c_w}{\rho} \tag{9}$$

At T_{lfw}, the fractional water pressure in the operational atmosphere ($P_{w,wa}$), could be obtained by multiplying the ϕ with the saturation pressure of the water vapour ($P_{sat,w}$), as indicated in the formulation below:

$$P_{w,wa} = \phi P_{sat,w} \tag{10}$$

The water vapour concentration in the operational atmosphere ($c_{w,wa}$) was determined using ρ_w calculated based on the fractional pressure and temperature. Then, the following equation could be used:

$$mf_{w,wa} = \frac{c_{w,wa}}{\rho} \tag{11}$$

where $mf_{w,wa}$: mass fraction of water vapour in the operational atmosphere.

The blowing factor (BF) was determined using the following definition to compute the adjusted mass transfer coefficient ($\overline{K}_{cor}$):

$$BF = \ln \frac{1 + \frac{mf_w - mf_{w,wa}}{mf_{w,wa} - 1}}{\frac{mf_w - mf_{w,wa}}{mf_{w,wa} - 1}} \tag{12}$$

The $\overline{K}_{cor}$ was adjusted using:

$$\overline{K}_{cor} = \overline{K}BF \tag{13}$$

Lastly, the water MFR resulting from evaporation was computed using the next equation:

$$\dot{m}_w = \overline{K}_{cor} A_{tot,wsu} (c_w - c_{w,wsu}) \tag{14}$$

where $A_{tot,wsu}$: the overall HT surface area of the WSU.

Establishing the water MFR through evaporation allowed for the determination of the temperature and ϕ_{wa} at the WSU exit. The cooled and humidified WA, upon leaving the spraying duct, was then guided to the counter-flow PHE to lower the temperature of the entering warm fresh air. Subsequently, the second phase of the simulation effort focused on examining the HT within the innovative plate-type PHE.

The PHE could be envisioned as a pair of straight channels accompanying the movement of fluid that are thermally linked. Assuming the ducts are of the same size (L) and transports the streams with heat capacity (c_j), the MFR of the streams is represented by ($\dot{m}_j$) that lower index and "sa" refers to the stale air and "fa" to fresh air. Therefore, $T_{sa}(x)$ and $T_{fa}(x)$ are the temperature profile to the streams and, here, x refers to the interval through the channel. The analyses were conducted under steady-state conditions, ensuring that the temperature profiles remained constant and were not dependent on time. Additionally, it was regarded as the exclusive HT from a limited quantity of fluid to the fluid particle inside the second channel at the equivalent location. The quantity of the HT within the channel due to temperature variations was disregarded. According to Newton's law of cooling, the rate of energy alteration within a limited section of the flow was directly linked to the temperature disparity between that segment and the corresponding component in the alternate duct:

$$\frac{\partial u_{sa}}{\partial t} = \xi(T_{fa} - T_{sa}) \tag{15}$$

$$\frac{\partial u_{sa}}{\partial t} = \xi(T_{sa} - T_{fa}) \tag{16}$$

Here, $u_j(x)$ is the ratio of the thermal energy to size; also, ξ means the thermal coupling constant divided to size among channels. This alteration in the internal energy led to a modification at T_{sa}. The temporal rate of variation for the stream element transported by the flow was as follows:

$$\frac{\partial u_{sa}}{\partial t} = \psi_{sa} \frac{dT_{sa}}{dx} \tag{17}$$

$$\frac{\partial u_{fa}}{\partial t} = \psi_{fa} \frac{dT_{fa}}{dx} \tag{18}$$

where $\psi_f = c_j \dot{m}_j$ means the thermal MFR. When rearranging the equations between (15) and (17) as well as (16) and (18), they could be rewritten as follows:

$$c_j \dot{m}_j \frac{dT_{sa}}{dx} = \xi(T_{fa} - T_{sa}) \tag{19}$$

$$c_j \dot{m}_j \frac{dT_{fa}}{dx} = \xi(T_{sa} - T_{fa}) \tag{20}$$

Looking at the rearranged equations above, it could be easily understood that there was no longer a time dependence because the system was essentially stationary. Moreover, given that the HT was nearly negligible along the extension of the pipe, it was not feasible to compute any secondary derivatives at any point x, as indicated by the heat equations, like in the next equations:

$$T_{sa} = \Omega_1 - \Omega_2 \frac{\sigma_1}{\sigma} e^{-\sigma x} \tag{21}$$

$$T_{fa} = \Omega_1 + \Omega_2 \frac{\sigma_1}{\sigma} e^{-\sigma x} \tag{22}$$

Therefore

$$\sigma_1 = \frac{\xi}{\psi_{sa}}, \quad \sigma_2 = \frac{\xi}{\psi_{fa}}, \quad \sigma = \sigma_1 + \sigma_2 \tag{23}$$

Also, Ω_1, as well as Ω_2, are in the mean of the integration constants. Now, to determine the mean temperatures, with the origin of the temperature points set at 0 and the channel size established as L, the following formulas arose:

$$\overline{T}_{sa} = \frac{1}{L} \int_0^L T_{sa} dx \tag{24}$$

$$\overline{T}_{fa} = \frac{1}{L} \int_0^L T_{fa} dx \tag{25}$$

$$T_{sa,0} = \Omega_1 - \Omega_2 \frac{\sigma_1}{\sigma} \tag{26}$$

$$T_{fa,0} = \Omega_1 + \Omega_2 \frac{\sigma_1}{\sigma} \tag{27}$$

$$T_{sa,l} = \Omega_1 - \Omega_2 \frac{\sigma_1}{\sigma} e^{-\sigma l} \tag{28}$$

$$T_{fa,l} = \Omega_1 + \Omega_2 \frac{\sigma_1}{\sigma} e^{-\sigma l} \tag{29}$$

$$\overline{T}_{sa} = \Omega_1 - \Omega_2 \frac{\sigma_1}{\sigma^2 l} (1 - e^{-\sigma L}) \tag{30}$$

$$\overline{T}_{fa} = \Omega_1 + \Omega_2 \frac{\sigma_2}{\sigma^2 l} (1 - e^{-\sigma L}) \tag{31}$$

Equations (27)–(31) did not matter at all. By choosing at least two of the temperatures, the integral constant could be eliminated and, thus, shed light on finding the other four temperatures. Additionally, the overall transfer of heat could be obtained using these next equations:

$$\frac{d\gamma_{sa}}{dt} = \int_0^L \frac{du_{sa}}{dt} dx = \Psi_{sa}(T_{sa,L} - T_{sa,0}) = \xi l(\overline{T}_{fa} - \overline{T}_{sa}) \tag{32}$$

$$\frac{d\gamma_{fa}}{dt} = \int_0^L \frac{du_{fa}}{dt} dx = \Psi_{fa}(T_{fa,L} - T_{fa,0}) = \xi l(\overline{T}_{sa} - \overline{T}_{fa}) \tag{33}$$

γ represents the overall transfer of heat, whilst u means the internal energy. As is known from the law of energy conservation, it could be understood from the above equations that the overall result of energy changes was zero. The above equation, given as the mean temperature differences, is defined in the literature as logarithmically average temperature difference, which is one of the measures of heat exchanger efficiency.

4. Results and Discussion

This section first begins with a theoretical evaluation of the effects of the WSU size, WA velocity and D_h on the amount of EWV. Accordingly, Figure 6a–c below was created for the theoretical evaluation.

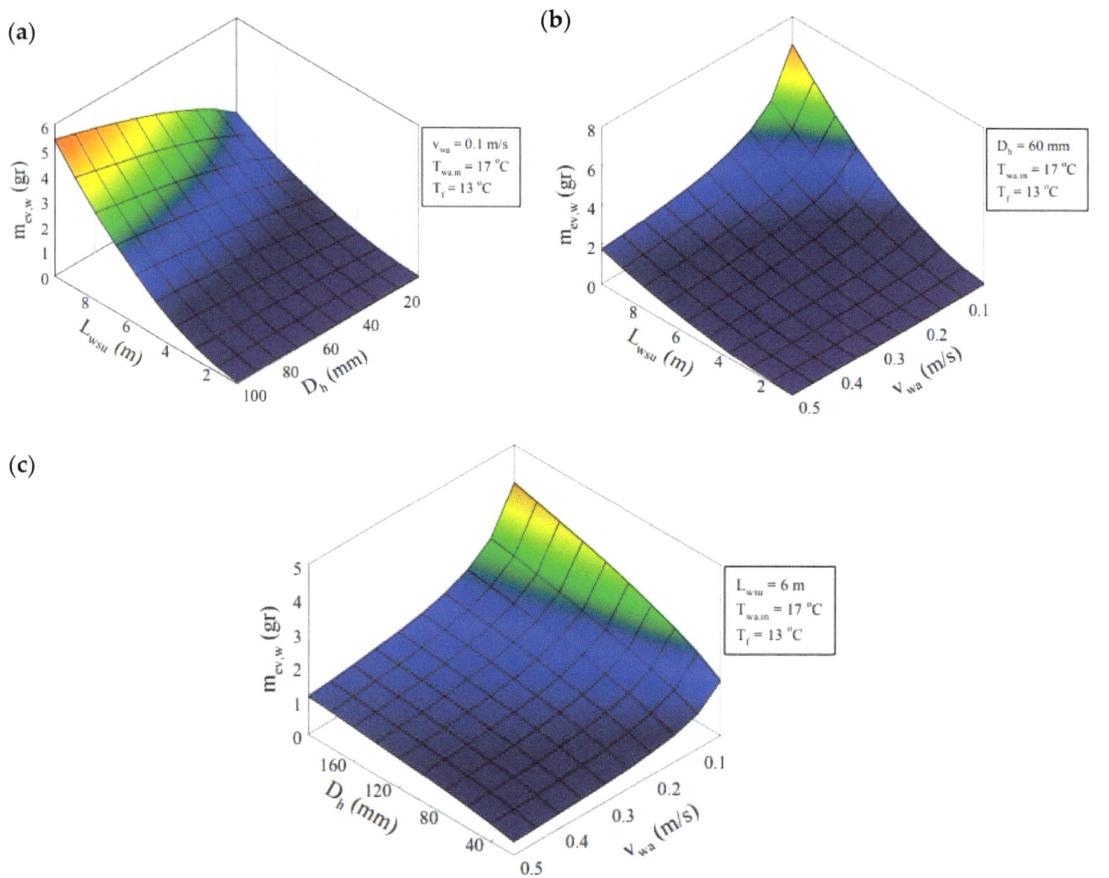

Figure 6. (**a**) Merged impacts of L_{wsu} and D_h onto the quantity of EWV. (**b**) Collective influence of L_{wsu} and the v_{wa} on the quantity of EWV. (**c**) Joint effects of D_h and v_{wa} on the quantity of EWV.

In an alternative investigation, the temperature without considering humidity at the exit of the WSU and the proportion of moisture in the WA was individually scrutinised at the WA inlet temperatures of 17 °C, 20 °C, 23 °C, 26 °C and 29 °C. The outcomes are depicted in Figure 7a–e, and the theoretical representation of the WSU size was provided.

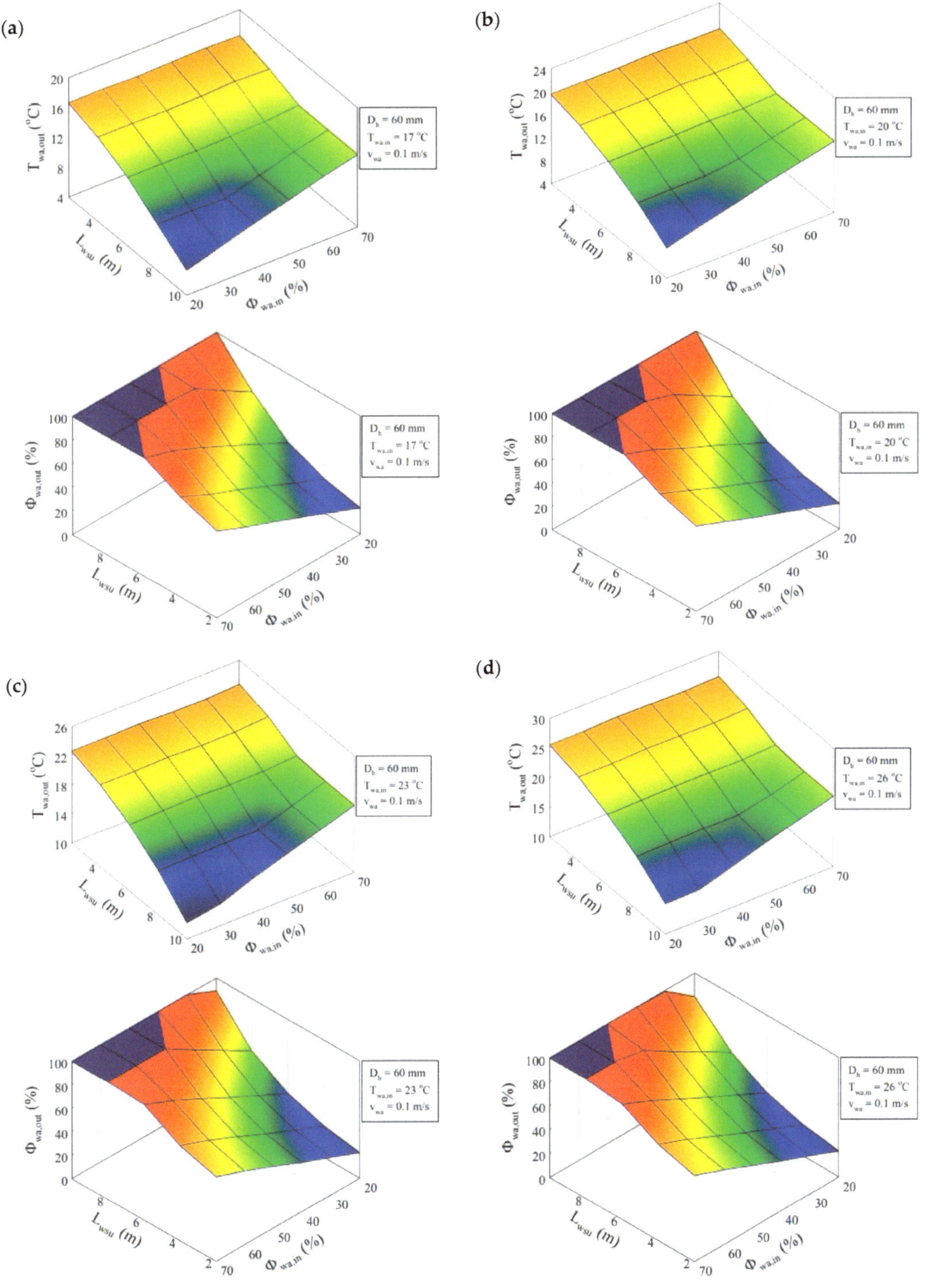

Figure 7. *Cont.*

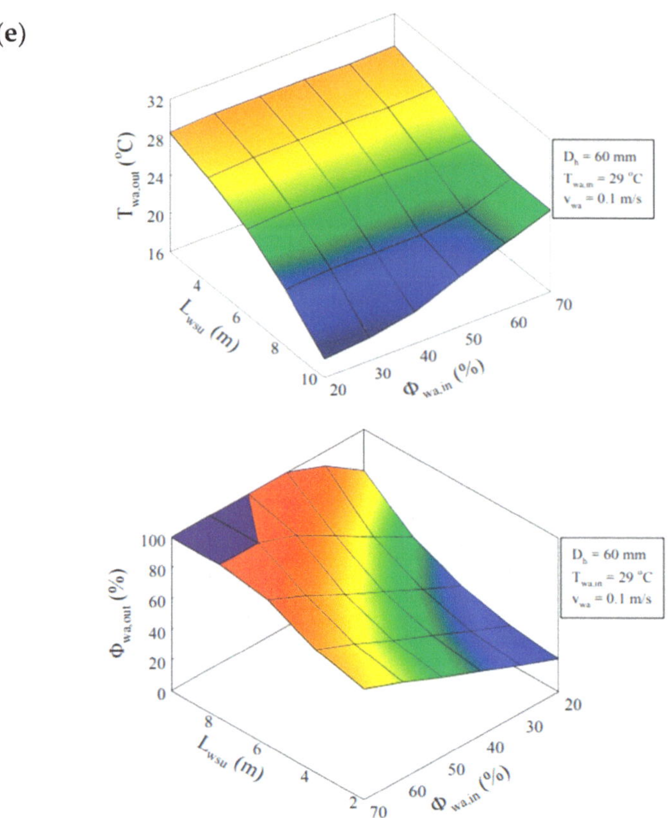

Figure 7. (**a**) Depicting the variation in the size of the WSU, the T_{db} and the ϕ of the WA at the exit of the WSU concerned a temperature of 17 °C, (**b**) at 20 °C, (**c**) at 23 °C, (**d**) at 26 °C and (**e**) 29 °C.

As indicated in the figures, it was evident that the temperature and ϕ of the WA, as extensively discussed in the literature, played a crucial role in defining the capacity of the evaporative cooling system. The findings suggested a trend of decreasing the T_{out} of the WA with the increasing size of the WSU. Moreover, it was observed that the ϕ of the WA attained saturation beyond a certain threshold for each size of the WSU, deviating from the peak cooling condition. Nevertheless, a WSU size of at least 8 m was the pivotal requirement to achieve an adiabatic saturation. Nevertheless, the outcomes indicated that with an increase in the inlet ϕ of the WA, both the temperature and ϕ of the outlet WA witnessed an upward trend.

In a different segment of the results, the theoretical computation was performed for the T_{out} and ϕ of the WA, considering a constant WSU size and air MFR (L_{wsu} = 6 m, v_{wa} = 0.1 m/s). The temperature range was scrutinised from 17 °C to 29 °C, at 3 °C increments, mirroring the approach taken in the previous analysis, as illustrated in Figure 8a–e. As evident from the illustrations below, how crucial the D_h is in influencing the WA outlet temperature became apparent. For instance, the observed $Tt_{wa,out}$ of less than 10 °C for a D_h of 0.3 m was an anticipated occurrence, clearly depicted in the images. However, it was deduced that the $\phi_{wa,in}$ did not surpass 50% at its best. As highlighted earlier, the $T_{out,wa}$ was markedly impacted by the $\phi_{wa,in}$. To illustrate, when the WA entered at a temperature of 17 °C with 70% ϕ, the $T_{out,wa}$ decreased to 13 °C, whereas, with 20% ϕ, it dropped below 10 °C.

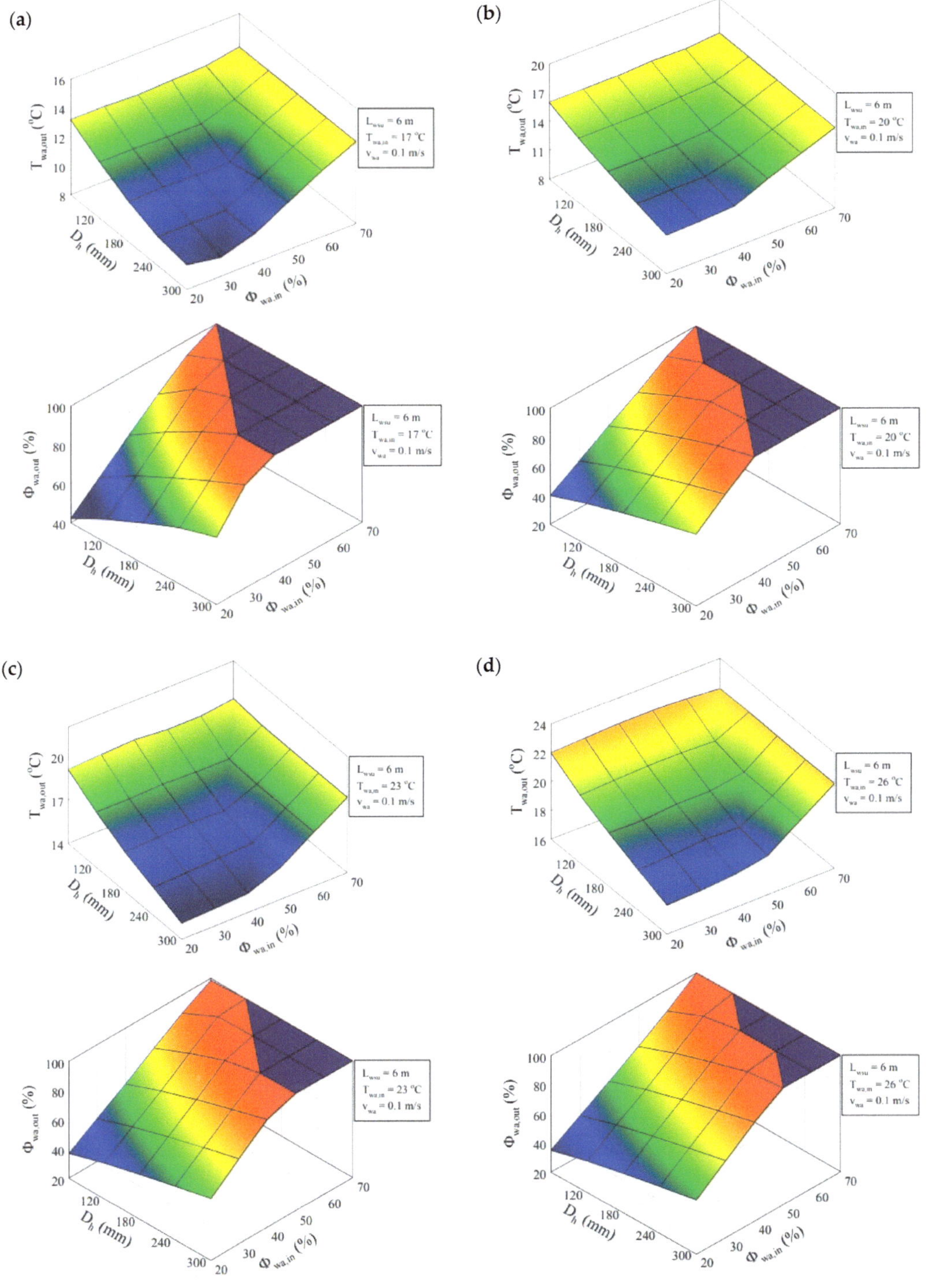

Figure 8. *Cont.*

(**e**)

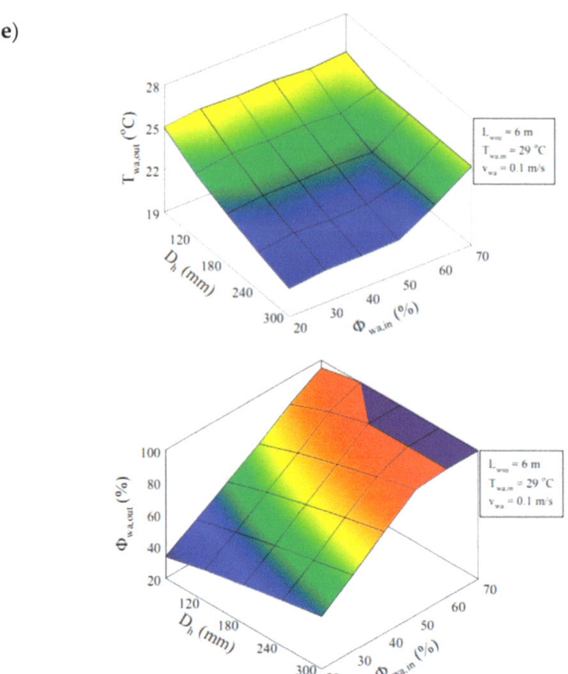

Figure 8. (**a**) As influenced by D_h, the T_{db} and the ϕ_{wa} at the WSU exit concerned $T_{wa,in}$ at 17 °C, (**b**) at 20 °C, (**c**) at 23 °C, (**d**) at 26 °C and (**e**) 29 °C.

Another aspect explored independently within the research involved considering the joint impacts of the MFR and ϕ_{wa} on the T_{out}, as depicted in Figure 9a–e below. The findings allowed us to observe that achieving a lower WA temperature in the outlet area required a significantly low air velocity, which could be attained by humidifying the environment through the addition of water vapour to the WA, albeit requiring a longer duration.

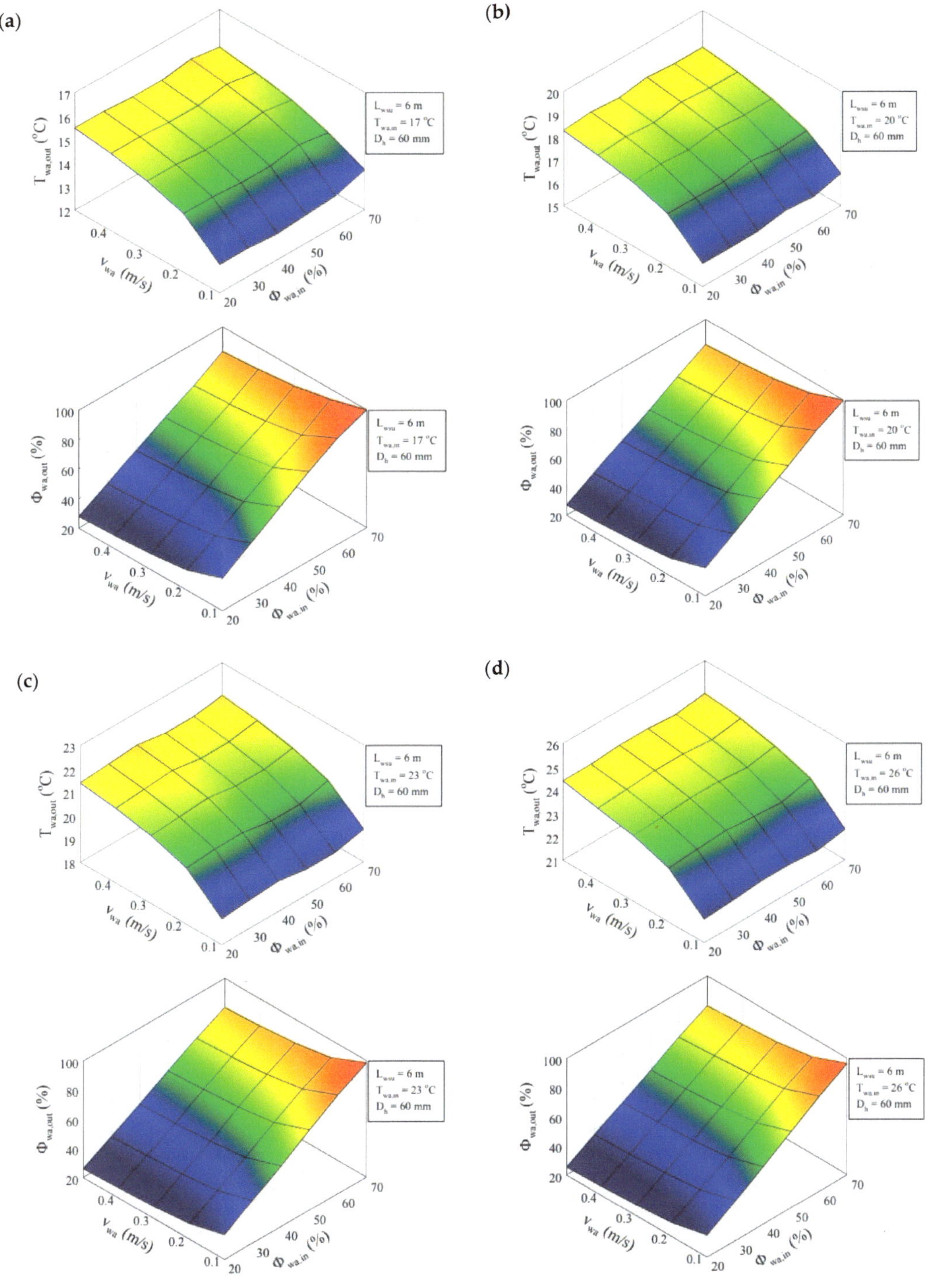

Figure 9. *Cont.*

(e)

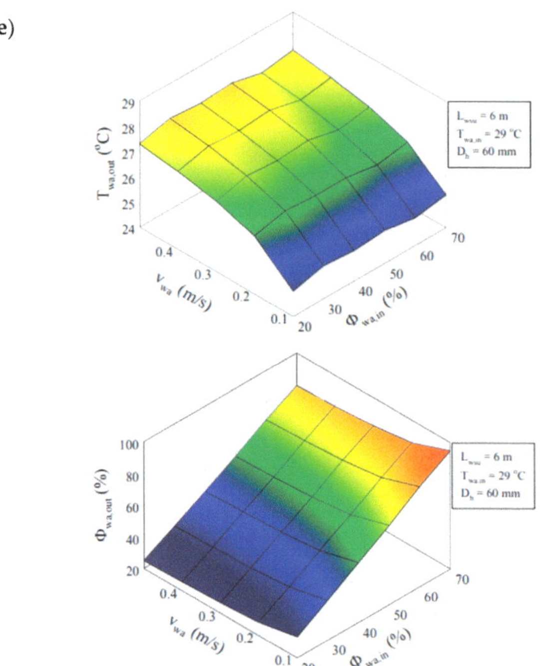

Figure 9. (**a**) In relation to the velocity of the WA, the T_{db} and the ϕ_{wa} at the exit of the WSU concerned $T_{wa,in}$ at 17 °C, (**b**) at 20 °C, (**c**) at 23 °C, (**d**) at 26 °C and (**e**) 29 °C.

Figure 10 illustrates a linear increase in the $T_{fa,in}$, correlating with both the $T_{fa,out}$ and the $T_{fa,wa}$.

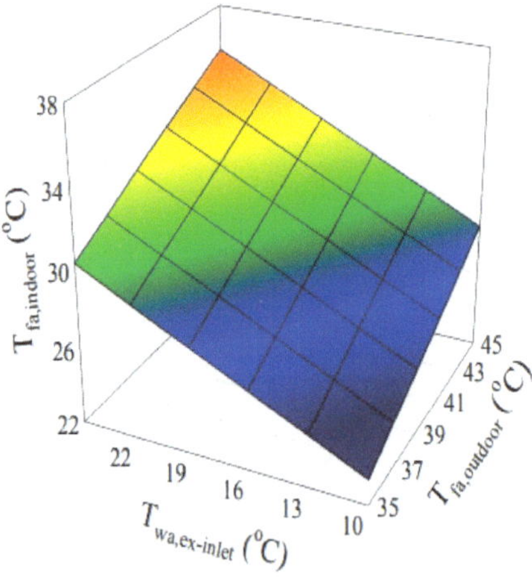

Figure 10. The inlet temperature of fresh air was influenced by both the $T_{fa,out}$ and the $T_{wa,in}$ of the PHE.

In the final analysis, we depicted the impact of $T_{fa,out}$ on $\phi_{fa,in}$ in Figure 11a–d. As per the results, if $\phi_{fa,out}$ was above 40% and the $T_{fa,out}$ exceeded 35 °C, it indicated that the thermal comfort range for $\phi_{fa,in}$ would not be met.

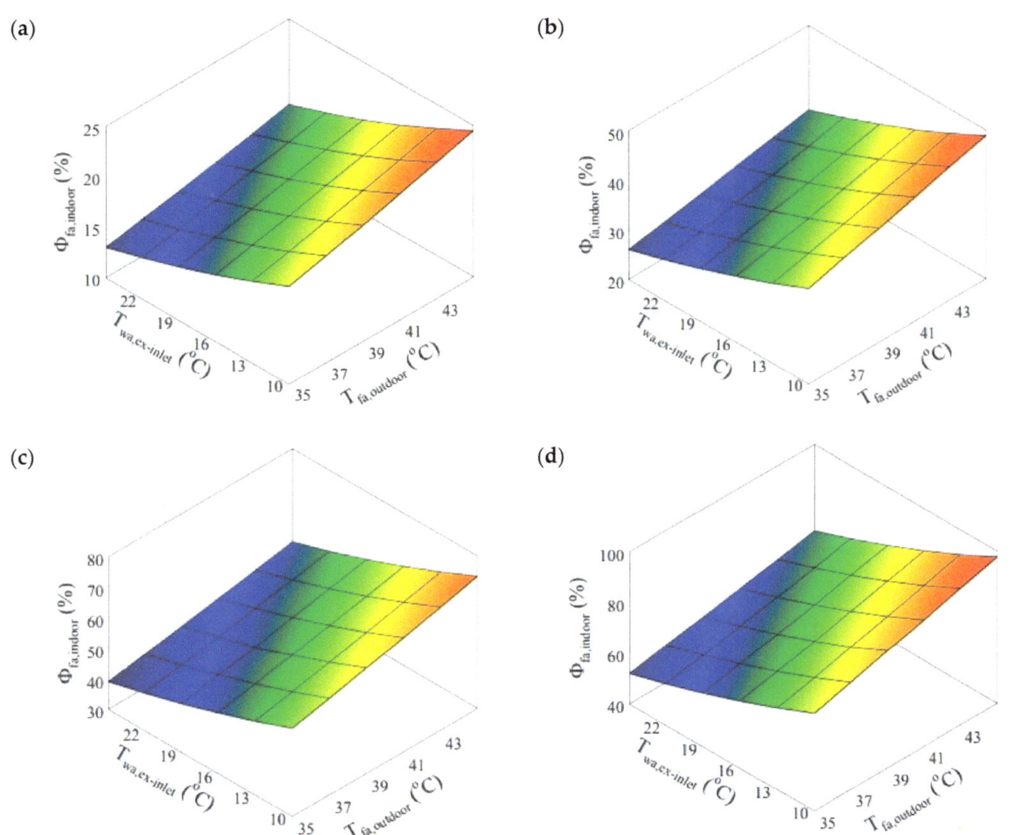

Figure 11. (**a**) $\phi_{fa,in}$ by $\phi_{fa,out}$, as well as WA temperature at the exchanger inlet at $\phi_{fa,out} = 10\%$ (**b**) at $\phi_{fa,out} = 20\%$, (**c**) at $\phi_{fa,out} = 30\%$, (**d**) at $\phi_{fa,out} = 40\%$.

5. Conclusions

- Enhancing the D_h and extending the size of the WSU could give rise to an increment in the quantity of water vapour.
- Conversely, accelerating the WA speed would contribute to a diminishment in water evaporation.
- When considered in terms of cooling efficiency, it became evident that the temperature of the WA was just as crucial as its ϕ. The findings demonstrated a significant reduction in the T_{out} of the WA with the rise in WSU size.
- On the flip side, it was proven that the ϕ in the WA attained adiabatic saturation after a certain value of the WSU size, which was the wanted status.
- Based on the findings from the optimisation analyses conducted with various inlet air temperatures, achieving the adiabatic saturation status required a WSU size of at least 8 m.
- In another analysis conducted with a similar approach, the focus shifted to diverse relative humidities of the WA. The results showed that an increase in the inlet ϕ of the WA led to higher temperatures and ϕ of the WA. This underscored the significance of selecting an optimal value whilst considering indoor thermal comfort conditions.
- Additionally, the outer temperature and ϕ of the WA were established for $L_{wsu} = 6$ m, $v_{wa} = 0.1$ m/s, contemplating severe temperatures of the WA at the interior. The cumulative impacts of D_h and ϕ were also assessed. We derived a conclusion that

the D_h had a notable impact on the T_{out} of the WA. For a D_h of 0.3 m, the $T_{wa,out}$ fell below 10 °C, which appeared to be highly favourable.
- It was stressed that, for the optimal circumstances to be achieved, the ϕ in the inlet WA should be maintained below 50%.
- This study also investigated the impacts of the MFR and ϕ on the T_{out}. It concluded that achieving the desired low WA temperature at the outlet required maintaining a low air velocity.
- $T_{fa,in}$ increased proportionally with both the temperature of the WA and the $T_{fa,out}$.
- The T_{out} significantly influenced the ϕ within the indoor environment. For instance, assuming $\phi_{fa,out} = 40\%$, if the $T_{fa,out}$ exceeded 35 °C, it indicated that $\phi_{fa,in}$ may not fall within the thermal comfort range, as demonstrated in this study.

Author Contributions: P.M.C.—Formal Analysis, Methodology, Investigation, Visualization, Validation, Writing—original draft, Writing—review & editing; E.C.—Visualization, Validation, Writing—original draft, Writing—review & editing; S.R.—Project administration, Supervision, Writing—review & editing. All authors have read and agreed to the published version of the manuscript.

Funding: This research received no external funding.

Data Availability Statement: The raw data supporting the conclusions of this article will be made available by the authors on request.

Conflicts of Interest: The authors declare no conflict of interest.

Nomenclature

A: Area [m²]	K: Mass transfer coefficient [m/s]
BF: Blowing factor	L: Length
c: Specific heat capacity [J/kgK]	mf: Mass fraction
D: Diameter [m]	ṁ: Mass flow rate [kg/s]
Dm: Mass diffusivity [m²/s]	P: Pressure [Pa]
HVAC: Heating, ventilating and air conditioning	Re: Reynolds number
ICES: Indirect-contact evaporative cooling system	u: Thermal energy
PHE: Polycarbonate heat exchanger	x: Distance along the duct
Sh: Sherwood number	Sc: Schmidt number
T: Temperature [°C]	EWV: Evaporated water vapour
MFR: Mass flow rate	DC: Direct contact
HT: Heat transfer	
SUBSCRIPTS	
cor: Corrected	lwf: Liquid water film
db: Dry bulb	out: Outside
f: Film	sa: Stale air
fa: Fresh air	sat: Saturated
h: Hydraulic	tot: Total
in: Inlet	w: Water vapour
j: Stream	wa: Working air
L: Length [m]	wsu: Water-spraying unit
GREEK LETTERS	
Φ: relative humidity	σ: Constant
ν: Kinematic viscosity [m²/s]	ξ: Thermal connection constant
μ: Dynamic viscosity [kg/ms]	p: Perimeter [m]
Ω: Constant	ρ: Density [kg/m³]
Ψ: Thermal mass flow rate	$\overline{w}$: Constant
γ: Overall transfer of heat [W]	

References

1. Zhang, L.; Xu, M.; Chen, H.; Li, Y.; Chen, S. Globalization, green economy and environmental challenges: State of the art review for practical implications. *Front. Environ. Sci.* **2022**, *10*, 870271. [CrossRef]
2. Al-Shetwi, A.Q. Sustainable development of renewable energy integrated power sector: Trends, environmental impacts, and recent challenges. *Sci. Total Environ.* **2022**, *822*, 153645. [CrossRef] [PubMed]
3. Cuce, P.M.; Cuce, E.; Alvur, E. Internal or external thermal superinsulation towards low/zero carbon buildings? A critical report. *Gazi Mühendislik Bilim. Derg.* **2024**, *9*, 435–442.
4. Cuce, E.; Cuce, P.M.; Wood, C.; Gillott, M.; Riffat, S. Experimental Investigation of Internal Aerogel Insulation Towards Low/Zero Carbon Buildings: A Comprehensive Thermal Analysis for a UK Building. *Sustain. Clean Build.* **2024**, *1*, 1–22.
5. Akram, M.W.; Hasannuzaman, M.; Cuce, E.; Cuce, P.M. Global technological advancement and challenges of glazed window, facade system and vertical greenery-based energy savings in buildings: A comprehensive review. *Energy Built Environ.* **2023**, *4*, 206–226. [CrossRef]
6. Ahmed, A.; Ge, T.; Peng, J.; Yan, W.C.; Tee, B.T.; You, S. Assessment of the renewable energy generation towards net-zero energy buildings: A review. *Energy Build.* **2022**, *256*, 111755. [CrossRef]
7. Li, L.; Lin, J.; Wu, N.; Xie, S.; Meng, C.; Zheng, Y.; Wang, X.; Zhao, Y. Review and outlook on the international renewable energy development. *Energy Built Environ.* **2022**, *3*, 139–157. [CrossRef]
8. Bertoldi, P.; Economidou, M.; Palermo, V.; Boza-Kiss, B.; Todeschi, V. How to finance energy renovation of residential buildings: Review of current and emerging financing instruments in the EU. *Wiley Interdiscip. Rev. Energy Environ.* **2021**, *10*, e384. [CrossRef]
9. Tsemekidi Tzeiranaki, S.; Bertoldi, P.; Diluiso, F.; Castellazzi, L.; Economidou, M.; Labanca, N.; Serrenho, T.R.; Zangheri, P. Analysis of the EU residential energy consumption: Trends and determinants. *Energies* **2019**, *12*, 1065. [CrossRef]
10. Drewniok, M.P.; Dunant, C.F.; Allwood, J.M.; Ibell, T.; Hawkins, W. Modelling the embodied carbon cost of UK domestic building construction: Today to 2050. *Ecol. Econ.* **2023**, *205*, 107725. [CrossRef]
11. Rashad, M.; Khordehgah, N.; Żabnieńska-Góra, A.; Ahmad, L.; Jouhara, H. The utilisation of useful ambient energy in residential dwellings to improve thermal comfort and reduce energy consumption. *Int. J. Thermofluids* **2021**, *9*, 100059. [CrossRef]
12. Cuce, P.M. Thermal performance assessment of a novel liquid desiccant-based evaporative cooling system: An experimental investigation. *Energy Build.* **2017**, *138*, 88–95. [CrossRef]
13. Chen, Q.; Ja, M.K.; Burhan, M.; Akhtar, F.H.; Shahzad, M.W.; Ybyraiymkul, D.; Ng, K.C. A hybrid indirect evaporative cooling-mechanical vapor compression process for energy-efficient air conditioning. *Energy Convers. Manag.* **2021**, *248*, 114798. [CrossRef]
14. Chang, C.K.; Yih, T.Y.; Othman, Z.B. Performance Assessment of Green Filtration System with Evaporative Cooling to Improve Indoor Air Quality (IAQ). *Int. J. Environ. Sci. Dev.* **2022**, *13*, 1–7.
15. Zhou, C.; Hu, Y.; Zhang, L.; Fang, J.; Xi, Y.; Hu, J.; Li, Y.; Liu, L.; Zhao, Y.; Yang, L.; et al. Investigation of heat transfer characteristics of deflected elliptical-tube heat exchanger in closed-circuit cooling towers. *Appl. Therm. Eng.* **2024**, *236*, 121860. [CrossRef]
16. Lai, L.; Wang, X.; Kefayati, G.; Hu, E. Evaporative cooling integrated with solid desiccant systems: A review. *Energies* **2021**, *14*, 5982. [CrossRef]
17. Blanco-Marigorta, A.M.; Tejero-Gonzalez, A.; Rey-Hernandez, J.M.; Gomez, E.V.; Gaggioli, R. Exergy analysis of two indirect evaporative cooling experimental prototypes. *Alex. Eng. J.* **2022**, *61*, 4359–4369. [CrossRef]
18. Mohammed, R.H.; El-Morsi, M.; Abdelaziz, O. Indirect evaporative cooling for buildings: A comprehensive patents review. *J. Build. Eng.* **2022**, *50*, 104158. [CrossRef]
19. Shi, W.; Min, Y.; Ma, X.; Chen, Y.; Yang, H. Performance evaluation of a novel plate-type porous indirect evaporative cooling system: An experimental study. *J. Build. Eng.* **2022**, *48*, 103898. [CrossRef]
20. Sun, T.; Tang, T.; Yang, C.; Yan, W.; Cui, X.; Chu, J. Cooling performance and optimization of a tubular indirect evaporative cooler based on response surface methodology. *Energy Build.* **2023**, *285*, 112880. [CrossRef]
21. Putra, N.; Sofia, E.; Gunawan, B.A. Evaluation of indirect evaporative cooling performance integrated with finned heat pipe and luffa cylindrica fiber as cooling/wet media. *J. Adv. Res. Exp. Fluid Mech. Heat Transf.* **2021**, *3*, 16–25.
22. Zhao, X.; Liu, S.; Riffat, S.B. Comparative study of heat and mass exchanging materials for indirect evaporative cooling systems. *Build. Environ.* **2008**, *43*, 1902–1911. [CrossRef]
23. Xu, P.; Ma, X.; Zhao, X.; Fancey, K.S. Experimental investigation on performance of fabrics for indirect evaporative cooling applications. *Build. Environ.* **2016**, *110*, 104–114. [CrossRef]
24. Maclaine-Cross, I.L.; Banks, P.J. A general theory of wet surface heat exchangers and its application to regenerative evaporative cooling. *J. Heat Transf.* **1981**, *103*, 579–585. [CrossRef]
25. Kettleborough, C.F.; Hsieh, C.S. The thermal performance of the wet surface plastic plate heat exchanger used as an indirect evaporative cooler. *J. Heat Transf.* **1983**, *105*, 366–373. [CrossRef]
26. Kettleborough, C.F.; Waugaman, D.G.; Johnson, M. The thermal performance of the cross-flow three-dimensional flat plate indirect evaporative cooler. *J. Energy Resour. Technol.* **1992**, *114*, 181–186. [CrossRef]
27. Wassel, A.T.; Mills, A.F. Design methodology for a counter-current falling film evaporative condenser. *J. Heat Transf.* **1987**, *109*, 784–787. [CrossRef]
28. Hsu, S.T.; Lavan, Z.; Worek, W.M. Optimization of wet surface heat exchanger. *Energy* **1989**, *14*, 757–770. [CrossRef]
29. Erens, P.J.; Dreyer, A.A. Modelling of indirect evaporative air coolers. *Int. J. Heat Mass Transf.* **1993**, *36*, 17–26. [CrossRef]

30. Tsay, Y.L. Analysis of heat and mass transfer in a countercurrent-flow wet surface heat exchanger. *Int. J. Heat Fluid Flow* **1994**, *15*, 149–156. [CrossRef]
31. Guo, X.C.; Zhao, T.S. A parametric study of an indirect evaporative air cooler. *Int. Commun. Heat Mass Transf.* **1998**, *25*, 217–226. [CrossRef]
32. Halasz, B. A general mathematical model of evaporative cooling devices. *Rev. Générale De Therm.* **1998**, *37*, 245–255. [CrossRef]
33. Riffat, S.B.; Zhu, J. Mathematical model of indirect evaporative cooler using porous ceramic and heat pipe. *Appl. Therm. Eng.* **2004**, *24*, 457–470. [CrossRef]
34. Adam, A.; Han, D.; He, W.; Amidpour, M.; Zhong, H. Numerical investigation of the heat and mass transfer process within a cross-flow indirect evaporative cooling system for hot and humid climates. *J. Build. Eng.* **2022**, *45*, 103499. [CrossRef]
35. Belarbi, R.; Ghiaus, C.; Allard, F. Modeling of water spray evaporation: Application to passive cooling of buildings. *Sol. Energy* **2006**, *80*, 1540–1552. [CrossRef]
36. Crafton, E.F.; Black, W.Z. Heat transfer and evaporation rates of small liquid droplets on heat horizontal surfaces. *Int. J. Heat Mass Transf.* **2004**, *47*, 1187–1200. [CrossRef]
37. Kaiser, A.S.; Lucas, M.; Viedma, M.; Zamora, B. Numerical model of evaporative cooling processes in a new type of cooling tower. *Int. J. Heat Mass Transf.* **2005**, *48*, 986–999. [CrossRef]
38. Maheshwari, G.P.; Al-Ragom, F.; Suri, R.K. Energy-saving potential of an indirect evaporative cooler. *Appl. Energy* **2001**, *69*, 69–76. [CrossRef]
39. Falkovich, G. *Fluid Mechanics*; Cambridge University Press: Cambridge, UK, 2011.
40. Cengel, Y.A.; Ghajar, A.J. *Heat and Mass Transfer: Fundamentals & Applications*, 4th ed.; The McGraw-Hill Companies: New York, NY, USA, 2011.
41. Frossling, N. Uber die Verdunstung fallender Tropfen. *Gerlands Beitr. Geophysik* **1938**, *52*, 170–216.
42. Bolz, R.E.; Tuve, G.L. *Handbook of Tables for Applied Engineering Sciemce*, 2nd ed.; CRC Press: New York, NY, USA, 1976.

Article

Influences of Heat Rejection from Split A/C Conditioners on Mixed-Mode Buildings: Energy Use and Indoor Air Pollution Exposure Analysis

Xuyang Zhong [1,2], Ming Cai [3], Zhe Wang [2,*], Zhiang Zhang [4] and Ruijun Zhang [5]

1   School of Architecture and Urban Planning, Chongqing University, Chongqing 400045, China; xuyangz@lsu.edu.cn
2   Department of Civil Engineering, Faculty of Engineering, Lishui University, Lishui 323000, China
3   State Grid Lishui Power Supply Company, Lishui 323000, China; mingcai09@hotmail.com
4   Department of Architecture and Built Environment, University of Nottingham Ningbo China, 199 East Taikang Road, Ningbo 315100, China; zhiang.zhang@nottingham.edu.cn
5   School of Architecture, Southeast University, 2 Sipailou, Nanjing 210096, China; ruijun_zhang@seu.edu.cn
*   Correspondence: wangzhe@lsu.edu.cn

Abstract: The heat rejected by outdoor units of split A/C conditioners can impact the ambient outdoor environment of mixed-mode buildings. Nevertheless, how this environmental impact may affect the space-conditioning energy use and indoor air pollution is poorly understood. By coupling EnergyPlus and Fluent, this study examines the effects of outdoor units' heat rejection on the building surroundings, building cooling load, and indoor $PM_{2.5}$ exposure of a six-storey mixed-mode building. The building had an open-plan room on each floor, with the outdoor unit positioned below the window. The coupled model was run for a selected day when the building was cooled by air conditioning and natural ventilation. Five mixed-mode cooling strategies were simulated, reflecting different window-opening schedules, airflow rates of outdoor units, and cooling set-points. The results indicate that compared with the always-air-conditioned mode, the mixed-mode operation could significantly mitigate the negative impact of heat rejection on space-cooling energy consumption. Increasing the airflow rate of outdoor units led to a lower increase in demand for space cooling and lower indoor $PM_{2.5}$ exposure. If one of the six rooms needs to be cooled to a lower temperature than the others; choosing the bottom-floor room helped achieve more energy savings and better indoor air quality.

Keywords: building simulation; split A/C conditioners; mixed-mode building; cooling loads; exposure to indoor $PM_{2.5}$

Citation: Zhong, X.; Cai, M.; Wang, Z.; Zhang, Z.; Zhang, R. Influences of Heat Rejection from Split A/C Conditioners on Mixed-Mode Buildings: Energy Use and Indoor Air Pollution Exposure Analysis. *Buildings* **2024**, *14*, 318. https://doi.org/10.3390/buildings14020318

Academic Editor: Eusébio Z.E. Conceição

Received: 25 December 2023
Revised: 10 January 2024
Accepted: 22 January 2024
Published: 23 January 2024

1. Introduction

In Hong Kong, split A/C conditioners are widely installed in buildings. A split A/C conditioner consists of a unit installed indoors and a unit installed outdoors. During cooling, outdoor units reject the heat from indoor units to outdoors. A building interacts with its ambient outdoor environment by heat convection between the ambient outdoor air and external surfaces, as well as by the exchange of air between the indoor and ambient outdoor environment of the building through ventilation and infiltration [1]. The outdoor air temperature can therefore significantly influence building energy consumption.

As the outdoor units' heat rejection can significantly influence ambient outdoor temperatures, the heat rejection data have been incorporated in several studies on building energy use. Chow and Lin [2] used a modelling approach to investigate the temperature profile of the air within a tall building re-entrant where the outdoor units were located; the model outputs indicate (1) that the heat rejected by the outdoor units of lower-floor rooms caused an increase in the temperatures of the air around upper-floor rooms and (2) that there was a higher amount of energy required for air conditioning due to this temperature increase. Many attempts have been made to reduce the negative effects caused by heat rejection on the energy efficiency of buildings. Chow et al. [3] examined the impacts of different re-entrant shapes on the space-cooling energy demand of residential buildings and suggested that outdoor units should be installed in T-shaped re-entrants to achieve better cooling efficiency. Research by Nada and Said [4] investigated the cooling performance of outdoor units arranged in different ways in a building shaft and found the layout that was most effective in terms of reducing space-conditioning energy consumption.

The above studies have shown the impact of heat rejection on the energy efficiency of sealed air-conditioned buildings, where the occupants kept the air conditioning on and windows closed. There has been, however, little research looking at how heat rejection may modify the energy consumption of mixed-mode buildings, which rely on both air conditioning and window-assisted natural ventilation to maintain occupant thermal comfort while avoiding a significant energy cost for the air conditioning [5]. Motivated by the need to cut carbon emissions from the building sector, the government of Hong Kong has started to encourage the implementation of mixed-mode cooling for existing buildings [6]. The effectiveness of mixed-mode cooling, however, is highly dependent on the ambient outdoor air temperatures. Opening windows cannot always help maintain indoor air temperatures at an acceptable level if the difference in temperature between ambient outdoor air and indoor air is too small and can even result in an energy penalty if the temperature of the ambient outdoor air is higher than that of the indoor air [7].

Another potential problem with mixed-mode cooling lies in diminished indoor air quality. Outdoor air pollutants can infiltrate into buildings via external walls, roofs, and open windows and therefore influence the level of indoor air pollution exposure, which can have negative impacts on occupant health [8]. Consequently, ambient outdoor air pollution is highly related to the risk of health problems for people living in mixed-mode buildings

There is a series of studies that analyse the characteristics of air pollutants around buildings There have been both field studies [9–11] indicating that buildings with different locations (e.g., urban, rural, and roadside) experienced different levels of ambient outdoor air pollution and modelling studies [12–14] showing different levels of ambient outdoor air pollution for flats on different floors of the containing building. Keshavarzian et al. [15] investigated the influences that different building cross-section shapes might have on the dispersion of air pollutants near a single building. Cui et al. [16] used a modelling approach to look at how the layout of buildings in an urban street canyon could affect the dispersion of pollutants around buildings. Xiong and Chen [17] assessed the impact of horizontal sunshields on the ambient outdoor air pollutant concentrations of rooms with single-sided ventilation.

Whereas previous research has examined the way in which the concentration of ambient outdoor air pollutants was influenced by factors including the location, height, building geometrics, and urban street canyon, there has been little research on how heat rejected by outdoor units can modify the concentration of the air pollutants around buildings. The majority of outdoor units are operated to discharge hot air into the outdoor environment. The high volumes of discharged hot air are likely to influence the airflow near the building, thereby influencing the dispersion of air pollutants near the building. Although some studies [18–20] have looked at changes in the airflow patterns around buildings as a result of heat rejection, their focus was centred on the relationship between the airflow pattern and the cooling efficiency of outdoor units. The impact of heat rejection on the level of air pollutants around buildings is still poorly understood.

The literature review concluded that: (1) more attention should be paid to the effect of heat rejection from outdoor units on the performance of mixed-mode buildings, where the air conditioning can be switched off and windows can be open under favourable indoor–outdoor temperature differences and (2) the effect of heat rejection from outdoor units on the concentrations of air pollutants in the proximity of mixed-mode buildings remains unclear. The objectives of this study are to examine how heat rejection may influence the ambient air temperatures and ambient fine particle (also known as $PM_{2.5}$) concentrations of a mixed-mode building and how these influences will change the space-cooling demand and occupant exposure to indoor $PM_{2.5}$. The building energy modelling tool EnergyPlus [21] was employed to estimate the space-cooling demand and occupant exposure to indoor $PM_{2.5}$, whilst the CFD software Fluent 2021R1 [22] was employed to estimate the airflow patterns and the concentrations of outdoor $PM_{2.5}$ around the building. Several mixed-mode cooling strategies were simulated in order to demonstrate how ambient outdoor environmental conditions could vary according to control variables including the window-opening schedule, airflow rate of outdoor units, and the cooling set-point. By assessing the building performance under different mixed-mode cooling strategies, the ways in which a mixed-mode building can be operated to reduce energy consumption and improve occupant health are discussed.

2. Methodology

2.1. Building Model

The building model was run for a city block that had nine six-storey buildings (Figure 1a). The nine buildings were arranged in a layout of 3×3 and were all identical in size, orientation, and construction. This type of city block is typical in the dense urban areas in Hong Kong according to the real estate data source [23]. The building in the middle was the mixed-mode building of interest. The building had an open-plan room (floor area: 80 m^2; ceiling height: 3.4 m) on every floor, the exception being the ground floor. Each room had a south-oriented window, which had a height of 1.8 m and was 5.0 m in width. Stairs, lifts, and the ground-floor space (Figure 1b) were not included in the analysis. The surrounding buildings were treated as shading and wind-blocking components. The street canyon was 10 m in width. The model parameters, including the environmental conditions, fabrics, occupancy pattern, cooling system, and mixed-mode cooling strategies, are described below.

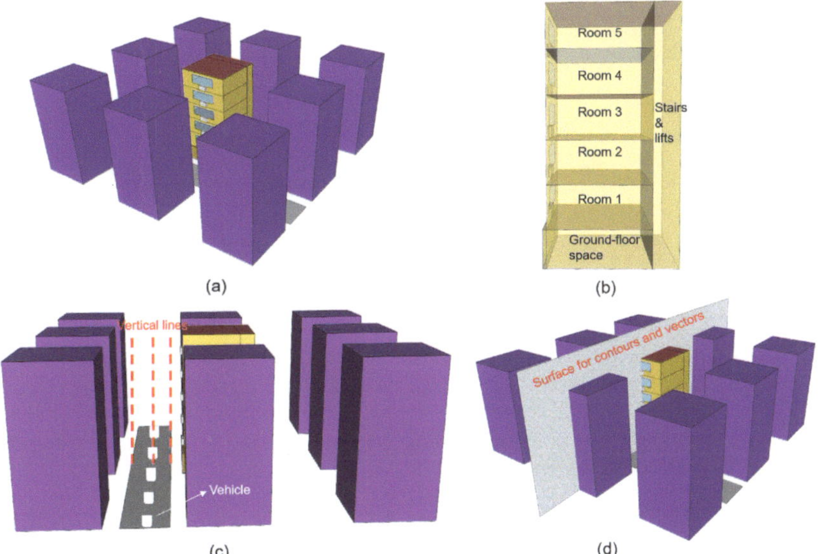

Figure 1. (**a**) The 3D view of the city block. (**b**) Section view of the mixed-mode building. (**c**) The 3D view of the street canyon. (**d**) The surface used for drawing contours and vectors.

2.1.1. Environmental Conditions

Data on weather conditions were provided by the Hong Kong Observatory 2021 database [24]. Data on the background levels of outdoor $PM_{2.5}$ were provided by the Hong Kong EPD 2021 database [25]. The model was run for a three-day period during which the mixed-mode building was cooled using both air-conditioning and natural ventilation via open windows. The first two days were treated as "warmup days" in order to facilitate the convergence of results. The outdoor temperatures, wind speeds, wind directions, and background levels of outdoor $PM_{2.5}$ for the third day are shown in Figure 2. The mean values of the weather variables of the selected three-day period can represent the annual average weather conditions for Hong Kong. The advantage of using a three-day period is the reduction in computational costs, especially for the Fluent model. Given that the wind mainly blew from the south towards the north, a southerly wind was simulated for simplicity. In addition to the southerly wind, a northerly wind was simulated by only changing the wind direction without modifying the other weather parameters so as to demonstrate the impact of the wind direction.

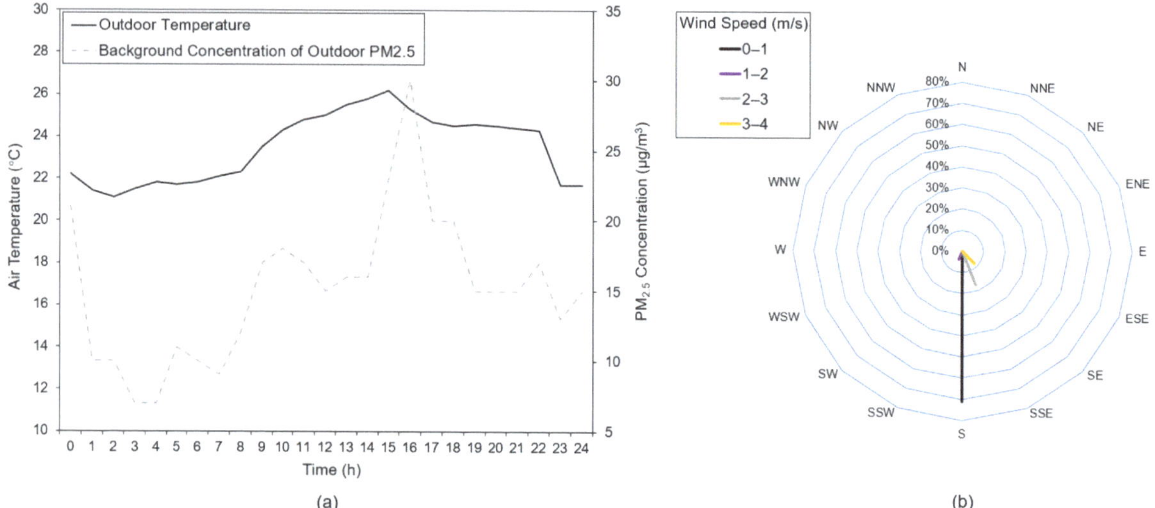

Figure 2. (**a**) Outdoor air temperatures and background outdoor $PM_{2.5}$ for the third day. (**b**) The wind rose diagram for the third day.

2.1.2. Fabrics

This paper used an existing study [26] for the building envelope data (Table 1). The thermal conductivity of each material was provided by the Buildings Department (BD) [27] It was assumed that there was no exchange of heat between a room and its adjoining rooms The airtightness of the building envelope was modelled using a permeability measurement (i.e., the rate of air leakage at 50 Pa indoor–outdoor pressure difference). The permeability was used to demonstrate the impact of different wind pressures on the infiltration rate. The infiltration rate of the building was first calculated using the ISO 13790 methodology [28] and then was changed into the permeability (11.5 $m^3/h/m^2$) based on the surface area and volume of the building.

2.1.3. Occupancy Pattern

The occupancy pattern determines both the internal heat gains and period of exposure The number of occupants in individual rooms was set as seven to comply with the appropriate design occupation density for offices [29]. Occupants were assumed to be present from 08:00 to 15:00. The generation rates of the heat from people (130 W/person), lights (12 W/m^2), and appliances (15 W/m^2) were provided by CIBSE tables [29]. The occupancy pattern was overlaid on the profile of indoor $PM_{2.5}$ concentrations to determine occupant exposure to indoor $PM_{2.5}$.

Table 1. The materials and thermal characteristics of individual building fabrics.

Type	Materials	U-Value (W/m²K)	Solar Absorptance	Longwave Emission Coefficient	Solar Heat Gain Coefficient (SHGC)
External walls	Mosaic tiles (5 mm) + Cement (10 mm) + Heavy concrete (100 mm) + Gypsum plaster (10 mm)	3.1	Front: 0.4 Back: 0.5	Front: 0.9 Back: 0.9	
Windows	Tinted glass (6 mm)	4.6			0.5
Roof	Concrete tiles (25 mm) + Asphalt (20 mm) + Cement (50 mm) + Polystyrene (50 mm) + Heavy concrete (150 mm) + Gypsum plaster (10 mm)	0.4	Front: 0.1 Back: 0.5	Front: 0.9 Back: 0.9	
Ground floor	Floor tiles (10 mm) + Gypsum plaster (10 mm) + Reinforced concrete (180 mm)	3.0	Front: 0.8 Back: 0.5	Front: 0.9 Back: 0.9	

2.1.4. Cooling Device

The cooling device in each room was modelled as a split A/C conditioner. The specification of the split A/C conditioner was taken from the engineering data provided by an A/C manufacturer [30]. The split A/C conditioner had a COP of 3.1. The outdoor unit, which took in air through the inlet on the back and discharged air through the outlet on the front, was positioned below the window of each room (Figure 1a). The air inlet and air outlet had areas of 1.1 m² and 0.9 m², respectively. To simplify analysis, the impact that the outdoor unit's air intake might have on the airflow pattern near the building was ignored. The outdoor unit could be set up to operate at a high airflow rate of 85 m³/min or a low airflow rate of 65 m³/min.

2.1.5. Mixed-Mode Cooling Strategies

The mixed-mode building of interest was neither only naturally ventilated nor fully air-conditioned but relied on a combination of different energy-efficiency cooling techniques. The techniques used to cool each room included air conditioning and natural ventilation via an open window. Four air-conditioning patterns were considered, including:

1. (*all rooms 27 °C + low airflow rate*): during occupied hours, all the rooms were cooled to a set-point of 27 °C. The outdoor units were set up to operate at a low airflow rate of 65 m³/min.
2. (*all rooms 27 °C + high airflow rate*): during occupied hours, all the rooms were cooled to 27 °C. The outdoor units were set up to operate at a high airflow rate of 85 m³/min.
3. (*room five 23 °C and the rest 27 °C + low airflow rate*): during occupied hours, the top-floor room (i.e., room 5) was cooled to 23 °C, whereas the rest were cooled to 27 °C. The outdoor units were set up to operate at a low airflow rate of 65 m³/min.
4. (*room one 23 °C and the rest 27 °C + low airflow rate*): during occupied hours, room 1 (i.e., the room on the bottom floor) was cooled to 23 °C, whereas the rest were cooled to 27 °C. The outdoor units were set up to operate at a low airflow rate of 65 m³/min.

Two ventilation patterns were considered, including:

A. (*no window opening*): all the windows in the building were closed. This reflects the ventilation pattern of a sealed air-conditioned building.
B. (*temperature-dependent window opening*): a large indoor temperature swing can occur when the window was open, especially when there is a large indoor–outdoor temperature difference. This ventilation pattern aimed to comply with ASHRAE 55-2017 [31], which specifies that in order to reduce the negative impact of a large indoor temperature swing on occupant thermal comfort, the change in the temperature of the indoor air during a four-hour period should not exceed 3.3 °C. To meet the ASHRAE 55-2017 requirement, the temperatures inside and outside each room were calculated throughout the simulation period. When the difference in temperature between indoors and outdoors was in the range of 0 and $T_{threshold}$, the window was open. When the indoor–outdoor delta temperature was larger than $T_{threshold}$, the window was closed. In both cases, the window was closed if the air conditioning was on, if natural ventilation

was not able to keep indoor temperatures below the cooling setpoint, or if no one was in the room. The value of $T_{threshold}$ was determined via a series of simulations with $T_{threshold}$ varying from 5.8 °C (i.e., the difference between the lowest temperature of the ambient outdoor air and the cooling set-point) to 0 in increments of −0.1 °C. Simulations stopped when the ASHRAE 55-2017 requirement was met; the value of $T_{threshold}$ was then determined.

Mixed-mode cooling strategies were developed by combining the air-conditioning and ventilation patterns described above. In total, there were five different mixed-mode cooling strategies to be simulated (Table 2).

Table 2. The five different mixed-mode cooling strategies.

Mixed-Mode Cooling Strategy	Air-Conditioning Pattern	Ventilation Pattern
Strategy 1A	1 (all rooms 27 °C + low airflow rate)	A (no window opening)
Strategy 1B	1 (all rooms 27 °C + low airflow rate)	B (temperature-dependent window opening)
Strategy 2B	2 (all rooms 27 °C + high airflow rate)	B (temperature-dependent window opening)
Strategy 3B	3 (room five 23 °C and the rest 27 °C + low airflow rate)	B (temperature-dependent window opening)
Strategy 4B	4 (room one 23 °C and the rest 27 °C + low airflow rate)	B (temperature-dependent window opening)

2.2. Simulations

The building model described in Section 2.1 was developed in both EnergyPlus (v9.6) and Fluent (v2021-r1). EnergyPlus was applied to calculate cooling loads using the equation for heat balance, model the indoor airflow using the AirflowNetwork model, and calculate concentrations of indoor $PM_{2.5}$ using the generic contaminant transport algorithm. Fluent was applied to model the turbulent airflow near the building using the governing equations for incompressible airflow and calculate the concentrations of ambient outdoor $PM_{2.5}$ through a stochastic tracking approach. To ensure the accuracy of the input data (i.e., the temperatures of ambient outdoor air, ambient outdoor $PM_{2.5}$ concentrations, wind pressure coefficients, and temperatures of exterior surfaces), this study adopted a quasi-dynamic coupling method for EnergyPlus—Fluent co-simulation. The methods and assumptions used in the simulations are detailed below.

2.2.1. EnergyPlus Simulations

The outputs from Energyplus simulations included exterior surface temperatures, cooling loads, and concentrations of indoor $PM_{2.5}$. These variables were calculated at an interval of 10 min and output hourly.

The temperature of an exterior surface was calculated using the equation:

$$q''_{asol} + q''_{LWR} + q''_{conv} - q''_{ko} = 0 \tag{1}$$

where q''_{asol} is the absorbed heat flux from solar radiation (W/m^2), q''_{LWR} is the exchange of long wavelength radiation flux with the air and the surroundings (W/m^2), q''_{conv} is the exchange of convective flux with the outside air (W/m^2), and q''_{ko} is the conduction heat flux (W/m^2).

The cooling load was estimated using the equation:

$$Q_{sun\_rad} + Q_{internal\ surface\_rad} + Q_{ven} + Q_{inf} + Q_{cond} + Q_{interal\ heat} - Q_{cooling} = 0 \tag{2}$$

where Q_{sun_rad} is the heat gain from solar radiation (W), $Q_{internal\ surface\_rad}$ is the transfer of radiative heat from the internal surfaces (W), Q_{ven} is the ventilation heat gain (W), Q_{inf} is the infiltration heat gain (W), Q_{cond} is the conduction heat gain (W), $Q_{interal\ heat}$ is the occupant and equipment heat gain (W), and $Q_{cooling}$ is the cooling load (W).

According to the previous study [32], the heat rejection rate was calculated by adding up the energy required to cool the indoor space and the energy required for the A/C operation. The outlet air temperature of the outdoor unit was estimated using the equation:

$$T_{air\_outlet} = \frac{Q_{rejected}}{M_{air}C_{p,air}} + T_{air\_inlet} \tag{3}$$

where $Q_{rejected}$ is the rate of heat rejection (kW), M_{air} is the outdoor unit airflow (kg/s), $C_{p,air}$ is the specific heat of air (kJ/kg·K), and T_{air_inlet} is the temperature of the air at the inlet of the outdoor unit (°C).

The indoor level of outdoor-sourced $PM_{2.5}$ was determined using the generic contaminant model [33]. The outdoor-sourced $PM_{2.5}$ was modelled using a combination of $PM_{2.5}$ from vehicles and the background levels of outdoor $PM_{2.5}$ for Hong Kong (described in Section 2.1.1). The emission rate of $PM_{2.5}$ from vehicles in the street canyon (Figure 1c) was 1.13×10^{-7} kg/s, according to a study on the air pollution caused by road traffic [34]. The ingress of the air from the ambient outdoor environment was modelled by assuming that there were cracks within the building fabrics (external walls, the roof and, when open, windows). The fraction of pollutant loss caused by the deposition of pollutants in the cracks (a.k.a. penetration factor) was set as 0.8 for closed windows and 1.0 for open windows [35]. Window opening was modelled assuming a two-way flow. The internal floors, ceilings, and walls were assumed to have cracks that allowed for the exchange of $PM_{2.5}$ between different rooms. The reference air mass flow coefficient of the cracks was assigned according to the area of the building surface and the permeability of the building envelope, with the exponent of air mass flow being 0.66 [36]. The rate of the deposition of outdoor-sourced $PM_{2.5}$ was set as 0.19 h^{-1} [37]. It should be noted that the penetration rate and deposition rate of a particle are strongly related to the size of the particle; however, for simplicity, all the particles in this study were assumed to have the same size.

2.2.2. Fluent Simulations

The outputs from Fluent simulations included the temperature of the air around the building, concentration of $PM_{2.5}$ around the building, and coefficient of wind pressure. Data for these outputs were obtained from the grid layer that was closest to the building envelope.

The flow field was modelled using a three-dimensional standard k-ε model, which has frequently been used by similar studies because of its robustness and relatively low computing costs [38,39]. As in other similar research [16,40], the airflow was assumed to be numerically stable and was therefore incompressible turbulent. The general form of the governing equation for the airflow that is incompressible turbulent is written as:

$$\frac{\partial}{\partial_t}(\partial) + \nabla \cdot (\overline{u}_\partial) - \nabla \cdot (\Gamma_\partial \nabla_\partial) = S_\partial \tag{4}$$

where $\overline{u}$ is the average velocity, S is the source term, Γ is the coefficient of effective diffusion, and ∂ is the scalar, which can be turbulent kinetic energy (k), rate of dissipation (ε), or velocity ingredients.

The velocity–pressure coupling was solved using the SIMPLE algorithm. The second-order method was used to solve the pressure, convective, and diffusive terms. A standard wall function was assigned to the regions near walls. The time step of the simulation was 1 s. The number of iterations was kept below 600 for each time step. The convergence of results was obtained when the values of scaled residuals were less than 10^{-5} for the continuity equation and less than 10^{-6} for other equations.

The distribution of outdoor $PM_{2.5}$ was determined via a stochastic tracking method that estimates the particle trajectories by solving the balance of forces acting on the particle. The equation of the force balance for a particle is as follows:

$$\frac{dv_p}{dt} = F_d(v - v_p) + \frac{g_x(\rho_p - \rho)}{\rho_p} + F_{add} \tag{5}$$

where v_p is the velocity of the particle, v is the velocity of air, $F_d(v - v_p)$ is the drag force for every unit mass of the particle, g_x is the acceleration due to gravity, ρ_p is the density of the particle, ρ is the density of air, and F_{add} is the additional acceleration term. No collision between particles was considered in this study.

A step-by-step report of particle trajectories was used to locate individual particles. The concentration of particles was determined using a particle-in-box approach. The equation for the particle-in-box approach is as follows:

$$C = \frac{M_p \sum_{i=1}^{n} N_k^i}{V_k} \tag{6}$$

where M_p is the particle mass, N_k^i is the ith particle that is in the box k, and V_k is the volume of the box k.

As suggested by COST [41] and AIJ [42], a computational domain of the flow field (Figure 3a) was developed, with the inlet being 5H (H: the building's height, which was 20 m) away from the city block, the outlet being 15H away from the city block, the sky being 5H away from the city block, and the lateral boundary being 5H away from the city block. To meet the size requirement of the validation carried out in Section 2.3, the entire model was scaled down by a factor of 25. For a reduced-scale model, the independence of the Reynolds number (R_e) should be assessed [16]. The results of pre-simulations show that the R_e values were 2.9×10^5 and 2.6×10^5 at building height and around outdoor units, respectively. Therefore, the independence requirement [43] that R_e at the height of the building or around the envelope feature should be greater than 1.0×10^4 was met.

The structured grids within the city block (especially near buildings) were dense, whereas those in the surrounding regions were coarse (Figure 3b). The grid layer closest to the building surfaces, outdoor units, and ground was 0.001 m in height. This meshing strategy led to a mean y^* of around 82, which was within the range of 30–300 required for a standard wall function [41]. To ensure there was a smooth transition in the size of the grids, the grids were drawn using an inflation ratio of no greater than 1.2.

An analysis of the grid sensitivity was conducted to keep the model outputs (including the normalised wind speeds and normalised $PM_{2.5}$ concentrations for ninety measurement points on the nominal vertical lines in the street canyon (Figure 1c)) independent of the grid size. The threshold for grid independence is that the root mean square error (RMSE) should be below 10%, as used in [15,44]. RMSE was determined using the equation:

$$RMSE = \sqrt{\frac{\sum_{i=1}^{n}(y_i - \hat{y}_i)^2}{n}} \tag{7}$$

where i is a measurement point up to n number of measurement points and y_i and $\hat{y}_i$ are the results of the measurement point i with two different grid configurations.

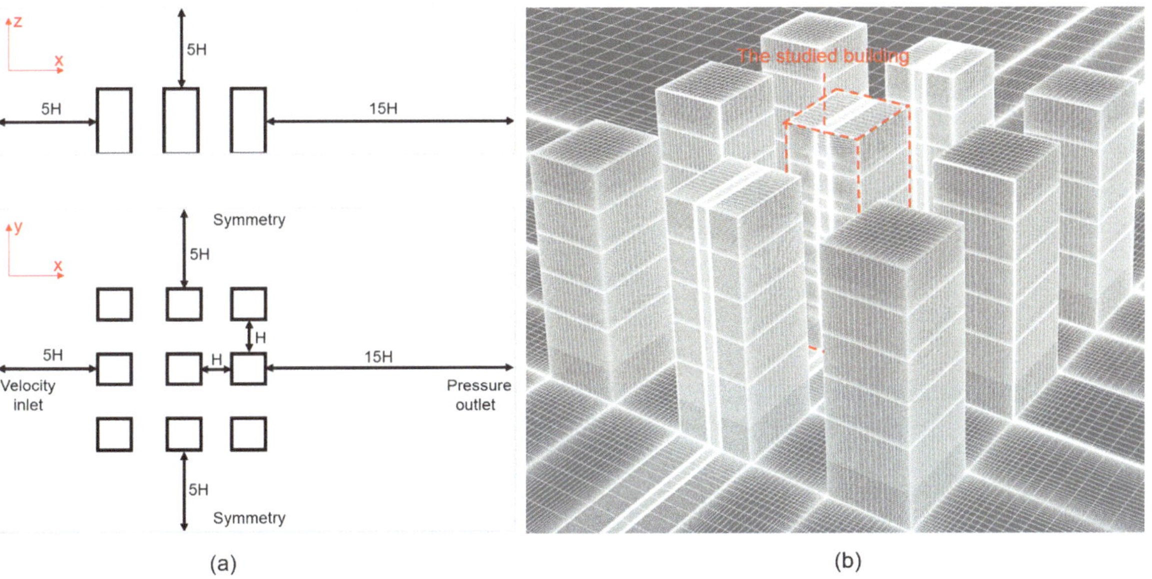

Figure 3. (**a**) The size of the flow field. (**b**) The mesh for the city block.

This study tested three different grid sizes (i.e., coarse, basic, and fine grids), with a refinement factor of 1.1 [41]. The number of cells for the coarse, basic, and fine grids were 1.9, 3.6, and 7.8 million, respectively. The results show that the RMSE between the coarse grids and basic grids was 23.1% and the RMSE between the basic grids and fine grids was only 4.6%. Therefore, the resolution of basic grids was considered sufficient and was applied to perform the validation performed in Section 2.3 and the parametric studies carried out in Section 3.

2.2.3. EnergyPlus and Fluent Co-Simulation

The quasi-coupling method was adopted for EnergyPlus–Fluent co-simulation. The coupling process is shown in Figure 4. First, EnergyPlus was run using the environmental data (shown in Section 2.1.1) to calculate the temperatures of the exterior surfaces and the outdoor units' heat rejection rates. Then, Fluent was run using the outputs from EnergyPlus (including the temperatures of exterior surfaces and the rates of heat rejection) as boundary conditions to calculate temperatures of ambient outdoor air, concentrations of ambient outdoor $PM_{2.5}$, and coefficients of wind pressure. Finally, EnergyPlus was run using the modified environmental data that included the temperatures of ambient outdoor air and concentrations of ambient outdoor $PM_{2.5}$ output from Fluent to calculate the cooling loads, concentrations of indoor $PM_{2.5}$, temperatures of exterior surfaces, and heat rejected by outdoor units. The data exchange between EnergyPlus and Fluent was handled by Matlab 2020b [45] and took place every hour.

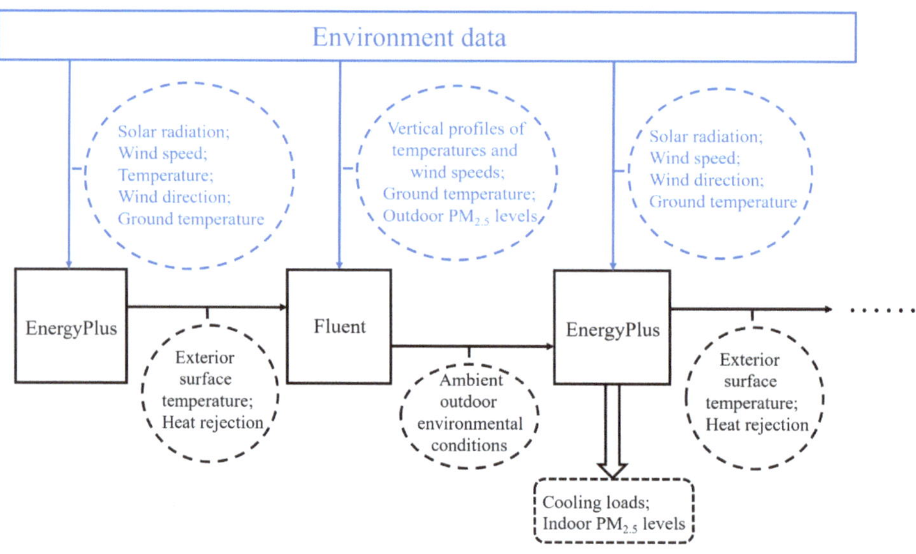

Figure 4. The coupling process: data exchange between EnergyPlus and Fluent.

Table 3 summarizes the Fluent simulation's boundary conditions.

Table 3. Boundary conditions of the CFD domain during the coupling process.

Boundary	Type	Conditions
Ground	Wall	Non-slip; surface temperatures based on the meteorological data.
Building envelope	Wall	Non-slip; exterior surface temperatures outputted by EnergyPlus.
Sky and non-inlet/outlet laterals	Wall	Non-slip; adiabatic.
Domain inlet	Velocity inlet	Wind speed profile: U_z; temperature profile: T_z; turbulence kinetic energy profile: k_z; and turbulence dissipation rate profile: ε_z.
Domain outlet	Pressure outlet	Gauge pressure of 0 pa; temperature profile T_z; turbulence profiles: k_z and ε_z.
Outlet of the outdoor unit	Velocity inlet	Airflow rate: 65 or 85 m^3/min; area of the air outlet: 0.9 m^2; and temperature profile of T_{air_outlet} (see Equation (3)).

It should be noted that U_z and T_z in Table 3 were calculated based on the following equations in order to match the wind speed and temperature patterns modelled in EnergyPlus:

$$U_{met}\left(\frac{\delta_{met}}{z_{met}}\right)^{\alpha_{met}}\left(\frac{z}{\delta}\right)^{\alpha} = U_z \tag{8}$$

$$T_{tro} + \frac{LEz}{E+z} - LH_{tro} = T_z \tag{9}$$

$$T_{tro} = T_{z,met} - L\left(\frac{Ez_{met}}{E+z_{met}} - H_{tro}\right) \tag{10}$$

where U_{met} is the measured velocity of the weather monitoring station (m/s), δ_{met} is the boundary layer thickness of the velocity profile of the weather monitoring station (m), z_{met} is the height of the velocity sensor of the weather monitoring station (m), α_{met} is the exponent of the velocity profile of the weather monitoring station, z represents the altitude, δ represents the thickness of the boundary layer of the velocity profile of the site, α is the velocity profile exponent of the site, L is the air temperature gradient (k/m), E is the Earth radius, z is the altitude, H_{tro} is the offset for the troposphere, T_{tro} is the ground-level air temperature for the troposphere (°C), $T_{z,met}$ is the measured outdoor air temperature of the weather station (°C), and z_{met} is the height of the air temperature sensor of the weather station (m).

The turbulence kinetic energy profile ($k_\mathcal{Z}$) mentioned in Table 3 was determined according to the following equation:

$$k_\mathcal{Z} = 0.5 \times \left(u'^2_{x,\mathcal{Z}} + u'^2_{y,\mathcal{Z}} + u'^2_{z,\mathcal{Z}} \right) \cong 1.5 \times u'^2_\mathcal{Z} \tag{11}$$

where $u'_{x,\mathcal{Z}}$, $u'_{y,\mathcal{Z}}$, and $u'_{z,\mathcal{Z}}$ are the root mean square of the velocity of the wind blowing in the x-axis, y-axis, or z-axis directions at a height of $\mathcal{Z}$ and $u'_\mathcal{Z}$ is the root mean square of the velocity of the streamwise wind at a height of $\mathcal{Z}$.

The turbulence dissipation rate profile ($\varepsilon_\mathcal{Z}$) mentioned in Table 3 was determined according to the following equation:

$$\varepsilon_\mathcal{Z} = C_\partial^{0.5} k_\mathcal{Z} \frac{U_{\text{reference}}}{\mathcal{Z}_{\text{reference}}} \alpha \left(\frac{\mathcal{Z}}{\mathcal{Z}_{\text{reference}}} \right)^{\alpha-1} \tag{12}$$

where C_∂ is the dimensionless constant of the standard k-ε model and $U_{\text{reference}}$ is the reference velocity of the wind at a reference height of $\mathcal{Z}_{\text{reference}}$.

2.3. Validation

The simulation tools used herein were validated according to the results from the literature review and filed measurements. In Sections 2.3.1 and 2.3.2, the grid size was determined according to the sensitivity analysis carried out in Section 2.2.2.

2.3.1. Airflow around Buildings

Data from a wind tunnel test carried out by the Architectural Institution of Japan (AIJ) [42] was used to assess the ability of Fluent to accurately model the flow field around buildings. In this wind tunnel test, nine cubes were arranged in a 3 × 3 array (Figure 5a). Each cube was 0.2 m in length. The street canyon was 0.2 m in width. As suggested by COST and AIJ (described in Section 2.2.2), the computational domain was developed using Fluent. The cube surface's grid layer height was 0.003 m. The mean y^* is about 75. The Fluent model used the boundary conditions of the wind tunnel test.

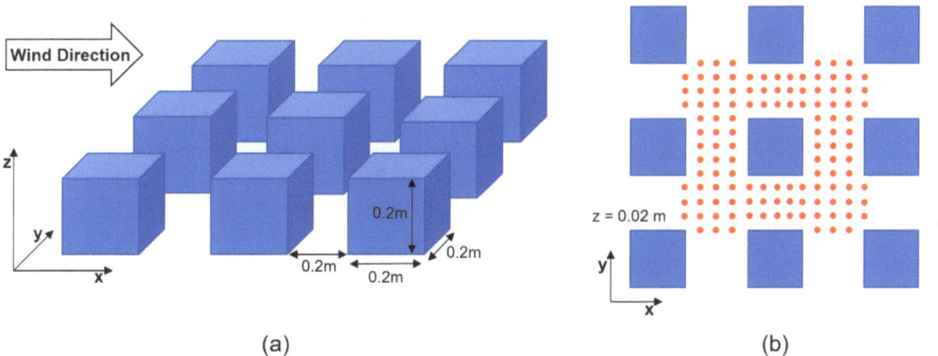

Figure 5. (**a**) The cubes used in the test carried out by AIJ. (**b**) The locations of points used to measure velocity.

The measurement point (Figure 5b) heights were 0.02 m to measure velocities around cubes. The normalised velocity was equal to the ratio of the velocity of a point (U_z^*) and the inflow velocity at the same height ($U_{z,\text{inlet}}$). The comparison between measurements and simulations is shown in Figure 6. The average of the percentage differences ($100\% \times$ (predicted value − measured value)/measured value) was 15.5%, which shows a good match between the predicted and measured velocities.

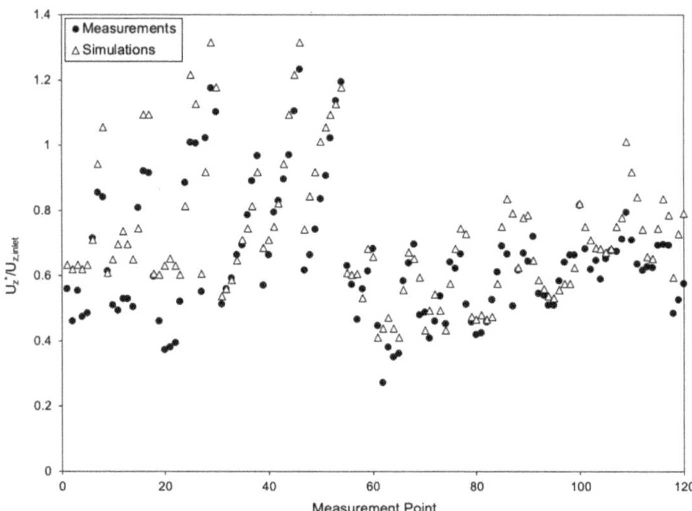

Figure 6. The simulated and measured normalised velocity for individual measurement points.

In addition, the model accuracy was assessed based on three dimensionless metrics, which were the factor of 2 of observations (FAC2), the fractional mean bias (FB), and the normalised mean square error (NMSE). The dimensionless metrics were calculated using:

$$FAC2 = \frac{1}{n}\sum_{i=1}^{n} n_i \text{ with } n_i = \begin{cases} 1, \text{if } 0.5 \leq \frac{s_i}{m_i} \leq 2 \\ 1, \text{if } m_i \leq W_F \text{ and } s_i \leq W_F \\ 0, \text{ else} \end{cases} \tag{13}$$

$$FB = \frac{2 \times \left(\sum_{i=1}^{n} m_i - \sum_{i=1}^{n} s_i\right)}{\sum_{i=1}^{n} m_i + \sum_{i=1}^{n} s_i} \tag{14}$$

$$NMSE = \frac{\sum_{i=1}^{n}(m_i - s_i)^2}{\sum_{i=1}^{n} m_i \times \sum_{i=1}^{n} s_i} \tag{15}$$

where n_i is the ith data point, s represents the simulation results, m represents the measurement results, and W_F is the allowable absolute difference [46]. A CFD model that is acceptable for use in urban scenarios should meet the following requirements: $FAC2 \geq 0.3$, $-0.67 \leq FB \leq 0.67$, and $NMSE \leq 6$ [47,48]. The CFD model developed in this section met the criteria, as FAC2 is 1, FB is -0.1, and NMSE is 0.03. Consequently, the final k-ε model in conjunction with the numerical setups could provide accurate estimates of the airflow patterns around buildings.

2.3.2. The Level of Air Pollution around Buildings

Data from a wind tunnel test carried out by the University of Hamburg [49] were used to assess the accuracy of ambient outdoor pollutant concentrations predicted by Fluent. In this wind tunnel test, 21 rectangular blocks (Figure 7a) were arranged in a 3 × 7 array. Each rectangular block had the same dimensions: 0.15 m length, 0.1 m width, and 0.125 m height. The street canyon was 0.1 m in width. The CFD domain was developed based on the suggestions from COST and AIJ (described in Section 2.2.2). Both the grid layers closest to the block surfaces and the ground have heights of 0.002 m (with y^* being about 67). The boundary conditions of the wind tunnel test were applied to the Fluent model. The air pollutant emission source is the block's bottom (Figure 7a). The emission velocity was 0.1 m/s, with the emission area being 4.6 cm^2.

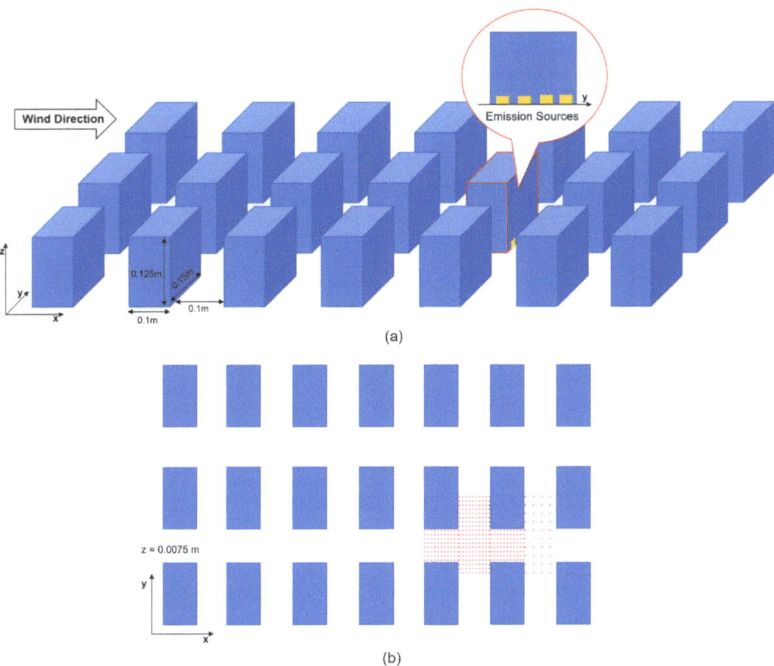

Figure 7. (**a**) The rectangular blocks. (**b**) The locations of measurement points.

The study used 437 measurement points (Figure 7b) with heights of 0.0075 m for pollutant concentration measurements. The concentration values were normalised to a dimensionless K value that was calculated based on the equation:

$$K = \frac{C_{measured}U_{ref}H^2}{C_{source}Q_{source}} \tag{16}$$

where $C_{measured}$ is the measured pollutant concentration (ppm), C_{source} is the pollutant concentration of the emission source (ppm), U_{ref} is the measured reference velocity at a height of 0.66 m (m/s), H is the block height (m), and Q_{source} is the emission rate of air pollutants (m^3/s).

The comparison between the simulations and measurements is shown in Figure 8, which shows there was a good accuracy between the simulation and the measurement results. This conclusion was further supported by the metrics that reveal that FAC2 is 1, FB is −0.1, and NMSE is 0.2. Consequently, the selected *k-ε* model and stochastic tracking model, in conjunction with the numerical setups, could provide reliable estimates of the pollutant concentrations within a city block.

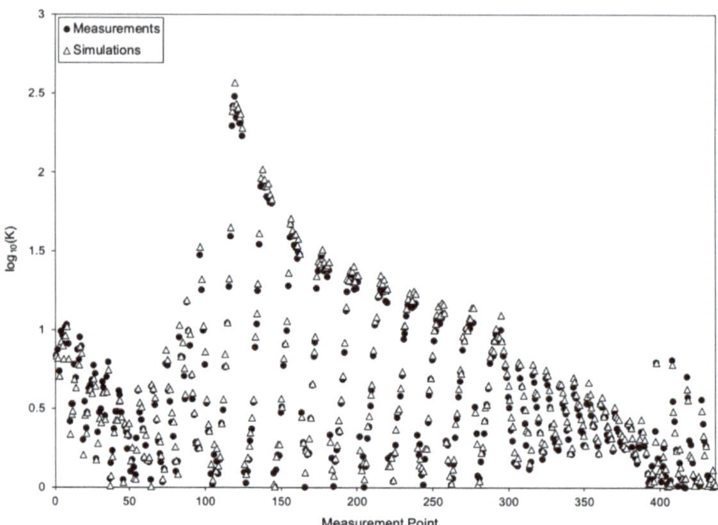

Figure 8. The simulated and measured normalised pollution levels of individual measurement points (the profile of the normalised concentrations is shown in the format of log due to sharp increases in concentrations relative to the strength of emission sources).

2.3.3. Space Cooling Demand and Indoor PM$_{2.5}$

Data from a field measurement was used to check the accuracy of space-cooling energy use and indoor pollutant levels predicted by EnergyPlus. A typical Hong Kong flat was used for field measurements (Figure 9a). The main bedroom of the flat had an air conditioner, with the COP being 2.4. The characteristics of the building fabrics were summarised as follows: (1) the rate of heat flow through the external walls, windows, roof, and ground floor were 3.1, 4.6, 0.42, and 0.54 W/m^{2}K, respectively; (2) the window had a SHGC of 0.76; and (3) the airtightness of the building at 50 Pa was 10.1 m^3/h/m^2. Windows and internal doors remained closed. During the measurement period, the air conditioner was run to maintain the indoor temperature of the main bedroom at 24 °C.

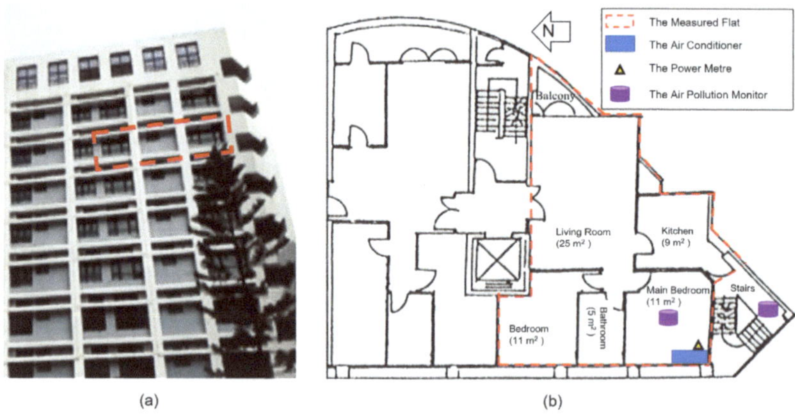

Figure 9. (**a**) The measured flat. (**b**) The layout of the measured flat.

Measurements were taken from 28–30 September 2017. A power metre (SP2; Broad-Link, Hangzhou, China) was applied to record the electric energy required to run the air conditioner in the main bedroom. The PM$_{2.5}$ level was measured via a calibrated air quality monitor (DUSTTRAK 8530 EP; TSI, Shoreview, MN, USA) that was accurate to $\pm 5\%$ and could measure the pollution level in the range from 0.001 mg/m^3 to 150 mg/m^3. The indoor PM$_{2.5}$ level was measured using the air quality monitor positioned in the middle of the main bedroom, whereas the outdoor PM$_{2.5}$ level was measured using the air quality monitor positioned on the stairs (Figure 9b). The nearby weather station (Kowloon City) data were used.

To understand the difference in results between measurements and simulations, the normalised mean bias error (NMBE) and the coefficient of variation of the root mean squared error (CVRMSE) were applied. ASHRAE Guideline 14-2014 [50] suggests that the simulation results are reliable if NMBE $< \pm 10\%$ and CVRMSE $< 30\%$. The two errors were calculated as follows:

$$\mathrm{NMBE}[\%] = 100 \times \frac{\sum_{i=1}^{n}\left(y_i - y_i^{*}\right)}{n \times \overline{y}} \tag{17}$$

$$\mathrm{CVRMSE}[\%] = 100 \times \frac{\sqrt{\sum_{i=1}^{n}\left(y_i - y_i^{*}\right)^{2}/n}}{\overline{y}} \tag{18}$$

where n is the number of data points, y is the measurement result, y^{*} is the simulation result, and $\overline{y}$ is the average of all the measurements.

An EnergyPlus model of the measured flat was constructed based on the geometry, building fabrics, air conditioner, weather conditions, and levels of outdoor PM$_{2.5}$ described above. Comparisons were made using the results for September 29, 2017. In general, the simulation results have good accuracy compared with the measurement results, as evidenced by Figure 10. The NMBE is 4.5% and the CVRMSE is 7.3%, which meets ASHRAE requirements. The simulated levels of indoor PM$_{2.5}$ were also in agreement with the measured ones, with the values of NMBE (8.5%) being less than 10% and CVRMSE (10.2%) being less than 30%. Consequently, the selected AirflowNetwork model and generic contaminant transport model, in conjunction with the numerical setups, could provide reliable estimates of energy demand for space-cooling and levels of indoor PM$_{2.5}$.

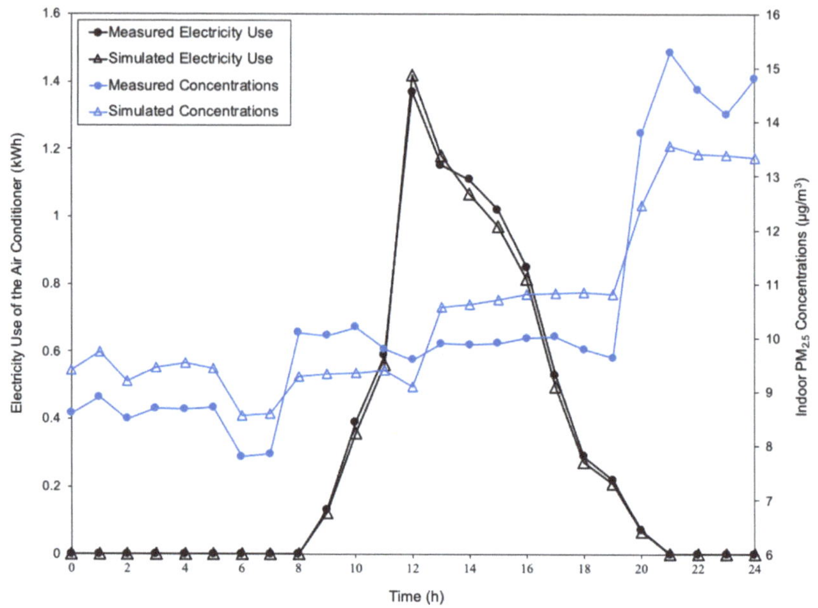

Figure 10. Comparisons between the measured and simulated electric energy needed for air conditioning and indoor $PM_{2.5}$ concentrations.

3. Results

The results presented herein show the ambient outdoor environmental conditions (i.e., temperatures, $PM_{2.5}$ concentrations, and airflow patterns), energy demand for space cooling, and indoor $PM_{2.5}$ exposure of the mixed-mode building operated using the cooling strategies described in Section 2.1.5. Based on the data taken from the surface shown in Figure 1d, the velocity vector, as well as the contours of temperatures and $PM_{2.5}$ concentrations, were drawn to help explain the results. A reference case with no heat rejected by the outdoor units was simulated so that variations in the surrounding outdoor environment because of heat rejection and the resulting changes in building performance could be estimated. There are two outdoor unit position scenarios: a leeward scenario, where the outdoor units are positioned on the leeward side, and a windward scenario, where the outdoor units are on the windward side (described in Section 2.1.1).

3.1. Ambient Outdoor Temperatures

Figure 11 shows the changes in the temperatures of the outdoor air around individual rooms due to heat rejection. The positive values represent that, compared with the reference case, the temperature increases.

Windward Scenario (unit: °C)

	Room 1	Room 2	Room 3	Room 4	Room 5
8:00–9:00	0.5	0.6	0.4	0.4	0.1
9:00–10:00	0.9	1.2	0.7	0.6	0.1
10:00–11:00	1.0	1.3	0.8	0.7	0.1
11:00–12:00	1.1	1.3	0.8	0.8	0.1
12:00–13:00	1.1	1.4	0.9	0.8	0.2
13:00–14:00	1.2	1.4	0.9	0.8	0.2
14:00–15:00	1.1	1.4	0.9	0.8	0.2
Average	**1.0**	**1.2**	**0.8**	**0.7**	**0.1**

(**a**) Strategy 1A (all rooms 27 °C + low airflow rate + no window opening)

Leeward Scenario (unit: °C)

	Room 1	Room 2	Room 3	Room 4	Room 5
8:00–9:00	1.3	1.5	1.6	1.7	1.9
9:00–10:00	1.4	1.6	1.7	1.8	2.0
10:00–11:00	1.6	1.7	1.9	2.0	2.2
11:00–12:00	1.7	1.8	2.0	2.1	2.4
12:00–13:00	1.7	1.8	2.0	2.2	2.5
13:00–14:00	1.7	1.9	2.1	2.2	2.6
14:00–15:00	1.7	1.9	2.1	2.2	2.6
Average	**1.6**	**1.7**	**1.9**	**2.1**	**2.3**

(**f**) Strategy 1A (all rooms 27 °C + low airflow rate + no window opening)

	Room 1	Room 2	Room 3	Room 4	Room 5
8:00–9:00	0	0	0	0	0
9:00–10:00	0	0	0	0	0
10:00–11:00	0	0	0	0	0
11:00–12:00	0.8	0.9	0.6	0.5	0.1
12:00–13:00	0.8	1.0	0.6	0.5	0.1
13:00–14:00	0.9	1.1	0.7	0.6	0.1
14:00–15:00	1.0	1.2	0.8	0.7	0.1
Average	**0.5**	**0.6**	**0.4**	**0.3**	**0.1**

(**b**) Strategy 1B (all rooms 27 °C + low airflow rate + temperature-dependent window opening)

	Room 1	Room 2	Room 3	Room 4	Room 5
8:00–9:00	0	0	0	0	0
9:00–10:00	0	0	0	0	0
10:00–11:00	0	0	0	0	0
11:00–12:00	1.4	1.5	1.7	1.8	2.0
12:00–13:00	1.4	1.5	1.7	1.8	2.1
13:00–14:00	1.4	1.5	1.7	1.8	2.2
14:00–15:00	1.4	1.5	1.7	1.8	2.2
Average	**0.8**	**0.9**	**1.0**	**1.0**	**1.2**

(**g**) Strategy 1B (all rooms 27 °C + low airflow rate + temperature-dependent window opening)

	Room 1	Room 2	Room 3	Room 4	Room 5
8:00–9:00	0	0	0	0	0
9:00–10:00	0	0	0	0	0
10:00–11:00	0	0	0	0	0
11:00–12:00	0.7	0.8	0.5	0.5	0.1
12:00–13:00	0.7	0.8	0.5	0.5	0.1
13:00–14:00	0.8	1.0	0.6	0.5	0.1
14:00–15:00	0.9	1.1	0.7	0.6	0.1
Average	**0.4**	**0.5**	**0.3**	**0.3**	**0.1**

(**c**) Strategy 2B (all rooms 27 °C + high airflow rate + temperature-dependent window opening)

	Room 1	Room 2	Room 3	Room 4	Room 5
8:00–9:00	0	0	0	0	0
9:00–10:00	0	0	0	0	0
10:00–11:00	0	0	0	0	0
11:00–12:00	1.1	1.2	1.3	1.4	1.6
12:00–13:00	1.2	1.2	1.4	1.5	1.7
13:00–14:00	1.2	1.2	1.4	1.5	1.7
14:00–15:00	1.2	1.2	1.4	1.5	1.8
Average	**0.7**	**0.7**	**0.8**	**0.8**	**1.0**

(**h**) Strategy 2B (all rooms 27 °C + high airflow rate + temperature-dependent window opening)

	Room 1	Room 2	Room 3	Room 4	Room 5
8:00–9:00	0.1	0.2	0.2	0.4	0.0
9:00–10:00	0.1	0.2	0.4	0.4	0.1
10:00–11:00	0.2	0.3	0.4	0.5	0.1
11:00–12:00	0.8	1.0	0.7	0.7	0.1
12:00–13:00	0.8	1.0	0.7	0.6	0.1
13:00–14:00	1.0	1.2	0.8	0.7	0.2
14:00–15:00	1.0	1.2	0.9	0.8	0.2
Average	**0.6**	**0.7**	**0.6**	**0.6**	**0.1**

(**d**) Strategy 3B (room five 23 °C and rest 27 °C + low airflow rate + temperature-dependent window opening)

	Room 1	Room 2	Room 3	Room 4	Room 5
8:00–9:00	0	0	0	0	1.3
9:00–10:00	0	0	0	0	1.3
10:00–11:00	0	0	0	0.1	1.5
11:00–12:00	1.4	1.5	1.7	1.8	2.2
12:00–13:00	1.4	1.5	1.7	1.8	2.3
13:00–14:00	1.4	1.5	1.7	1.9	2.4
14:00–15:00	1.4	1.5	1.7	1.9	2.4
Average	**0.8**	**0.9**	**1.0**	**1.1**	**1.9**

(**i**) Strategy 3B (room five 23 °C and rest 27 °C + low airflow rate + temperature-dependent window opening)

	Room 1	Room 2	Room 3	Room 4	Room 5
8:00–9:00	0.4	0	0	0	0
9:00–10:00	0.5	0	0	0	0
10:00–11:00	0.5	0	0	0	0
11:00–12:00	0.9	1.1	0.6	0.5	0.1
12:00–13:00	1.0	1.1	0.6	0.5	0.1
13:00–14:00	1.0	1.2	0.7	0.6	0.1
14:00–15:00	1.0	1.3	0.8	0.7	0.1
Average	**0.8**	**0.7**	**0.4**	**0.3**	**0.1**

(**e**) Strategy 4B (room one 23 °C and rest 27 °C + low airflow rate + temperature-dependent window opening)

	Room 1	Room 2	Room 3	Room 4	Room 5
8:00–9:00	1.1	0.4	0.3	0.3	0.1
9:00–10:00	1.2	0.6	0.5	0.3	0.1
10:00–11:00	1.3	1.4	0.8	0.4	0.2
11:00–12:00	1.6	1.7	1.7	1.9	2.1
12:00–13:00	1.6	1.7	1.8	1.9	2.2
13:00–14:00	1.6	1.7	1.8	2.0	2.3
14:00–15:00	1.6	1.7	1.8	1.9	2.3
Average	**1.4**	**1.3**	**1.2**	**1.2**	**1.3**

(**j**) Strategy 4B (room one 23 °C and rest 27 °C + low airflow rate + temperature-dependent window opening)

Figure 11. Changes in the temperature of the outdoor air around individual rooms caused by heat rejection from outdoor units under different mixed-mode cooling strategies (the grey background indicates that the outdoor unit was on).

3.1.1. Windward Scenario

Strategy 1A reflects those from previous modelling studies [17,18] examining the influence of heat rejection on the energy efficiency of sealed air-conditioned buildings, with heat rejection resulting in higher temperatures for the outdoor air around the building during the entire working period (Figure 11a). The increase in temperature for the outdoor air around individual rooms ranged from 0.1 °C to 1.4 °C. Averaged over the working hours, room 2 saw the greatest temperature increase. This was because whereas the hot air from each outdoor unit moved upwards because of the buoyancy effect, the conflict between the downward wind movement and the upward hot air movement kept the heat accumulated at second-floor height (Figure 12a). Room 5 saw the lowest temperature increase, since the downward wind took away the heat rejected by its outdoor unit (Figure 12a).

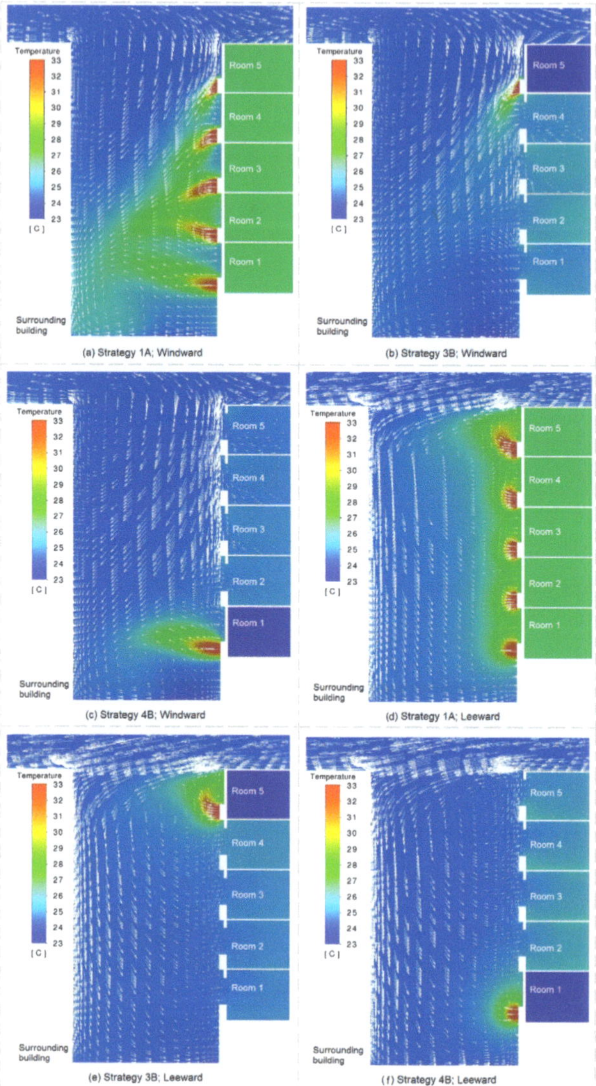

Figure 12. The contours of outdoor temperatures and velocity vectors within the street canyon at 10:00 with different mixed-mode cooling strategies and wind directions.

Compared with Strategy 1A, Strategy 1B shows that the temperature increase was eliminated or reduced at certain times during the working period (Figure 11b). No changes in the temperature of the outdoor air around individual rooms were observed from 08:00 to 11:00, during which all the outdoor units were switched off. During the period from 11:00 to 15:00, the increase in the temperature of the outdoor air around individual rooms was reduced to a range of 0.1 °C to 1.2 °C. This range was brought down further by Strategy 2B (Figure 11c), largely due to the fact that the outlet air temperature of the outdoor unit decreases as the airflow rate of the outdoor unit increases (according to Equation (1)).

When modelled under Strategy 3B, the temperature of the outdoor air around each room increased during the period from 08:00 to 11:00 (Figure 11d). This result is in line with Figure 12b. This is because the hot air from room 5's outdoor unit was dispersed by the downward wind and caused the outdoor air temperature to increase around rooms

1 to 4. The results for Strategy 4B show that the temperature of the outdoor air around room 1 increased during the period from 08:00 to 11:00 (Figure 11e). This was caused by the trapped heat of the heat rejection from the outdoor unit of room 1 due to the downward wind (Figure 12c).

3.1.2. Leeward Scenario

The leeward scenario predicted an increase in the temperature of the outdoor air around individual rooms operating under Strategy 1A (Figure 11f). However, the leeward scenario shows a 1.6 °C greater temperature increase than the windward scenario. This result was because of two reasons: (1) under the windward scenario, wind dispersed part of the discharged heat (Figure 12a), and (2) under the leeward scenario, most of the heat was accumulated around the building (Figure 12d). The higher-floor rooms saw a higher temperature increase because the outdoor units' rejected heat moved upwards.

For rooms 1 to 4, there was no significant difference between Strategy 1B and Strategy 3B on the temperature increase (Figure 11g,i). Room 5, on the other hand, saw a greater temperature increase under Strategy 3B than under Strategy 1B. This result was due to two factors: (1) more heat was rejected for room 5 as the cooling set-point was lower and (2) the rejected heat of room 5's outdoor unit moved upwards (Figure 12e). An increase in the temperature of the outdoor air around individual rooms was predicted by Strategy 4B during the period from 08:00 to 11:00 (Figure 11j) because of the rejected heat of room 1's outdoor unit moved upwards (Figure 12f). It should also be noted that under Strategy 4B, the outdoor unit of room 2 was switched on one hour earlier (10:00–11:00) because the hot air coming from the room 1's outdoor unit caused higher ambient outdoor air temperatures.

3.2. Ambient Outdoor $PM_{2.5}$ Concentrations

The changes in the concentrations of outdoor $PM_{2.5}$ around individual rooms caused by heat rejection are shown in Figure 13, where the positive values represent higher concentrations compared with the reference case, whereas the negative values represent lower concentrations.

Windward Scenario (unit: µg/m³)

(a) Strategy 1A (all rooms 27 °C + low airflow rate + no window opening)

	Room 1	Room 2	Room 3	Room 4	Room 5
8:00–9:00	-9.5	-6.4	-2.7	-0.8	0
9:00–10:00	-10.0	-6.7	-2.9	-0.8	0
10:00–11:00	-10.5	-7.1	-3.0	-0.8	0
11:00–12:00	-10.7	-7.2	-3.1	-0.8	0
12:00–13:00	-10.8	-7.3	-3.1	-0.9	0
13:00–14:00	-11.2	-7.5	-3.2	-0.9	0
14:00–15:00	-11.4	-7.6	-3.2	-0.9	0
Average	-10.6	-7.1	-3.0	-0.8	0.0

(b) Strategy 1B (all rooms 27 °C + low airflow rate + temperature-dependent window opening)

	Room 1	Room 2	Room 3	Room 4	Room 5
8:00–9:00	0	0	0	0	0
9:00–10:00	0	0	0	0	0
10:00–11:00	0	0	0	0	0
11:00–12:00	-10.3	-6.8	-3.0	-0.8	0
12:00–13:00	-10.4	-6.9	-3.0	-0.8	0
13:00–14:00	-10.7	-7.1	-3.1	-0.8	0
14:00–15:00	-10.9	-7.3	-3.1	-0.8	0
Average	-6.0	-4.0	-1.7	-0.5	0.0

(c) Strategy 2B (all rooms 27 °C + high airflow rate + temperature-dependent window opening)

	Room 1	Room 2	Room 3	Room 4	Room 5
8:00–9:00	0	0	0	0	0
9:00–10:00	0	0	0	0	0
10:00–11:00	0	0	0	0	0
11:00–12:00	-12.1	-8.1	-3.5	-0.9	0
12:00–13:00	-12.3	-8.1	-3.5	-1.0	0
13:00–14:00	-12.6	-8.4	-3.6	-1.0	0
14:00–15:00	-12.9	-8.6	-3.7	-1.0	0
Average	-7.1	-4.7	-2.1	-0.6	0.0

(d) Strategy 3B (room five 23 °C and rest 27 °C + low airflow rate + temperature-dependent window opening)

	Room 1	Room 2	Room 3	Room 4	Room 5
8:00–9:00	0	0	0	-0.3	0
9:00–10:00	0	0	0	-0.3	0
10:00–11:00	0	0	0	-0.4	0
11:00–12:00	-10.3	-6.8	-3.0	-0.8	0
12:00–13:00	-10.4	-6.9	-3.0	-0.8	0
13:00–14:00	-10.7	-7.1	-3.1	-0.8	0
14:00–15:00	-10.9	-7.3	-3.1	-0.8	0
Average	-6.0	-4.0	-1.7	-0.6	0.0

(e) Strategy 4B (room 23 °C and rest 27 °C + low airflow rate + temperature-dependent window opening)

	Room 1	Room 2	Room 3	Room 4	Room 5
8:00–9:00	-5.6	-3.1	-2.0	-0.5	0
9:00–10:00	-5.9	-3.3	-2.2	-0.5	0
10:00–11:00	-6.1	-3.4	-2.2	-0.5	0
11:00–12:00	-10.5	-7.0	-3.0	-0.8	0
12:00–13:00	-10.6	-7.1	-3.1	-0.8	0
13:00–14:00	-10.9	-7.3	-3.1	-0.9	0
14:00–15:00	-11.2	-7.4	-3.2	-0.9	0
Average	-8.7	-5.5	-2.7	-0.7	0.0

Leeward Scenario (unit: µg/m³)

(f) Strategy 1A (all rooms 27 °C + low airflow rate + no window opening)

	Room 1	Room 2	Room 3	Room 4	Room 5
8:00–9:00	11.1	-4.7	-3.5	-2.6	-1.3
9:00–10:00	11.7	-5.0	-3.7	-2.7	-1.3
10:00–11:00	12.7	-5.4	-4.0	-2.9	-1.4
11:00–12:00	13.3	-5.7	-4.2	-3.1	-1.5
12:00–13:00	13.8	-5.9	-4.3	-3.2	-1.6
13:00–14:00	14.4	-6.1	-4.5	-3.4	-1.6
14:00–15:00	14.8	-6.3	-4.6	-3.5	-1.7
Average	13.1	-5.6	-4.1	-3.1	-1.5

(g) Strategy 1B (all rooms 27 °C + low airflow rate + temperature-dependent window opening)

	Room 1	Room 2	Room 3	Room 4	Room 5
8:00–9:00	0	0	0	0	0
9:00–10:00	0	0	0	0	0
10:00–11:00	0	0	0	0	0
11:00–12:00	12.8	-5.4	-4.0	-2.9	-1.4
12:00–13:00	13.3	-5.6	-4.2	-3.0	-1.5
13:00–14:00	13.8	-5.8	-4.4	-3.2	-1.5
14:00–15:00	14.2	-6.0	-4.5	-3.3	-1.6
Average	7.7	-3.3	-2.4	-1.8	-0.8

(h) Strategy 2B (all rooms 27 °C + high airflow rate + temperature-dependent window opening)

	Room 1	Room 2	Room 3	Room 4	Room 5
8:00–9:00	0	0	0	0	0
9:00–10:00	0	0	0	0	0
10:00–11:00	0	0	0	0	0
11:00–12:00	12.0	-6.3	-4.8	-3.4	-1.8
12:00–13:00	12.5	-6.6	-5.0	-3.6	-1.9
13:00–14:00	12.9	-6.9	-5.2	-3.7	-1.9
14:00–15:00	13.0	-7.1	-5.3	-3.8	-2.0
Average	7.2	-3.8	-2.9	-2.1	-1.1

(I) Strategy 3B (room five 23 °C and rest 27 °C + low airflow rate + temperature-dependent window opening)

	Room 1	Room 2	Room 3	Room 4	Room 5
8:00–9:00	0	0	0	0	0.4
9:00–10:00	0	0	0	0	0.5
10:00–11:00	0	0	0	0	0.5
11:00–12:00	12.8	-5.4	-4.0	-2.8	-1.4
12:00–13:00	13.3	-5.6	-4.2	-2.9	-1.4
13:00–14:00	13.8	-5.8	-4.4	-3.0	-1.5
14:00–15:00	14.2	-6.0	-4.5	-3.1	-1.5
Average	7.7	-3.3	-2.4	-1.7	-0.6

(J) Strategy 4B (room one 23 °C and rest 27 °C + low airflow rate + temperature-dependent window opening)

	Room 1	Room 2	Room 3	Room 4	Room 5
8:00–9:00	10.9	-4.8	-3.5	-2.6	-1.2
9:00–10:00	11.6	-5.0	-3.7	-2.8	-1.3
10:00–11:00	11.8	-5.3	-4.1	-2.9	-1.5
11:00–12:00	12.8	-5.5	-4.1	-3.0	-1.5
12:00–13:00	13.6	-5.7	-4.3	-3.1	-1.5
13:00–14:00	14.4	-6.1	-4.5	-3.3	-1.6
14:00–15:00	14.8	-6.3	-4.6	-3.4	-1.7
Average	12.8	-5.5	-4.1	-3.0	-1.5

Figure 13. Changes in the concentrations of outdoor $PM_{2.5}$ around individual rooms caused by heat rejection from outdoor units under different mixed-mode cooling strategies (the grey background indicates that the outdoor unit was on).

3.2.1. Windward Scenario

The results of Strategy 1A indicate a reduction in the concentrations of outdoor $PM_{2.5}$ around rooms 1 to 4 during the entire working period, with the average reduction for each room ranging from 0.8 µg/m³ to 10.6 µg/m³ (Figure 13a). This reduction was attributed in large part to the variation in the characteristics of the airflow within the street canyon. The air flow pattern at the bottom of the street canyon is affected when considering the hot air coming from outdoor units in the simulation. The airflow vortex was moved to the leeward side of the surrounding building (Figure 14a,b). This means that the fine particles released from the middle of the street canyon were prone to be blown to the surrounding buildings by the wind. A reduction in the levels of outdoor $PM_{2.5}$ around rooms 1 to 4 was predicted by Strategy 1B during the period from 11:00 to 15:00 (Figure 13b); this was again related to the movement of the airflow vortex at the street canyon's bottom. Strategy 2B led to a greater reduction in the ambient outdoor $PM_{2.5}$ concentrations than Strategy 1B (Figure 13b,c), largely because the position of the airflow vortex was closer to the surrounding buildings, so the airflow rate of the outdoor units increased (Figure 14c).

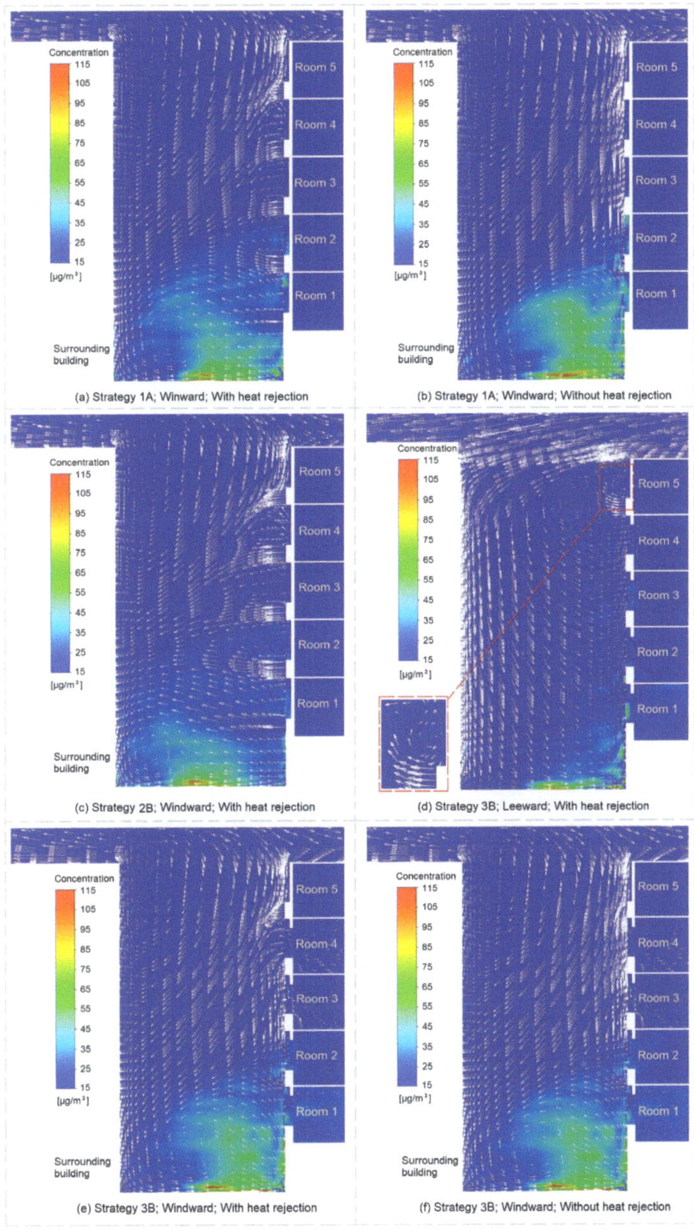

Figure 14. *Cont.*

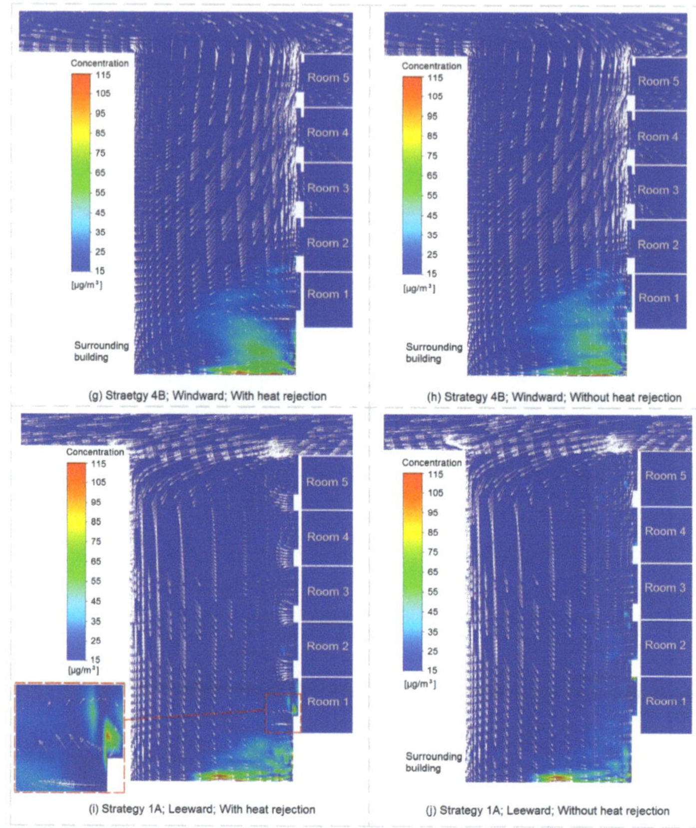

(g) Straetgy 4B; Windward; With heat rejection

(h) Strategy 4B; Windward; Without heat rejection

(i) Strategy 1A; Leeward; With heat rejection

(j) Strategy 1A; Leeward; Without heat rejection

Figure 14. The contours of the concentrations of outdoor $PM_{2.5}$ and velocity vectors within the street canyon at 10:00 with different mixed-mode cooling strategies and wind directions and with and without considering heat rejection from outdoor units.

The results of Strategy 3B show no obvious changes in the levels of outdoor $PM_{2.5}$ around the building (the exception being room 4, which saw a slight concentration decrease) during the period from 08:00 to 11:00 (Figure 13d). This was because of two reasons: (1) the heat rejection from room 5 was found to only change the pattern of the airflow near room 4 and room 5 (Figure 14e,f) and (2) the $PM_{2.5}$ from the middle of the street hardly moved to room 5 when there were no winds blowing them upwards. A reduction in the concentrations of outdoor $PM_{2.5}$ around rooms 1 to 4 could be seen under Strategy 4B during the period from 08:00 to 11:00 (Figure 13e). This was because the heat rejected by room 1 interfered with the upward wind that transported $PM_{2.5}$ from the street to the surroundings of the building (Figure 14g,h).

3.2.2. Leeward Scenario

The results of Strategy 1A show an increase in the concentration of outdoor $PM_{2.5}$ around room 1 and a reduction in the concentrations around rooms 2 to 5 during the entire working period (Figure 13f). When heat rejection was absent, $PM_{2.5}$ from the street was blown to the surroundings of individual rooms by the upward wind (Figure 14j). Above room 1's outdoor unit there was a clockwise airflow vortex if heat rejection was considered in simulation (Figure 14i). This vortex could trap $PM_{2.5}$ that was transported by upward winds, resulting in an increase in the levels of outdoor $PM_{2.5}$ around room 1. Rooms 2 to 5 saw lower levels of ambient outdoor $PM_{2.5}$, likely attributable to more $PM_{2.5}$ from the street being contained near room 1. Similarly, higher levels of outdoor $PM_{2.5}$ around room 1 and lower levels of outdoor $PM_{2.5}$ around rooms 2 to 5 could be seen under Strategies 1B, 2B, and 3B during the period from 11:00 to 15:00 or under Strategy 4B during the entire working period (Figure 13g–j). When modelled under Strategy 3B, no change was seen in the concentrations of outdoor $PM_{2.5}$ around rooms 1 to 4 during the period from 08:00 to 11:00 (Figure 13i). However, during the same period, room 5 experienced higher ambient outdoor $PM_{2.5}$ concentrations due to the clockwise vortex above its outdoor unit (Figure 14d).

3.3. Cooling Loads

The building cooling loads were determined by summing the cooling loads of rooms 1 to 5. The changes in the building cooling loads due to heat rejection are shown in Figure 15, where the positive values represent higher cooling loads compared with the reference case.

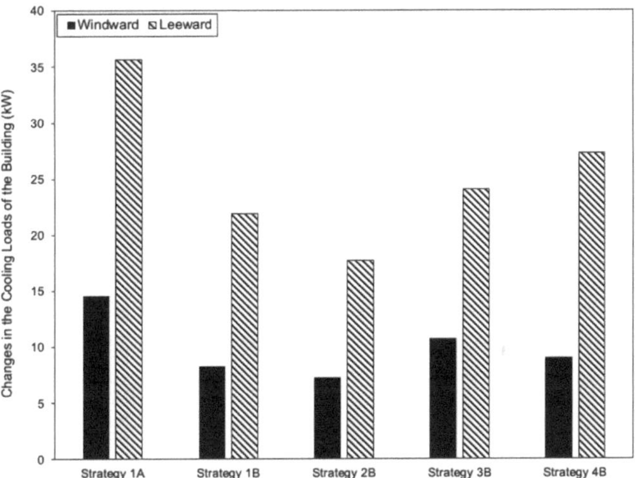

Figure 15. Changes in building cooling loads because of heat rejection under different mixed-mode cooling strategies.

3.3.1. Windward Scenario

As expected, the increased ambient outdoor temperatures described in Section 3.1.1 resulted in higher building cooling loads. Compared with Strategy 1A, Strategy 1B led to a 43.2% reduction in the increase in cooling loads (Figure 15). This result is in line with Figure 11a,b, showing that each room experienced lower ambient outdoor temperatures under Strategy 1B than under Strategy 1A. Similarly, the building cooling load predicted by Strategy 4B was 15.9% lower than that predicted by Strategy 3B (Figure 15) because of the increase in the temperature of outdoor air around rooms 2 to 5 being lower under Strategy 4B than under Strategy 3B (Figure 11d,e). Strategy 2B showed energy saving benefits, as the predicted building cooling load increase was 13.3% lower than Strategy 1B (Figure 15). This means increasing the airflow rate of outdoor units can improve energy efficiency.

3.3.2. Leeward Scenario

Higher outdoor temperatures meant greater building cooling loads, which is similar to the windward scenario. The trend between different strategies was: Strategy 1A resulted in the greatest cooling-load increase, followed Strategy 4B, then Strategy 3B, then Strategy 1B, and finally Strategy 2B (Figure 15). The cooling-load increase predicted by the windward scenario was on average 60.6% smaller than that predicted by the leeward scenario. The difference between Strategy 3B and Strategy 4B was the opposite of that observed under the windward scenario, with Strategy 3B leading to a 11.7% lower increase in cooling loads than Strategy 4B.

3.4. Indoor Levels of $PM_{2.5}$ Exposure

The results of exposure to indoor $PM_{2.5}$ were averaged over the working hours. Changes in the exposure of each room caused by heat rejection are shown in Figure 16, where the positive values represent a greater exposure in comparison with the reference case and the negative values represent lower exposures.

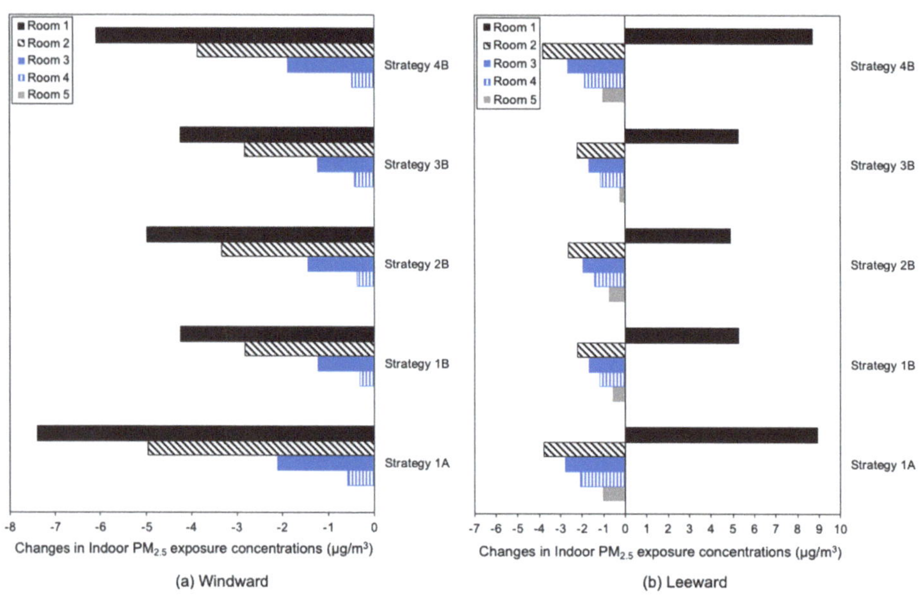

Figure 16. Changes in the exposure concentrations of individual rooms caused by heat rejection under different mixed-mode cooling strategies.

3.4.1. Windward Scenario

The results of Strategy 1A indicate that people in rooms 1 to 4 had reduced exposure to indoor $PM_{2.5}$ (Figure 16a), which was attributable to the reduced levels of ambient outdoor $PM_{2.5}$ (Figure 13a). No change in exposure was seen for room 5 (Figure 16a) as the concentrations of outdoor $PM_{2.5}$ around room 5 remained the same (Figure 13a). Compared with Strategy 1B, Strategy 2B led to lower indoor $PM_{2.5}$ exposures for rooms 1 to 4 (Figure 16a), showing the potential health benefit of increasing the airflow rates of outdoor units. Strategy 3B led to higher indoor $PM_{2.5}$ exposure than Strategy 4B (Figure 16a) mainly because the heat rejection from room 1 had a much stronger modifying effect on the concentrations of outdoor $PM_{2.5}$ around the building than the heat rejection room 5 (Figure 14e–h).

3.4.2. Leeward Scenario

Higher levels of indoor $PM_{2.5}$ exposure for room 1 were seen under each strategy (Figure 16b). This is because, as shown in (Figure 14i), there were greater concentrations of ambient outdoor $PM_{2.5}$ due to the formation of a vortex above the outdoor unit of room 1. People in rooms 2 to 5, on the other hand, experienced lower levels of indoor $PM_{2.5}$ exposure because of the reduction in the concentrations of outdoor $PM_{2.5}$ around rooms 2 to 5 (Figure 13f–j). Strategy 2B led to lower indoor $PM_{2.5}$ exposures for each room than Strategy 1B (Figure 16b), showing the potential health benefit of increasing the airflow rates of outdoor units.

4. Discussion

4.1. Main Findings

The simulation results indicate that the building had higher cooling loads because of the higher temperatures of ambient outdoor air due to heat rejection. This adverse energy effect varied in response to the wind direction, with simulation results showing that the cooling-load increase predicted by the windward scenario was 60.6% lower than that predicted by the leeward scenario. Under the leeward scenario, the heat rejected by the lower-level outdoor units could even result in the increased use of air-conditioning for rooms on higher levels. Averaged over the windward and leeward scenarios, the cooling-load increase due to heat rejection was 40.9% lower when windows were open based on temperature difference than when windows remained closed. This finding means that the influence of heat rejection on space-cooling energy consumption can vary significantly between sealed air-conditioned and mixed-mode buildings. Although an increase in the airflow rate of outdoor units could reduce the rate of ambient outdoor temperature increase due to heat rejection and thus lead to a smaller increase in cooling loads, this may not be possible in all buildings, for example those with concerns about noise. Additionally, the increased costs of running fans may reduce or even offset this energy benefit.

The position of the room that was cooled to a lower temperature than the others had a major influence on the cooling-load increase due to heat rejection, with the results of the windward scenario showing that the cooling-load increase predicted by Strategy 4B (room 1 being cooled to 23 °C and rooms 2 to 5 being cooled to 27 °C) was 15.9% lower than that predicted by Strategy 3B (room 5 being cooled to 23 °C and rooms 1 to 4 being cooled to 27 °C). The results for the windward scenario also indicate that Strategy 4B led to a lower indoor $PM_{2.5}$ exposure than Strategy 3B. An implication of these two findings is that under the windward scenario, an activity (e.g., meeting) that requires the indoor space to be cooled to a relatively low set-point temperature should take place in room 1 (i.e., the room on the bottom floor) rather than room 5 (i.e., the room on the top floor) in order to reduce space-conditioning energy consumption while improving indoor air quality.

Under the windward scenario, occupants experienced lower levels of exposure to indoor $PM_{2.5}$ since the heat rejection led to lower levels of ambient outdoor $PM_{2.5}$. The results for the leeward scenario show that occupants in room 1 faced higher levels of exposure to indoor $PM_{2.5}$ when heat rejection was present. In addition, it was found that the windward scenario led to lower cooling-load increases than the leeward scenario. Therefore, the location of outdoor units is a critical design decision. Outdoor units should be placed on the windward side of a building. Under both wind direction scenarios, higher airflow rates for outdoor units led to lower ambient outdoor $PM_{2.5}$ concentrations and thus contributed to better indoor air quality. Therefore, in addition to user preference, ease of installation, and cost, the decision to choose an outdoor unit may also come down to the rated airflow rate.

4.2. Limitations and Future Research

This study carries with it several limitations. Whereas the typical arrangement of outdoor units in Hong Kong was applied to the modelled building, future research on buildings that have different arrangements of outdoor units (e.g., outdoor units installed in re-entrants) is worth investigating. In addition, placing outdoor units on a single side is an important modelling simplification and must be acknowledged. Further work is ongoing to simulate the scenario that outdoor units are installed on both the windward and leeward sides. Previous studies [17,19] have suggested that the airflow pattern in a street canyon can change significantly between different aspect ratios. Further work is required to test more different aspect ratios to validate the sensitivity of the model. Indoor-sourced fine particles play an important role in indoor $PM_{2.5}$ exposure but are not considered herein. Heat rejection can influence the air exchange between the indoors and outdoors, thereby affecting the ability of indoor-sourced fine particles to exfiltrate. Fine particles from both indoor and outdoor sources will be modelled in future research. Outdoor units and envelope features such as external shading devices and balconies could be closely connected because they are usually positioned next to each other. The presence of external shading devices or balconies is likely to affect the flow pattern of the air discharged from outdoor units. Therefore, the combined effects of outdoor units and envelope features on the indoor environmental conditions should be investigated in future research.

Caution is also needed when generalising simulation results to the buildings that have the same geometrics as the modelled building. This study assigned a single orientation, a single occupancy pattern, and a single vertical profile of wind speeds to the model. However, it is acknowledged that variations in these model inputs may significantly influence the model outcomes. In terms of orientation, due to a greater exposure to the sun, a room with south-facing windows generally has higher indoor temperatures compared with its counterparts with non-south-facing windows [51]. This means that people in rooms with non-south-facing windows may air condition their rooms less frequently to maintain indoor thermal comfort. A decreased use of A/C can reduce the effects of heat rejection on space-cooling demands and indoor $PM_{2.5}$ exposure and thus influences the study results. The occupancy pattern determines both the internal heat gains and the periods when rooms are occupied and thus could have great impacts on space-cooling demands and indoor $PM_{2.5}$ exposure. The simulations were run based on an occupancy schedule representative of offices, and further research on whether similar results can be obtained from different occupancy schedules (e.g., household occupancy) is required. All simulations were run with a vertical profile of wind speeds that reflect a dense urban environment. The hot air discharged by outdoor units can influence the airflow pattern within the street canyon and therefore changes the conditions of the outdoor air around the building. The impact of the hot air on the airflow pattern within the street canyon, however, is likely to vary when wind speeds are modified to reflect a less dense urban environment.

5. Conclusions

By using a coupled EnergyPlus–Fluent modelling approach, this work has been able to investigate how heat rejection from outdoor units influenced the ambient outdoor environment of a mixed-mode building operating under different cooling strategies and how these influences affected space-cooling demands and indoor $PM_{2.5}$ exposure. The main outcomes of this study are:

1. The mixed-mode building had higher cooling loads because of the increase in ambient outdoor temperatures due to heat rejection. This adverse energy effect was more significant when windows remained closed than when windows were open based on temperature difference;
2. Placing outdoor units on the windward side is beneficial to disperse the rejected heat from outdoor units, whereas the leeward scenario may "trap" the heat. Therefore, the windward scenario had 60.6% lower cooling load increase than the leeward scenario;
3. In the windward scenario, $PM_{2.5}$ from the street was kept away from the buildings due to the airflow vortex generated by the heat rejection of the outdoor units, so the indoor $PM_{2.5}$ was lower. Under the leeward scenario, the bottom-floor room saw higher ambient outdoor $PM_{2.5}$ concentrations due to heat rejection; occupants in the bottom-floor room thus experienced greater exposure to indoor $PM_{2.5}$;
4. The combination of outcomes (2) and (3) indicates that outdoor units should be placed on the windward side of a building in order to reduce both the space-cooling demands and exposure to indoor $PM_{2.5}$;
5. An increase in the airflow rate of outdoor units offers the co-benefits of energy savings and occupant health under both the windward and leeward scenarios;
6. Under the windward scenario, if one room needs to be cooled to a lower temperature than the others, the bottom-floor room is a better choice than the top-floor room for energy savings and occupant health.

Author Contributions: Conceptualization, X.Z., Z.Z. and R.Z.; methodology, X.Z. and Z.Z.; software, Z.Z.; validation, X.Z., Z.Z. and R.Z.; formal analysis, X.Z. and Z.Z.; investigation, X.Z., Z.Z. and M.C.; resources, Z.Z. and M.C.; data curation, Z.Z. and Z.W.; writing—original draft preparation, X.Z. and Z.Z.; writing—review and editing, X.Z., Z.Z., R.Z. and Z.W.; visualization, X.Z. and Z.Z.; supervision, Z.Z. and Z.W.; project administration, Z.Z. and Z.W.; funding acquisition, X.Z. and Z.Z. All authors have read and agreed to the published version of the manuscript.

Funding: This research was supported by Zhejiang Provincial Natural Science Foundation of China under Grant No. LQ22E080026 and Major Science and Technology Programme of Ningbo Science and Technology Bureau under Project Code 2022Z161.

Data Availability Statement: The raw data supporting the conclusions of this article will be made available by the authors on request.

Acknowledgments: The authors would like to thank the Department of Architecture and Built Environment, University of Nottingham Ningbo China, and the School of Energy and Environment, City University of Hong Kong, for providing materials used for experiments and simulations.

Conflicts of Interest: Author Ming Cai was employed by the company State Grid Lishui Power Supply Company. The remaining authors declare that the research was conducted in the absence of any commercial or financial relationships that could be construed as a potential conflict of interest.

References

1. Hong, T.; Xu, Y.; Sun, K.; Zhang, W.; Luo, X.; Hooper, B. Urban microclimate and its impact on building performance: A case study of San Francisco. *Urban Clim.* **2021**, *38*, 100871. [CrossRef]
2. Chow, T.T.; Lin, Z. Prediction of on-coil temperature of condensers installed at tall building re-entrant. *Appl. Therm. Eng.* **1999**, *19*, 117–132. [CrossRef]
3. Chow, T.T.; Lin, Z.; Wang, Q.W. Effect of building re-entrant shape on performance of air-cooled condensing units. *Energy Build.* **2000**, *32*, 143–152. [CrossRef]
4. Nada, S.A.; Said, M.A. Performance and energy consumptions of split type air conditioning units for different arrangements of outdoor units in confined building shafts. *Appl. Therm. Eng.* **2017**, *123*, 874–890. [CrossRef]
5. Engelmann, P.; Kalz, D.; Salvalai, G. Cooling concepts for non-residential buildings: A comparison of cooling concepts in different climate zones. *Energy Build.* **2014**, *82*, 447–456. [CrossRef]
6. Liu, S.; Kwok, Y.T.; Lau, K.K.-L.; Ouyang, W.; Ng, E. Effectiveness of passive design strategies in responding to future climate change for residential buildings in hot and humid Hong Kong. *Energy Build.* **2020**, *228*, 110469. [CrossRef]
7. Bamdad, K.; Matour, S.; Izadyar, N.; Omrani, S. Impact of climate change on energy saving potentials of natural ventilation and ceiling fans in mixed-mode buildings. *Build. Environ.* **2022**, *209*, 108662. [CrossRef]
8. Gens, A.; Hurley, J.F.; Tuomisto, J.T.; Friedrich, R. Health impacts due to personal exposure to fine particles caused by insulation of residential buildings in Europe. *Atmos. Environ.* **2014**, *84*, 213–221. [CrossRef]
9. Jones, N.C.; Thornton, C.A.; Mark, D.; Harrison, R.M. Indoor/outdoor relationships of particulate matter in domestic homes with roadside, urban and rural locations. *Atmos. Environ.* **2000**, *34*, 2603–2612. [CrossRef]
10. Canha, N.; Almeida, S.M.; Freitas, M.d.C.; Trancoso, M.; Sousa, A.; Mouro, F.; Wolterbeek, H.T. Particulate matter analysis in indoor environments of urban and rural primary schools using passive sampling methodology. *Atmos. Environ.* **2014**, *83*, 21–34. [CrossRef]
11. Semmens, E.O.; Noonan, C.W.; Allen, R.W.; Weiler, E.C.; Ward, T.J. Indoor particulate matter in rural, wood stove heated homes. *Environ. Res.* **2015**, *138*, 93–100. [CrossRef] [PubMed]
12. Taylor, J.; Davies, M.; Mavrogianni, A.; Shrubsole, C.; Hamilton, I.; Das, P.; Jones, B.; Oikonomou, E.; Biddulph, P. Mapping indoor overheating and air pollution risk modification across Great Britain: A modelling study. *Build. Environ.* **2016**, *99*, 1–12. [CrossRef]
13. Taylor, J.; Shrubsole, C.; Symonds, P.; Mackenzie, I.; Davies, M. Application of an indoor air pollution metamodel to a spatially-distributed housing stock. *Sci. Total Environ.* **2019**, *667*, 390–399. [CrossRef] [PubMed]
14. Taylor, J.; Shrubsole, C.; Davies, M.; Biddulph, P.; Das, P.; Hamilton, I.; Vardoulakis, S.; Mavrogianni, A.; Jones, B.; Oikonomou, E. The modifying effect of the building envelope on population exposure to PM2.5 from outdoor sources. *Indoor Air* **2014**, *24*, 639–651. [CrossRef] [PubMed]
15. Keshavarzian, E.; Jin, R.; Dong, K.; Kwok, K.C.S. Effect of building cross-section shape on air pollutant dispersion around buildings. *Build. Environ.* **2021**, *197*, 107861. [CrossRef]
16. Cui, D.; Li, X.; Liu, J.; Yuan, L.; Mak, C.M.; Fan, Y.; Kwok, K. Effects of building layouts and envelope features on wind flow and pollutant exposure in height-asymmetric street canyons. *Build. Environ.* **2021**, *205*, 108177. [CrossRef]
17. Xiong, Y.; Chen, H. Effects of sunshields on vehicular pollutant dispersion and indoor air quality: Comparison between isothermal and nonisothermal conditions. *Build. Environ.* **2021**, *197*, 107854. [CrossRef]
18. Hsieh, C.-M.; Aramaki, T.; Hanaki, K. Managing heat rejected from air conditioning systems to save energy and improve the microclimates of residential buildings. *Comput. Environ. Urban Syst.* **2011**, *35*, 358–367. [CrossRef]

19. Adelia, A.S.; Yuan, C.; Liu, L.; Shan, R.Q. Effects of urban morphology on anthropogenic heat dispersion in tropical high-density residential areas. *Energy Build.* **2019**, *186*, 368–383. [CrossRef]
20. Yuan, C.; Zhu, R.; Tong, S.; Mei, S.; Zhu, W. Impact of anthropogenic heat from air-conditioning on air temperature of naturally ventilated apartments at high-density tropical cities. *Energy Build.* **2022**, *268*, 112171. [CrossRef]
21. Crawley, D.B.; Lawrie, L.K.; Winkelmann, F.C.; Buhl, W.F.; Huang, Y.J.; Pedersen, C.O.; Strand, R.K.; Liesen, R.J.; Fisher, D.E.; Witte, M.J.; et al. EnergyPlus: Creating a new-generation building energy simulation program. *Energy Build.* **2001**, *33*, 319–331. [CrossRef]
22. Ansys, Inc. *Ansys Fluids: Computational Fluid Dynamics (CFD) Simulation Software*, version 2021; Ansys, Inc.: Canonsburg, PA, USA, 2021.
23. EMPORIS. Building Types Hong Kong. 2017. Available online: https://www.emporis.com/city/101300/hong-kong-china (accessed on 16 September 2023).
24. HKO. Climatological Information Services. 2021. Available online: https://www.hko.gov.hk/en/cis/climat.htm (accessed on 16 September 2023).
25. EPD. Environmental Protection Interactive Centre: Air Quality Data—Download/Display. 2021. Available online: https://cd.epic.epd.gov.hk/EPICDI/air/station/?lang=en (accessed on 16 September 2023).
26. Wan, K.S.Y.; Yik, F.H.W. Representative building design and internal load patterns for modelling energy use in residential buildings in Hong Kong. *Appl. Energy* **2004**, *77*, 69–85. [CrossRef]
27. BD. *Guidelines on Design and Construction Requirements for Energy Efficiency of Residential Buildings*; Buildings Department Hong Kong: Hong Kong, China, 2014.
28. *ISO 13790*; Energy Performance of Buildings, Calculation of Energy Use for Space Heating and Cooling. International Organization for Standardization (ISO): Geneva, Switzerland, 2008.
29. CIBSE. *CIBSE Guide A: Environmental Design*; The Chartered Institution of Building Services Engineers: London, UK, 2006.
30. Daikin. *Engineering Data—Daikin AC*; Daikin: Osaka, Japan, 2022.
31. *Standard 55-2017*; Thermal environmental conditions for human occupancy. American Society of Heating, Refrigerating and Air Conditioning Engineers (ASHRAE): Peachtree Corners, GA, USA, 2017.
32. Kikegawa, Y.; Genchi, Y.; Yoshikado, H.; Kondo, H. Development of a numerical simulation system toward comprehensive assessments of urban warming countermeasures including their impacts upon the urban buildings' energy-demands. *Appl. Energy* **2003**, *76*, 449–466. [CrossRef]
33. Taylor, J.; Shrubsole, C.; Biddulph, P.; Jones, B.; Das, P.; Davies, M. Simulation of pollution transport in buildings: The importance of taking into account dynamic thermal effects. *Build. Serv. Eng. Res. Technol.* **2014**, *35*, 682–690. [CrossRef]
34. Meng, M.-R.; Cao, S.-J.; Kumar, P.; Tang, X.; Feng, Z. Spatial distribution characteristics of PM2.5 concentration around residential buildings in urban traffic-intensive areas: From the perspectives of health and safety. *Saf. Sci.* **2021**, *141*, 105318. [CrossRef]
35. Taylor, J.; Mavrogianni, A.; Davies, M.; Das, P.; Shrubsole, C.; Biddulph, P.; Oikonomou, E. Understanding and mitigating overheating and indoor PM2.5 risks using coupled temperature and indoor air quality models. *Build. Serv. Eng. Res. Technol.* **2015**, *36*, 275–289. [CrossRef]
36. Jones, B.; Das, P.; Chalabi, Z.; Davies, M.; Hamilton, I.; Lowe, R.; Milner, J.; Ridley, I.; Shrubsole, C.; Wilkinson, P. The Effect of Party Wall Permeability on Estimations of Infiltration from Air Leakage. *Int. J. Vent.* **2013**, *12*, 17–30. [CrossRef]
37. Long, C.M.; Suh, H.H.; Catalano, P.J.; Koutrakis, P. Using Time- and Size-Resolved Particulate Data to Quantify Indoor Penetration and Deposition Behavior. *Environ. Sci. Technol.* **2001**, *35*, 2089–2099. [CrossRef]
38. Liu, J.; Heidarinejad, M.; Pitchurov, G.; Zhang, L.; Srebric, J. An extensive comparison of modified zero-equation, standard k-ε, and LES models in predicting urban airflow. *Sustain. Cities Soc.* **2018**, *40*, 28–43. [CrossRef]
39. Zhang, R.; Mirzaei, P.A.; Jones, B. Development of a dynamic external CFD and BES coupling framework for application of urban neighbourhoods energy modelling. *Build. Environ.* **2018**, *146*, 37–49. [CrossRef]
40. Zheng, J.; Tao, Q.; Chen, Y. Airborne infection risk of inter-unit dispersion through semi-shaded openings: A case study of a multi-storey building with external louvers. *Build. Environ.* **2022**, *225*, 109586. [CrossRef] [PubMed]
41. Franke, J.; Baklanov, A. *Best Practice Guideline for the CFD Simulation of Flows in the Urban Environment: COST Action 732 Quality Assurance and Improvement of Microscale Meteorological Models*; COST European Cooperation in Science and Technology: Hamburg, Germany, 2007.
42. Tominaga, Y.; Mochida, A.; Yoshie, R.; Kataoka, H.; Nozu, T.; Yoshikawa, M.; Shirasawa, T. AIJ guidelines for practical applications of CFD to pedestrian wind environment around buildings. *J. Wind Eng. Ind. Aerodyn.* **2008**, *96*, 1749–1761. [CrossRef]
43. Snyder, W. *Guideline for Fluid Modeling of Atmospheric Diffusion*; U.S. Environmental Protection Agency: Washington, DC, USA, 1981.
44. Shirzadi, M.; Tominaga, Y. CFD evaluation of mean and turbulent wind characteristics around a high-rise building affected by its surroundings. *Build. Environ.* **2022**, *225*, 109637. [CrossRef]
45. Gdeisat, M.; Lilley, F. 1—MATLAB®Integrated Development Environment. In *Matlab by Example*; Gdeisat, M., Lilley, F., Eds.; Elsevier: Amsterdam, The Netherlands, 2013; pp. 1–20. [CrossRef]
46. Roache, P.J. Perspective: A Method for Uniform Reporting of Grid Refinement Studies. *J. Fluids Eng.* **1994**, *116*, 405–413. [CrossRef]
47. Tominaga, Y.; Stathopoulos, T. CFD simulations of near-field pollutant dispersion with different plume buoyancies. *Build. Environ.* **2018**, *131*, 128–139. [CrossRef]

48. Hanna, S.; Chang, J. Acceptance criteria for urban dispersion model evaluation. *Meteorol. Atmos. Phys.* **2012**, *116*, 133–146. [CrossRef]

49. Leitl, B.; Schatzmann, M.C. *Compilation of Experimental Data for Validation of Microscale Dispersion Models*; Meteorological Institute of the University of Hamburg: Hamburg, Germany, 2010.

50. *Guideline 14–2014*; Measurement of Energy, Demand, and Water Savings. American Society of Heating, Refrigerating, and Air Conditioning Engineers (ASHRAE): Peachtree Corners, GA, USA, 2014.

51. Zhong, X.; Zhang, Z.; Wu, W.; Ridley, I. Comprehensive evaluation of energy and indoor-PM2.5-exposure performance of residential window and roller blind control strategies. *Energy Build.* **2020**, *223*, 110206. [CrossRef]

 *buildings*

Article

A Hybrid Residential Short-Term Load Forecasting Method Using Attention Mechanism and Deep Learning

Xinhui Ji [1,2,3], Huijie Huang [1,2], Dongsheng Chen [3], Kangning Yin [3], Yi Zuo [1,2], Zhenping Chen [1,2,*] and Rui Bai [4]

[1] School of Electronic and Information Engineering, Suzhou University of Science and Technology, Suzhou 215009, China
[2] Suzhou Smart City Research Institute, Suzhou University of Science and Technology, Suzhou 215009, China
[3] Kashi Institute of Electronics and Information Industry, Kashi 844199, China
[4] State Grid Suzhou Power Supply Company, Suzhou 215004, China
* Correspondence: zhpchen@usts.edu.cn

Abstract: Development in economics and social society has led to rapid growth in electricity demand. Accurate residential electricity load forecasting is helpful for the transformation of residential energy consumption structure and can also curb global climate warming. This paper proposes a hybrid residential short-term load forecasting framework (DCNN-LSTM-AE-AM) based on deep learning, which combines dilated convolutional neural network (DCNN), long short-term memory network (LSTM), autoencoder (AE), and attention mechanism (AM) to improve the prediction results. First, we design a T-nearest neighbors (TNN) algorithm to preprocess the original data. Further, a DCNN is introduced to extract the long-term feature. Secondly, we combine the LSTM with the AE (LSTM-AE) to learn the sequence features hidden in the extracted features and decode them into output features. Finally, the AM is further introduced to extract and fuse the high-level stage features to achieve the prediction results. Experiments on two real-world datasets show that the proposed method is good at capturing the oscillation characteristics of low-load data and outperforms other methods.

Keywords: residential short-term load forecasting; deep learning; dilated convolutional neural network; long and short-term memory network; attention mechanism

Citation: Ji, X.; Huang, H.; Chen, D.; Yin, K.; Zuo, Y.; Chen, Z.; Bai, R. A Hybrid Residential Short-Term Load Forecasting Method Using Attention Mechanism and Deep Learning. *Buildings* **2023**, *13*, 72. https://doi.org/10.3390/buildings13010072

Academic Editor: Jan Kośn

Received: 10 November 2022
Revised: 14 December 2022
Accepted: 22 December 2022
Published: 28 December 2022

1. Introduction

With the development of smart cities and smart homes, the daily electricity units of the residents are becoming increasingly numerous, resulting in a more complicated power system. As shown in Figure 1, a typical residential power supply includes biomass power generation, photovoltaic power generation, hydropower generation, and wind power generation. The residential electricity demand poses a huge potential threat to maintaining the stability of the power system. However, the existing energy supply structure is still dominated by thermal power generation, and the carbon emissions of thermal power generation will lead to climate warming, which, on the contrary, leads to an increase in energy demand. Hence, accurate residential power load forecasting is important. Generally, power load forecasting is divided into three categories: short-term load forecasting, medium-term load forecasting, and long-term load forecasting [1]. Short-term load forecasting can predict the sum of energy consumption within a few minutes or hours, which can optimize the energy dispatch and reduce the micro-grid primary energy loss [2]. Further, residents can reduce electricity costs by formulating electricity utilization strategies with the current pricing scheme. In addition, short-term load forecasting provides a judgment basis for some abnormal power information and guarantees the safety of people's lives and property. Therefore, in this paper, we mainly consider how to design a suitable short-term load forecasting method to improve forecasting accuracy.

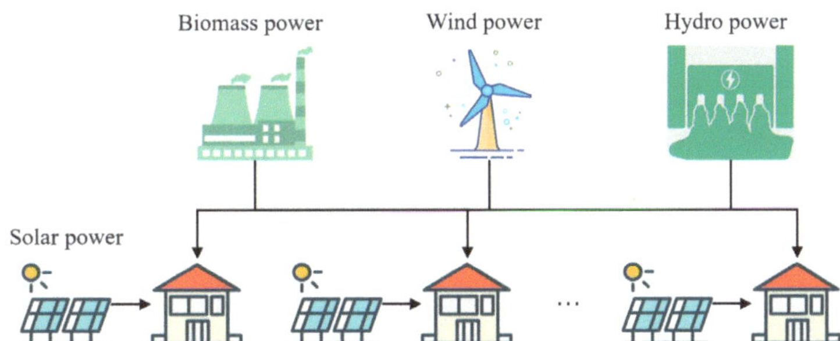

Figure 1. Structure of residential energy supply.

Short-term load forecasting is a time series forecasting problem. The factors that influence residential short-term load forecasting are complex and significant, including social events, electricity price adjustments, human behavior, and other uncertainties. Initially, researchers relied on artificial feature analysis to obtain empirical models, which is time-consuming and inaccurate [3]. Due to the widely different distribution of residential electricity consumption, load forecasting tends to utilize machine learning and deep learning to optimize the results. Support vector machine (SVM) [4], autoregressive integrated moving average (ARIMA) [5], extreme gradient boosting regressor (XGBoost) [6], artificial neural network (ANN) [7], and long short-term memory network (LSTM) [8] are the most commonly used methods [9]. Machine learning methods exhibit some benefits: (1) they have high computational efficiency; (2) they are highly interpretable in their basic form [10]. Although machine learning methods perform well in load forecasting, deep learning for load forecasting can obtain better results. First, deep learning does not require the creation of feature engineering in machine learning. Second, deep learning has a good generalization [11].

Currently, many studies have focused on the accuracy of short-term load forecasting. However, they have ignored trend tracking in oscillating data, especially in valley data. The oscillating data are likely to correspond to the operation of some power consumption units. If this situation cannot be predicted accurately, then for future fault monitoring and other studies, the system's stability will be challenging to achieve because of the increased likelihood of misjudgment. Therefore, this paper proposes a hybrid model, i.e., DCNN-LSTM-AE-AM, for residential short-term load forecasting. Figure 2 shows the architecture of the proposed DCNN-LSTM-AE-AM. By broadening the temporal horizon, we utilize the dilated convolutional neural network to extract temporal features from the time series. Then, the LSTM-AE is applied to mine the electricity consumption characteristics thoroughly. Finally, an attention mechanism (AM) is used to reflect the importance of behaviors in load prediction. The main contributions of this paper are listed as follows:

- Considering that individual data loss may still occur due to various conditions, we propose the T-nearest neighbors (TNN) algorithm to solve the problem of missing values, which can estimate the missing load according to the load data of adjacent similar days.
- We propose a hybrid short-term residential electricity load forecasting model (DCNN-LSTM-AE-AM). This proposed model focuses on the trend tracking of oscillating data, which can be captured with almost no delay, and provides a technical basis for predicting power failures in advance. Compared with other methods, DCNN-LSTM-AE-AM can capture the valley load data, which improves the prediction accuracy.
- The proposed DCNN-LSTM-AE-AM model is validated on two real-world datasets and compared with the existing methods. Experimental results show that this model improves the prediction results and has a good generalization.

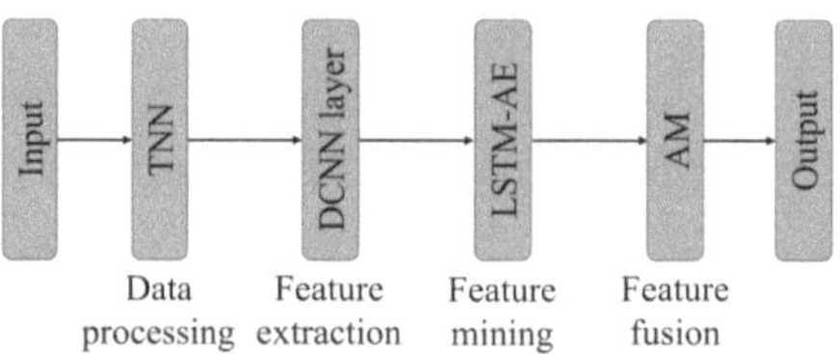

Figure 2. Architecture of the proposed DCNN-LSTM-AE-AM.

The rest of this paper is organized as follows. Section 2 reviews the related work. Section 3 proposes the main results on how to design a hybrid short-term residential electricity load forecasting model. In Section 4, the experimental results are demonstrated, and some comparisons with the other three models are made. Finally, Section 5 concludes the paper and future work.

2. Related Work

In this section, we discuss the previous research on short-term load forecasting. The existing approaches for feature extraction can be divided into three: manual feature screening, traditional machine learning, and deep earning.

In the early days, load data were less affected by uncertain factors such as human behavior and climate change. Researchers first processed the data by relying on expert experience. Then, with the application of some statistical methods, they tried to build a prediction model. For example, Taylor analyzed the seasonal cycle within the day and week as the key features and processed them with five statistical methods to obtain an optimal forecasting result [12]. Statistical methods are rich in theory and highly interpretable for some specific characteristics. However, due to the growth in the living standard, the approaches are difficult for experts to interpret the complex data.

With the advance of big data and artificial intelligence, data-driven load forecasting technologies have received attention extensively [13–17]. The computing capacity of hardware devices has greatly influenced short-term load forecasting. He et al. [18] utilized ARIMA to design a high-frequency short-term load forecasting model. The model divided the inputs by season and then used hourly load data to predict energy consumption for the next month. Cao et al. [19] divided the dataset by the characteristics of similar daily meteorological conditions and applied ARIMA for prediction. Although the above models predicted well in the same season, such models based on seasons are extremely dependent on manual screening and do not have the ability to generalize. Cai et al. [20] proposed an energy prediction model that combined the K-means and data mining methods to analyze the energy consumption of 16,000 residential buildings. Mohammadi et al. [21] proposed a hybrid model based on sliding window empirical mode decomposition (SWEMD) to predict the power consumption of small buildings. They also proposed an algorithm to optimize the model parameters. Chauhan et al. [22] designed a hybrid model based on SVM and ensemble learning for prediction, where load data were processed separately through hourly and daily resolutions. Experiments in Aguilar Madrid and Antonio [6] showed that the XGBoost could obtain the best performance from the machine learning algorithms set. Massaoudi et al. [23] proposed an ensemble-based LGBM-XGB-MLP hybrid model to improve prediction performance. Although traditional machine learning models can achieve good results in load forecasting, they all require more effort in the feature selection and parameter optimization process [11].

Deep learning has received explosive growth in various fields. Some typical networks for deep learning, such as ANN [24–26], convolutional neural network (CNN) [27], recurrent neural network (RNN) [28], and LSTM [29], have been widely used in load forecasting Chen et al. [30] added periodic features and utilized a deep residual network (ResNet) to predict the hourly residential load, whose evaluation performance had been improved From the aspect of demand-side management (DSM), Kong et al. [31] proposed an improved deep belief network (DBN) method to forecast 1-h-level loads. Compared with others, this method could significantly improve both the day-ahead and week-ahead load forecasting results. Dong et al. [32] proposed a distributed deep belief network (DDBN) with Markov switching topology, which improved distributed communication stability and prediction accuracy.

It should be noted that energy consumption is generated by a variety of complex factors, so how to utilize the nonlinear models to improve prediction accuracy is the main consideration. Recently, CNN has injected vitality into short-term load forecasting [33–35]. Amarasinghe et al. [36] tried to apply CNN for prediction on the residential load dataset [37]. Experimental results show that this method is effective, but the prediction accuracy needs to be improved. Sadaei et al. [38] proposed a hybrid algorithm combining CNN and fuzzy time series (FTS) for forecasting. This method converted multivariate time series into multi-channel images. The proposed model overcame some advanced time series models for short-term load forecasting with better results and solved the problem of over-fitting.

RNN, LSTM, and gated recurrent neural network (GRU) are designed for time-series data, which is sensitive to temporal features. They can fully mine time-related features between adjacent data. Rahman et al. [39] utilized RNN to forecast the 1-h commercial and residential load data in the medium and long-term forecasting and thus realized the load trend tracking. However, it is challenging for RNNs to achieve convergence on data within a long time interval. Kong et al. [8] relied on the multi-layer LSTM to predict individual energy consumption. Compared with others, Kong et al. [8] not only solved the problem of RNN but also improved the prediction accuracy. Li et al. [40] proposed an improved GRU to dynamically capture temporal correlations within the forecast period, enabling the model to adapt to different datasets.

In the field of load forecasting, it is more inclined to utilize the feature extraction capabilities of the hybrid model to improve forecasting accuracy. Shi et al. [41] proposed a pooling method based on the LSTM network, which further improved the prediction accuracy of LSTM through backpropagation and pooling. Jiang et al. [42] proposed a hybrid model based on CNN and LSTM to predict household energy consumption. Unlike other combined methods, this method divided the input into two parts: long and short data. Then, they obtained a result with sufficient feature interaction through data fusion. Yue et al. [43] combined ensemble empirical mode decomposition (EEMD), permutation entropy (PE), feature selection (FS), LSTM, and bayesian optimization algorithm (BOA) to optimize the prediction accuracy. Further, some reasonable explanations were made for the reconstructed subsequences. Lin et al. [44] proposed an AM-based auto-encoder structure for LSTM, which provided superior prediction accuracy and had a good generalization. Wei et al. [45] proposed detrend singular spectrum fluctuation analysis (DSSFA) to extract trend and periodic components and then input these components into LSTM to improve short-term forecasting accuracy. Laouafi et al. [46] proposed an adaptive hybrid ensemble method named CMKP-EG-SVR and optimized the result of the mixture model through a gaussian-based error correction strategy.

3. Model Architecture of Residential Short-Term Load Forecasting

The goal of residential short-term load forecasting is to improve residents' electricity experience. It is an essential part of energy supply management. In this section, we mainly introduce a complete forecasting architecture that can improve residential load forecasting accuracy.

Figure 3 shows the process of residential short-term load forecasting, DCNN-LSTM-AE-AM. In Figure 3, the missing data are first processed based on the TNN algorithm, and the DCNN layer extracts the initial features. Then, the LSTM-AE is used to extract the spatiotemporal feature information. Finally, the AM is introduced to analyze the importance of the extracted features and outputs the final prediction result.

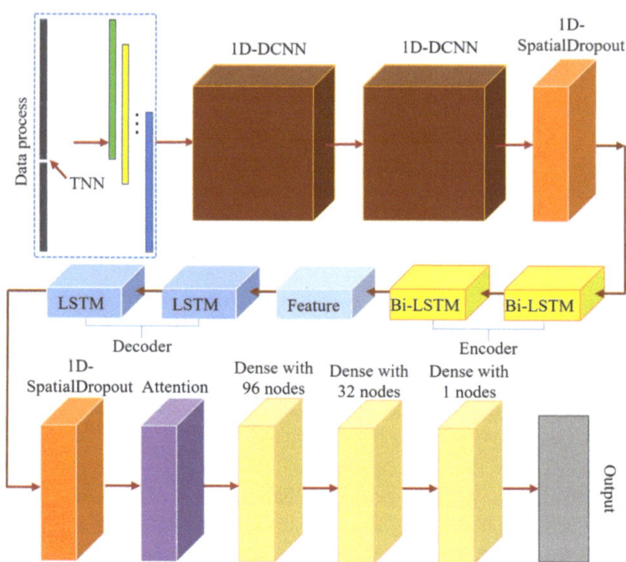

Figure 3. Process of residential short-term load forecasting.

3.1. Data Processing

Residential energy consumption data are collected by smart meters. Smart meters can accurately collect and transmit data to the data management center through various communication networks. Considering that the communication network is susceptible to interference by multiple factors, data loss is inevitable. Therefore, the missing data needs to be processed by some methods. This paper proposes the T-nearest neighbors (TNN) algorithm to fill in the missing data. At time t, TNN is defined as follows:

$$I_t \leftarrow \frac{1}{K}\left(I_{t-\frac{K}{2}T} + I_{t-(\frac{K}{2}-1)T} + \cdots + I_{t-T} + I_{t+T} + \cdots + I_{t+(\frac{K}{2}-1)T} + I_{t+\frac{K}{2}T'}\right) \tag{1}$$

where K represents the number of selected adjacent values; T represents the interval period; I_t is the output of TNN at time t. The algorithm can solve the problem that the duration of missing data is relatively long. When the duration of the missing data is too long, the missing data will be ignored.

Moreover, the collected load data usually has a small amount of singular data, which will affect the overall model. Therefore, it is necessary to scale these data to some fixed range so that they will conform to a certain distribution. In this paper, we use a linear normalization, i.e., the max-min normalization, to process the load data, which is defined as follows:

$$\theta_{norm} = \frac{\theta - \theta_{\min}}{\theta_{\max} - \theta_{\min}}, \tag{2}$$

where θ_{norm} represents the normalized output; θ represents the current input; $\theta_{\max}$ and $\theta_{\min}$ represent the upper and lower bounds of the current sequence input, respectively.

3.2. Dilated Convolutional Neural Network

CNN has been widely used in image processing ever since it was proposed [36]. Recently, time series data has also tried to use CNN to deal with short-term load forecasting. The core of CNN is weight sharing. Each CNN has a convolution kernel, which shares different weights according to the convolution operation. However, CNN will lose this feature information with long-term regularity, such as valley oscillation load [47]. To broaden the horizon of CNN, in this paper, we transform the convolution computation of continuous data into the convolution computation of skipping data, which is called the dilated convolutional neural network (DCNN) [48,49]. For a τ-dimension input vector $v \in \mathbb{R}^\tau$ and a kernel $w : \{0, \ldots, k-1\} \in \mathbb{R}$, the dilated convolution operation on element s of input vector v is defined as:

$$y_s = \sum_{i=0}^{k-1} v[s + r \cdot i] \cdot w[i],$$

(3)

where the dilated convolution adjustment rate r is expressed as the interval step size for selecting input data; k represents the kernel size. In addition, τ represents the time step.

Figure 4a,b illustrate the internal structures of CNN and DCNN, respectively. In Figure 4, CNN has one dimension, a kernel of one, and a dilation rate of one, while DCNN has one dimension, a kernel of one, and a dilation rate of one. The first layer is the original input layer, the second layer is the hidden layer, and the third layer is the output layer. In addition, after adding the convolution layer, the activation function and pooling layer are added appropriately to help the backpropagation of the gradient.

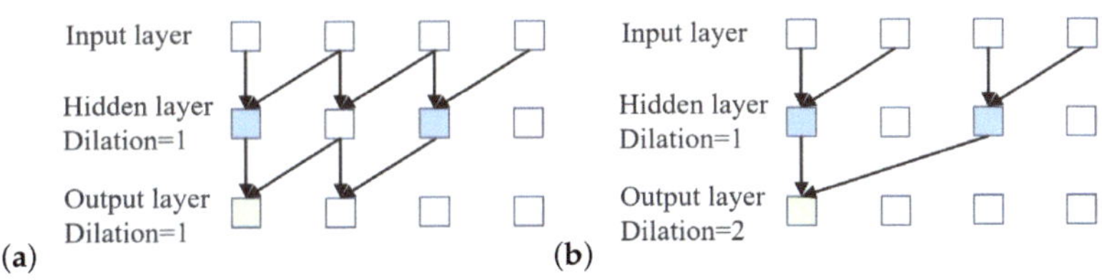

Figure 4. Structure of CNN and DCNN: (**a**) CNN; (**b**) DCNN.

3.3. LSTM-Based Autoencoder

3.3.1. Long Short-Term Memory Network

RNN is a forward-propagating sequential neural network. When it deals with long-duration data, it usually faces some challenges, such as gradient disappearance and gradient explosion. Hochreiter and Schmidhuber [29] proposed an improved network based on the RNN structure and named it LSTM. The internal structure of the LSTM is shown in Figure 5, where the memory cell can retain information from a long time ago, and the forget gate can choose to discard some feature information. Backpropagation in the LSTM strengthens the interaction ability of context information and reserves more useful spatiotemporal feature information.

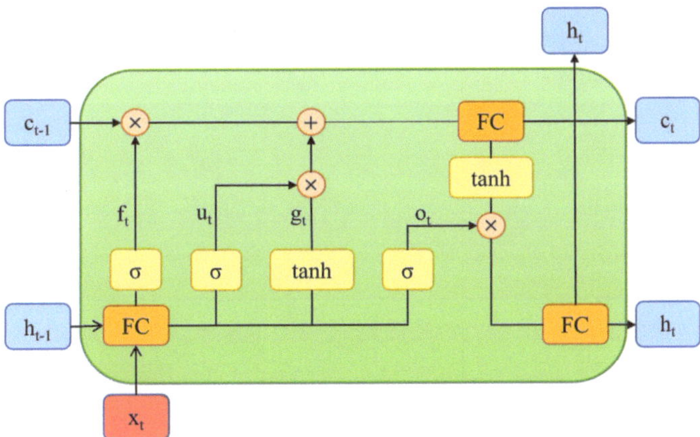

Figure 5. Internal structure of LSTM.

The update principles of the LSTM are defined as:

$$f_t = \sigma\left(W_{ft}x_t + W_{fh}h_{t-1} + b_f\right),\tag{4}$$

$$u_t = \sigma(W_{tx}x_t + W_{uh}h_{t-1} + b_u),\tag{5}$$

$$g_t = \tanh\left(W_{gx}x_t + W_{gh}h_{t-1} + b_g\right),\tag{6}$$

$$o_t = \sigma(W_{ox}x_t + W_{oh}h_{t-1} + b_o),\tag{7}$$

$$c_t = g_t \odot u_t + c_{t-1} \odot f_t,\tag{8}$$

$$h_t = \tanh(c_t) \odot o_t,\tag{9}$$

where f_t, u_t, g_t, o_t in Equations (4)–(7) represent the information at the forget gate, input gate, input node, and output gate at time t; σ, tanh are the multiplication calculations and activation functions; b_f, b_u, b_g, b_o are the bias parameters of the corresponding processing units; c_t, h_t in Equations (8) and (9) represent memory cells; W_{ft}, W_{fh}, W_{tx}, W_{uh}, W_{gx}, W_{gh}, W_{ox}, and W_{oh} are the weight matrices of the corresponding processing units; $\odot$ represents element multiplication; FC represents the fully connected layer. These units use functions σ and *tanh* to continuously compress the input x_t to a smaller range.

3.3.2. Bidirectional Long Short-Term Memory Network

The bidirectional long short-term memory network (BiLSTM) is a variant of LSTM. As shown in Figure 6, BiLSTM is a special network structure formed by superimposing two LSTM layers. The LSTMs synchronously train the input data at the same time step. These two LSTM layers differ in that one input the data in a positive temporal order, and the other processes it in a reverse temporal order. This structure not only utilizes the information of the previous moment but also relies on the information of the latter moment [50,51].

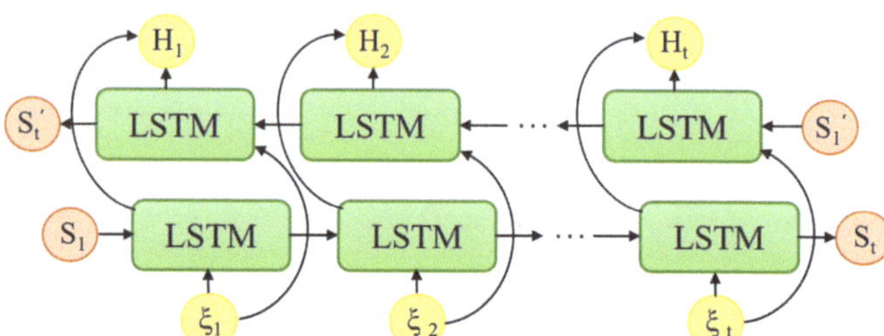

Figure 6. Standard structure of BiLSTM.

Let $\overrightarrow{h_t}$ and $\overleftarrow{h_t}$ be the hidden states of forward and backward propagation, respectively. Then, $\overrightarrow{h_t}, \overleftarrow{h_t}$, and the output H_t of BiLSTM are calculated as follows:

$$\overrightarrow{h_t} = \overrightarrow{LSTM}(\xi_t, S_{t-1}), t \in [1, \mathbb{T}],\tag{10}$$

$$\overleftarrow{h_t} = \overleftarrow{LSTM}(\xi_t, S_{t+1}), t \in [\mathbb{T}, 1],\tag{11}$$

$$H_t = FC\left(\overrightarrow{h_t}, \overleftarrow{h_t}\right).\tag{12}$$

where ξ_t represents the current input at current time t; S_t represents the internal state in the LSTM, that is, the memory cells and the hidden state; $\mathbb{T}$ represents the time steps.

3.3.3. Autoencoder

Autoencoder (AE), as an artificial neural network (ANN), is a conceptual network structure with an encoder and decoder. The AE aims to find an optimal set of connection weights by minimizing the reconstruction error between the original input and the output [52]. For any AE, there is a n-dimensional input vector φ_t and an output vector ϵ_t with random dimensions. The input φ_t can be mapped to output ϵ_t according to the following mapping functions:

$$\Theta = f\left(\tilde{W} \cdot \varphi_t + \alpha\right), \tag{13}$$

$$\epsilon_t = \rho\left(\hat{W} \cdot \Theta + \beta\right), \tag{14}$$

where Θ is the mapping output of the encoding layer; $\tilde{W}$ and $\hat{W}$ are two weight matrices; f and ρ are two activation functions; α and β are the bias parameters of the encoder and decoder, respectively.

Obviously, simple nonlinear AE is difficult for time series data feature extraction. As shown in Figure 3, in this paper, the LSTM and BiLSTM are combined to construct the AE. In order to learn more feature information and increase the sensitivity of the contextual information connection, the encoder part adopts two BiLSTM layers. In addition, the increase in the depth of the neural network structure is conducive to the extraction and fusion of load features. In the decoder part, we only need to add two LSTM layers as the feature analysis layer to reduce the unnecessary network computing burden.

3.3.4. Attention Mechanism

The AM pays more attention to the important parts of reconstructing the cognitive world instead of making an average judgment on the whole. Figure 7 shows the structure of the AM. In Figure 7, FC is the fully connected layer, which fuses with the output of the AM. The output of the AM can be calculated as follows:

$$\mu_t = \sum \Omega_t \odot \eta_t. \tag{15}$$

where $\eta_t = \{h_1, \cdots, h_\chi\}$ is the output decoded by LSTM-AE and is a χ-dimensional hidden state vector at time t; $\Omega_t = \{\lambda_1, \cdots, \lambda_\chi\}$ is a weight matrix.

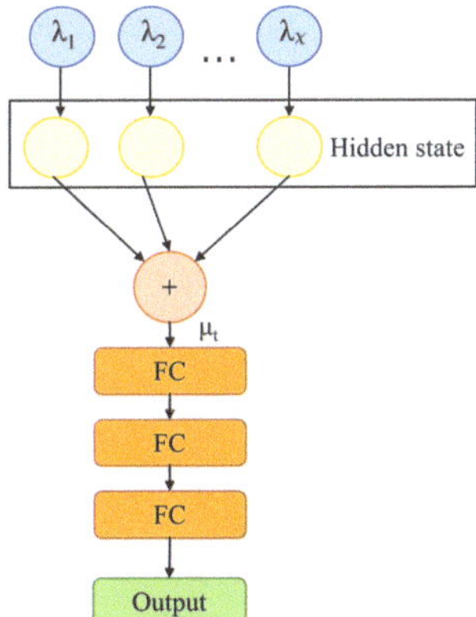

Figure 7. Structure of attention mechanism.

The matrix Ω_t in (15) can be implemented according to the following procedure. First, the output η_t represents the input of the AM. Then, the alignment model $a(\cdot)$ aligns the input with the output vector $\phi_t = \{\varepsilon_1, \cdots, \varepsilon_\chi\}$. The alignment score ϕ_t is calculated as follows:

$$\phi_t = a(\delta_{t-1}, \eta_t). \tag{16}$$

In this study, the alignment model $a(\delta_{t-1}, \eta_t)$ represents $\tanh(\delta_{t-1} \odot \eta_t + \gamma)$, where the cell state δ_{t-1} decoded by LSTM-AE represents the χ-dimensional hidden state vectors at time $t-1$; γ is a vector of bias parameters. Finally, each element λ_j is computed by applying a softmax operation:

$$\lambda_j = \frac{\exp(\varepsilon_j)}{\sum_{i=1}^{X} \exp(\varepsilon_i)}, \tag{17}$$

where i and j represent the i-th and j-th elements in ϕ_t. After injecting the AM, we also apply multiple fully connected layers as the output layer to complete the final load forecasting.

4. Experiment Results and Analysis

In this section, we will show the effectiveness of our proposed DCNN-LSTM-AE-AM load forecasting method through some experiments.

4.1. Experiment Settings

4.1.1. Dataset Selection

The experiments use two real-world power load datasets to test our proposed method's robustness and generalization. The household electricity consumption dataset from the UCI machine learning library (IHEPC) [37] records the energy consumption information of a house from 2006 to 2010. It contains multiple attributes: date, time, global active power, global reactive power, voltage, current, and active power of three-room types: kitchen, bathroom, and bedroom. There are 2,075,269 1-min-level data, including 25,979 missing values. Table 1 shows detailed information on IHEPC. The global active power represents the actual power consumption, so this paper only takes the global active power as the input. Another dataset is from the smart grid and smart city (SGSC) [53] projects carried out by the Australian government and industry consortium Ausgrid. Data in the SGSC dataset are collected from 10,000 households and some retail stores in New South Wales (NSW) from 2010 to 2014. In this paper, the IHPEC is organized as the sum of energy consumption within 15 min and 1 h. Since the SGSC only provides electricity consumption per half hour, the SGSC is organized as the sum of energy consumption within 30 min and 1 h. Figure 8 shows the randomly selected data of IHPEC at a 15-min resolution and the randomly selected data of one household from SGSC at a 30-min resolution.

Table 1. Information from the IHEPC dataset.

Variable	Description
Data	Recorded date
Time	Current moment
Global active power	Sum of active power per minute
Global reactive power	Sum of reactive power per minute
Voltage	Voltage per minute
Global intensity	Sum of current per minute
Sub metering1	Power used by kitchen per minute
Sub metering2	Power used by room per minute
Sub metering3	Power used by bathroom per minute

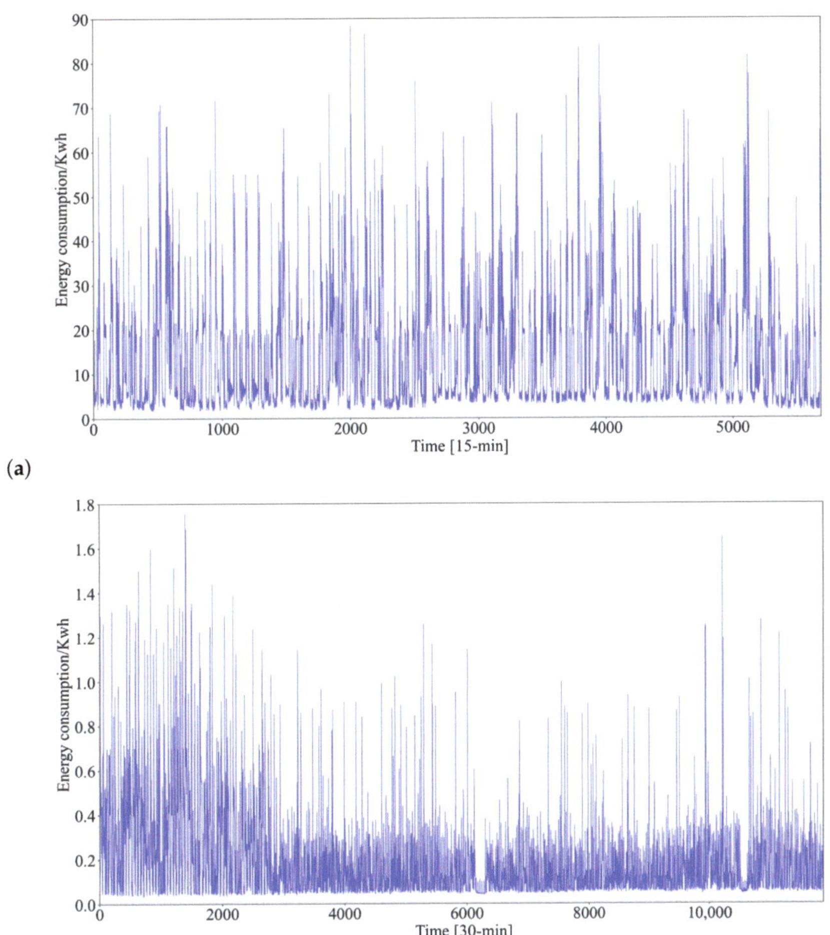

Figure 8. Visualization of selected datasets: (**a**) IHEPC at a 15-min resolution; (**b**) SGSC at a 30-min resolution.

4.1.2. Experiment Setup

All networks are built based on Python3.6, Keras2.2, and TensorFlow2.0. The device is configured with a 2.6 GHz intel i9 CPU and a 16GB NVIDIA TESLA T4 GPU.

According to the findings in [8], some rules of thumb for hyperparameter selection are adopted. Since hyperparameter selection is a time-consuming task, in this paper, we try to use different combinations of parameters to obtain the optimal performance in MSE. Table 2 lists the hyperparameter settings of the proposed DCNN-LSTM-AE-AM. The first 1D-Conv layer uses the DCNN with a kernel of 3, 12 filters, and a dilation rate of 2 to extract features and ReLU as output results. The number of the second 1D-Conv layer's filters is upgraded to 24. These two SpatialDropout layers randomly zero the parameters with probabilities 0.1 and 0.2, respectively. In the LSTM-AE, 32 units are used in all four temporal models. In the final output layer, the AM uses 32 units, and the three Dense layers use 96, 32, and 1 unit, respectively. We set the maximum training epoch to 50. All other methods are also tested on the same equipment, hyperparameter configuration, and environment to allow horizontal comparisons. In addition, XGBoost obtains optimal performance with the number of estimators set to 30 after constantly changing the number of estimators.

In this paper, we split the dataset into a training set and a test set, whose ratios are designed as 0.67 and 0.33. We fill the missing values in the training set according to the TNN algorithm. The missing values in the test set are ignored to prevent prior knowledge leakage

Table 2. Hyperparameter settings.

Network	Hyperparameters
1D-Conv	The convolution kernel size is 3, the number of filters is 12, the dilation rate is 2, and the activation function is ReLU
1D-Conv	The convolution kernel size is 3, the number of filters is 24, the dilation rate is 2, and the activation function is ReLU
1D-SpatialDropout	0.1
BiLSTM	32 units
BiLSTM	32 units
LSTM	32 units
LSTM	32 units
1D-SpatialDropout	0.2
Attention	32 units
Dense	96 units
Dense	32 units
Dense	1 unit

4.1.3. Evaluation Metric

It is well known that the classification task can be evaluated by accuracy in percentage However, this kind of accuracy is not appropriate for evaluating any regression task In this paper, in order to objectively evaluate the fairness and integrity of the methods, we use four evaluation metrics: MAE, RMSE, MSE, and MAPE. The MAE is a method of averaging quantization errors. Compared with MAE, the other methods have made some improvements. MSE pays more attention to the influence of outliers on the overall prediction effect. RMSE performs arithmetic square root on the overall basis of the MSE, which amplifies the difference of the MSE. MAPE can focus on the gap between the error and the actual value. The formulations for the MAE, RMSE, MSE, and MAPE are listed as follows:

$$MAE = \frac{1}{N} \sum_{i=1}^{N} \left| \psi_{pred} - \psi \right|, \tag{18}$$

$$MSE = \frac{1}{N} \sum_{i=1}^{N} (\psi_{pred} - \psi)^2, \tag{19}$$

$$RMSE = \sqrt{\frac{1}{N} \sum_{i=1}^{N} (\psi_{pred} - \psi)^2}, \tag{20}$$

$$MAPE = \frac{100\%}{N} \sum_{i=1}^{N} \left| \frac{\psi_{pred} - \psi}{\psi} \right|. \tag{21}$$

where N represents the total amount of the load data; ψ and ψ_{pred} represent the actual load and the predicted load, respectively. For all those evaluation metrics, we have that the smaller the value, the more accurate the model.

4.2. Influence of Hyperparameters

This subsection sets various optimized hyperparameters to achieve the best performance. We evaluate the performance of the proposed model under these hyperparameters by the MSE. To ensure the fairness of the experiments, we compare the influence of batch size, optimizer, and learning rate on the MSE with other parameters that remain unchanged Figure 9 shows the MSE of the proposed method under different hyperparameter settings. From Figure 9, we can have that our method has a certain degree of sensitivity to hyperparameters. According to the experimental results in Figure 8, we choose the batch size of 64, the optimizer as Adam, and the learning rate as 0.001.

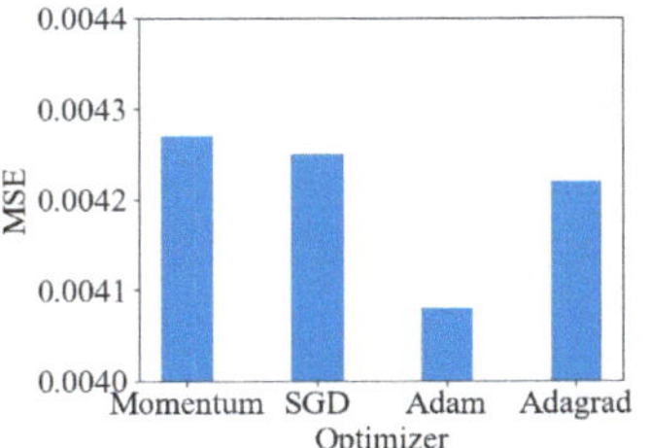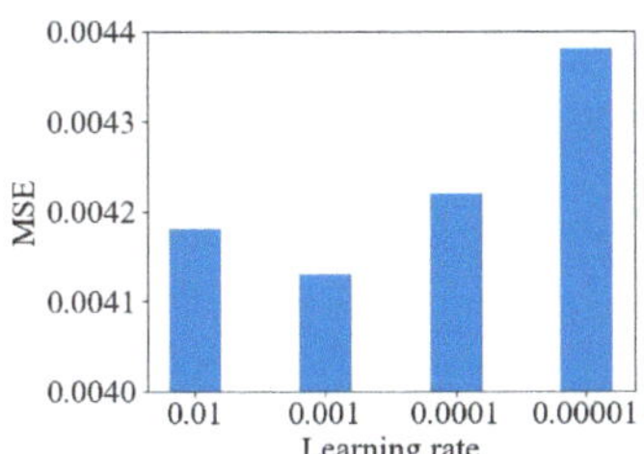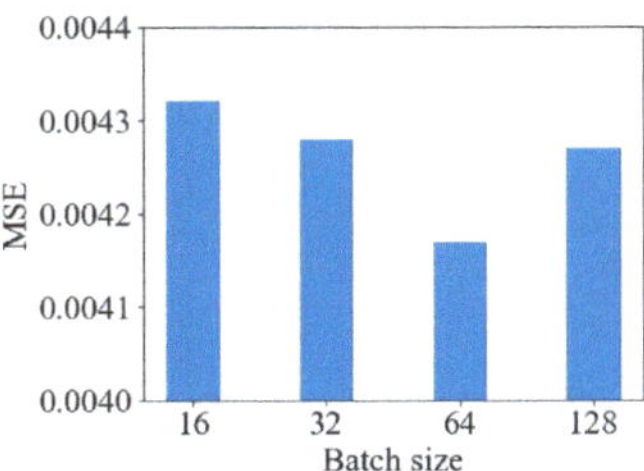

Figure 9. MSE of the proposed model under different hyperparameters.

4.3. Influence of Time Step

Time step τ, i.e., the length of the input data, is one of the factors affecting the robustness and accuracy of the proposed short-term load forecasting model. In deep networks, long time-series data often results in overfitting. Hence, the data length also affects the performance of our proposed method. The MSE, RMSE, MAE, and MAPE values under different data lengths are shown in Table 3. The evaluation metrics with lengths between 8 and 14 have very little difference. Therefore, in these following experiments, we let the data length be 12.

Table 3. Performance of our proposed method under different data lengths.

Length	MSE	RMSE	MAE	MAPE
6	0.00434	0.0659	0.0351	0.7599
8	0.00420	0.0648	0.0332	0.8521
10	0.00418	0.0647	0.0344	0.7662
12	0.00411	0.0641	0.0331	0.6175
14	0.00415	0.0644	0.0334	0.6757
16	0.00426	0.0653	0.0338	0.7421
18	0.00444	0.0666	0.0348	0.7952
20	0.00450	0.0671	0.0357	0.7853
22	0.00458	0.0677	0.0366	0.8322
24	0.00446	0.0668	0.0370	0.8964

4.4. Performance Evaluation on IHEPC Dataset

The performance of the proposed DCNN-LSTM-AE-AM is compared with some sole models and some hybrid models. The sole models include XGBoost, DCNN, LSTM [8], and AM; while the hybrid models include LSTM-AE [54], CNN-LSTM [42], DCNN-AM, DCNN-LSTM-AE, and LSTM-AE-AM. The evaluation metrics comparison among different methods at 15-min and 1-h resolutions are shown in Table 4. From Table 4, we have that: (1) our proposed method obtains the best metrics of 0.0041 in MSE, 0.0640 in RMSE, 0.0333 in MAE, and 0.6757 in MAPE at the 15-min resolution; (2) while for the dataset at the 1-h resolution, our proposed method outperforms others with metrics of 0.0086 in MSE, 0.0926 in RMSE, 0.0667 in MAE, and 0.7257 in MAPE at the 1-h resolution. (3) Compared with the existing methods, the overall prediction accuracy of the DCNN-LSTM-AE-AM at different resolutions has obtained a little improvement. (4) Performances of those sole models are worse than those of hybrid models. It is worth noting that there is not much difference between the performance of DCNN-AM, LSTM-AE-AM, and DCNN-LSTM-AE-AM.

Table 4. Prediction performance comparisons on the SGSC dataset.

Method	Resolution	MSE	RMSE	MAE	MAPE
XGBoost	15 min	0.0463	0.2152	0.1541	0.8683
	1 h	0.0410	0.2025	0.1120	0.7268
DCNN	15 min	0.0403	0.2007	0.1192	0.6869
	1 h	0.0412	0.2030	0.1132	0.7284
LSTM	15 min	0.0491	0.2216	0.1263	1.0384
	1 h	0.0548	0.2341	0.1356	1.4831
AM	15 min	0.0663	0.2575	0.1228	1.4286
	1 h	0.0687	0.2621	0.1346	1.3788
LSTM-AE	15 min	0.0179	0.1338	0.0855	0.7863
	1 h	0.0232	0.1523	0.0878	0.8265
CNN-LSTM	15 min	0.0158	0.1257	0.0712	0.6930
	1 h	0.0197	0.1403	0.0997	0.7556
DCNN-AM	15 min	0.0081	0.0900	0.0452	0.6938
	1 h	0.0091	0.0954	0.0466	0.7028
DCNN-LSTM-AE	15 min	0.0222	0.1490	0.0709	0.7380
	1 h	0.0295	0.1718	0.0823	0.7428
LSTM-AE-AM	15 min	0.0043	0.0656	0.0378	0.6854
	1 h	0.0095	0.0975	0.0682	0.7530
DCNN-LSTM-AE-AM	15 min	0.0041	0.0640	0.0333	0.6757
	1 h	0.0086	0.0927	0.0667	0.7257

To show the effect of individual data in the model, we also use four box-figures to show the performance of different models in MSE, RMSE, MAE, and MAPE. As shown in Figure 10, four subfigures are used to show the performance of the IHEPC dataset at a 15-min resolution. Since we have given the average results in Table 4, the average results are no longer marked in the subfigures, and only the median results are marked From Figure 10, one can have that: (1) the proposed method outperforms others in all the evaluation metrics; (2) the median of the metrics is lower than that of other methods; (3) the MSE, RMSE, and MAE obtain significant improvement, while the increase in MAPE is relatively small.

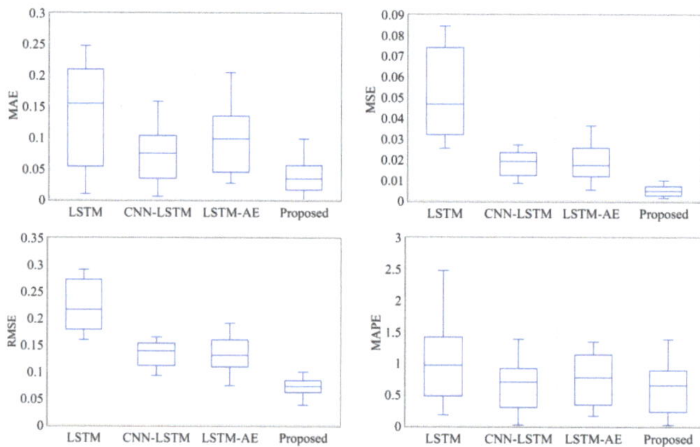

Figure 10. Prediction results of different methods on the IHEPC dataset at a 15-min resolution.

Figure 10 also shows that all the mentioned forecasting methods are still far from accurate prediction, which is the direction we need to focus on. Further, CNN-LSTM, LSTM-AE, and our proposed method all inherit the feature extraction capability of LSTM. To demonstrate our proposed model's trend-tracking capability, we show the prediction results of the mentioned four sole models and the six hybrid models in Figures 11 and 12. From Figure 11, it can be seen that the tracking trend of deep learning methods, except for the AM, can outperform traditional machine learning. DCNN is always trying to match the low-load data. The LSTM has outstanding temporal feature extraction and good trend tracking, but the error with respect to the actual load is large. Although the overall performance of using only AM is not good, the AM is suitable for forecasting low-load data. The performance of AM benefits from local feature fusion capability.

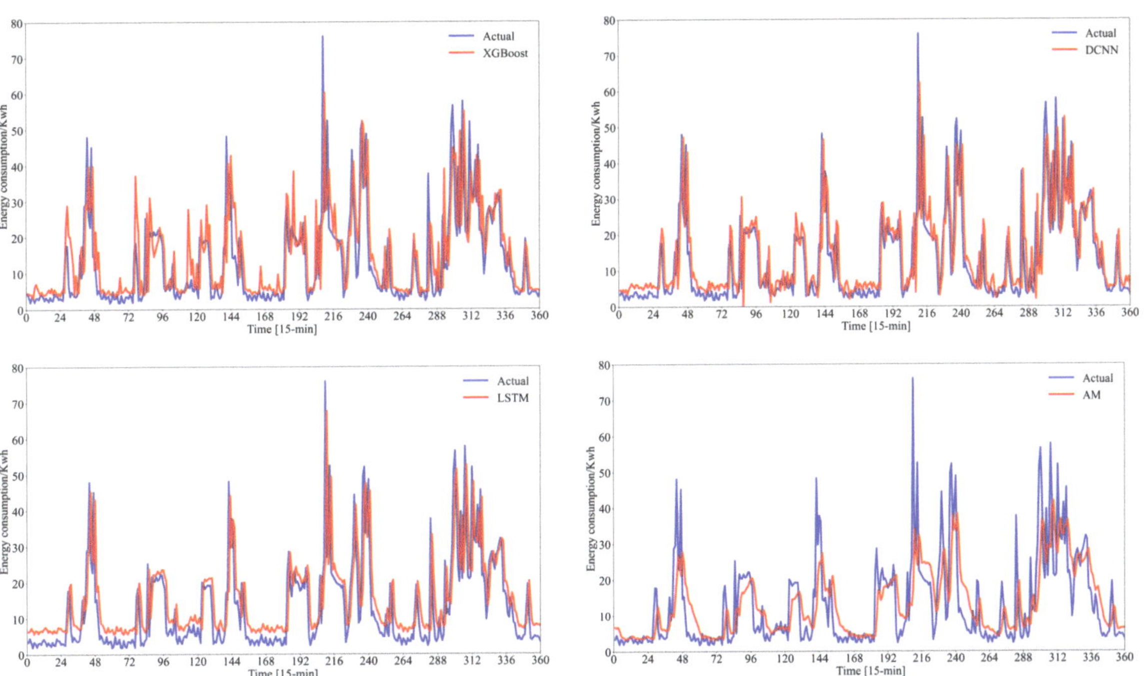

Figure 11. Prediction results of some sole models on the IHEPC dataset at a 15-min resolution.

From Figure 12, one can have that: (1) Different hybrid models have different prediction results. (2) The LSTM-AE is similar to the LSTM, which further narrows the numerical gap. (3) Although the CNN-LSTM is sensitive to data with large fluctuations in value and predicts the result with a small error, there is a large error when capturing the valley values. This is one of the reasons for the large value of MAPE. (4) DCNN-AM and LSTM-AE-AM are increasingly perceptive to the low-load data. The addition of AM in the last layer is definitely the main contributor to improving the prediction performance of the hybrid model on the low-load data. (5) Compared with the other five mentioned hybrid models, DCNN-LSTM-AE-AM can quickly capture the data in this situation and predict the valley data very well. This benefits from the broadened time horizon of the DCNN. The introduction of AM makes the weights closer to the actual load, which significantly reduces the prediction error. In addition, the results of DCNN-LSTM-AE suggest that no combination of hybrid models can improve the prediction results.

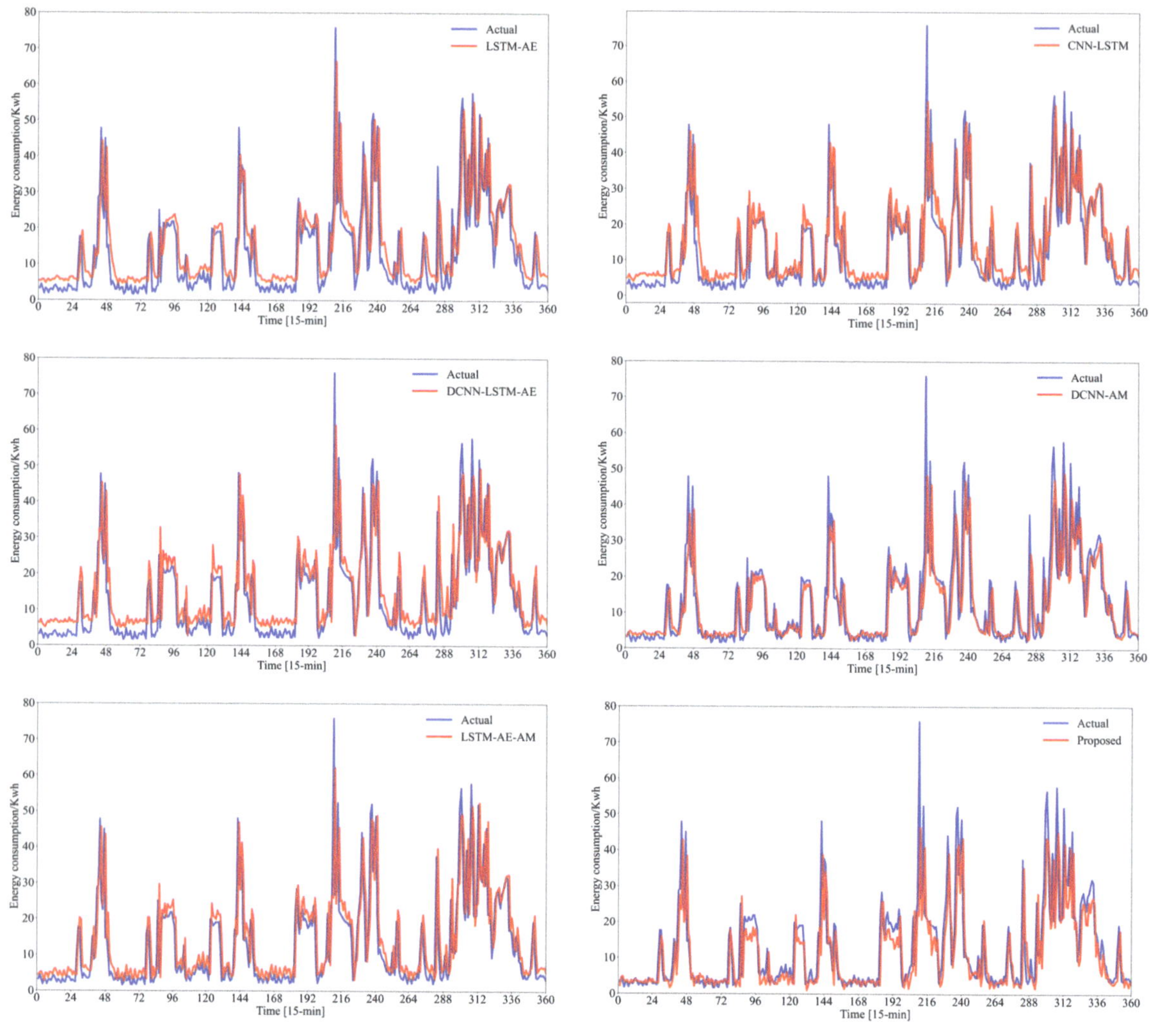

Figure 12. Prediction results of some hybrid models on the IHEPC dataset at a 15-min resolution.

4.5. Performance Evaluation on SGSC Dataset

Our proposed method is also tested on the SGSC dataset, where it shows excellent accuracy and has a good generalization. Table 5 shows the evaluation metrics comparisons of different forecasting methods on the SGSC dataset. The performance of the proposed method on the SGSC dataset is similar to that on the IHEPC dataset, and both have obtained the best results for each evaluation metric. The MSE of the proposed method obviously gets improved compared with other methods. The other evaluation metrics also have a small increase.

As shown in Figure 13, we compare the actual load with the prediction results. In this study, a residential user is arbitrarily selected, and the proposed DCNN-LSTM-AE-AM can easily predict the load at the next moment. The proposed method can capture the overall trend of the actual load with small time offsets and numerical errors. In addition, the method also has good predictability for the valley data. This confirms that DCNN and AM have a good ability to capture long-term regular data.

Table 5. Prediction performance comparisons on the SGSC dataset.

Method	Resolution	MSE	RMSE	MAE	MAPE
XGBoost	30 min	0.0456	0.2135	0.1203	0.9298
	1 h	0.0403	0.2007	0.0980	0.8296
DCNN	30 min	0.0433	0.2081	0.1329	0.6796
	1 h	0.0465	0.2156	0.1366	0.7862
LSTM	30 min	0.0483	0.2198	0.1298	1.1001
	1 h	0.0522	0.2285	0.1401	1.5131
AM	30 min	0.0652	0.2553	0.1893	1.6235
	1 h	0.0689	0.2625	0.2006	1.7692
LSTM-AE	30 min	0.0166	0.1288	0.0799	0.7567
	1 h	0.0218	0.1476	0.0762	0.7992
CNN-LSTM	30 min	0.0142	0.1192	0.0804	0.6728
	1 h	0.0182	0.1349	0.0991	0.7256
DCNN-AM	30 min	0.0083	0.0911	0.0489	0.6118
	1 h	0.0089	0.0943	0.0496	0.6412
DCNN-LSTM-AE	30 min	0.0242	0.1556	0.0756	0.7128
	1 h	0.0289	0.1700	0.0862	0.7196
LSTM-AE-AM	30 min	0.0048	0.0693	0.0384	0.6203
	1 h	0.0056	0.0748	0.0696	0.6495
DCNN-LSTM-AE-AM	30 min	0.0041	0.0640	0.0329	0.5901
	1 h	0.0081	0.0900	0.0657	0.6336

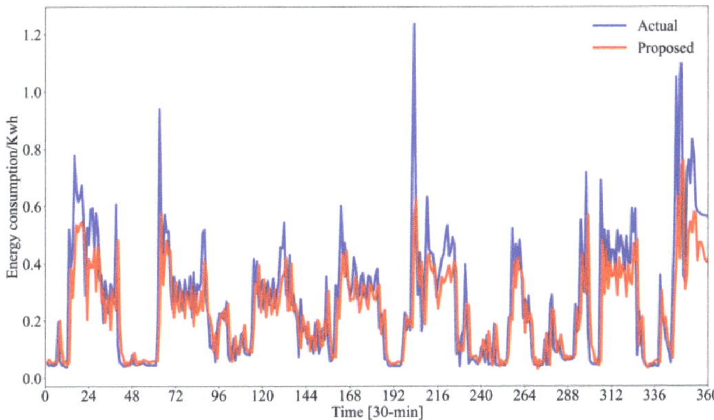

Figure 13. Prediction results of DCNN-LSTM-AE-AM on the SGSC dataset at a 30-min resolution.

5. Conclusions

This paper proposes a short-term load forecasting model to predict residential energy consumption. A hybrid electric load forecasting model, i.e., DCNN-LSTM-AE-AM, is constructed with the introduction of existing deep learning methods. We use multiple similar-days data in the vicinity of missing values as the basis for inferring the original data and use the TNN algorithm to fill in the missing data. In the initial feature-extraction stage, DCNN broadens the time horizon to retain the load features. LSTM-AE is used to improve the analysis capability of features. In the feature fusion stage, the importance of features in each period is summarized based on AM, which enhances the final prediction accuracy. The validity of the proposed method is verified on two real-world datasets. Experimental results show that the proposed method improves the accuracy of residential load forecasting and can capture low-load data features.

In future work, we need to improve the accuracy of residential load forecasting by exploiting residential lifestyle features. Moreover, how to achieve real-time forecasting through methods such as online learning is another work.

Author Contributions: Conceptualization, X.J.; methodology, K.Y.; software, H.H.; validation, Y.Z and H.H.; formal analysis, X.J. and R.B.; investigation, X.J.; resources, R.B.; data curation, D.C.; writing—original draft preparation, X.J.; writing—review and editing, Z.C.; visualization, Y.Z.; supervision, Z.C. and R.B.; project administration, D.C.; funding acquisition, Z.C. All authors have read and agreed to the published version of the manuscript.

Funding: This research was funded by the National Natural Science Foundation of China (NSFC) under grant nos. 51874205; Jiangsu Qinglan Project.

Institutional Review Board Statement: Not applicable.

Informed Consent Statement: Not applicable.

Data Availability Statement: The figures and tables used to support the findings of this study are included in the article.

Conflicts of Interest: The authors declare no conflict of interest.

References

1. Jacob, M.; Neves, C.; Vukadinović Greetham, D. *Forecasting and Assessing Risk of Individual Electricity Peaks*; Springer Nature: Berlin/Heidelberg, Germany, 2020.
2. Hernandez, L.; Baladron, C.; Aguiar, J.M.; Carro, B.; Sanchez-Esguevillas, A.J.; Lloret, J.; Massana, J. A survey on electric power demand forecasting: Future trends in smart grids, microgrids and smart buildings. *IEEE Commun. Surv. Tutor.* **2014**, *16*, 1460–1495 [CrossRef]
3. Ghofrani, M.; Hassanzadeh, M.; Etezadi-Amoli, M.; Fadali, M.S. Smart meter based short-term load forecasting for residential customers. In Proceedings of the 2011 North American Power Symposium, Boston, MA, USA, 4–6 August 2011; pp. 1–5.
4. Ullah, I.; Ahmad, R.; Kim, D. A prediction mechanism of energy consumption in residential buildings using hidden markov model. *Energies* **2018**, *11*, 358. [CrossRef]
5. Lee, Y.S.; Tong, L.I. Forecasting time series using a methodology based on autoregressive integrated moving average and genetic programming. *Knowl.-Based Syst.* **2011**, *24*, 66–72. [CrossRef]
6. Aguilar Madrid, E.; Antonio, N. Short-term electricity load forecasting with machine learning. *Information* **2021**, *12*, 50. [CrossRef]
7. Fumo, N.; Biswas, M.R. Regression analysis for prediction of residential energy consumption. *Renew. Sustain. Energy Rev.* **2015**, *47*, 332–343. [CrossRef]
8. Kong, W.; Dong, Z.Y.; Jia, Y.; Hill, D.J.; Xu, Y.; Zhang, Y. Short-term residential load forecasting based on LSTM recurrent neural network. *IEEE Trans. Smart Grid* **2017**, *10*, 841–851. [CrossRef]
9. Ahmad, A.S.; Hassan, M.Y.; Abdullah, M.P.; Rahman, H.A.; Hussin, F.; Abdullah, H.; Saidur, R. A review on applications of ANN and SVM for building electrical energy consumption forecasting. *Renew. Sustain. Energy Rev.* **2014**, *33*, 102–109. [CrossRef]
10. Arik, S.Ö.; Pfister, T. Tabnet: Attentive interpretable tabular learning. In Proceedings of the AAAI Conference on Artificial Intelligence, Virtual, 2–9 February 2021; Volume 35, pp. 6679–6687.
11. Janiesch, C.; Zschech, P.; Heinrich, K. Machine learning and deep learning. *Electron. Mark.* **2021**, *31*, 685–695. [CrossRef]
12. Taylor, J.W. Short-term load forecasting with exponentially weighted methods. *IEEE Trans. Power Syst.* **2011**, *27*, 458–464 [CrossRef]
13. Hammad, M.A.; Jereb, B.; Rosi, B.; Dragan, D. Methods and models for electric load forecasting: A comprehensive review. *Logist. Supply Chain Sustain. Glob. Chall.* **2020**, *11*, 51–76. [CrossRef]
14. Liu, X.; Zhang, Z.; Song, Z. A comparative study of the data-driven day-ahead hourly provincial load forecasting methods: From classical data mining to deep learning. *Renew. Sustain. Energy Rev.* **2020**, *119*, 109632. [CrossRef]
15. Panda, S.K.; Ray, P.; Salkuti, S.R. A Review on Short-Term Load Forecasting Using Different Techniques. In *Recent Advances in Power Systems*; Springer, Singapore, 2022; pp. 433–454.
16. Vanting, N.B.; Ma, Z.; Jørgensen, B.N. A scoping review of deep neural networks for electric load forecasting. *Energy Inform.* **2021**, *4*, 49. [CrossRef]
17. Haben, S.; Arora, S.; Giasemidis, G.; Voss, M.; Greetham, D.V. Review of low voltage load forecasting: Methods, applications, and recommendations. *Appl. Energy* **2021**, *304*, 117798. [CrossRef]
18. He, H.; Liu, T.; Chen, R.; Xiao, Y.; Yang, J. High frequency short-term demand forecasting model for distribution power grid based on ARIMA. In Proceedings of the 2012 IEEE International Conference on Computer Science and Automation Engineering (CSAE), Zhangjiajie, China, 25–27 May 2012; Volume 3, pp. 293–297.
19. Cao, X.; Dong, S.; Wu, Z.; Jing, Y. A data-driven hybrid optimization model for short-term residential load forecasting. In Proceedings of the 2015 IEEE International Conference on Computer and Information Technology; Ubiquitous Computing and Communications; Dependable, Autonomic and Secure Computing; Pervasive Intelligence and Computing, Liverpool, UK, 26–28 October 2015; pp. 283–287.
20. Cai, H.; Shen, S.; Lin, Q.; Li, X.; Xiao, H. Predicting the energy consumption of residential buildings for regional electricity supply-side and demand-side management. *IEEE Access* **2019**, *7*, 30386–30397. [CrossRef]

21. Mohammadi, M.; Talebpour, F.; Safaee, E.; Ghadimi, N.; Abedinia, O. Small-scale building load forecast based on hybrid forecast engine. *Neural Process. Lett.* **2018**, *48*, 329–351. [CrossRef]

22. Chauhan, M.; Gupta, S.; Sandhu, M. Short-Term Electric Load Forecasting Using Support Vector Machines. *ECS Trans.* **2022**, *107*, 9731. [CrossRef]

23. Massaoudi, M.; Refaat, S.S.; Chihi, I.; Trabelsi, M.; Oueslati, F.S.; Abu-Rub, H. A novel stacked generalization ensemble-based hybrid LGBM-XGB-MLP model for short-term load forecasting. *Energy* **2021**, *214*, 118874. [CrossRef]

24. Lee, K.; Cha, Y.; Park, J. Short-term load forecasting using an artificial neural network. *IEEE Trans. Power Syst.* **1992**, *7*, 124–132. [CrossRef]

25. Bisht, B.S.; Holmukhe, R.M. Electricity load forecasting by artificial neural network model using weather data. *IJEET Trans. Power Syst.* **2013**, *4*, 91–99.

26. Kuo, P.H.; Huang, C.J. A high precision artificial neural networks model for short-term energy load forecasting. *Energies* **2018**, *11*, 213. [CrossRef]

27. Krizhevsky, A.; Sutskever, I.; Hinton, G.E. Imagenet classification with deep convolutional neural networks. *Adv. Neural Inf. Process. Syst.* **2012**, *60*, 25–31. [CrossRef]

28. Sherstinsky, A. Fundamentals of recurrent neural network (RNN) and long short-term memory (LSTM) network. *Phys. D Nonlinear Phenom.* **2020**, *404*, 132306. [CrossRef]

29. Hochreiter, S.; Schmidhuber, J. Long short-term memory. *Neural Comput.* **1997**, *9*, 1735–1780. [CrossRef] [PubMed]

30. Chen, K.; Chen, K.; Wang, Q.; He, Z.; Hu, J.; He, J. Short-term load forecasting with deep residual networks. *IEEE Trans. Smart Grid* **2018**, *10*, 3943–3952. [CrossRef]

31. Kong, X.; Li, C.; Zheng, F.; Wang, C. Improved deep belief network for short-term load forecasting considering demand-side management. *IEEE Trans. Power Syst.* **2019**, *35*, 1531–1538. [CrossRef]

32. Dong, Y.; Dong, Z.; Zhao, T.; Li, Z.; Ding, Z. Short term load forecasting with markovian switching distributed deep belief networks. *Int. J. Electr. Power Energy Syst.* **2021**, *130*, 106942. [CrossRef]

33. Deng, Z.; Wang, B.; Xu, Y.; Xu, T.; Liu, C.; Zhu, Z. Multi-scale convolutional neural network with time-cognition for multi-step short-term load forecasting. *IEEE Access* **2019**, *7*, 88058–88071. [CrossRef]

34. Aouad, M.; Hajj, H.; Shaban, K.; Jabr, R.A.; El-Hajj, W. A CNN-Sequence-to-Sequence network with attention for residential short-term load forecasting. *Electr. Power Syst. Res.* **2022**, *211*, 108152. [CrossRef]

35. Zhang, G.; Bai, X.; Wang, Y. Short-time multi-energy load forecasting method based on CNN-Seq2Seq model with attention mechanism. *Mach. Learn. Appl.* **2021**, *5*, 100064. [CrossRef]

36. Amarasinghe, K.; Marino, D.L.; Manic, M. Deep neural networks for energy load forecasting. In Proceedings of the 2017 IEEE 26th International Symposium on Industrial Electronics (ISIE), Edinburgh, UK, 19–21 June 2017; pp. 1483–1488.

37. UCI. Individual Household Electric Power Consumption Data Set. Available online: https://archive.ics.uci.edu/ml/datasets/Individual+household+electric+power+consumption (accessed on 4 March 2020).

38. Sadaei, H.J.; e Silva, P.C.d.L.; Guimarães, F.G.; Lee, M.H. Short-term load forecasting by using a combined method of convolutional neural networks and fuzzy time series. *Energy* **2019**, *175*, 365–377. [CrossRef]

39. Rahman, A.; Srikumar, V.; Smith, A.D. Predicting electricity consumption for commercial and residential buildings using deep recurrent neural networks. *Appl. Energy* **2018**, *212*, 372–385. [CrossRef]

40. Li, D.; Sun, G.; Miao, S.; Gu, Y.; Zhang, Y.; He, S. A short-term electric load forecast method based on improved sequence-to-sequence GRU with adaptive temporal dependence. *Int. J. Electr. Power Energy Syst.* **2022**, *137*, 107627. [CrossRef]

41. Shi, H.; Xu, M.; Li, R. Deep learning for household load forecasting—A novel pooling deep RNN. *IEEE Trans. Smart Grid* **2017**, *9*, 5271–5280. [CrossRef]

42. Jiang, L.; Wang, X.; Li, W.; Wang, L.; Yin, X.; Jia, L. Hybrid multitask multi-information fusion deep learning for household short-term load forecasting. *IEEE Trans. Smart Grid* **2021**, *12*, 5362–5372. [CrossRef]

43. Yue, W.; Liu, Q.; Ruan, Y.; Qian, F.; Meng, H. A prediction approach with mode decomposition-recombination technique for short-term load forecasting. *Sustain. Cities Soc.* **2022**, *85*, 104034. [CrossRef]

44. Lin, J.; Ma, J.; Zhu, J.; Cui, Y. Short-term load forecasting based on LSTM networks considering attention mechanism. *Int. J. Electr. Power Energy Syst.* **2022**, *137*, 107818. [CrossRef]

45. Wei, N.; Yin, L.; Li, C.; Wang, W.; Qiao, W.; Li, C.; Zeng, F.; Fu, L. Short-term load forecasting using detrend singular spectrum fluctuation analysis. *Energy* **2022**, *256*, 124722. [CrossRef]

46. Laouafi, A.; Laouafi, F.; Boukelia, T.E. An adaptive hybrid ensemble with pattern similarity analysis and error correction for short-term load forecasting. *Appl. Energy* **2022**, *322*, 119525. [CrossRef]

47. Khan, N.; Haq, I.U.; Khan, S.U.; Rho, S.; Lee, M.Y.; Baik, S.W. DB-Net: A novel dilated CNN based multi-step forecasting model for power consumption in integrated local energy systems. *Int. J. Electr. Power Energy Syst.* **2021**, *133*, 107023. [CrossRef]

48. Chen, L.C.; Papandreou, G.; Schroff, F.; Adam, H. Rethinking atrous convolution for semantic image segmentation. *arXiv* **2017**, arXiv:1706.05587.

49. Bai, S.; Kolter, J.Z.; Koltun, V. Convolutional sequence modeling revisited. In Proceedings of the 2018 International Conference of Learning Representation Workshop, Vancouver, BC, Canada, 30 April–3 May 2018. Available online: https://openreview.net/forum?id=BJEX-H1Pf (accessed on 26 June 2022).

50. Graves, A.; Fernández, S.; Schmidhuber, J. Bidirectional LSTM networks for improved phoneme classification and recognition. In Proceedings of the International Conference on Artificial Neural Networks, Warsaw, Poland, 11–15 September 2005; Springer Berlin/Heidelberg, Germany, 2005; pp. 799–804.
51. Jia, M.; Huang, J.; Pang, L.; Zhao, Q. Analysis and research on stock price of LSTM and bidirectional LSTM neural network. In Proceedings of the 3rd International Conference on Computer Engineering, Information Science & Application Technology (ICCIA 2019), Chongqing, China, 30–31 May 2019; Atlantis Press: Paris, France, 2019; pp. 467–473.
52. Wang, T.; Lai, C.S.; Ng, W.W.; Pan, K.; Zhang, M.; Vaccaro, A.; Lai, L.L. Deep autoencoder with localized stochastic sensitivity for short-term load forecasting. *Int. J. Electr. Power Energy Syst.* **2021**, *130*, 106954. [CrossRef]
53. Smart-Grid Smart-City Customer Trial Data. Available online: https://data.gov.au/data/dataset/smart-grid-smart-city-customer-trial-data (accessed on 26 June 2022).
54. Khan, Z.A.; Hussain, T.; Ullah, A.; Rho, S.; Lee, M.; Baik, S.W. Towards efficient electricity forecasting in residential and commercial buildings: A novel hybrid CNN with a LSTM-AE based framework. *Sensors* **2020**, *20*, 1399. [CrossRef] [PubMed]

Article

Deep Forest-Based DQN for Cooling Water System Energy Saving Control in HVAC

Zhicong Han [1,2], Qiming Fu [1,2,*], Jianping Chen [2,3,*], Yunzhe Wang [1,2], You Lu [1,2], Hongjie Wu [1] and Hongguan Gui [4]

1 School of Electronic and Information Engineering, Suzhou University of Science and Technology, Suzhou 215009, China
2 Jiangsu Province Key Laboratory of Intelligent Building Energy Efficiency, Suzhou University of Science and Technology, Suzhou 215009, China
3 School of Architecture and Urban Planning, Suzhou University of Science and Technology, Suzhou 215009, China
4 Data Grand Information Technology, Co., Ltd., Shanghai 200120, China
* Correspondence: fqm_1@mail.usts.edu.cn (Q.F.); alanjpchen@aliyun.com (J.C.)

Abstract: Currently, reinforcement learning (RL) has shown great potential in energy saving in HVAC systems. However, in most cases, RL takes a relatively long period to explore the environment before obtaining an excellent control policy, which may lead to an increase in cost. To reduce the unnecessary waste caused by RL methods in exploration, we extended the deep forest-based deep Q-network (DF-DQN) from the prediction problem to the control problem, optimizing the running frequency of the cooling water pump and cooling tower in the cooling water system. In DF-DQN, it uses the historical data or expert experience as a priori knowledge to train a deep forest (DF) classifier, and then combines the output of DQN to attain the control frequency, where DF can map the original action space of DQN to a smaller one, so DF-DQN converges faster and has a better energy-saving effect than DQN in the early stage. In order to verify the performance of DF-DQN, we constructed a cooling water system model based on historical data. The experimental results show that DF-DQN can realize energy savings from the first year, while DQN realized savings from the third year. DF-DQN's energy-saving effect is much better than DQN in the early stage, and it also has a good performance in the latter stage. In 20 years, DF-DQN can improve the energy-saving effect by 11.035% on average every year, DQN can improve by 7.972%, and the model-based control method can improve by 13.755%. Compared with traditional RL methods, DF-DQN can avoid unnecessary waste caused by exploration in the early stage and has a good performance in general, which indicates that DF-DQN is more suitable for engineering practice.

Keywords: HVAC; cooling water system; reinforcement learning; DF-DQN

Citation: Han, Z.; Fu, Q.; Chen, J.; Wang, Y.; Lu, Y.; Wu, H.; Gui, H. Deep Forest-Based DQN for Cooling Water System Energy Saving Control in HVAC. *Buildings* **2022**, *12*, 1787. https://doi.org/10.3390/buildings12111787

Academic Editors: Shi-Jie Cao, Dahai Qi, Junqi Wang and Gwanggil Jeon

Received: 21 August 2022
Accepted: 17 October 2022
Published: 25 October 2022

Publisher's Note: MDPI stays neutral with regard to jurisdictional claims in published maps and institutional affiliations.

1. Introduction

In order to achieve the goal of carbon neutrality, countries around the world are committed to energy saving and emission reduction. Building energy consumption accounts for a large part of energy consumption around the world [1], and heating, ventilation, and air-conditioning (HVAC) systems occupy a major part, reaching more than half of energy consumption. The cooling water system is an essential subsystem of the HVAC system, which mainly consists of cooling water pumps, cooling towers, and chiller condensers [2]. The operation of the cooling water system has an important influence on the entire HVAC system, and optimal control of the cooling water system can effectively reduce energy consumption of the HVAC system. Thus, the optimal control of the cooling water system is crucial.

In HVAC systems, optimal control policies are often used to reduce operation costs and to ensure the thermal comfort of occupants [3,4]. Optimal control policies can be classified into traditional control policies and advanced control policies in intelligent buildings, where the former one contains sequencing control (rule-based control) and process control, and the latter one includes soft-computing control policies, hard-computing control policies, and hybrid control policies [5]. Many optimal control methods have been tried for cooling water system control, such as proportional-integral (PI) controllers, proportional integral derivative (PID) controllers, and model predictive control (MPC) controllers. These methods heavily rely on the system model, various sensors, and controllers in the system, so the disadvantages of these methods are also obvious. Model-based methods often require a perfect model of the system, while system modeling is usually difficult in real applications even if we can attain enough data from different sensors. According to Zhu et al., the uncertainties of the model have a serious impact on the control performance [6]. In the actual system operation, the aging of the equipment or the renewal of some equipment may lead to inconsistency between the system model and the actual system [7]. Even if the initially established model is accurate enough, continuous changes in the actual system over time lead to an unavoidable decrease in the performance of the control method.

To avoid the impact of the imperfect system model on control policies, data-driving methods in artificial intelligence have received too much attention in HVAC control problems recently. Reinforcement learning (RL) is a kind of classical data-driven and model-free method in artificial intelligence. In recent years, RL has attracted increasing attention for building energy efficiency control problems [8,9], because it can provide a simple framework by learning from interaction with the environment directly. In these studies, RL methods can provide a model-free framework for achieving energy saving, but they often fail to achieve a good control effect in the early stage, or can be even worse than some baseline control policies, which are mainly caused by the agent's exploration of the environment. Moreover, in the exploration process by the trial-and-error mechanism, they may also cause a certain degree of damage to the equipment, which may directly lead to an increase in cost. These two problems severely limit the practical use of RL in the field of HVAC optimization applications. Therefore, in order to maintain RL control effectiveness and achieve the maximum possible energy savings, it is necessary to reduce the time of this process in some way so that the RL control policy converges more quickly to reduce unnecessary costs.

In this paper, we tried to use DF-DQN to tackle this problem. Due to the introduction of DF, we mapped the original action space to a smaller one, and then combined the label of DF to attain the final control action, which directly reduced the output action of DQN. Moreover, the label of DF had the guidance of a priori knowledge, which not only ensured a good control effect, but also can realize energy saving in the early stage. The main contributions of this paper are as follows:

- We extended our previously proposed DF-DQN from the prediction problem to the control problem. The introduction of DF mapped the original DQN output action space into a new smaller action space, which could accelerate the convergence speed of DQN;
- We used DF-DQN to control a cooling water system in HVAC and to realize energy savings from the early stage. A priori knowledge was introduced as a deep forest classifier, which can not only reduce the action space, but also reduce the exploration of the agent. The experimental results show that DF-DQN can save energy from the first year, while DQN can achieve similar energy saving from the third year;
- We verified the performance of DF-DQN in an environment based on the modeling of a real cooling water system, so as to ensure the credibility of DF-DQN. The data that DF-DQN and other compared methods used were collected from a real-world system, and the simulation environment was built based on this system. The code and the experimental data are available at: https://github.com/H-Phoebe/DF-DQN-for-energy-saving-control (accessed on 20 August 2022).

2. Related Works

In recent years, more and more researchers have tried to solve practical problems with RL methods. In the applications of the HVAC system, the complexity and lag of the HVAC system directly lead to an increase in modeling cost, while RL can provide model-free control and have good control performance. However, RL generally takes a relatively long time to learn a better control policy, and this process may lead to some unnecessary energy wastage and cost increases, so some researchers try to avoid this wastage by speeding up the convergence of RL algorithms, which can achieve more energy saving at the same time.

Applications of RL in HVAC. Lork et al. [10] used Q-learning to achieve a balance between comfort and energy savings in rooms. They used a Bayesian convolutional neural network combined with data from all rooms to construct a temperature and air conditioning power prediction model to reduce uncertainty. This model was then adapted to individual rooms and the temperature set point was controlled using Q-learning. Qiu et al. [8] used Q-learning to obtain optimal control of the cooling water system in HVAC, wherein the RL controller can save 11% of the system energy, more than the 7% saved by the local feedback controller. Ahn et al. [11] used DQN to achieve a model-free optimal control policy in HVAC and the results proved that DQN can reduce energy consumption, and provided model-free optimal control. Brandi et al. [12] used DQN to control the water supply temperature set point of the heating system terminal unit, which can achieve a heating energy saving ranging between 5% and 12%. Yan et al. [13] applied DDPG to generate an optimal control policy for a multi-zone residential HVAC system, which can greatly reduce energy consumption while ensuring comfort. In addition, the DDPG-trained agent can intelligently balance different optimization objectives with generalization ability and adaptability to unknown environments. Ding et al. [14] used RL algorithms to control the indoor temperature of a residential HVAC system, which can achieve energy conservation while maintaining indoor thermal comfort. Qiu et al. [15] used three multi-agent RL algorithms to control the condenser system in HVAC. The experimental results showed that the interaction multi-agent RL algorithm can achieve better energy-saving effects compared to the other two algorithms. Amasyali et al. [16] used the deep RL controller to control the power cost of electric water heaters in residential buildings. The experimental results showed that this method does not cause discomfort to users, and can save 19–35% of the power cost compared with the baseline control.

Improve RL convergence speed. In engineering applications, the convergence time of RL methods may be several months or even years, which directly leads to an increase in cost. Therefore, some researchers have tried to shorten this time to reduce the cost of practical applications. Li et al. [17] controlled the HVAC system in order to control energy consumption and ensure comfort, and they put forward multi-grid Q-learning to solve the problem of slow convergence rate in RL. Yu [18] et al. developed an exploration policy for the RL controller using a priori knowledge, which can guide the RL controller to explore the action space, thus reducing the training time. Fu et al. [19] used a multi-agent RL to realize the collaborative control optimization of multiple devices in the HVAC system. The experimental results showed that the method converges faster than single-agent RL method. In [20], the authors mention that adding a priori knowledge can help the RL controller reduce training time.

3. Preliminaries

3.1. MDP

RL, as a class of control techniques in machine learning, has been explored for its potential in HVAC systems. In RL, the problem can often be considered as a sequential decision-making case, and the agent can learn by interacting with the environment directly, as shown in Figure 1.

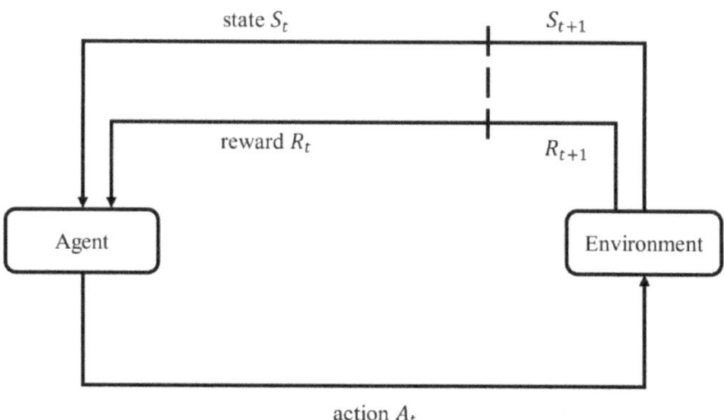

Figure 1. The interaction progress between agents and environments in RL.

Markov's decision process (MDP), a classical formalization of sequential decision-making is often used to model RL problems. An MDP can be defined as a tuple $\langle S, A, T, R, \gamma \rangle$, where S is the collection of states, A is the collection of actions, T is a transition function, R is an immediate reward function, and γ is a discount factor. In the interactive process, for some states, the agent can select an action to act on the environment, and receive a scalar reward and a next state [21]. The final goal of RL is to maximize a cumulative numerical reward R_t, as is shown in Equation (1).

$$R_t = \sum_{k=0}^{\infty} \gamma^k r_{t+k+1} \tag{1}$$

where $\gamma \in [0, 1]$ is the discount factor, k represents k time steps after time step t, and r_{t+k+1} represents the immediate reward of the corresponding time step. The agent selects an action $a \in A$ by policy π, then the agent moves to the next state s_{t+1}, then the agent obtains the immediate reward r_{t+1} from the environment. In RL, the action value Q is used to represent the exception of a cumulative discounted reward, which is starting from state s and taking action a. The action value Q is shown in Equation (2).

$$Q_\pi(s,a) = E_\pi \left[\sum_{k=0}^{\infty} \gamma^k R_{t+k+1} | s_t = s, a_t = a \right] \tag{2}$$

The optimal policy π_* can be achieved by evaluating the action value function:

$$Q_*(s,a) = \max Q_\pi(s,a) = E \left[R_{t+1} + \gamma \max_{a'} Q_* (s_{t+1}, a') | s_t = s, a_t = a \right] \tag{3}$$

Finally, the optimal policy π_* can be obtained.

3.2. Deep Forest

Deep forest (DF) [22] is a decision tree ensemble approach and can be applied to classification tasks. DF can obtain good performance in most cases, even with different data in different domains, which mainly benefits from two techniques, namely multi-grained scanning and the cascade forest structure.

Multi-grained scanning uses sliding windows of various sizes for sampling to obtain more feature sub-samples, so as to obtain more and richer feature relationships. Then, a certain amount of the random forest and cascade forest are trained with the obtained feature sub-samples to obtain the feature vector.

The cascade forest structure is used to enhance the representation learning ability of DF. In a cascade forest, each level receives the characteristic information processed by the previous level, then the processing results in inputs, which are then output to the next level. The first level's input of a cascade forest is the feature vector after multi-granularity scanning transformation. The final prediction result is obtained at the last level and expressed as an aggregate value.

In addition, the training process of the deep forest is efficient, and it can operate normally even if the training data scale is small. The structure of DF is shown in Figure 2.

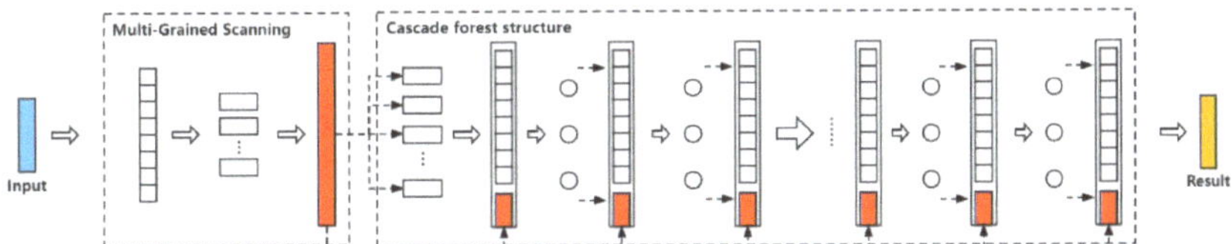

Figure 2. Structure of deep forest.

3.3. DQN

Traditional methods in RL, such as SARSA and Q-learning, can effectively solve problems with a small state and action space by establishing Q-table. However, when the state space is large enough or continuous, such as practical problems in HVAC, these methods may fail to achieve a control policy. DQN, a method proposed by Google's DeepMind in 2015 [23], has been applied in HVAC controls in recent years. Different from SARSA and Q-learning, DQN can solve problems with large or continuous state space [24], mainly benefiting from its two specific techniques.

Firstly, DQN uses the mechanism of experience replay to eliminate the correlation of network inputs. This means storing the transfer samples (s, a, r, s') while the agent interacts with the environment and samples randomly to train the agent. Secondly, there are two networks in DQN, where one is the Q-network, and the other is the target network. These two networks have the same structure, but have different parameters. The Q-network outputs the current Q value, and the target network outputs the target Q value. After some iterations, the parameters of the Q-network are copied to the target network. The loss function is shown in Equation (4).

$$L(\theta_i) = E\left[\left(r + \gamma \max_{a'} Q(s', a' \,|\theta_i^-) - Q(s, a \,|\theta_i)\right)^2\right] \tag{4}$$

where a' is the action selected in state s', and θ_i and θ_i^- are the parameters of Q-network and target network, respectively.

4. Environment and Modeling

4.1. Cooling Water System Layout

In this paper, we tried to control the cooling water system to reduce the energy consumption of the HVAC system. The cooling water system is an important part of HVAC, including chillers, cooling water pumps, cooling towers, and some other necessary equipment. To achieve the goal of energy saving, it is important to enable this equipment to be controlled more efficiently. In another word, we should try to find an optimal policy to coordinate this equipment. Based on a real application, we constructed a cooling water system platform, which contained four chillers, three cooling water pumps, and seven cooling towers (the same type of equipment has the same settings), as shown in Figure 3.

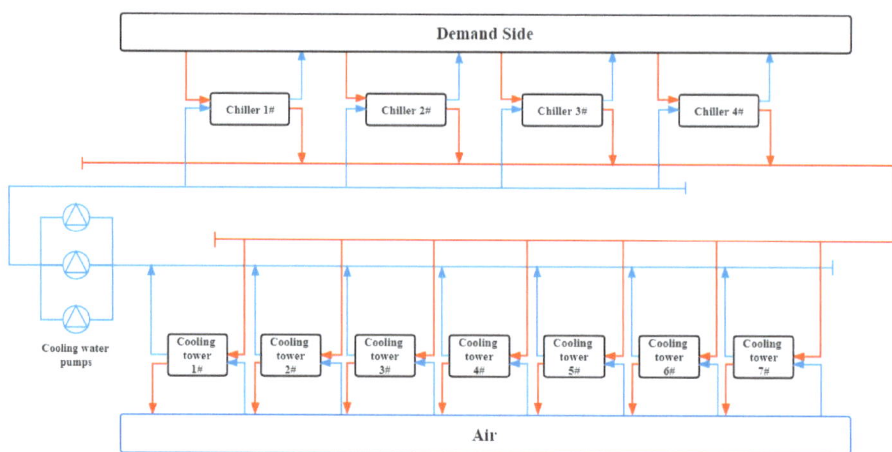

Figure 3. The layout of the cooling water system.

To measure the effect of the control policy, we adopted the system coefficient of performance (COP), which is often used to measure the energy-saving performance of HVAC systems. The system COP is defined in Equation (5).

$$COP = \frac{CL_{system}}{\sum P_{chillers} + \sum P_{towers} + \sum P_{pumps}} \tag{5}$$

where $\sum P_{chillers}$ is the total power of all chillers (kW), $\sum P_{towers}$ is the total power of all cooling towers (kW), and $\sum P_{pumps}$ is the total power of all cooling water pumps (kW). CL_{system} is the system cooling load, which is defined in Equation (6).

$$CL_{system} = C_p \times \rho \times F_{chw} \times (T_{chwr} - T_{chws}) \div 3600 \text{ s/h} \tag{6}$$

where C_p is the specific heat capacity of water (4.2 kJ/(kg·K)), ρ is the water density (1000 kg/m^3), F_{chw} is the chilled water flowrate (m^3/h), T_{chwr} is the inlet chilled water temperature of chillers (°C), and T_{chws} is the outlet chilled water temperature of chillers (°C).

4.2. System Simulation Modeling

For the system simulation, some real data and parameters were collected, but some others could not be achieved directly, so we tried to use the regression method to attain them.

We regressed the chiller model with historical data, which could be used to attain the chiller's COP, and further, we calculated the chiller's power, as shown in Equation (7).

$$P_{chiller} = CL \, / \, COP_{chiller} \tag{7}$$

where $COP_{chiller}$ is obtained by Equation (8).

$$COP_{chiller} = chiller \; model(CL, \, T_{cwr}, \, T_{chws}, \, F_{chw}) \tag{8}$$

where T_{cwr} is the inlet cooling water temperature of chillers (°C). Some other related parameters are shown in Equations (9)–(11) [8].

$$T_{cws} = T_{cwr} + (P_{chiller} + CL) \div \frac{C_p \times F_{cw} \times \rho}{3600 \text{ s/h}} \tag{9}$$

$$T_{chws} = \max\left[T_{chwsset}, \, T'_{chwr} - CC \div \frac{C_p \times F_{chw} \times \rho}{3600 \text{ s/h}} \right] \tag{10}$$

$$T_{chwr} = T_{chws} + CL \div \frac{C_p \times F_{chw} \times \rho}{3600 \text{ s/h}} \tag{11}$$

where $T_{chws_{set}}$ is the T_{chws} set point of a chiller, T'_{chws} is the T_{chws} of last time step, F_{cw} is the cooling water flowrate (m^3/h), and CC is the chiller cooling capacity.

The power of the cooling water pump model is calculated by Equations (12) and (13).

$$K = \frac{f_{pump_{actual}}}{f_{pump_{rated}}} \tag{12}$$

$$P_{pump} = a + b \times K + c \times K^2 + d \times K^3 \tag{13}$$

where $f_{pump_{actual}}$ and $f_{pump_{rated}}$ are the actual running frequency and rated running frequency of the real cooling water pump, and a, b, c, d are determined by the regression of historical data.

The cooling tower model is defined as Equations (14) and (15).

$$P_{tower} = a + b \times f_{tower_{actual}} + c \times f_{tower_{actual}}^2 + d \times f_{tower_{actual}}^3 \tag{14}$$

$$T_{cwr} = tower\ model\left(T_{cws},\ f_{tower_{actual}},\ T_{wb},\ F_{cw}\right) \tag{15}$$

In Equation (14), $f_{tower_{actual}}$ is the actual running frequency of cooling tower, a, b, c, d are determined by the regression of historical data. In Equation (15), T_{cwr} is the inlet cooling water temperature of chillers (°C), T_{cws} is the outlet cooling water temperature of chillers (°C), T_{wb} is ambient wet-bulb temperature (°C), and F_{cw} is the cooling water flowrate (m^3/h).

For each model, we randomly selected 80% of the collected data set for training and 20% of the data for testing, using MAPE (mean absolute percentage error) and CVRMSE (the coefficient of variation of the root mean square error) as the error metrics to evaluate the accuracy of the models. All models had a MAPE of less than 5% and CVRMSE of less than 10%, which indicates that the accuracy of each model was within the acceptable range.

The controller controlled the on and off states of this equipment and the operating frequency. An iterative process was as follows: Firstly, T_{cws} was obtained from the CL and the switching state of the chiller model; then, F_{cw} was obtained by combining the operating frequency and T_{cws}; finally, T_{cwr} was obtained by combining the cooling tower model with F_{cw} and the T_{wet}. All the parameters were iterated until T_{cwr} converged (i.e., the difference of T_{cwr} between two successive iterations was less than 0.1 °C). If the T_{cwr} did not converge within 50 iterations, the last result of T_{cwr} was adopted and the iteration was stopped [8]. The specific process is shown in Figure 4.

4.3. Data Collection

We used the data collected from the actual system to verify our proposed method. The details of this actual system corresponded with our simulation environment.

We collected real data from 1 July 2021 to 10 October 2021, 102 days in total, where the sample interval was half an hour. The CL is shown in Figure 5, and the wet-bulb temperature is shown in Figure 6.

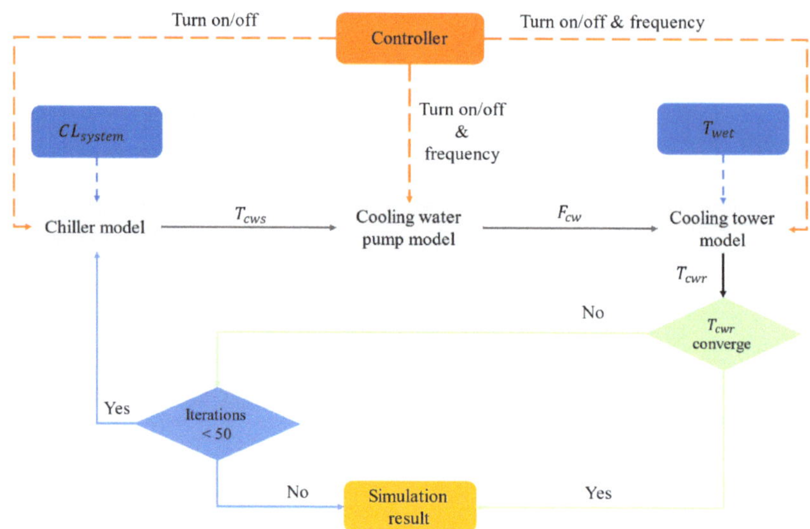

Figure 4. Simulation process.

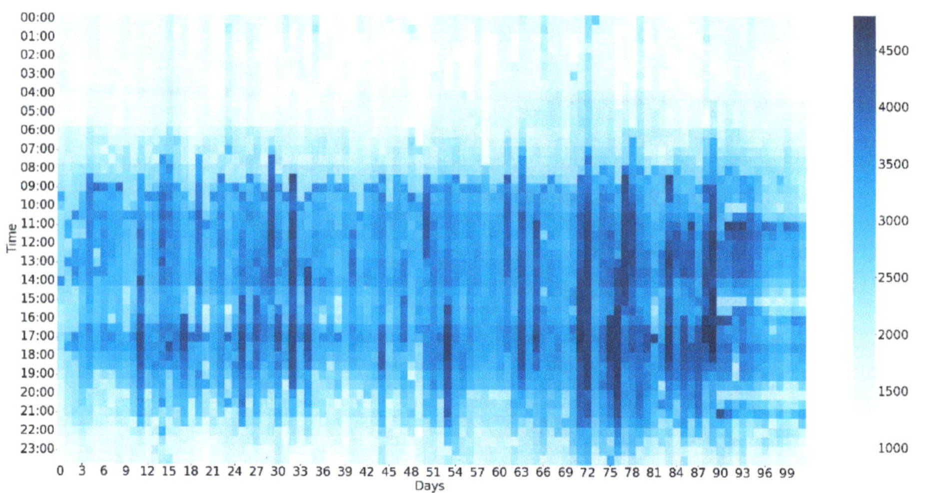

Figure 5. The temporal distribution of the cooling load. The deeper the color, the heavier the load.

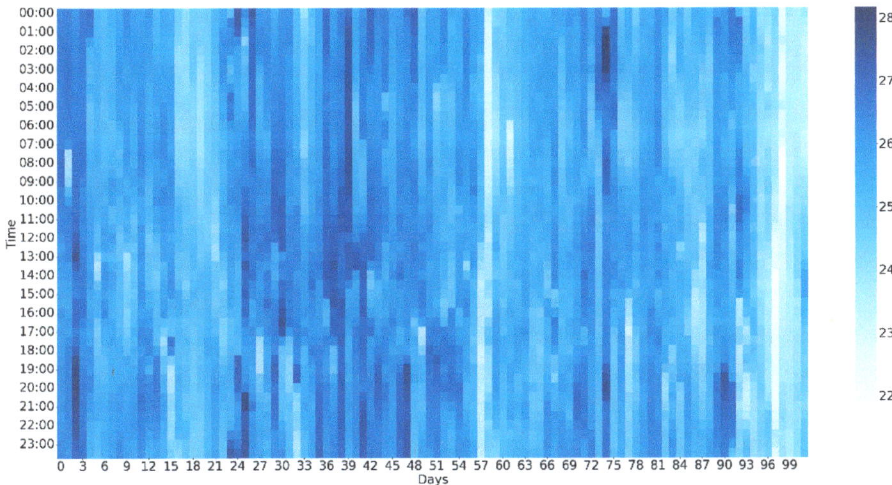

Figure 6. The temporal distribution of the wet-bulb temperature. The deeper the color, the higher the temperature.

As shown in Figures 5 and 6, the darker the color, the greater the value of CL and the wet-bulb temperature (T_{wet}). From Figure 5, we can find that the main cooling demand of the system was concentrated between 6:00 and 23:00 every day.

5. Methodology

5.1. MDP Modeling

Using RL methods for control problem requires MDP modeling of the environment. The details of the modeling are represented as follows:

(a) State

In this paper, we took the combination of ambient the wet-bulb temperature (T_{wet}) and system cooling load (CL_{system}) as state. There were two reasons for using these two variables:

(1) The operation of the system has no influence of these variables;

(2) CL_{system} is a component factor of COP, which is related to the operation of cooling water system.

(b) Action

In this paper, operating frequencies of cooling tower fans and cooling water pumps were taken as the action (e.g., [*pump_action* : 35 hz, *tower_action* : 35 hz]). In addition, the action was discretized and the control accuracy was 1 hz. In order to protect the equipment, the action needed to be limited within a reasonable range. We limited the action frequency within [20, 50] for both the cooling tower and cooling water pump, so there were 31 actions in total for each one.

(c) Reward

COP was taken as the reward in this paper. In the case of the same $CL_{sysytem}$, the higher the COP value is, the sum of power is the lowest, which reflects the purpose of energy saving. The reward is shown in Equation (16).

$$Reward = COP = \frac{CL_{system}}{\sum P_{chillers} + \sum P_{towers} + \sum P_{pumps}} \quad (16)$$

5.2. DF-DQN for Control

Figure 7 depicts the overall framework of DF-DQN for control in cooling water system. Firstly, we labeled the collected state data, including cooling load and wet-bulb temperature, according to the a priori knowledge. The label was the running frequency of the equipment more or less than the value of the base number under this state. If the operating frequency of the equipment under this state was less than base number, the label was '0'; otherwise, the label was '1'. The labeled state data were used to train the deep forest classification model, of which 80% was used for training and 20% was used to test the accuracy of the trained model.

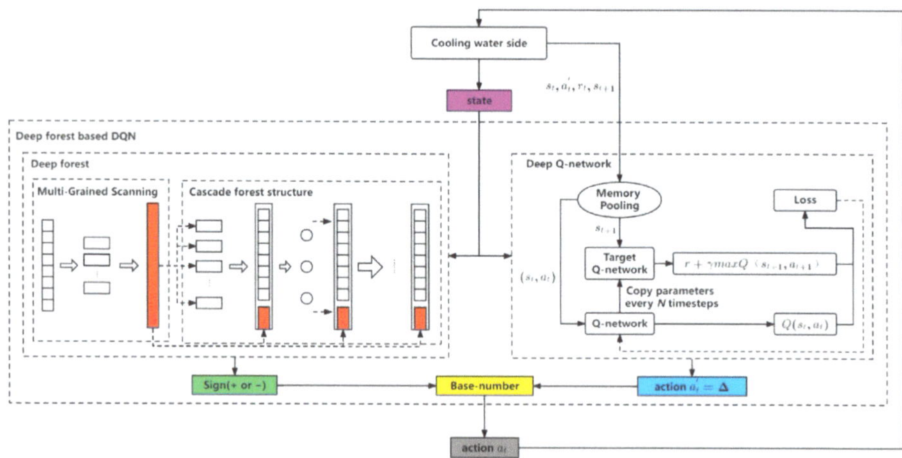

Figure 7. Overall framework of DF-DQN for cooling water system.

After training, the DF classification model can output a label for the new state, which represents the relationship between the actual frequency of equipment operation and the base number, and this label can be converted into a sign to shrink the action space thereafter. Figure 8 gives more details.

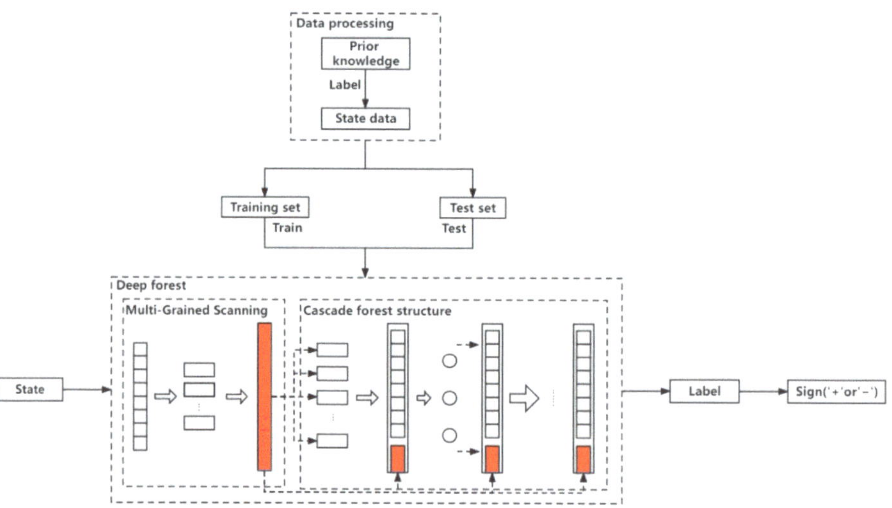

Figure 8. The process of training a DF classifier.

Secondly, we obtained the relationship between the actual frequency of the equipment operation and base number under some states, namely the sign. We also needed the difference between the operating frequency of the actual equipment and the base number.

In this part, we trained DQN agent, which output an action a'_t, the absolute value of the difference between the true action and base number, namely Δ.

At last, according to the actual data we collected, the DF classifier output a positive sign or negative sign ('+' & '−'), and DQN output Δ. Based on sign, Δ, and base number, we could obtain the actual equipment running frequency, namely action a_t. The actual action is calculated according to Equations (17) and (18).

$$Sign = DF_{classifier}(state) \tag{17}$$

$$a_t = base_{number} + Sign(a'_t) \tag{18}$$

where *sign* is output by DF, a'_t is output by DQN in DF-DQN.

5.3. Theoretical Analysis of Shrink Action

The DF classifier labeled each state. It could replace the original action space with a smaller action space combined with the label, so as to realize the reduction of the action space.

Owing to the introduction of DF in DQN, the original action space of each equipment was reduced from 31 to 16, so the original combined action pace was reduced from 31×31 to 16×16. Therefore, the action space of each equipment was reduced by nearly half, while the combined action of the two equipment reduced the action space by nearly 3/4 with the increase in equipment types. The introduction of DF could make the combined action space decrease exponentially. Figure 9 presents more details.

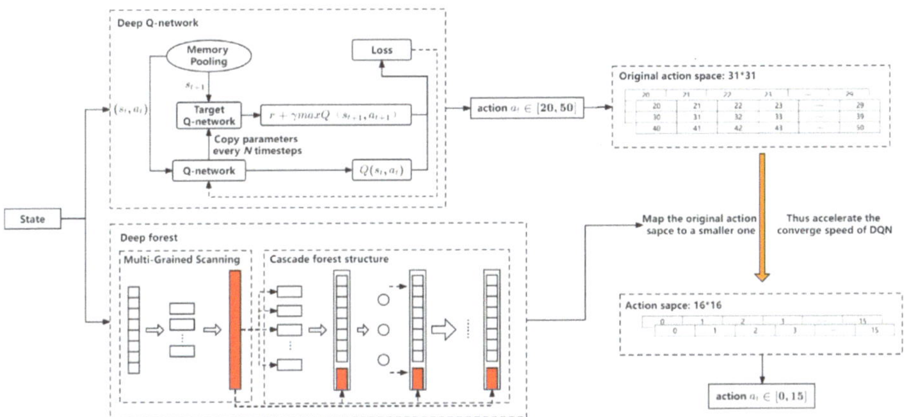

Figure 9. DF reduce action space.

Theoretically, if DF can divide each action into M categories with the same number, and the final combined actions include N kinds, the reduced action space of DF-DQN can follow Equation (19).

$$\frac{(Action\ space)_{DF-DQN}}{Original\ action\ space} \approx \left(\frac{1}{M}\right)^N \tag{19}$$

Based on the above analysis, when dealing with the problem with large action space or multiple action combinations, DF can significantly shrink the scale of the original action space, which can reduce the complexity of the problem to a certain extent finally.

In this paper, $M = 2$, $N = 2$, so the combined action of the two equipment was shrunk into about 1/4 of the original action space.

5.4. DF-DQN Algorithm

The details of DF-DQN for the cooling water system is shown in Algorithm 1.

Algorithm 1. DF-DQN for cooling water system

Initialize replay memory D to capacity N
Initialize action value function Q with random weights θ
Detect and replace outliers in training set
Split the training set (80% for training, 20% for testing)
Train the deep forest classifier F
For episode = 1, M do
    Attain initial state s_t of the cooling water system
    For t = 1, T do
        Select a random action a'_t with probability ε, otherwise $a'_t = \max Q(s_t, a; \theta)$
        Attain positive or negative sign through F
        Combine base number, sign ('+' or '−'), a'_t to derive a_t (the true running frequency of cooling
water system)
        Execute action a_t in cooling water system
        Observe reward r_t and state s_{t+1} from the simulation system
        Store transition $(s_t, a'_t, r_t, s_{t+1})$ in D
        Sample random minibatch of transitions $(s_j, a'_j, r_j, s_{j+1})$ from D

$$\text{Set } y_j = \begin{cases} r_j & for\ terminal\ state\ s_{t+1} \\ r_j + \gamma \max_{a'} Q\left(s_{j+1}, a'; \theta\right) & otherwise \end{cases}$$

        Update Q function using $\left(y_i - Q\left(s_j, a_j; \theta\right)\right)^2$
        Copy parameters every J steps
        Update state $s_t \leftarrow s_{t+1}$
    End for
End for

---

6. Experiment and Result

To verify the performance of DF-DQN, we compared it with three other benchmark methods. In addition, we presented some experiments about the effect of DF accuracy on the performance of DF-DQN.

6.1. Compare Methods

1. DF-DQN: DF-DQN is the method we proposed before [25], which has been used to solve prediction problem. We extended DF-DQN to control problems in this paper;
2. DQN: In this paper, DQN and DF-DQN share the same parameter settings in the DQN part. For the cooling water system, the action space was small and discrete, and its state space was large enough, so usually DQN can provide a good control policy according to paper [24];
3. Baseline control: The PID control was selected as the baseline method, which is often used in real HVAC control applications. This method selects the action by approaching the difference between T_{cws} and T_{cwr}. We took the baseline control method for comparison because it is the original control method in this system;
4. Model-based control: The model-based control is the best method among all methods, and can select the best action in each situation, but this method is heavily dependent on the model. In this paper, we traversed the best action in each state as the model-based control. Actually, it is often impossible to deploy the model-based control method in real applications, but in this paper, based on our simulation model, the model-based control method provided the best policy. We used the model-based control method for comparison because it has the best control performance of all methods in this system.

6.2. Parameters Setting

We used DF-DQN and three other methods to control this system for comparison, including DQN, a baseline control, and a model-based control.

The agent in DF-DQN took the $\varepsilon - greedy$ policy to select the action. At the beginning, we ensured the agent could explore the environment as much as possible, so we set $\varepsilon_{inital} = 1$. We used a liner decay during the process, and set $\Delta\varepsilon = 0.0001$, $\varepsilon_{min} = 0.01$. In order to make the agent take more focus on the current COP, we set $\gamma = 0.01$. The agent's policy network and target network were both composed of two hidden layers. The minibatch was set to 32. The capacity of memory pooling ($Memory_{capacity}$) was set to 1000, and we set C_{step} (Copy steps) to 100, $C_{step} = 100$. The learning rate was set to 0.01, $\alpha = 0.01$. All parameters are shown in Table 1.

Table 1. Parameters of DF-DQN and DQN.

Parameters	Value
ε_{inital}	1
$\Delta\varepsilon$	0.0001
ε_{min}	0.01
γ	0.01
$Memory_{capacity}$	1000
C_{step}	100
α	0.01

In this system, the equipment contained chillers, cooling water pumps, and cooling towers. We used RL to control the cooling water pumps and cooling towers. As for the chillers, we used a sequence control [26] to reduce unnecessary refrigerating capacity, which can protect chillers at the same time. The workflow of this system is shown in Appendix A.

6.3. Experimental Result

In this paper, we used the model-based control to attain the best action in each state, so that we could attain the label of the cooling water pump and cooling tower's action under each state. We used DF for two classifications to judge whether the frequency of the cooling water pump and cooling tower was more or less than the base number in each state. If it was more than the base number, we labeled it as 1 (represent '+'); otherwise, we labeled it as 0 (represent '−'). The accuracy of DF can reach 97.319% and 99.694%. DF-DQN combined DF and DQN, where DF output sign '+' or '−', and DQN of DF-DQN output Δ, and then we combined them with base number to attain the final action of the cooling water pump and cooling tower.

Cumulative reward in an episode was taken to prove the convergence of DF-DQN. With the increase in episode, when the value of cumulative reward fluctuated less, we believe that the method converged. One of the comparison methods, DQN, also used the same method. The reward was defined by Equation (16), namely COP, and the higher reward not only conveyed that it had better converge, but also represented that the method had better energy-saving performance.

We explain the experimental results from two aspects: one is the influence of DF's accuracy on DF-DQN, and the other is control performance of DF-DQN.

6.3.1. Influence of DF's Accuracy on DF-DQN

The accuracy of DF affected the performance of DF-DQN. In order to better explain the influence of DF accuracy on the performance of DF-DQN, we made a test with a low accuracy case. We used DF-DQN (false label) and DF-DQN to control the system for 20 years in our simulation environment. We randomly generated labels to replace the original labels that DF generated, so that we could analyze the impact of DF accuracy on the performance of DF-DQN. The accuracy of the randomly generated labels was 50% of the original labels. We compared the experimental results of DF-DQN (false label) with the DF-DQN from three aspects: COP, cumulative power, and energy-saving effect.

Before comparison, we needed to ensure that both methods could converge. The convergence of the two methods is shown in Figure 10, where one episode in the training process is one year.

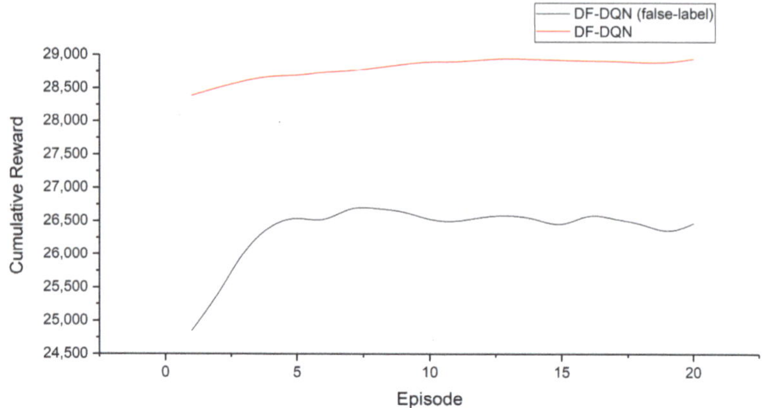

Figure 10. Comparison of cumulative reward between DF-DQN (false label) and DF-DQN.

DF-DQN (false label) and DF-DQN both converged at last. Although the wrong label was used, DQN still learned the control policy under the wrong labels and converged. However, the cumulative reward of DF-DQN was higher than that of DF-DQN (false label) on the whole, which also reflected the better performance of DF-DQN. In addition, the performance of DF-DQN (false label) decreased a lot due to the false labels.

As shown in Figure 11, the COP of DF-DQN (false label) was lower than that of DF-DQN in 20 years, which indicates that the control performance of DF-DQN (false label) was worse than that of DF-DQN in each year, and the energy-saving effect decreased accordingly, which also can be found in the cumulative power comparison in Figure 12.

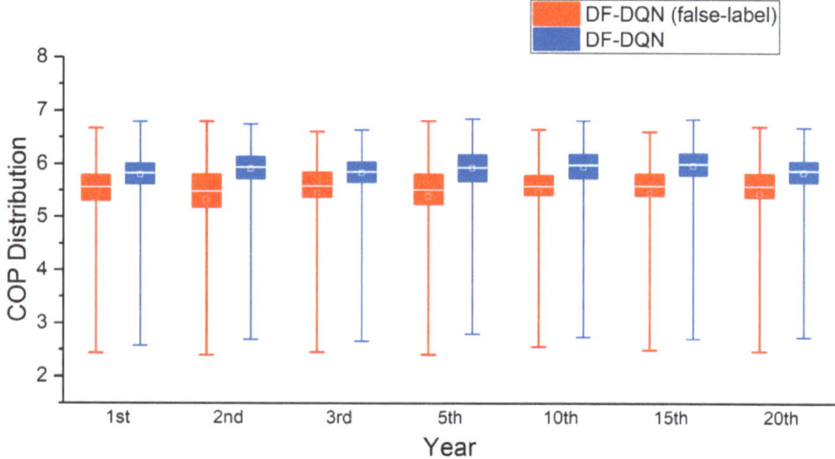

Figure 11. Comparison of COP between DQN and DF-DQN.

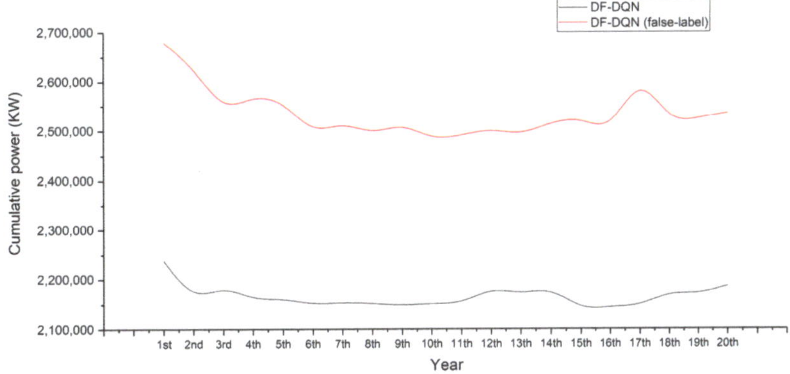

Figure 12. Comparison of cumulative power between DF-DQN and DF-DQN (false label).

We compared the energy-saving effect of these two methods, and used the baseline control method of the system as a benchmark. The partial energy-saving effect comparison can be found in Table 2.

Table 2. Partial energy-saving effect comparison result.

	Energy Saving (Compared to Baseline Control)	
Year	**DF-DQN**	**DF-DQN (False Label)**
1st	8.074%	−10.022%
2nd	11.337%	−8.037%
3rd	10.086%	−4.224%
5th	11.157%	−5.191%
10th	11.580%	−1.934%
15th	12.168%	−3.754%
20th	10.177%	−4.094%
Average (20 years)	11.035%	−4.104%

Regardless of the comparison of COP, the cumulative power, or the energy-saving effect, DF-DQN (false label) was worse than DF-DQN. The direct reason for this result was the wrong labels. From the comparison result, we found that the accuracy of DF directly affected the performance of DF-DQN, and the low accuracy of DF led to a decrease in the performance of DF-DQN. Therefore, for DF-DQN control in this problem, it was crucial to improve the accuracy of DF as much as possible.

6.3.2. Performance of DF-DQN Compared with DQN, Baseline Control, and Model-Based Control

DF-DQN and DQN both converged at last, but in the beginning episodes, DF-DQN achieved a higher cumulative reward. The difference between DF-DQN and DQN can be found in Figure 13 more clearly.

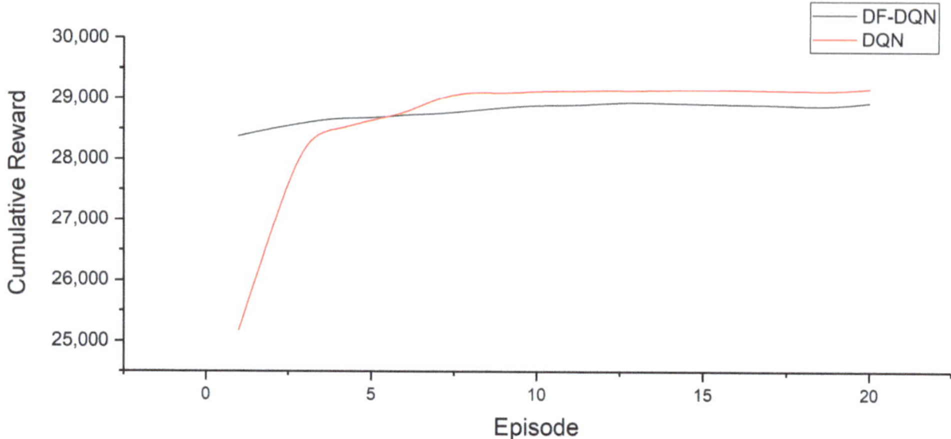

Figure 13. Comparison of cumulative reward between DQN and DF-DQN.

As shown in Figure 13, the cumulative reward of DF-DQN was much greater than that of DQN before the fifth episode, which reflects that the control performance of DF-DQN was much better than DQN in the early stage. After the fifth episode, DQN outperformed DF-DQN, which was due to the accuracy of DF. However, the performance of DF-DQN was almost approaching DQN.

In order to compare the performance of the baseline control method, DQN, DF-DQN, and the model-based control method, we used them to control the system for 20 years in our simulation environment. We compared their performance in three aspects: the COP, the cumulative power, and the energy-saving effect.

(a)   COP

The COP is shown in Figure 14. The model-based control method was the best method among these methods in theory, and its COP was the highest in practice. The baseline control method is a relatively poor control method compared with others, and its COP was the lowest in most of the years.

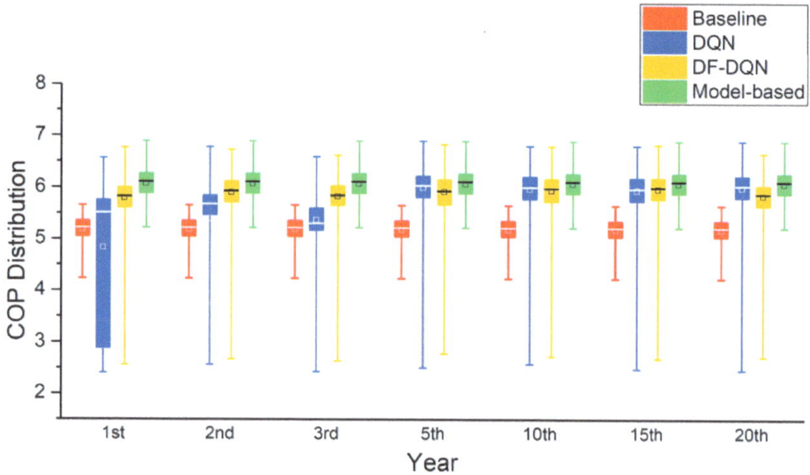

Figure 14. COP comparison of each method.

From Figure 14, we found that the COP obtained by DQN and DF-DQN was gradually becoming higher, indicating that their energy-saving effect was gradually becoming better, which is just as we have mentioned before, and the higher system COP, the better the energy-saving performance. The distribution of COP in the first year reflects that the control effect of DQN was much worse than that of DF-DQN, but it gradually became better in the later years. The COP reflected its better energy-saving performance to a certain extent, but not absolutely. In addition, the minimum COP obtained by DQN and DF-DQN was relatively small, which was due to the poorly selected actions in a few states.

As for DQN, its COP was less than the baseline control method in the first year, and the distribution of COP in the first year was also relatively scattered, but in the second year, the distribution of COP became concentrated, and its COP was more than the baseline control method, which meant that DQN's control performance became better in the second year. Finally, DQN's COP became stable, which means that the control policy of DQN was becoming stable and convergent.

In contrast with DQN, DF-DQN's COP was between the baseline control method and model-based control method from the first year and this trend remained in the following 20 years, which reflected that DF-DQN can obtain a better control effect from the beginning, and the control effect was better than DQN in the early stage. Moreover, the performance of DF-DQN was more stable than DQN in 20 years.

The performance of DQN was not good in the early stage, and its COP was not between the baseline control method and model-based control method until the policy converged. In contrast, DF-DQN met this condition not only after the convergence of the control policy, but also from the very beginning, which reflected that DF-DQN converged faster than DQN, and had a better performance than DQN in the early stage.

(b)   Cumulative power

In order to intuitively analyze the energy-saving effect of these four methods, we compared the annual cumulative power under these four methods' control policies, and the comparison results are shown in Figure 15.

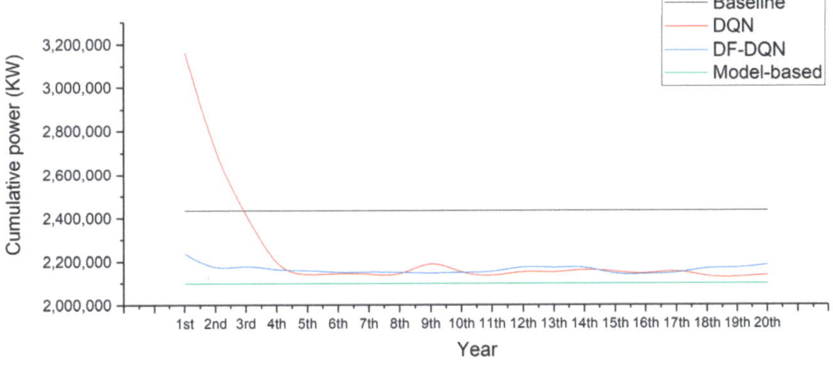

Figure 15. Cumulative power of each method in 20 years.

In Figure 15, the model-based control method had the lowest cumulative power, which is consistent with the intuition. The baseline control method had a relatively poor energy-saving effect, and its cumulative power was relatively more than others.

The magnitude of DQN's cumulative power was larger than that of the model-based control method and less than the baseline control method from the third year. In Figure 14, though DQN's COP was more than the baseline control method in the second year, its cumulative power was still more than the baseline control method. In Figure 15, the cumulative power of DQN in the second year was less than that in the first year, and it continued to decrease until the fourth year, and then remained stable. As we mentioned in the previous part, the lower COP did not absolutely mean a better energy-saving effect, but from the experimental results, we found that the second year's policy was better than the first year's, which is also be conveyed in Figure 15.

DF-DQN's cumulative power was much lower than the baseline control method from the first year, and this trend continued to decrease until the second year and then remained stable. In addition, DF-DQN's cumulative power was also much less than DQN's in the first three years, and after that, it almost approached DQN. It is obvious that DF-DQN can not only achieve a good energy-saving effect, but can also save energy from an early stage. Compared with DQN, the energy-saving effect of DF-DQN in the early stage was much better.

(c) Energy saving

Taking the baseline control method as the benchmark, we compared the other three methods' energy-saving effects in each year. The partial comparison results are represented in Table 3, and the complete comparison results are shown in Appendix B.

Table 3. Partial comparison effect results.

	Energy Saving (Compared to Baseline Control)		
Year	DQN	DF-DQN	Model-Based Control
1st	−29.996%	8.074%	13.755%
2nd	−9.843%	11.337%	13.755%
3rd	0.798%	10.086%	13.755%
5th	12.195%	11.157%	13.755%
10th	11.908%	11.580%	13.755%
15th	11.461%	12.168%	13.755%
20th	12.094%	10.177%	13.755%
Average (20 years)	7.972%	11.035%	13.755%

There is no doubt that the model-based control method had the best energy-saving effect, reaching 13.775%. The energy-saving effect of DQN and DF-DQN both had a growth process before the convergence.

According to the experimental results, DQN could not achieve the goal of energy saving until the third year. In particular, in the first year, its energy-saving effect was 29.996% worse than the baseline control method. In the second year, DQN's saving effect became much better than the first year, but was still 9.843% worse than the baseline control. Until the third year, DQN's saving effect was 0.798% better than the baseline control method, and began to remain stable from the fourth year, and was able to achieve a 10–12% energy-saving effect each year. DQN's energy-saving effect was not good in the early stage, but it became better and better with training. After 20 years control, its average energy-saving effect reaches 7.972%.

In contrast, DF-DQN could achieve the goal of energy saving from the first year, and remained with a 10–11% energy-saving effect. In the first year, it could achieve 8.074% better than the baseline control method, and kept becoming better in the second year, reaching 11.337%. After 20 years of the control, its average energy-saving effect reached about 11.035%.

DQN may have a better energy-saving effect in the later years, but it has to explore the environment before converging, which led to its worse performance in the early stage. Considering of the service life of the equipment, DF-DQN may have a better energy-saving effect than DQN in general, and our experimental results also proved this.

7. Conclusions and Future Work

In this paper, we extended DF-DQN from the prediction problem to the control problem, which was used to achieve the goal of energy saving with respect to the cooling water system control in HVAC. We compare its performance with DQN, baseline control method and the model-based control method. The experimental results show that since the a priori knowledge was introduced as a deep forest classifier, DF-DQN's action space could be mapped to a smaller one. DF-DQN did not need to spend a lot of on exploring the environment, so it converged much faster than DQN, which is the main reason that DF-DQN shows a better performance in the early stage compared to DQN. In the latter stage, the performance of DF-DQN was always slightly worse than DQN, and the reason is that the DF classifier may have output some wrong labels in a few states, which directly affected the result and DF-DQN's performance. Compared with the model-based control method, DF-DQN performed slightly worse in saving energy, but it did not require any complete system model, thus avoiding the unnecessary cost of modeling, which was valuable in the engineering practice.

DF-DQN had obvious energy-saving effects in the early stage and the overall energy-saving effect was also good, but its performance was directly affected by DF, which relied on historical data or expert experience. Thus, it is particularly important to train a DF classifier with excellent performance. DF-DQN has a good energy-saving effect in engineering applications, and is more practical than traditional RL methods, but it is not suitable for systems lacking historical data or expert experience. In addition, in this paper, we only considered two controllable equipment, but if more equipment need to be controlled, for example, more than 10 equipment, the performance of DF-DQN might decrease, which is limited by the DQN. Thus, for the future works, we will focus on following two aspects: (1) improving the accuracy of DF classifier or constructing a new classifier with higher accuracy, which could improve the final control performance in the current DF-DQN framework. (2) When more equipment of different types is involved, multi-agent reinforcement learning method can be adopted into the DF-DQN framework.

Author Contributions: Conceptualization, Z.H.; data curation, Z.H., Q.F. and Y.L.; formal analysis, Z.H. and Q.F.; funding acquisition, Q.F., J.C. and Y.W.; investigation, J.C., Y.W., Y.L., H.W. and H.G.; methodology, Z.H. and Q.F.; project administration, J.C.; software, Z.H. and Q.F.; supervision, Q.F., J.C., Y.W., Y.L., H.W. and H.G.; validation, Z.H. and Q.F.; writing—original draft, Z.H.; writing—review and editing, Z.H., Q.F., Y.W., Y.L., H.W. and H.G. All authors have read and agreed to the published version of the manuscript.

Funding: This work was financially supported by National Key R&D Program of China (No. 2020YFC2006602), National Natural Science Foundation of China (No. 62172324, No. 62072324, No. 61876217, No. 61876121), University Natural Science Foundation of Jiangsu Province (No. 21KJA520005), Primary Research and Development Plan of Jiangsu Province (No. BE2020026), Natural Science Foundation of Jiangsu Province (No. BK20190942).

Institutional Review Board Statement: Not applicable.

Informed Consent Statement: Not applicable.

Data Availability Statement: The experiment results are available at: https://github.com/H-Phoebe/DF-DQN-for-energy-saving-control (accessed on 20 August 2022).

Acknowledgments: The authors appreciate the support of Shunian Qiu.

Conflicts of Interest: The authors declare no conflict of interest.

Appendix A

The workflow of this system:

The workflow can be described as following steps:

A.   In time step t, the agent observes the state s_t, and decides to turn the system on or off according to CL. This process is shown in the right-hand part of Figure A1. The details of this process are shown below:

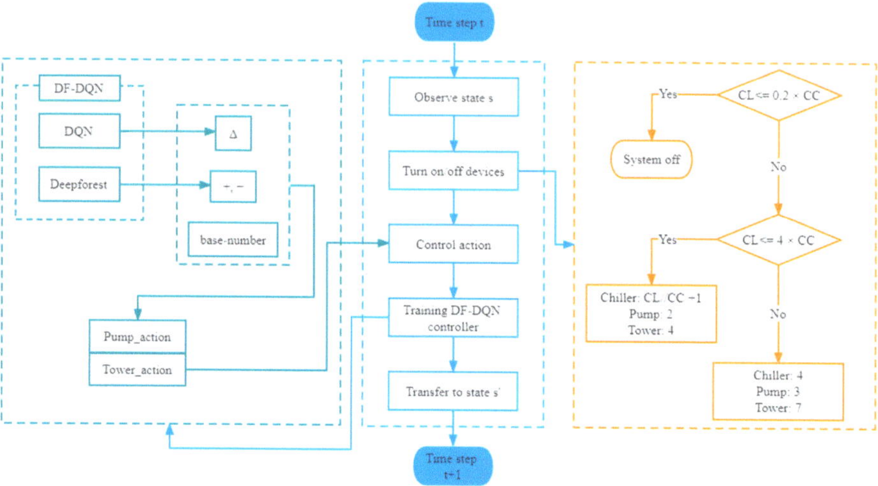

Figure A1. The workflow of system.

(1)   If CL is less than 20% of the chiller cooling capacity (CC) that one chiller can offer, the system will be turned off;

(2)   If CL is more than 20% of the rated refrigerating capacity that one chiller can offer and less than the refrigerating capacity that all the chillers can offer, namely $4 \times CC$, we will turn on the system, and the number of chillers is decided by the minimum x, which can make $x \times CC \geq CL$, and x is the number of chillers we turn on. x can be calculated by Equation (A1).

$$x = CL \; // \; CC + 1 \tag{A1}$$

where "$//$" represents exact division. No matter how many chillers we turn on, the CL assigned to each chiller is the same. As for cooling water pumps and the cooling towers, we turn on 2 and 4, respectively.

(3)   If CL is more than $4 \times CC$, we turn on all the chillers, cooling water pumps, and cooling towers, namely 4, 3, 7, respectively.

B.   We use the DF-DQN controller to control cooling water pumps and cooling towers, select the frequency of them, and combine them into an action ($pump\_action$, $tower\_action$) The system COP, reward in RL, can be observed after executing the action. The action is selected by $\varepsilon - greedy$ policy;

C.   Then we train our DF-DQN agent;

D.   Transfer to next state s_{t+1};

E.   End the current learning and move to step (A).

Appendix B

The energy-saving effects obtained by all methods in this paper are shown in Table A1.

Table A1. Energy-saving effect of each method compared with baseline control.

	Energy Saving (Compared to Baseline Control)			
Year	DQN	DF-DQN	DF-DQN (False Label)	Model-Based Control
1st	−29.996%	8.074%	−10.022%	13.755%
2nd	−9.843%	11.337%	−8.037%	13.755%
3rd	0.798%	10.086%	−4.224%	13.755%

Table A1. *Cont.*

	Energy Saving (Compared to Baseline Control)			
Year	**DQN**	**DF-DQN**	**DF-DQN (False Label)**	**Model-Based Control**
4th	11.362%	11.273%	−5.659%	13.755%
5th	12.195%	11.157%	−5.191%	13.755%
6th	11.752%	11.677%	−2.374%	13.755%
7th	11.879%	11.480%	−3.465%	13.755%
8th	12.503%	11.578%	−2.349%	13.755%
9th	8.957%	11.757%	−3.350%	13.755%
10th	11.908%	11.580%	−1.934%	13.755%
11th	12.440%	11.636%	−2.274%	13.755%
12th	11.195%	10.299%	−2.866%	13.755%
13th	11.763%	10.879%	−2.266%	13.755%
14th	10.893%	10.311%	−3.364%	13.755%
15th	11.461%	12.168%	−3.754%	13.755%
16th	12.042%	11.855%	−2.420%	13.755%
17th	10.967%	11.850%	−7.476%	13.755%
18th	12.583%	10.639%	−3.268%	13.755%
19th	12.490%	10.893%	−3.704%	13.755%
20th	12.094%	10.177%	−4.094%	13.755%
Average	7.972%	11.035%	−4.104%	13.755%

References

1. Cao, X.; Dai, X.; Liu, J. Building energy-consumption status worldwide and the state-of-the-art technologies for zero-energy buildings during the past decade. *Energy Build.* **2016**, *128*, 198–213. [CrossRef]
2. Taylor, S.T. *Fundamentals of Design and Control of Central Chilled-Water Plants*; ASHRAE Learning Institute: Atlanta, GA, USA, 2017.
3. Wang, S.; Ma, Z. Supervisory and optimal control of building HVAC systems: A review. *HVAC&R Res.* **2008**, *14*, 3–32. [CrossRef]
4. Wang, J.; Hou, J.; Chen, J.; Fu, Q.; Huang, G. Data mining approach for improving the optimal control of HVAC systems: An event-driven strategy. *J. Build. Eng.* **2021**, *39*, 102246. [CrossRef]
5. Gholamzadehmir, M.; Del Pero, C.; Buffa, S.; Fedrizzi, R.; Aste, N. Adaptive-predictive control strategy for HVAC systems in smart buildings–A review. *Sustain. Cities Soc.* **2020**, *63*, 102480. [CrossRef]
6. Zhu, N.; Shan, K.; Wang, S.; Sun, Y. An optimal control strategy with enhanced robustness for air-conditioning systems considering model and measurement uncertainties. *Energy Build.* **2013**, *67*, 540–550. [CrossRef]
7. Heo, Y.; Choudhary, R.; Augenbroe, G.A. Calibration of building energy models for retrofit analysis under uncertainty. *Energy Build.* **2012**, *47*, 550–560. [CrossRef]
8. Qiu, S.; Li, Z.; Li, Z.; Li, J.; Long, S.; Li, X. Model-free control method based on reinforcement learning for building cooling water systems: Validation by measured data-based simulation. *Energy Build.* **2020**, *218*, 110055. [CrossRef]
9. Claessens, B.J.; Vanhoudt, D.; Desmedt, J.; Ruelens, F. Model-free control of thermostatically controlled loads connected to a district heating network. *Energy Build.* **2018**, *159*, 1–10. [CrossRef]
10. Lork, C.; Li, W.T.; Qin, Y.; Zhou, Y.; Yuen, C.; Tushar, W.; Saha, T.K. An uncertainty-aware deep reinforcement learning framework for residential air conditioning energy management. *Appl. Energy* **2020**, *276*, 115426. [CrossRef]
11. Ahn, K.U.; Park, C.S. Application of deep Q-networks for model-free optimal control balancing between different HVAC systems. *Sci. Technol. Built Environ.* **2020**, *26*, 61–74. [CrossRef]
12. Brandi, S.; Piscitelli, M.S.; Martellacci, M.; Capozzoli, A. Deep reinforcement learning to optimise indoor temperature control and heating energy consumption in buildings. *Energy Build.* **2020**, *224*, 110225. [CrossRef]
13. Du, Y.; Zandi, H.; Kotevska, O.; Kurte, K.; Munk, J.; Amasyali, K.; Mckee, E.; Li, F. Intelligent multi-zone residential HVAC control strategy based on deep reinforcement learning. *Appl. Energy* **2021**, *281*, 116117. [CrossRef]
14. Ding, Z.; Fu, Q.; Chen, J.; Wu, H.; Lu, Y.; Hu, F. Energy-efficient control of thermal comfort in multi-zone residential HVAC via reinforcement learning. *Connect. Sci.* **2022**, *34*, 2364–2394. [CrossRef]
15. Qiu, S.; Li, Z.; Li, Z.; Wu, Q. Comparative Evaluation of Different Multi-Agent Reinforcement Learning Mechanisms in Condenser Water System Control. *Buildings* **2022**, *12*, 1092. [CrossRef]
16. Amasyali, K.; Munk, J.; Kurte, K.; Kuruganti, T.; Zandi, H. Deep reinforcement learning for autonomous water heater control. *Buildings* **2021**, *11*, 548. [CrossRef]
17. Li, B.; Xia, L. A multi-grid reinforcement learning method for energy conservation and comfort of HVAC in buildings. In Proceedings of the 2015 IEEE International Conference on Automation Science and Engineering (CASE), Gothenburg, Sweden, 24–28 August 2015; pp. 444–449. [CrossRef]

18. Yu, Z.; Yang, X.; Gao, F.; Huang, J.; Tu, R.; Cui, J. A Knowledge-based reinforcement learning control approach using deep Q network for cooling tower in HVAC systems. In Proceedings of the 2020 Chinese Automation Congress, CAC 2020, Shanghai, China, 6–8 November 2020; pp. 1721–1726. [CrossRef]
19. Fu, Q.; Chen, X.; Ma, S.; Fang, N.; Xing, B.; Chen, J. Optimal control method of HVAC based on multi-agent deep reinforcement learning. *Energy Build.* **2022**, *270*, 112284. [CrossRef]
20. Yang, L.; Nagy, Z.; Goffin, P.; Schlueter, A. Reinforcement learning for optimal control of low exergy buildings. *Appl. Energy* **2015**, *156*, 577–586. [CrossRef]
21. Sutton, R.; Barto, A. *Reinforcement Learning: An Introduction*, 2nd ed.; MIT Press: Cambridge, MA, USA; London, UK, 2018.
22. Zhou, Z.H.; Feng, J. Deep forest. *Natl. Sci. Rev.* **2019**, *6*, 74–86. [CrossRef]
23. Mnih, V.; Kavukcuoglu, K.; Silver, D.; Rusu, A.A.; Veness, J.; Bellemare, M.G.; Graves, A.; Riedmiller, M.; Fidjeland, A.K.; Ostrovski, G.; et al. Human-level control through deep reinforcement learning. *Nature* **2015**, *518*, 529–533. [CrossRef]
24. Fu, Q.; Han, Z.; Chen, J.; Lu, Y.; Wu, H.; Wang, Y. Applications of reinforcement learning for building energy efficiency control: A review. *J. Build. Eng.* **2022**, *50*, 104165. [CrossRef]
25. Fu, Q.; Li, K.; Chen, J.; Wang, J. A Novel Deep-forest-based DQN method for Building Energy Consumption Prediction. *Buildings* **2022**, *12*, 131. [CrossRef]
26. Li, Z.; Huang, G.; Sun, Y. Stochastic chiller sequencing control. *Energy Build.* **2014**, *84*, 203–213. [CrossRef]

Article

Machine Learning-Based Method for Detached Energy-Saving Residential Form Generation

Haixu Guo [1], Ding Duan [1], Jincheng Yan [1], Keyuan Ding [1], Fengkui Xiang [2] and Ran Peng [1,*]

[1] School of Civil Engineering and Architecture, Wuhan Institute of Technology, Wuhan 430074, China
[2] China Gezhouba Group Co., Ltd., Wuhan 430022, China
* Correspondence: pengran@wit.edu.cn

Citation: Guo, H.; Duan, D.; Yan, J.; Ding, K.; Xiang, F.; Peng, R. Machine Learning-Based Method for Detached Energy-Saving Residential Form Generation. *Buildings* **2022**, *12*, 1504. https://doi.org/10.3390/buildings12101504

Academic Editors: Shi-Jie Cao, Dahai Qi, Junqi Wang and Gwanggil Jeon

Received: 24 August 2022
Accepted: 17 September 2022
Published: 22 September 2022

Abstract: In recent years, machine learning has gradually been applied to building energy-saving designs to reduce the time consumption of the optimization screening stage. However, since most of the existing research scholars come from the fields of computers and engineering, the application of machine learning technology mostly involves complex programming as well as software in the field of engineering, which requires multiple software to be coupled to achieve. In view of the differences between disciplines and the high learning threshold, these theories are difficult to apply and promote in practical work in the field of architecture. In this regard, this paper focuses on the improvement of methods, based on the Grasshopper platform, proposes a detached energy-saving residential form generation design method and process, to explore the optimal energy-saving building form in a more concise and efficient way. Based on this new method, on the basis of verifying its feasibility through a residential building case, two machine learning algorithms, neural network (ANN) and support vector machine (SVM), are compared and studied, and the applicability of these two algorithms in different building performance indicators is further discussed. The results show that the ANN model has the highest accuracy and is more suitable for the prediction of building energy consumption; in view of the simple and fast operation of SVM, it is more suitable for comfort prediction with relatively low accuracy requirements. By combining the above two machine learning methods, work efficiency can be improved while satisfying the prediction of relevant performance indicators. This method can help architects quickly search for the best building energy-saving form design scheme in the scheme design stage and provide data support and information feedback for architects in design conception and deepening.

Keywords: detached house; energy efficient building; machine learning; multi-objective optimization; Grasshopper

1. Introduction

With the rapid growth of China's economy after the reform and opening up, residential buildings have flourished in the past decades and become one of the most important building types for national construction development and solving people's livelihood problems [1]. Residential buildings consume a large amount of primary energy, increase a large amount of carbon emissions, and become an important cause of global warming. Therefore, low-carbon residential construction is the most important task to achieve the goal of "double carbon".

In recent years, the multi-objective optimization method has been more widely used in the retrofitting and generation of energy-efficient buildings. Di Wu et al. [2] incorporated building cost into the optimization index and used the Design Builder software platform to optimize a comprehensive screening of building energy consumption and economics relying on the NSGA-II algorithm. Shao Teng et al. [3] took the regional climate as the entry point and used the particle swarm algorithm (PSO) and Hooke-Jeeves algorithm to work together with the coupling of multiple software such as GENOPT and Energy-Plus to carry out the design optimization study of the building envelope as a variable.

Zhang Yong et al. [4] incorporated comfort into the optimization index and proposed a decomposition-based multi-objective evolutionary optimization (MOEA/D) method for building energy efficiency design, which is comparatively better than the typical NSGA-II algorithm in terms of distributivity and convergence, and raised the issue of how to combine machine learning to reduce computational time consumption in the outlook. Yuan Gao et al. [5,6] relied on the Grasshopper platform to find the optimal design for winter heating energy consumption and comfort with building planning and single-unit parameters as variables, and incorporated the TOPSIS comprehensive evaluation method to filter the Pareto solution set. Chingwen Xue et al. [7] based on MATLAB platform with the help of C programming language, through the building performance of 1000 samples simulations and used neural networks (ANN) to establish mapping relationships between variables and performance parameters to speed up the convergence of the optimization screening phase.

Through combing the literature, in the study of multi-objective optimization of buildings, in order to achieve better convergence and distribution, most research focuses on the improvement of algorithms. However, since the optimization process requires the intervention of performance simulation software to simulate the corresponding building variables and then the performance parameters obtained are optimized and filtered, taking the Honeybee plug-in as an example, its simulation time is 1–20 min per time. In the process of multi-objective optimization, hundreds or thousands of performance simulations are often required, so most of the time in the optimization process is consumed in the performance simulation, which can take up to several hours. It can be seen that it is difficult to solve the above problem by algorithmic improvement alone.

Although the newly proposed agent model idea [8–12] reduces the time consumption in the performance simulation link through machine learning, it is mostly used for building energy consumption prediction and building reconstruction, and there is little research on the building form generation link. In addition, most of the scholars in the current research come from the computer and engineering fields, so the application of machine learning technology mostly involves complex programming and software in the engineering field and needs to be realized by coupling multiple software. In view of the differences between disciplines and the high learning threshold, these theories are difficult to apply and popularize in practical work in the field of architecture.

This paper focuses on the improvement of the method and takes Rhino-Grasshopper (GH), a popular parametric design software in the field of architectural design, as the operation platform, and adds innovative modules of data sampling, performance simulation drive, and accuracy evaluation in the Octopus plug-in through visual programming and improves the optimization algorithm and process equipped with it from the perspective of architects. The optimization algorithms and processes are improved to achieve a set of independent energy-efficient residential form generation design methods that are simple to operate and can be completed on only one software platform, which successfully solves the above problems. The applicability of two machine learning methods, ANN and SVM, for different prediction target values is further analyzed. The method can help architects quickly find the optimal energy-efficient building form design in the schematic design stage and provide data support and feedback to architects in design conceptualization and refinement.

2. Literature Review

As mentioned in the introduction of this paper, the application of machine learning in building energy consumption was originally used to solve the problem of excessive time consumption in the multi-objective optimization stage. Therefore, this section, in order to improve the efficiency of the multi-objective optimization problem for the origin, from algorithm improvement to the intervention of machine learning algorithms, then reviews the literature on the application of machine learning in building energy consumption and building generation, so as to analyze the advantages and disadvantages of existing technologies and methods, so as to help explain the process of the research problems proposed in this paper and the value and significance of the research methods proposed in this paper.

2.1. Optimization

As an important branch of artificial intelligence, the theoretical prototype of optimization algorithms, "heuristic search", can be traced back to Herbert Simon's "The Sciences of the Artificial" book in 1969 [13]. With the increasing demand for design optimization in architectural design, more and more optimization algorithms are embedded in the architectural design platform, which provides a variety of available algorithms for architects to solve different design optimization problems. In recent years, multi-objective optimization algorithms have also been widely used in rural reconstruction. The research can be divided into the following two directions: the first is the improvement of the optimization algorithm, and the second is the optimization results and information feedback (Table 1).

Table 1. The development of optimization algorithm research problems.

Research Direction	Research Problem	Reference
Improvements to the Optimization Algorithm	Galapagos plug-in based on genetic algorithm and annealing algorithm, but only for single-objective optimization.	[14]
	Octopus plug-in based on GH platform, this plug-in introduces SPEA-2 and Hyper-Volume Evolutionary Algorithm (HypE), which can perform multi-objective optimization compared to the previous generation Galapagos plug-in.	[15]
	The application of NSGA-II algorithm in Design Builder, MATLAB, and other software platforms is better than SPEA-2 algorithm in terms of convergence and distribution.	[2]
	The MOEA/D algorithm uses the information of adjacent problems to update the individual position, avoiding the population falling into local optimum, and has great advantages over the previous generation algorithm in maintaining the solution distribution.	[16,17]
	The introduction of machine learning solves the time-consuming problem of multi-objective optimization by establishing a "surrogate model".	[8–12]
Optimization results and information feedback	The optimization results appear in the visual form of the Pareto solution set, which is convenient for the architect to conduct intuitive screening, but the number of the generated solution sets is often large, which brings certain difficulties for the architect to analyze the optimization results.	[18,19]
	By grouping and clustering the distribution of design variants in the result space, architects can compare only the differences between different clusters, and avoid redundant information from analyzing a large number of similar design variants.	[20]
	The K-means clustering algorithm is used to group the design variants in the optimization results in the target result space, and the performance characteristics of each group are analyzed and compared through parallel coordinates.	[21]
	The Pareto solution set is further evaluated and screened by the TOPSIS comprehensive evaluation method, which reduces the workload of architects in the solution set screening process.	[5]

The improvement of the optimization algorithm helps to improve its convergence and distribution. The algorithm that first appeared on the GH platform was the Galapagos plug-in developed by David Rutten [14]. The plug-in includes the genetic algorithm and the annealing algorithm. Among them, the genetic algorithm is based on the standard genetic algorithm and controls the genetic similarity of the two parent individuals selected in the hybridization process through the "breeding" parameter; the appearance of this plug-in is also an important step in the era of computer optimization for architectural design. It is widely used in building performance optimization and urban climate improvement. However, since Galapagos can only be optimized for a single objective, Vierlinger et al. [15] developed an Octopus plugin capable of multi-objective optimization. The plug-in introduces SPEA-2 and Hyper-Volume Evolutionary Algorithm (HypE) to allow users to optimize multiple objectives and visualize the individuals found during the optimization process through a 3D coordinate system, which can help architects filter suitable design variants in the Pareto solution set. The emergence of Octopus has a greater improvement than Galapagos in terms of algorithm improvement and software interaction design. With the introduction of the NSGA-II algorithm in software platforms such as Design Builder and MATLAB, its comprehensiveness is better than SPEA-2. To improve, Wu Di [2] and other scholars use this algorithm to screen passive energy-saving technologies and comprehensive energy-saving technologies by means of multiple software couplings, but there are still problems such as slow convergence speed and easy local convergence. Then the MOEA/D (multi-objective evolutionary algorithm based on decomposition) algorithm appeared, which has great advantages in maintaining the solution distribution [16,17]; at the same time, by using the information of adjacent problems to update the individual position, it avoids the population falling into a local optimum. Zhang Yong [4] and other scholars established a building model in SketchUp, ran the MOEA/D algorithm in MATLAB to generate a new solution, and then used the Visual C++ interface program to decode the new solution to Energy-Plus and output the energy of the building. The index value of time consumption and uncomfortable time is proposed, and in its outlook, it proposes how to combine machine learning and other methods to solve the problem of computational time consumption.

Although the improvement of the new algorithm can improve the optimization efficiency to a certain extent because the performance simulation software is required to intervene in the optimization process to simulate the corresponding building variables, and then the obtained performance parameters are optimized and screened, so most of the time it will be simulated by the performance. It is difficult to solve the above problems by the improvement of the algorithm alone; with the development of computer technology in recent years, machine learning algorithms have been introduced into the multi-objective optimization design of buildings; that is, by establishing a "surrogate model" to reduce the performance in the optimization process. The time consumed by the simulation will be discussed in detail in Section 2.2.

At the level of optimization results and information feedback, the processing and visualization of optimization results are also important means to help architects extract design information from them. In the results of multi-objective optimization, they often appear in the form of Pareto solution sets. However, the number of generated solution sets is It is often larger, which brings certain cognitive difficulties for architects in analyzing and optimizing results [18,19], so the processing of solution sets is particularly important. Some researchers propose to post-process the optimization results; that is, to extract a small number of representative design variants from the optimization results containing a large number of design variants to help architects quickly grasp the core information reflected in the optimization results. For example, scholars such as Negendahl [20] carried out a design optimization experiment of building a facade curtain wall with energy consumption, sunshine, cost, and heat load as multiple objectives. In order to extract key information from the Pareto front surface of design variants, the authors chose to visualize the optimization results in an "objective space" composed of a coordinate system and group and cluster these design variants according to their distribution in the resulting space. Through clustering, architects can compare only the differences between different clusters and avoid redundant information from analyzing a large number of similar design variants. Then, Chen [21] and other scholars took the space cooling system of the building as the design optimization object and carried out multi-objective optimization for electricity consumption, cost, and sunshine. The target result space is grouped, and the performance characteristics of each group are analyzed and compared by parallel coordinates.

In addition, scholars such as Gao Yuan [5] further evaluated and screened the Pareto solution based on the GH platform through the TOPSIS (technology for order preference by similarity to an ideal solution) comprehensive evaluation method, which reduced the workload of architects in the process of solution selection.

2.2. Application of Machine Learning in Architectural

In recent years, with the development of computer technology, digital design has gradually become the mainstream of architectural design, and the mutual penetration between disciplines has also brought the application of many new technologies to architectural design. Machine learning technology originated from artificial intelligence and statistics, and is widely used in image recognition, manufacturing, and other fields. In recent years, it has been used in the field of architectural design. Literature review according to 5 main directions of application of machine learning in architectural design (Table 2).

Table 2. The development of research problems in the application of machine learning in architectural design.

Research Direction	Research Problem	References
	The time-consuming problem in multi-objective optimization is solved by establishing a "surrogate model".	[8–12,22]
Building Performance Prediction	Based on the Rhino platform, a real-time visualization modeling platform is established by using GAN neural network, which can help architects to analyze the impact of urban wind and heat environment intuitively.	[23]
	Generating 3D Wind Field Data Based on SOM Algorithm.	[24]
	Based on the idea of "surrogate model", scholars such as Thomas Wortmann developed an optimization plug-in (Opossum) based on the GH platform. Compared with the traditional optimization algorithm, this plug-in can obtain performance optimization results more quickly.	[25]
	Based on the GH platform, a three-dimensional convolutional neural network (3dCNN) is created. By transferring the three-dimensional shape of the building, the network can identify three architectural features.	[26]
Architectural Form Generation	On the basis of the former research, the adjacency information of each sampling point is added, and the translated building shape information is handed over to the autoencoder for processing.	[27]
	Compared with the previous generation method, the three-dimensional model of the building is segmented in different directions, and the image of the cut surface is used as the learning information of the input of the neural network, and then the GAN neural network is used to generate the building form.	[28]
	Generative Design Method Based on GAN Neural Network.	[29–31]
Building Plan Generation	The machine learning method based on graph structure (Top-view representations) is a kind of deep learning algorithm. Compared with the GAN neural network algorithm, this method has more advantages in expressing the spatial topology relationship of buildings.	[32–35]
	Building facade generation based on GAN neural network.	[36]
Building Renovation	Building Performance Improvement Based on ANN Neural Network.	[7]

Machine learning has demonstrated its potential as a "pattern recognition" tool for applied research in building performance prediction. Among them, in solving the problem of slow convergence in the multi-objective optimization stage mentioned in Section 2.1, the application of machine learning technology in this stage is mainly used to establish the relationship between input and output variables, also known as the surrogate model; that is, the machine learning algorithm learns the relationship between input and output variables from multi-dimensional sample data, so as to establish a model of the correlation between input and output values, so this method is widely used in building energy consumption prediction [8–12], architectural environment simulation [22] and other building performance related fields. In solving the urban-scale environmental performance simulation, Duering S, Chronis, and other scholars established a real-time visualization modeling platform based on the Rhino platform using the GAN neural network [23]. The relationship between the graphs can help architects to intuitively analyze the influence of urban wind and heat environment, thereby optimizing the urban form. In addition, the SOM algorithm also has corresponding applications in generating 3D wind field data [24]. Based on the idea of a "surrogate model" in machine learning, scholars such as Thomas Wortmann [25] developed an optimization plug-in (Opossum) based on the GH platform. Compared with the traditional optimization algorithm, the plug-in can obtain performance optimization results more quickly, but due to the random number sampling method being used, results in a low model accuracy.

At the level of architectural form generation, in recent years, the focus has been on how to transfer the 3D shape of the building for machine learning to call, and it has been applied in the 3D convolutional neural network (3dCNN), that is, the 3D model of the building is pixelated. By making a three-dimensional distribution of pixels, the corresponding neural network can be created. At present, David Newton [26] created a three-dimensional convolutional neural network based on the GH platform, which can identify three architectural features, and then scholars such as Jaime de Miguel [27] added the adjacency information of each sampling point on this basis, The translated architectural shape information is handed over to the autoencoder for processing, and the architectural shape information is finally compressed into a vector in a latent space through four hidden layers. Scholars such as STEINFELD [28] adopted another method to establish a method of building shape generation based on machine learning. That is, another idea was adopted in the translation of building shapes, and the three-dimensional model of the building was divided in different directions. The image is used as the learning information at the input of the neural network, and then the GAN neural network is used to generate new architectural shapes.

The research on building plan generation is mainly divided into two different paths. The first is the generative design method based on the GAN neural network, that is, training in the form of image pixels to obtain a new building plan scheme [29–31]. In recent years, this method has also been widely used in the reconstruction and generation of building facades, but the problem is that it cannot be accurately exchange data with modeling software. The second is a machine learning method based on top-view representations, which is a kind of deep learning algorithm. This method has more advantages than the GAN neural network algorithm in expressing the spatial topology relationships of buildings [32–35].

With the saturation of urban buildings nowadays, building renovation gradually becomes a new research hotspot, and machine learning techniques are applied here. Thanks to the advantages of GAN neural networks in processing two-dimensional image data, similar to building planes, GAN neural networks are mainly used in the study of building façade renovation, which can assist architects in rapid batch pattern design in old neighborhoods and township renovation [36], eliminating the need to manually perform a large number of design patterns based on "mechanical" operations. In building performance retrofitting [7], neural networks (ANNs) are widely used because of their outstanding performance in terms of model accuracy, mostly using the MATLAB platform to invoke performance simulation software through programming, and then a certain number of simulation data samples are handed over to neural networks for learning to establish "agent models". Finally, the best combination of building parameters is selected through a multi-objective optimization algorithm, which can also be regarded as a further application of machine learning technology in the area of building energy consumption.

Through the review of related research status and problems and the reviews of "optimization algorithm" and "application of machine learning technology in architectural design", it can be seen that machine learning technology has been widely used in architectural design. Although SVM, ANN, and other traditional machine learning algorithms do not have the same "magic" as deep learning, they have great potential in the fields of building performance and building renovation, and the performance feedback in multi-objective optimization is accelerated by machine learning technology, which is also more in line with "intelligent" building needs. However, since most of the scholars in the existing research are from computer and engineering fields, the application of machine learning techniques mostly involves complex programming and software in the engineering field and requires multiple software coupling to achieve, which is difficult to promote in the practical work in the field of architecture due to the difference between disciplinary fields and high learning threshold.

3. Detached Energy-Efficient Housing Generation Method

In contrast to the diversity and complexity of urban architecture, detached housing is often characterized by modality and lightness, and passive energy efficiency tools such as form adaptation are particularly important. With the development of computer technology, the relationship between computers and architecture has become closer and closer. From the early days when the two fields were unrelated, to the 1990s when architects began to use computers for simple drafting and modeling, and nowadays when computers begin to assist architects in simple design tasks, architectural design has gradually differentiated into an independent research system.

Building design generation as a broad computer-generated concept covers many building generation ideas and methods. Specifically, it can be divided into the following two categories: the first is the "top-down" generation approach, i.e., the design generator builds a parametric model and then hands it over to the solver for filtering using optimization algorithms. This method is usually used in cases where there are few design elements and the relationship between them is simple, such as residential buildings with few building blocks and simple functions, or the generation of building façade components; the second type is the "bottom-up" operational design optimization, which integrates the solution and generation process of the design problem into a unified algorithm, which can solve the layout problem among multiple buildings, but the optimization algorithm is difficult to intervene [13–15]. Therefore, it is appropriate to adopt the first type of "top-down" approach in this study.

This paper regards building energy conservation as an important index of building form generation. The genetic algorithm carried in Octopus is used to optimize and screen the energy conservation indicators of buildings, so as to obtain the building form with the lowest energy consumption and relatively high comfort. As mentioned in the introduction, in the previous solution process, this stage often took a lot of time, and the intervention of the machine learning algorithm greatly shortened the time consumed in this process.

The parametric model part mainly includes parameter setting, LHS, performance simulation, machine learning, and accuracy evaluation, which is the core part of this study; the optimization algorithm part uses the Octopus solver for optimization screening. The detailed process is shown in Figure 1.

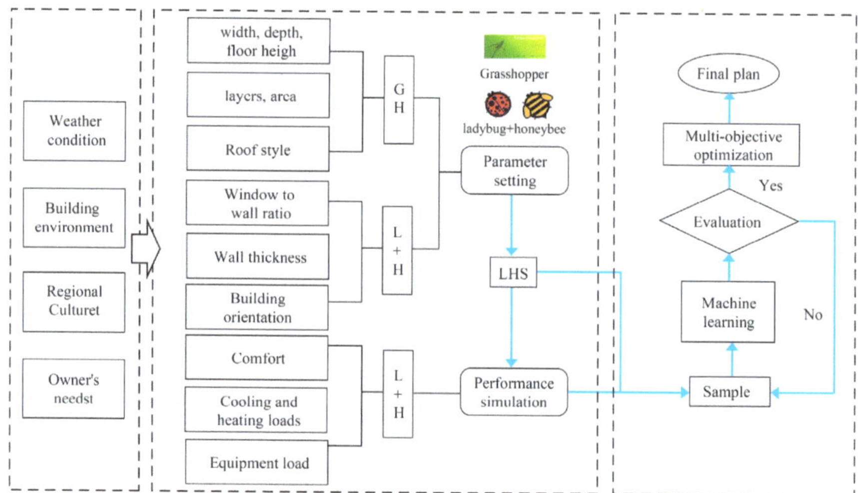

Figure 1. Flow chart of generation and optimization of detached energy-saving houses.

3.1. Parameter Setting

In the parameter setting, the architect needs to use field research and environmental monitoring to analyze the meteorological conditions, cultural customs, the owner's needs, and building codes of the countryside to set the number, type, and value range of the corresponding variables. The basic form parameters such as width, depth, height, and area of the building can be controlled by native commands such as domain box and number slider in GH; the parameters such as window to wall ratio and building orientation need to be controlled by the corresponding battery in the create module of Honeybee.

3.2. Latin Hypercube Sampling

Honeybee is a full-featured and highly accurate performance simulation plug-in based on the GH platform. Its internal operator can directly invoke the Energy-Plus calculation kernel to complete simulation analysis of annual thermal and cooling loads, comfort, etc. It is important to note that there are usually hundreds of thousands of variable parameter values arranged and combined in buildings, and as mentioned in the introduction of this paper, it would take a long time to perform computational simulations if Octopus is used directly for optimization screening at this stage. Therefore, it is necessary to sample the parameter values to generate a small and representative sample, and then input the sample variables into Honeybee separately for performance simulation, and finally establish the relationship between the design variables and the target values by means of machine learning.

Latin hypercube sampling (LHS) is a special stratified sampling method [37,38]. Its characteristic is that the collected sample data has good distribution and avoids the problem of sample data aggregation caused by the traditional random sampling method. Therefore, it is also widely used in computer experiments. However, due to the lack of plug-ins for sample collection in octopus, and the existing studies mostly using Python or R language programming to sample [6,8], the operation is cumbersome and requires other software intervention. Therefore, this paper supplements this cluster in GH through visual programming (Figure 2).

With a total of m variables $x_1, x_2, \ldots x_m$, N samples need to be collected. According to the sampling principle of LHS can be summarized as the following three steps: segmentation, taking values, and disruption. Step 1 (Segmentation): Use Divide Domain in GH to decompose each of the m variables into the same N small intervals. The number of small intervals is the number of required samples. Step 2 (take value): Use the Random command to select a random value in each interval. If you want to adjust the number of decimal places to be retained in this stage, you can enter the corresponding expression in Evaluate. Step 3 (Disorder): Finally, use the Jitter command to disorder each of the N values in the m variables and then combine them with the values in the other variables.

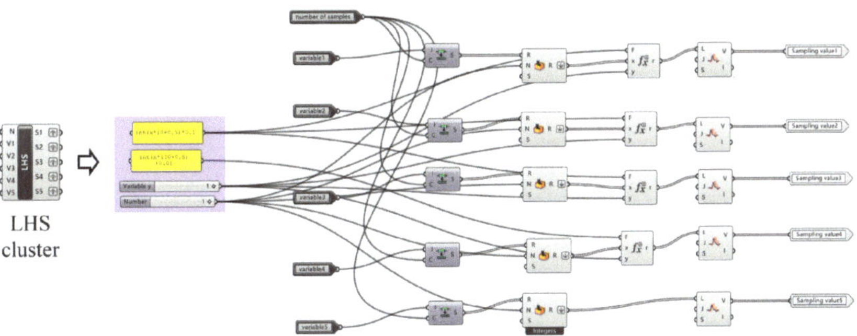

Figure 2. The battery diagram of the Latin hypercube sampling method based on GH.

3.3. Performance Simulation

Before the performance simulation, we need to clarify the performance target; generally, the more target values need to be simulated, the longer it takes. Relying on the power of Honeybee, the target values can be obtained through simulation calculation, such as cooling and heating load, equipment load, comfort, solar radiation value, and other performance targets, which can be set flexibly according to the demand in specific project practice. After setting the performance targets of the building, it is necessary to import the parameter samples obtained by LHS sampling into Honeybee for performance simulation, but the number of samples is often hundreds and it is difficult to perform manually, so it is necessary to build a driver module in GH that can import the data from the sample library into Honeybee simulation in turn. In this paper, the simulated parameters in the sample data are exported sequentially by data record and timer commands, and the parameter values to be simulated are assigned to the corresponding cell groups sequentially by list item (Figure 3). When all the parameter samples are simulated, the simulated values collected in the data record are checked and verified for later use in machine learning.

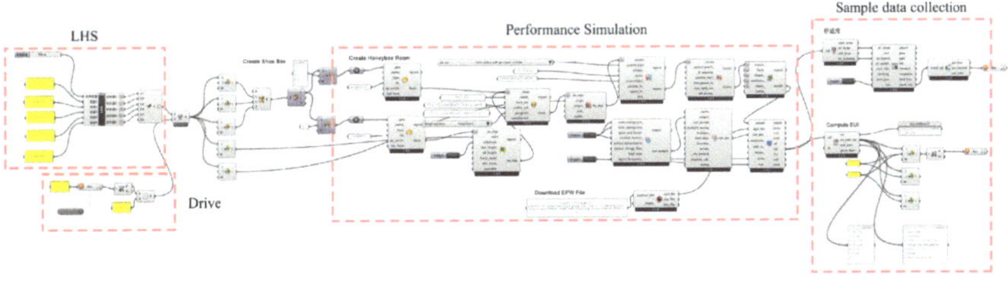

Figure 3. Performance simulation computing battery pack.

3.4. Machine Learning and Evaluation
3.4.1. Machine Learning

Machine learning originated in the fields of artificial intelligence and statistics. In recent years, it has been used in the field of building energy consumption prediction to speed up optimization. That is, an input-output correlation model is established through machine learning, and new data is input into the trained model. Predictive models can predict the corresponding data outputs, a process known as "surrogate models" or "metamodels" [39].

As mentioned in Section 2.2, there are many types of machine learning algorithms; however, due to the different characteristics and principles of the algorithms, they have their own areas of expertise; for example, the GAN neural network in deep learning is good at identifying image data, so it is mostly used in the field of architecture for the generation of plans and facades. In this study, the predicted indexes are the performance parameters of detached houses, including building energy consumption and comfort, which are regression problems with data labels, and the application of traditional machine learning algorithms has more potential in the field of building energy consumption and renovation, therefore, this paper unfolds with ANN and SVM learning algorithms.

In addition, relevant machine learning plug-ins have appeared in GH in recent years, such as Lunchbox, Octopus, Owl, and other plug-ins that package algorithms into different operators, among which Octopus is equipped with machine learning and multi-objective optimization modules, which have the characteristics of simple operation and relatively comprehensive functions; therefore, it is appropriate to choose Octopus as the operating platform for the machine learning part.

An ANN is a mathematical model that simulates a biological neural network for information processing, referred to as a neural network. It has the characteristics of high classification accuracy, strong learning ability, and good prediction and classification ability for untrained data [40].

The emergence of the MP-neuron model laid the foundation for the development of neural networks, however, the model with only one layer of functional neurons has limited learning ability and can only solve linearly separable problems for the nonlinear problem in Figure 4a cannot be fitted accurately. The neural network composed of multiple neurons and hidden layers enables ANN to deal with nonlinear problems better (Figure 4b). As shown in Figure 4, the greater the number of neurons, the smoother the fitted curve and the higher the accuracy of the model, but if the number of neurons exceeds a certain value, overfitting will occur and the accuracy decreases. In the problem of building performance prediction, the output values of energy consumption and indoor comfort are continuous variables, which can be regarded as a regression problem. The neural network can build a prediction model with high accuracy by adjusting the number of hidden layers and the number of neurons, so it is widely used in the field of building performance prediction and is therefore one of the machine learning algorithms focused on in this paper.

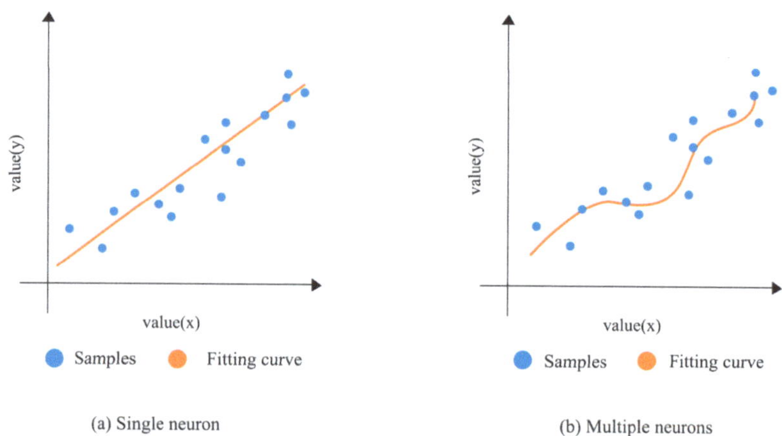

Figure 4. Neural network fitting curve comparison.

Its operation in Octopus relies mainly on the following two parts: the input module (network learning) and the output module (network evaluate). The input of the learning module corresponds to the sample parameter values obtained from LHS; the output corresponds to the target values after simulating the performance of the sample parameter values; since the input and output values in this battery module only identify the values within $[-1, 1]$, the sample data needs to be mapped to the interval $[-1, 1]$ and then given to network learning for neural network training. The infrastructure of a neural network is composed of multiple layers of parallel units called neurons, so the number of hidden layers and the number of neurons inside are important factors that affect the accuracy of a neural network, and if the neurons are underestimated, the ANNs ability to store information may be reduced. On the contrary, overestimation may lead to unnecessary learning and even overfitting of the neural network, and the "trial-and-error" method is commonly used to find the right number of hidden layers and neurons [7]. In Octopus, the number of hidden layers and neurons can be adjusted by controlling the parameter values of layers and nodes at the input side of the network learning battery. After the training is completed, the trained neural network is connected to the network evaluate battery, and the target value after prediction through the neural network is obtained by inputting the parameter values that need to be simulated on the I side (Figure 5).

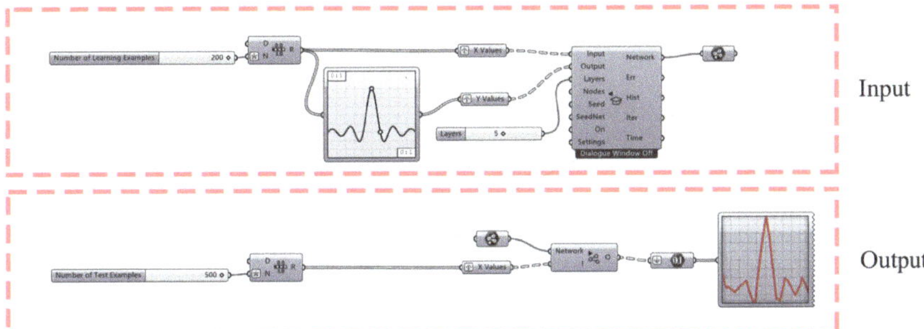

Figure 5. ANN battery pack.

SVM is a supervised binary classifier based on the statistical VC dimension theory and the principle of structural risk minimization. Cortes and Vapnik formally proposed a support vector machine (SVM) in 1995 [41]. SVM was first developed from statistical theory and is also known as SVC (support vector classify) for classification problems and SVR (support vector regression) for regression problems. SVMs usually separate samples by a linear function for linearly separable binary classification problems, but there is not only one but an infinite number of straight lines that can separate the data, so the support vector machine corresponds to the straight line that can correctly divide the data with the largest interval. However, such a classifier is still a weak classifier and still cannot solve some linear indistinguishable problems in reality, as shown in Figure 6 in the original space where the data are located. The kernel function in SVM solves this problem by mapping the data from the binary plane to a higher-dimensional space and finding an optimal plane in that space to separate these two types of data. From a mathematical point of view, let the curve in Figure 6, which separates orange and blue, be a circle with the equation $x^2 + y^2 = 1$. The principle of the kernel function is seen as mapping x^2 to X and y^2 to Y. The equation of the hyperplane becomes $X + Y = 1$, which makes the nonlinear separable problem in the original space become a linear separable problem in the new space through the mapping effect of the kernel function.

In dealing with the regression problem, the principle is similar to that of SVC. First of all, it is necessary to ensure that the fitted line should reflect all the sample data as much as possible, so the distance between the hyperplane and the farthest sample point in the regression task should be as large as possible, but the situation shown in Figure 7a will occur when the distance is guaranteed to be the largest, so a limit is added to the interval in SVR. The deviation of the model $f(x)$ from y should be less than or equal to ε. The deviation range is also called the ε pipeline, and the correct fitted curve shown in Figure 7b can be achieved by adding restrictions to the sample data interval. In the case of nonlinear fitting, the same principle as in the classification problem is used to transform the nonlinear fit into a linear one by mapping the sample data to a higher dimensional space with a kernel function (Figure 8).

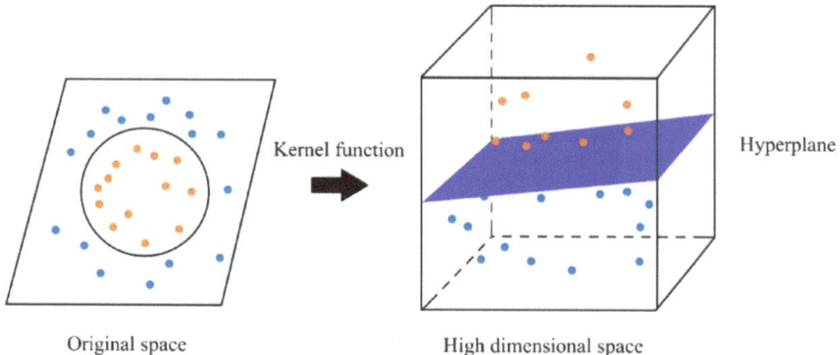

Figure 6. The principle analysis of the kernel function in the SVM linear inseparable problem.

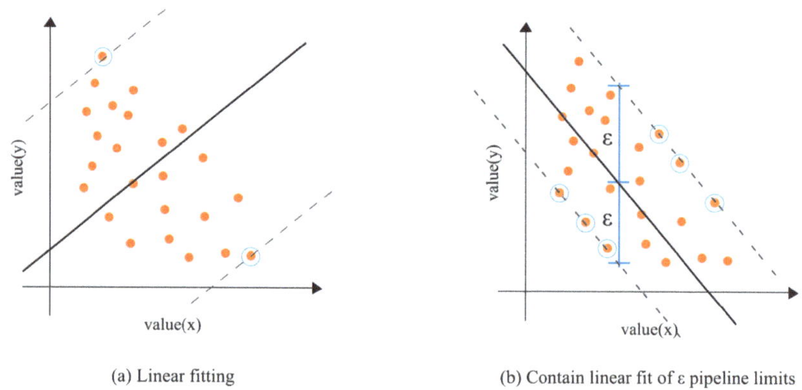

Figure 7. SVR linear fitting schematic.

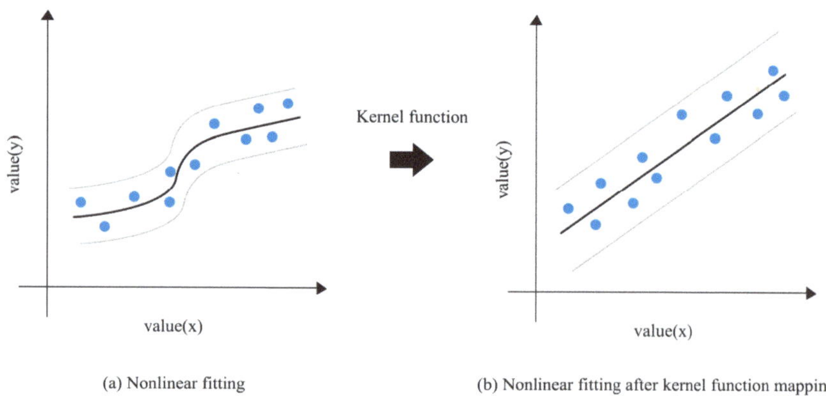

Figure 8. SVR nonlinear fitting schematic diagram.

In Octopus, there is a similar operation logic to ANN, which is also composed of an input module (SVM learning) and an output module (SVM Evaluate). Unlike ANN, the input sample value does not need to be mapped. Because SVM has a strict statistical theory and mathematical foundation, unlike ANN, which needs to rely on the experience and knowledge of designers, it needs to adjust fewer parameter values, and its operation difficulty is lower than ANN. (Figure 9).

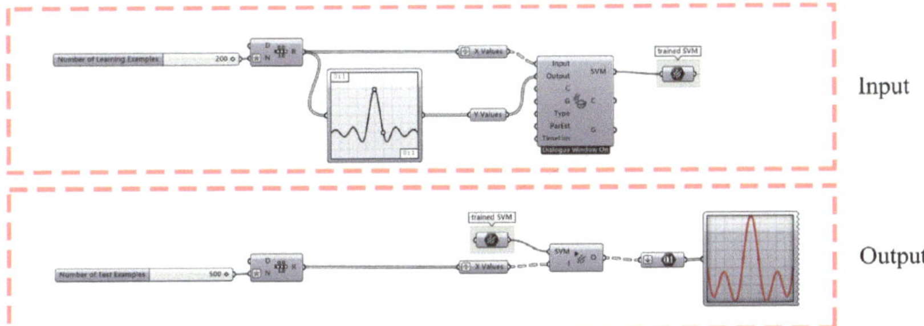

Figure 9. SVM battery pack.

3.4.2. Evaluation

The two machine learning algorithms are not superior or inferior in nature, and the more suitable machine learning method should be selected when dealing with different sample data; for example, SVM can be used for small samples with high dimensionality, while ANN can be used for samples with a large number of samples and noisy data. However, in the specific practice process, the operator needs to have strong data sensitivity and certain operating experience, otherwise, it is difficult to judge the nature of the sample data, so this paper proposes a classification attempt, that is, the sample data are imported into two machine learning algorithms for training, and the accuracy of machine learning is evaluated by the following two common model evaluation methods: root mean square error (RMSE) and coefficient of determination (R^2) [41]. The machine learning method with the better evaluation grade is used to establish the relationship between the variables requiring higher model accuracy and the performance simulation target value. Among them, the smaller the root mean square error indicates, the better the model prediction, the larger the coefficient of determination, and the higher the model accuracy. Its calculation formula is as follows:

$$\text{RMSE} = \sqrt{\frac{1}{n} \cdot \sum_{i=1}^{n}(y_i - \hat{y}_i)^2} \tag{1}$$

$$R^2 = 1 - \frac{\sum_{i=1}^{n}(y_i - \hat{y}_i)^2}{\sum_{i=1}^{n}(y_i - \overline{y}_i)^2} \tag{2}$$

where y_i represents the predicted value generated by machine learning, $\overline{y}_i$ represents the average value of y_i, and $\hat{y}_i$ represents the simulated value of performance generated by Honeybee calculation. In the "classification attempt" process, 70% of the sample data are usually used for machine learning and 30% for accuracy evaluation. In this paper, the above evaluation method is implemented in GH using a native cell written in the math module (Figure 10).

In addition, if the accuracy of the model is too low for both machine learning methods, it may be caused by factors such as insufficient sample data or improper operation and needs to be returned for verification and modification.

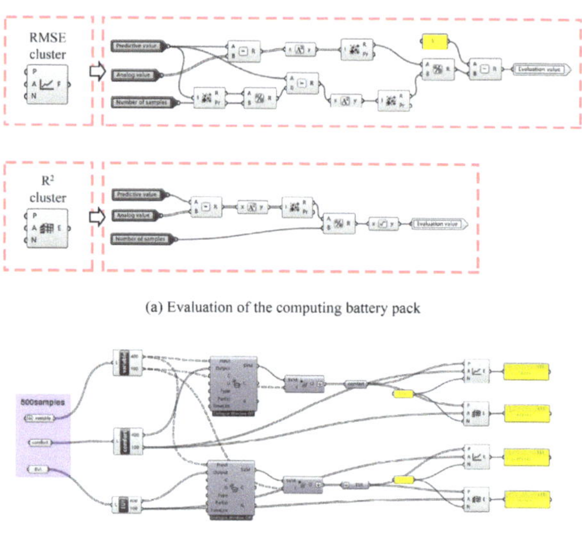

(a) Evaluation of the computing battery pack

(b) Evaluation module usage example

Figure 10. Evaluation of modular battery packs and usage examples.

3.5. Multi-Objective Optimization

Two important indicators of building energy conservation are energy consumption and comfort. In the optimization screening, multi-objective optimization of the residential building form parameters is performed using the SPEA-2 algorithm equipped with Octopus through a machine learning module in order to find the building form with the lowest energy consumption and the highest comfort level within a specific range of parameters (Figure 11). Since many-objective optimization parameters are contradictory in nature, a solution may be better for one objective but worse for others, so it is necessary to perform a specific analysis in the Pareto solution set [42,43], taking into account the actual situation, e.g., the user of the house spends more time in the house and needs to take more into account the level of indoor comfort, so the Pareto solution set can be chosen to favor the comfort level. For example, in areas where energy is relatively scarce, building energy efficiency is more important, so solutions with lower energy consumption can be selected in the Pareto solution set.

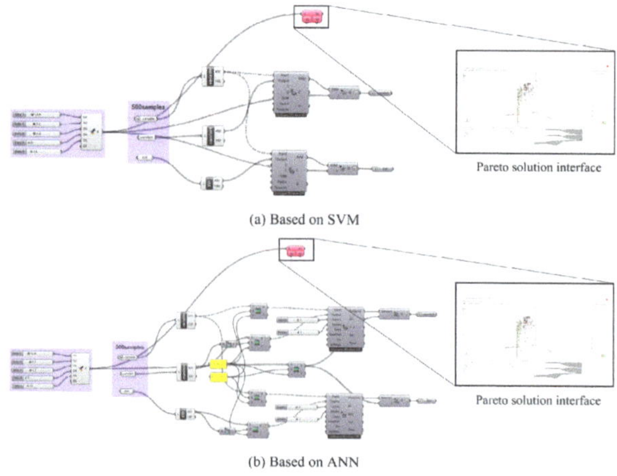

(a) Based on SVM

(b) Based on ANN

Figure 11. Multi-objective optimization operation battery pack.

4. Experiment and Results

In this section, the above operation process is practically performed so as to verify the effectiveness of this method, and the ANN and SVM algorithms are compared and studied to derive a machine learning approach more suitable for detached residential buildings for practical reference.

4.1. Sampling Method Comparison Experiment

With variables x_1, x_2, which both take values in the range [0, 1], 10 sample data are sampled in the interval using the LHS operation process proposed in this paper with GHs native random number command, respectively, and the generated data are displayed using scatter plots. The comparison of the sampling points in (Figure 12) easily shows that the sampled data after the operation using the above method exhibits the characteristics of uniform LHS sampling distribution, while the sampled data in the traditional random number command has the disadvantages of overlapping data and uneven sampling. This experiment proves the effectiveness of the sampling method in this paper.

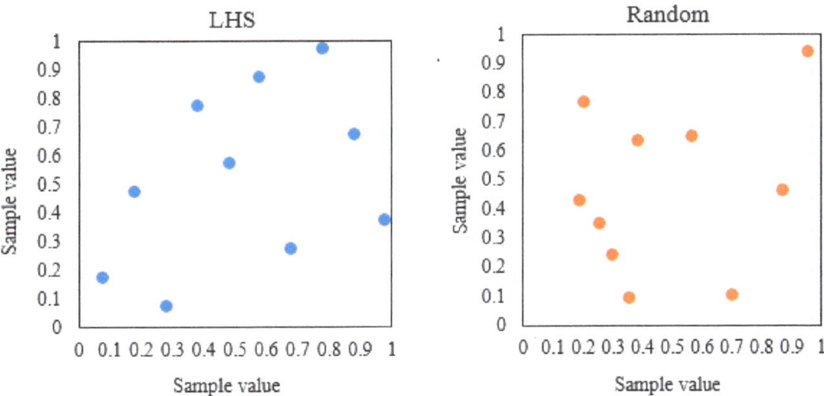

Figure 12. Comparison of LHS and Random sampling.

4.2. A Comparative Study of Machine Learning Algorithms

Two machine learning algorithms, ANN and SVM, are selected for comparative study. The two algorithms use the same learning samples and test data to ensure the objectivity of the experiment. When generating the sample data, the weather data of Jinan City is selected as an example for the reference independent house building shown in this paper. It adopts the form of a single-layer flat roof. The building is a steel frame structure as a whole, the exterior walls and roof are equipped with insulation layers, and the ventilation rate is set to $0.5\,\text{h}^{-1}$. The windows are made of double-layer hollow Low-e glass with an argon-filled air layer in the middle. The thermal transmittance (U-value) is $0.8\,\text{w}/(\text{m}^2{\cdot}\text{K})$, and the solar heat gain coefficient (SHGC) is 0.65. Tables 3 and 4 provide details of the structural and material properties of the building envelope and the internal heat gain.

In this paper, five main parameter variables affecting the form and performance of buildings are set (Table 5), and two important indicators in building energy efficiency design, building energy consumption (EUI) and comfort, are taken as optimization target values. Through the Honeybee plug-in, 500 performance simulations were conducted; 400 samples were used for machine learning, and the remaining 100 samples were used for testing.

Table 3. Structural and material properties of the building envelope.

Construction Component	Layers	Thickness (mm)	Conductivity (W/(m²·K))	Density (kg/m³)
Roof	Roof membrane	25	0.71	1856
	Typical insulation	-	-	-
	Metal roof surface	0.8	45	7824
	Generic insulation	50	0.03	43
	HW concrete	200	1.95	2240
	Ceiling air gap	100	0.55	1.28
	Acoustic tile	20	0.06	368
Exterior wall	Stucco	25	0.71	1856
	Gypsum board	15	0.16	800
	Typical insulation	-	-	-
	Gypsum board	15	0.16	800
Exterior window	Low-e glass	6	0.99	2528
	Argon cavity	12	0.017	1.78
	Low-e glass	6	0.99	2528

Table 4. Internal heat gains.

Source of Internal Heat Gain	Heat Gain	Schedule	
		Weekday	Weekend
Owner	100 W/Owner	00:00–8:00 and 18:00–24:00	Always at home
Lighting	6 W/m²	19:00–22:00	
Refrigerator	150 W	Always on	
Television	60 W	18:00–22:00	8:00–10:00 and 18:00–22:00

Table 5. Variable parameter settings.

Variable Name	Value Range	Step
Width/m	[12, 15]	0.1
Depth/m	[5, 6]	0.1
Floor height/m	[3, 4]	0.1
Building orientation/(°)	[0, 180]	1
Building window–wall ratio	[0.3, 0.5]	0.1

The first step of training is to determine the number of hidden layers and the number of neurons in the ANN by using the "trial and error" method. The RMSE value reaches the minimum and the R^2 value reaches the maximum when the number of hidden layers is 4, and then the RMSE value starts to increase and the R^2 value becomes smaller when the number of hidden layers increases. After determining the number of hidden layers to 4 and continuing to test the optimal number of neurons for the ANN, the RMSE and the R^2 values start to level off when the number of neurons increases to 7. Continuing to increase the number of neurons does not significantly improve the accuracy of the model but increases the training time. Therefore, in order to balance the training accuracy and training time, a hidden layer of 4 and a number of neurons of 7 were selected to build the neural network, and the RMSE and R^2 values of the model were 0.20 and 0.99, respectively. The model has the highest accuracy and takes the least time when the number of hidden layers is 3 and the number of neurons is 10 (Figure 13).

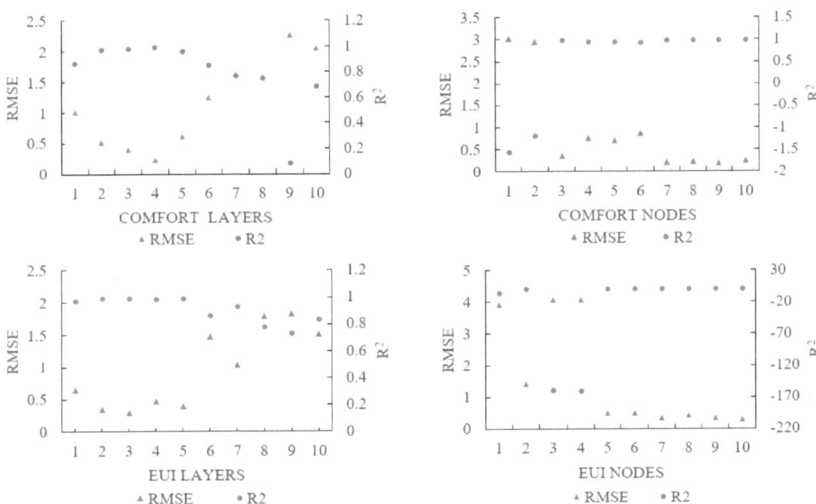

Figure 13. Test map of hidden layer and number of neurons.

The same sample data as ANN was used to import SVM for machine learning and use 100 samples for the evaluation test, and the accuracy of the obtained model is shown in Table 6. The comparison of the accuracy of the two machine learning algorithms in Table 6 shows that ANN is better than the SVM algorithm in the prediction accuracy of building energy consumption and comfort for both sample data. The RMSE has the same units as the output variable and can be considered as the average error of the model; the error of SVM in this target value of the annual comfort time is about 3 days, for which the error is within the acceptable range and can be used. SVM algorithm; however, in the prediction of building energy consumption, the error of 1.17 kwh/m^2 is too large for this target value because the area of detached houses is generally around 90–200 m^2, and it is more appropriate to use the machine learning algorithm of ANN.

Table 6. Accuracy comparison of machine learning methods.

Sample Name	Machine Learning Algorithm	RMSE	R^2
Comfort	ANN	0.20	0.99
	SVM	1.08	0.89
EUI	ANN	0.29	0.99
	SVM	1.17	0.90

In previous studies, the accuracy of the model was often used as the only criterion to evaluate the applicability of machine learning algorithms, and only a single algorithm was used in practical applications. This section further compares ANN and SVM algorithms widely used in dealing with regression problems through a simple detached house case on the basis of verifying the feasibility of the proposed method. Considering the complexity and accuracy of machine learning algorithm operation, SVM is simple and fast in operation, so it can be used for year-round comfort prediction with relatively low model accuracy; while ANN is more complicated in operation, but the model is more accurate, so it is more suitable for the model that requires a high forecast of building energy consumption. In summary, two machine learning methods, ANN and SVM, can be integrated and used in the generation of detached residential forms so as to achieve the purpose of efficient operation and accurate prediction.

4.3. Building Form Generation Result

The trained machine learning model is used for multi-objective optimization. After 86 iterations, the optimization results converge to form a Pareto front solution set. Figure 14 shows the EUI and comfort values of 500 database cases and Pareto solution sets used for machine learning. Each point in the figure is a solution associated with a set of decision variables representing a design scenario. The Pareto-front solutions yielded better building performance as far as the two optimization objectives are concerned.

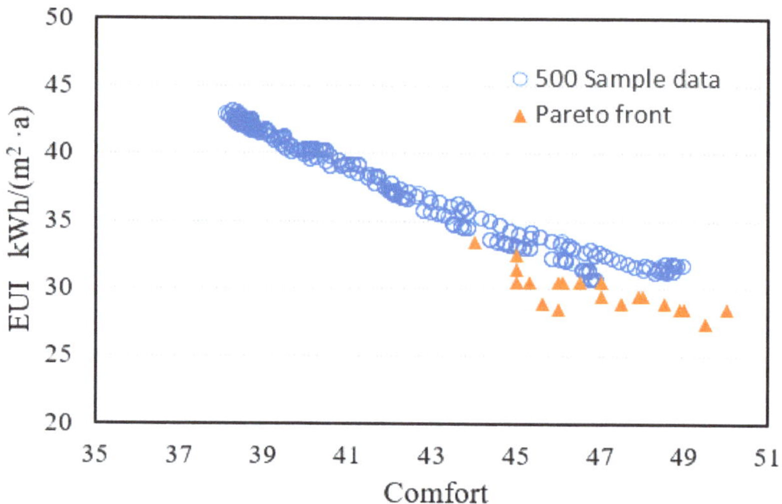

Figure 14. EUI and comfort value of 500 sample data and Pareto-front solutions.

Another advantage of the method proposed in this paper is that all the operational processes are completed on one software platform, and the overall consistency is good. The solutions generated from the Pareto solution set can be displayed in Rhino-Grasshopper in real-time. Figure 15 shows nine building form generation cases selected in the Pareto frontier solution set, in which the orange block is the generated building form, the white block is the surrounding environment of the site, and the red line represents the red line of building land. As described in Section 3.5 Multi-Objective Optimization in this paper, the two performance indexes optimized are often contradictory, so they need to be analyzed in detail in practical applications. For example, in the generated case shown in Figure 15, considering the integration of the building and the surrounding sites, Case5, Case6, Case8, and Case9 have improved the comfort of the whole year while ensuring low energy consumption, but they are obviously less integrated with the surrounding sites and exceed the scope of land use. Although Case7 has slightly higher energy consumption than Case4, the comfort of the whole year has been greatly improved; therefore, it is more suitable as a preliminary architectural form solution.

With the support of machine learning and a multi-objective optimization algorithm, this method takes building energy conservation as the goal to explore the optimal building volume. The generated building form model can be intuitively expressed in Rhino-Grasshopper, which can provide data support and intuitive form visual feedback for architects at the conception stage and facilitate further deepening and analysis of the scheme.

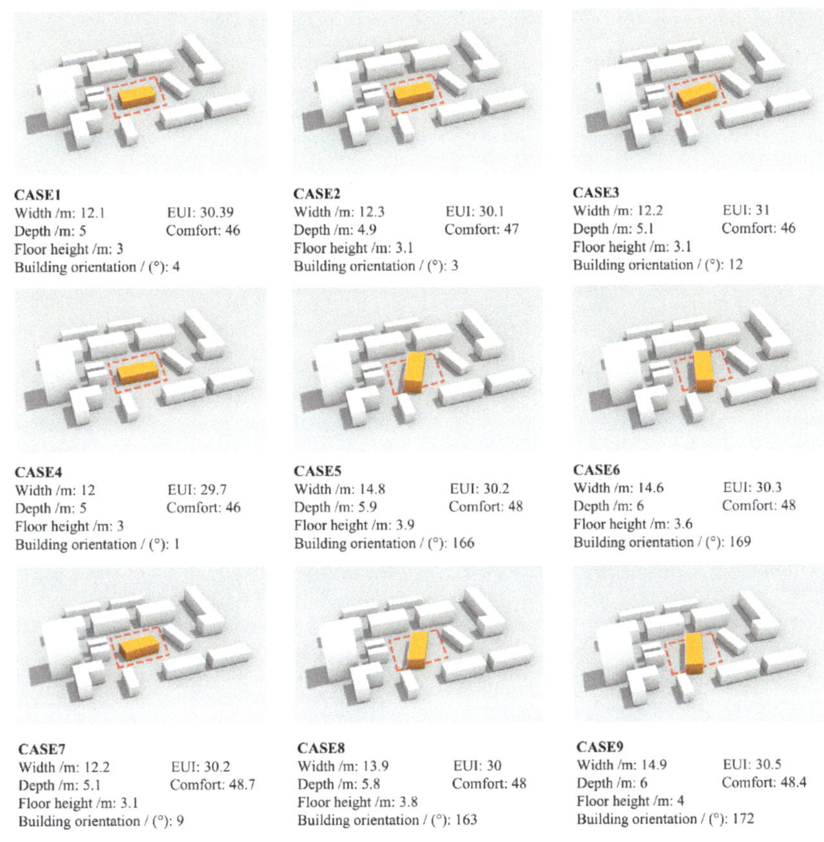

CASE1
Width /m: 12.1   EUI: 30.39
Depth /m: 5   Comfort: 46
Floor height /m: 3
Building orientation / (°): 4

CASE2
Width /m: 12.3   EUI: 30.1
Depth /m: 4.9   Comfort: 47
Floor height /m: 3.1
Building orientation / (°): 3

CASE3
Width /m: 12.2   EUI: 31
Depth /m: 5.1   Comfort: 46
Floor height /m: 3.1
Building orientation / (°): 12

CASE4
Width /m: 12   EUI: 29.7
Depth /m: 5   Comfort: 46
Floor height /m: 3
Building orientation / (°): 1

CASE5
Width /m: 14.8   EUI: 30.2
Depth /m: 5.9   Comfort: 48
Floor height /m: 3.9
Building orientation / (°): 166

CASE6
Width /m: 14.6   EUI: 30.3
Depth /m: 6   Comfort: 48
Floor height /m: 3.6
Building orientation / (°): 169

CASE7
Width /m: 12.2   EUI: 30.2
Depth /m: 5.1   Comfort: 48.7
Floor height /m: 3.1
Building orientation / (°): 9

CASE8
Width /m: 13.9   EUI: 30
Depth /m: 5.8   Comfort: 48
Floor height /m: 3.8
Building orientation / (°): 163

CASE9
Width /m: 14.9   EUI: 30.5
Depth /m: 6   Comfort: 48.4
Floor height /m: 4
Building orientation / (°): 172

Figure 15. Building form generation case (from Pareto solution).

5. Conclusions

From the perspective of architects, this paper relies on Grasshopper, a widely used parametric design software in architecture, as a technical platform, and adds innovative modules of data sampling, performance simulation drive, and accuracy evaluation to the Octopus plug-in through visual programming and improves and perfects the algorithm and optimization process it carries. It solves the problem of complex programming and coupling of multiple software programs in the previous study of energy-efficient building design optimization and further improves work efficiency. The applicability of SVM and ANN in different building performance indicators is further analyzed through a residential building case. The experimental results show that ANN and SVM can be used to predict the energy consumption and comfort level of buildings respectively in the design of detached energy-saving houses, which can improve work efficiency while ensuring the accuracy of the model. This method can help architects quickly search for the best building energy-saving form design scheme in the scheme design stage and provide data support and information feedback for architects in design conception and deepening.

Architecture is a unique discipline, combining technology and art. In addition to meeting the requirements of building performance, such as energy-saving optimization, the process of building form generation is also the process of shaping spatial experience and modeling art, so how to effectively balance the relationship between building performance and design aesthetics in the context of future artificial intelligence is a further problem to be solved.

Author Contributions: Data curation, H.G.; Formal analysis, D.D. and F.X.; Funding acquisition, D.D.; Investigation, K.D.; Methodology, D.D. and F.X.; Project administration, R.P.; Resources, J.Y.; Software, K.D.; Visualization, J.Y.; Writing – original draft, H.G.; Writing—review & editing, H.G. and R.P. All authors have read and agreed to the published version of the manuscript.

Funding: This research was funded by "Teaching Research Project of Wuhan Institute of Technology, grant number x2021019"; "MOE (Ministry of Education in China) Liberal Arts and Social Sciences Foundation, grant number 19YJC760079" and "Graduate Innovative Fund of Wuhan Institute of Technology, grant number CX2020092".

Data Availability Statement: Not applicable.

Conflicts of Interest: The authors declare no conflict of interest.

References

1. Building Energy Conservation Research Center of Tsinghua University. *Building Energy Efficiency Annual Development Research Report 2021 (Special Topic of Rural Housing)*; China Construction Industry Press: Beijing, China, 2021. (In Chinese)
2. Wu, D.; Liu, L.; Li, X.; Liu, C. Research on passive low energy consumption building technology based on multi-objective optimization—Taking Residential Buildings in cold areas as an example. *J. South China Univ. Technol. (Nat. Sci. Ed.)* **2018**, *46*, 98–104+120. (In Chinese)
3. Shao, T.; Jin, H. Research on energy-saving optimization design of rural housing in severe cold area based on optimization algorithm. *Archit. Sci.* **2019**, *35*, 99–107. (In Chinese) [CrossRef]
4. Zhang, Y.; Liang, X.; Yuan, L. Energy saving oriented multi-objective optimization method for building design. *J. Huazhong Univ. Sci. Technol. (Nat. Sci. Ed.)* **2021**, *49*, 107–112. (In Chinese) [CrossRef]
5. Gao, Y.; Chi, J.; Luo, S.; Hu, K.; Yuan, J.; Yue, X. Multi objective optimization technology framework for passive energy saving design of rural residential buildings in Tianjin Based on octopus+honeybee. *J. Xi'an Univ. Technol.* **2021**, 1–11. Available online: http://kns.cnki.net/kcms/detail/61.1294.N.20210911.1252.002.html (accessed on 23 January 2022). (In Chinese)
6. Gao, Y.; Luo, S.; Chi, J.; Yuan, J. Multi objective optimization evaluation method for low-carbon transformation of rural existing residential buildings. *South. Archit.* **2021**, 1–9. Available online: http://kns.cnki.net/kcms/detail/44.1263.TU.20211224.1252.008.html (accessed on 23 January 2022). (In Chinese)
7. Xue, Q.; Wang, Z.; Chen, Q. Multi-objective optimization of building design for life cycle cost and CO_2 emissions: A case study of a low-energy residential building in a severe cold climate. *Build. Simul.* **2021**, *15*, 83–98. [CrossRef]
8. Fan, C.; Sun, Y.; Zhao, Y.; Song, M.; Wang, J. Deep learning-based feature engineering methods for improved building energy prediction. *Appl. Energy* **2019**, *240*, 35–45. [CrossRef]
9. Si, B.; Wang, J.; Yao, X.; Shi, X.; Jin, X.; Zhou, X. Multi-objective optimization design of a complex building based on an artificial neural network and performance evaluation of algorithms. *Adv. Eng. Inform.* **2019**, *40*, 93–109. [CrossRef]
10. Zhang, J.; Li, Y.; Li, H.; Wang, X. Sensitivity Analysis of Thermal Performance of Granary Building based on Machine Learning. In Proceedings of the 24th CAADRIA Conference, Wellington, New Zealand, 15–18 April 2019; pp. 665–674.
11. Nguyen, A.T.; Reiter, S.; Rigo, P. A review on simulation-based optimization methods applied to building performance analysis. *Appl. Energy* **2014**, *113*, 1043–1058. [CrossRef]
12. Xie, Y.; Ishida, Y.; Hu, J.; Mochida, A. A backpropagation neural network improved by a genetic algorithm for predicting the mean radiant temperature around buildings within the long-term period of the near future. *Build. Simul.* **2021**, *15*, 473–492. [CrossRef]
13. Simon, H.A. *The Sciences of the Artificial*, 3rd ed.; MIT Press: Cambridge, MA, USA, 1969; Volume 33.
14. Rutten, D. Galapagos: On the Logic and Limitations of Generic Solvers. *Archit. Des.* **2013**, *83*, 132–135. [CrossRef]
15. Vierlinger, R. Multi Objective Design Interface. Master's Thesis, Technischen Universität Wien Fakultät für Bauingenieurwesen, Wien, Austria, 2013.
16. Okudan, G.E.; Tauhid, S. Concept selection methods—A literature review from 1980 to 2008. *Int. J. Des. Eng.* **2008**, *1*, 243. [CrossRef]
17. Li, S.; Liu, L.; Peng, C. A Review of Performance-Oriented Architectural Design and Optimization in the Context of Sustainability: Dividends and Challenges. *Sustainability* **2020**, *12*, 1427. [CrossRef]
18. Scheibehenne, B.; Greifeneder, R.; Todd, P.M. Can There Ever Be Too Many Options? A Meta-Analytic Review of Choice Overload. *J. Consum. Res.* **2010**, *37*, 409–425. [CrossRef]
19. Yousif, S.; Yan, W. Clustering Forms for Enhancing Architectural Design Optimization. In *CAADRIA 2018: Learning, Adapting and Prototyping, Proceedings of the 23rd Annual Conference of The Association for Computer-Aided Architectural Design Research in Asia (CAADRIA), Beijing, China, 17–19 May 2018*; Tsinghua University: Beijing, China, 2018.
20. Negendahl, K.; Nielsen, T.R. Building energy optimization in the early design stages: A simplified method. *Energy Build.* **2015**, *105*, 88–99. [CrossRef]
21. Chen, K.W. Architectural Design Exploration of Low-Exergy (LowEx) Buildings in the Tropics. Doctoral Thesis, ETH Zurich, Zurich, Switzerland, 2016.

22. Lu, S.; Li, J.; Li, J. Auxiliary method for performance evaluation and modification of building form based on machine learning. *J. Archit.* **2019**, 31–37. [CrossRef]
23. Duering, S.; Chronis, A.; Koenig, R. Optimizing Urban Systems: Integrated Optimization of Spatial Configurations. In Proceedings of the SimAUD 2020, Vienna, Austria, 25–27 May 2020; pp. 503–509.
24. Zaghloul, M. Machine-Learning aided Architectural Design—Synthesize Fast CFD by Machine-Learning. Doctoral Thesis, ETH Zurich, Zurich, Switzerland, 2017.
25. Wortmann, T.; Natanian, J. Multi-Objective Optimization for Zero-Energy Urban Design in China: A Benchmark. In Proceedings of the SimAUD 2020, Vienna, Austria, 25–27 May 2020; pp. 203–210.
26. Newton, D. Multi-Objective Qualitative Optimization (MOQO) in Architectural Design. *Appl. Constr. Optim.* **2018**, *1*, 187–196.
27. de Miguel, J.; Villafañe, M.E.; Piškorec, L.; Sancho-Caparrini, F. Deep Form Finding Using Variational Autoencoders for deep form finding of structural typologies. In *Blucher Design Proceedings*; Editora Blucher: São Paulo, Spain, 2019; pp. 71–80.
28. Steinfeld, K.; Park, A.; Walker, S. Fresh Eyes A framework for the application of machine learning to generative architectural design, and a report of activities at Smartgeometry 2018. In *"Hello, Culture!", Proceedings of the 18th International Conference, CAAD Futures 2019, Daejeon, Korea, 26–28 June 2019*; Lee, J.-H., Ed.; Springer: Singapore, 2019; p. 22. ISBN 978-89-89453-05-5.
29. Huang, W.; Zheng, H. Architectural drawings recognition and generation through machine learning. In Proceedings of the 38th Annual Conference of the Association for Computer Aided Design in Architecture, ACADIA 2018, Mexico City, Mexico, 18 October 2018; pp. 156–165.
30. Newton, D. Deep Generative Learning for the Generation and Analysis of Architectural Plans with Small Datasets. In *Blucher Design Proceedings*; Editora Blucher: São Paulo, Spain, 2019; Volume 2, pp. 21–28.
31. Chaillou, S. AI +Architecture. Towards a New Approach. 2019, pp. 1–188. Available online: https://www.academia.edu/39599650/AI_Architecture_Towards_a_New_Approach (accessed on 14 June 2020).
32. Azizi, V.; Usman, M.; Patel, S.; Schaumann, D.; Zhou, H.; Faloutsos, P.; Kapadia, M. Floorplan Embedding with Latent Semantics and Human Behavior Annotations. In Proceedings of the SimAUD 2020, Vienna, Austria, 25–27 May 2020; pp. 337–344.
33. Ferrando, C.; Dalmasso, N.; Mai, J.; Llach, D.C. Architectural Distant Reading:Using Machine Learning to Identify Typological Traits across Multiple Buildings. In Proceedings of the 18th International Conference on CAAD Futures, Daejeon, Korea, 26–28 June 2019; pp. 204–217.
34. Abi, W.; Alymani, A. Graph Machine Learning using 3D Topological Models. In Proceedings of the SimAUD 2020, Vienna, Austria, 25–27 May 2020; pp. 421–428.
35. Bdelrahman, M.M.; Chong, A.; Miller, C. Build2Vec:Building Representation in Vector Space. In Proceedings of the SimAUD 2020, Vienna, Austria, 25–27 May 2020; pp. 101–104.
36. Tang, P.; Wang, X.; Hua, H. Decoding history—Digital technology and practice in the protection and renewal of the historical features of Dingshu ancient South Street in Yixing. *J. Archit.* **2021**, 24–30. (In Chinese) [CrossRef]
37. Janssen, H. Monte-Carlo based uncertainty analysis: Sampling efficiency and sampling convergence. *Reliab. Eng. Syst. Saf.* **2013**, *109*, 123–132. [CrossRef]
38. Zhu, C.; Tian, W.; Shi, J.; Yin, B. Research on the influence of sampling method and sample size on the accuracy of building energy consumption machine learning model. *Build. Sci.* **2020**, *36*, 199–207. [CrossRef]
39. Ma, C.; Zhu, S.; Wang, M. Research on the application of machine learning technology in architectural design. *South. Archit.* **2021**, 121–131. [CrossRef]
40. Yang, J.; Qiao, P.; Li, Y.; Wang, N. A survey of machine learning classification problems and algorithms. *Stat. Decis. Mak.* **2019**, *35*, 36–40. [CrossRef]
41. Etherm, A. *Introduction to Machine Learning*; Fan, M.; Min, H.; Niu, C., Translators; China Machine Press: Beijing, China, 2009.
42. Chiandussi, G.; Codegone, M.; Ferrero, S.; Varesio, F.E. Comparison of multi-objective optimization methodologies for engineering applications. *Comput. Math. Appl.* **2012**, *63*, 912–942. [CrossRef]
43. Gou, S.; Nik, V.M.; Scartezzini, J.-L.; Zhao, Q.; Li, Z. Passive design optimization of newly-built residential buildings in Shanghai for improving indoor thermal comfort while reducing building energy demand. *Energy Build.* **2018**, *169*, 484–506. [CrossRef]

Article

Leakage Diagnosis of Air Conditioning Water System Networks Based on an Improved BP Neural Network Algorithm

Rundong Liu [1,2,3,*], **Yuhang Zhang** [4] **and Zhengwei Li** [4]

[1] School of Environmental Science and Engineering, Suzhou University of Science and Technology, Suzhou 215009, China

[2] Jiangsu Province Key Laboratory of Intelligent Building Energy Efficiency, Suzhou University of Science and Technology, Suzhou 215009, China

[3] National and Local Joint Engineering Laboratory of Municipal Sewage Resource Utilization Technology, Suzhou 215009, China

[4] School of Mechanical and Energy Engineering, Tongji University, Shanghai 200092, China; 1932711@tongji.edu.cn (Y.Z.); zhengwei_li@tongji.edu.cn (Z.L.)

[*] Correspondence: lrd@mail.usts.edu.cn

Abstract: Compared with traditional pipe networks, the complexity of air conditioning water systems (ACWSs) and the alternation of cooling and heating are more likely to cause pipe network leakage. Pipe leakage failure seriously affects the reliability of the air conditioning system, and can cause energy waste or reduce human comfort. In this study, a two-stage leakage fault diagnosis (LFD) method based on an Adam optimization BP neural network algorithm, which locates leakage faults based on the change values of monitoring data from flow meters and pressure sensors in air conditioning water systems, is proposed. In the proposed LFD method, firstly, the ACWS network's hydraulic model is built on the Dymola platform. At the same time, a cuckoo algorithm is used to identify the pipe network's characteristics to modify the model, and the experimental results show that the relative error between the model-simulated value and the actual values is no more than 1.5%. Secondly, all possible leakage conditions in the network are simulated by the model, and the dataset is formed according to the change rate of the observed data, and is then used to train the LFD model. The proposed LFD method is verified in a practical project, where the average accuracy of the first-stage LFD model in locating the leaking pipe is 86.96%; The average R^2 of the second-stage LFD model is 0.9028, and the average error between the predicted location and its exact location with the second-stage LFD model is 6.3% of the total length of the leaking pipe. The results show that the proposed method provides a feasible and convenient solution for timely and accurate detection of pipe network leakage faults in air conditioning water systems.

Keywords: BP neural network; air conditioning water systems; leakage fault diagnosis

Citation: Liu, R.; Zhang, Y.; Li, Z. Leakage Diagnosis of Air Conditioning Water System Networks Based on an Improved BP Neural Network Algorithm. *Buildings* **2022**, *12*, 610. https://doi.org/10.3390/buildings12050610

Academic Editors: Shi-Jie Cao, Dahai Qi, Junqi Wang and Gwanggil Jeon

Received: 31 March 2022
Accepted: 5 May 2022
Published: 6 May 2022

Publisher's Note: MDPI stays neutral with regard to jurisdictional claims in published maps and institutional affiliations.

1. Introduction

With people's aspirations and pursuit of a better life, the requirements for indoor air quality have gradually increased [1], and centralized or semi-centralized central air conditioning systems are being more and more widely used in today's buildings. As an important part of air conditioning systems, the pipe network not only plays an important role in connecting the unit, the air conditioning terminal device, and the cooling side, but also undertakes the key task of transporting and distributing the cold or heat to each terminal device. However, as the use of air conditioning systems grows over the years, many factors—such as the pipe material, surrounding environment, laying method, construction quality, and operation and maintenance management—affect the reliability of the air conditioning water system (ACWS) pipe network, and pipe leakage has become one of the most frequent failures in the whole life cycle of air conditioning systems [2].

Usually, ACWSs are equipped with water make-up devices, so a small leakage in the pipe network is often not obvious, and is often overlooked. However, pipe leakage failure often causes immeasurable damage to the actual running of the system. Figure 1 shows the common forms of damage to ACWS pipe networks. The "Practical Heating and Air Conditioning Design Manual" [3] stipulates the hourly leakage and replenishment of the ACWS. If a large public building is designed according to this standard with a floor area of 20,000 m^2, the hourly make-up water to the ACWS due to pipe network leakage is 800 L. At the same time, the make-up water is cold or hot water treated by high-priced softening; in addition to the large waste of water resources and energy consumption, pipe network leakage may also cause the system to deviate from the best working operating point [4], thus affecting human comfort. Pipe network leakage has become the most common but difficult-to-deal-with problem in ACWS failures. In addition, the vast majority of accidents in the pipe network do not occur suddenly, but are gradual, slow processes [5]. Pipe network leakage is a precursor to major accidents, such as long-term neglect of the leakage problem which, in the event of an accident, can seriously threaten people's lives and the safety of their property. Therefore, if the pipe leakage can be detected and solved in a timely manner, one can not only avoid the waste of water resources and energy consumption by the system, but also prevent the problem before it occurs, effectively reducing the possibility of safety accidents such as pipe bursts.

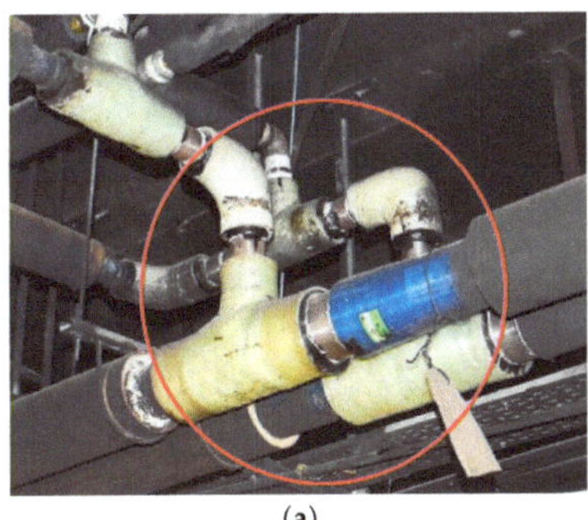

(a)   (b)

Figure 1. Common forms of damage to ACWS pipe networks: (**a**) leakage causes ACWS performance degradation; (**b**) ACWS water leakage causes mildew.

The key means of solving leakage problems in pipe networks is to use scientific and effective pipe network leakage detection methods to accurately locate the location of the leak(s) in the faulty pipe network. Discussion has been made of various pipeline fault detection methods, viz., vibration analysis, pulse-echo methodology, acoustic techniques, negative-pressure-wave-based leak detection systems, support-vector machine (SVM)-based pipeline leakage detection, interferometric-fiber-sensor-based leak detection, filter diagonalization method (FDM), etc. It was found that these methods have been applied for specific fluids, such as oil, gas, and water [6]. In air conditioning systems, the popularity of temperature, pressure, and flow sensors has led many scholars to propose diagnostic methods for pipe leakage based on statistical or analytical methods, including the pressure gradient method [7], pressure point analysis [8], negative pressure wave method [9], and extended Kalman filter [10] (the leakage position is determined using the 9-DOF IMU (3D accelerometer, 3D gyroscope, and 3D magnetometer) sensor data in the extended Kalman filter), which were found to have limited application scenarios based on examples, or to be more appropriate as auxiliary diagnostic tools.

With the development of artificial neural network technology, more and more scholars are exploring the possibilities of this technology for identifying pipe network leakage faults [11–13]. Lei et al. [14] established a BP-neural-network-based heating pipe network leakage fault diagnosis model based on artificial neural network, and introduced the idea of hierarchy into pipe network leakage diagnosis. Duan et al. [15] established an adaptive neuro-fuzzy inference system (ANFIS)-based LFD model for centralized heating pipe networks. The example verified that the ANFIS-based heat network LFD model is stable and has high diagnostic accuracy. Xue et al. [16] proposed an XGBoost-based district heating network leakage diagnosis method, using the rate of change of observed data from the flow and pressure sensors installed in the system to locate the leaky pipe section. A series of studies have shown that the technology applied to district heating networks (DHNs) has good diagnostic accuracy and stability—especially BP neural networks, which show a strong ability of nonlinear fitting and self-sample learning. However, with the increasing volume and complexity of ACWS networks, the traditional manual inspection or hardware-based network leakage diagnosis methods struggle to meet the requirements of "quick, accurate and steady" diagnosis of leakage faults in ACWS networks [14].

While there are a few studies on ACWS leakage diagnosis, this study focuses on the actual diagnosis effect of a BP neural network algorithm on the ACWS leakage problem. In addition, most of the pipe network LFD methods mentioned above regard the leaking pipe's number as the final diagnosis result. When the leaking pipe is long, this causes problems with the diagnostic efficiency. Leakage diagnosis of pipelines can prevent environmental and financial losses [17], and instability of air conditioning water systems can lead to lower energy efficiency of air conditioning systems [18]. This study aims to diagnose pipes' exact leakage location after diagnosing the identity of the leaking pipe. Compared with DHNs, ACWSs are more directly customer-facing, and affect the customer experience more quickly when they leak. Therefore, compared with other networks, it is more necessary to study the leakage diagnosis of ACWS networks, and to explore the feasibility of BP neural network technology for practical applications.

This paper proposes the possibility of applying an Adam-optimized BP neural network algorithm to the diagnosis of pipe leakage in ACWSs, establishes a simulated hydraulic model of pipe network leakage based on the Dymola platform (a kind of engineering modeling platform, detailed in Section 2.2.1), adopts the cuckoo search algorithm to identify the characteristics of the pipe network so as to ensure the reliability and accuracy of the simulation model, proposes a two-stage leakage diagnosis method with the Adam-optimized BP neural network—which not only locates the leaky pipe section, but also locates the exact leakage point on the pipe—and demonstrates the application of the method based on a practical project.

The rest of this paper is organized as follows: Section 2 briefly describes the research methodology, with practical examples. Section 3 shows and analyzes the actual results of the method. Section 4 discusses the case results and implications. Section 5 presents the main conclusions of this study.

2. Materials and Methods

2.1. Research Overview

Figure 2 shows the research path and design ideas of this paper, and the methods used are detailed in the subsequent section.

Figure 3 shows the detailed diagnosis process. When the ACWS is stable, the leakage diagnosis is triggered by monitoring the make-up water's flow rate (if the pipe network system is not installed with a make-up water flow sensor, the difference between the water supply's main flow rate and the return's main flow rate is used as the make-up water flow rate) when the set threshold is exceeded in the duration (which can be set). The real-time data (i.e., flow rate and pressure) monitored by the system are provided to the pipe network model and the BP neural network for second-stage diagnosis, respectively, and the first-stage diagnosis is performed by the sample set of leak conditions from the Dymola pipe network water system model, which outputs the identification number of its leaky pipe section. At the same time, to determine whether the length of the pipe section exceeds the set threshold length, if the length of the pipe section is shorter than the set threshold, the second-stage diagnosis is skipped and the network fault information is output directly; if the length of the pipe section is longer than the set threshold, the second-stage diagnosis is triggered, and the exact location of the leakage and the identity of the leaky pipe section are output as the results of the fault information to complete the accurate identification of the leakage fault in the ACWS.

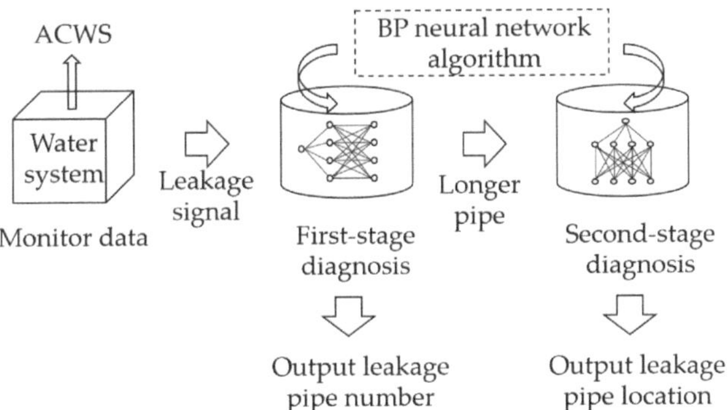

Figure 2. Research path and design ideas.

The ACWS network hydraulic model was built on the Dymola platform (detailed in Section 2.2.1); the identification of network resistance characteristics by the cuckoo algorithm (detailed in Section 2.2.2) helps to improve the hydraulic model and ensure that the model is as close as possible to the actual system. Meanwhile, the two-stage leakage diagnosis methods of the pipe network were all built using an Adam-optimized BP neural network, with different parameters set according to the different output variables of the two stages (detailed in Section 2.2.3).

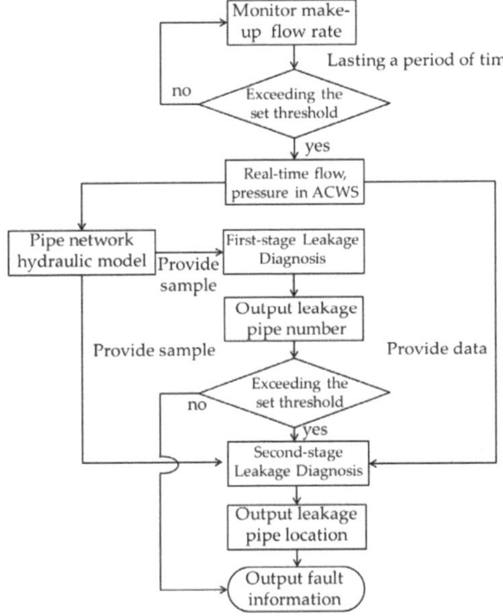

Figure 3. Research diagram of two-stage leakage diagnosis in ACWSs.

2.2. Method Description

2.2.1. Hydraulic Model of the ACWS Network under Simulated Leakage Conditions

The Dymola platform is a multidisciplinary system modeling and simulation tool based on the Modelica language (an object-oriented physical modeling language) [19] It has a library of models and simulation specialties suitable for multiple engineering domains, and has proven its applicability in the engineering field, while "Modelica.Fluid" provides the basis for developing the network hydraulic model. However, the existing components cannot meet the simulation under leakage conditions, so for this paper we developed a hydraulic model that can simulate different leakage conditions based on the Dymola platform.

The model is shown in Figure 4. In the event that a leak occurs at a certain point on the pipe, the simulated leakage is divided into two pipes—"pipe1" and "pipe2"—and the two pipes are connected by a tee junction without pressure loss, and then connected by a "negative flow source" (corresponding to the source module in Figure 4), which can be set by the user. All of the above components are "packaged" into a new element.

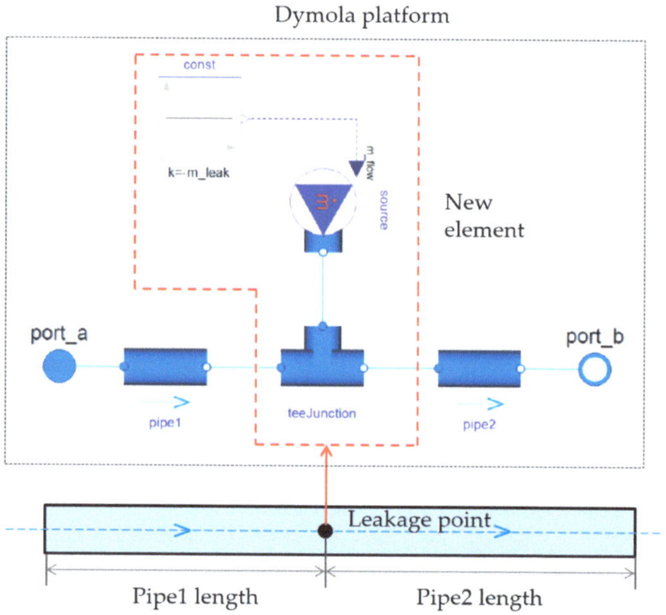

Figure 4. Hydraulic model of pipe leakage on the Dymola platform.

By setting the flow rate of the "negative flow source" and the ratio of the length of each pipe to the total pipe length (i.e., the simulated pipe leakage volume and leakage location), the model is built under different leakage conditions. In order to verify the accuracy of the hydraulic model of the pipe leakage, the classical theoretical calculation method and the simulation method based on the hydraulic model are used to calculate the pressure difference between the inlet and outlet of the same pipe, and the model is verified by comparing the calculation results (details in Appendix A).

2.2.2. Cuckoo Search Algorithm for the Identification of Network Resistance Characteristics

The hydraulic model established using Dymola described in Section 2.2.1 is not sufficient to simulate the actual pipe network system. In the actual pipe network, there are factors such as pipe corrosion, internal wall scaling, etc. The resistance characteristics of the pipe network inevitably deviate from the initial design calculation values, and the key to establishing an accurate hydraulic model of the pipe network lies in the identification of the resistance characteristics of the pipe network. Optimization algorithms are a common method for the identification of pipe network resistance characteristics. The performance of the cuckoo search algorithm was compared with that of the particle swarm algorithm, differential evolution algorithm, artificial bee swarm algorithm, and other algorithms based on various test functions [20], showing that the cuckoo search algorithm has fewer parameters, simple operation, easy implementation, generality and robustness, and excellent local and global search capabilities with comprehensive advantages. Meanwhile, the object of this paper is similar to the research objects in the literature on the identification of pipe resistance characteristics, so the cuckoo search algorithm was used in this study to help improve the hydraulic model.

Before applying the optimization algorithm, it is necessary to determine the objective function of the identification of the resistance characteristics of the pipe network in this study. The purpose is that the final identification parameters optimized by the algorithm can make the simulated parameters of the hydraulic model as close as possible to the actual monitored values. In this study, the sum of the absolute value of the relative error between the actual monitored values and the model-simulated values of each sensor (pressure and flow rate) in the pipe network system is used as the objective function, and the formula is shown in Equation (1):

$$obj(S) = \sum_{z=1}^{Z} \left(\sum_{i=1}^{N_P} \left| \frac{P_s - P_m}{P_m} \right| + \sum_{j=1}^{N_Q} \left| \frac{Q_s - Q_m}{Q_m} \right| \right) \tag{1}$$

where Z is the number of conditions involved in the calibration; N_P and N_Q represent the numbers of pressure and flow sensors installed in the network system, respectively; P_m, and P_s represent the monitored and simulated pressures, respectively; and Q_m and Q_s represent the monitored and simulated pipe section flow rates, respectively. There are also implicit constraints between the simulated pressure P_s and the simulated pipe flow Q_s, i.e., the hydraulic balance equations (i.e., nodal continuity equation, basic loop energy equation, and Bernoulli's equation) of the network itself need to be met.

The process of identifying the resistance characteristics of the ACWS network based on the cuckoo search algorithm is as follows:

Step 1: Build a pipe network simulation model based on deterministic parameters.

Step 2: Set parameters such as population size, discovery probability P_a, maximum number of iterations N of the algorithm, and random initialization of bird's nest locations, with each set of bird's nest locations representing a set of pipe resistance characteristic coefficients to be identified.

Step 3: Substitute each nesting position into the pipe network simulation model to calculate the objective function value of each nesting position (i.e., each set of pipe resistance characteristic coefficients), and compare them to obtain the current optimal nesting position and the optimal objective function value.

Step 4: Keep the optimal nest location in the previous generation, update the nest locations other than the optimal nest using Lévy flight, and calculate the corresponding objective function value, compare the obtained objective function value with the current optimal value, and update the current optimal objective function value [21].

Step 5: Compare the random number r with the discovery probability P_a. If $r > P_a$, change the nest location once randomly; if not, keep it the same. Finally, retain the optimal set of nest locations.

Step 6: If the maximum number of iteration generations has been reached or the search precision requirement has been met, continue to the next step; otherwise, return to Step 4

Step 7: Output the global optimal nest location, which is the optimal resistance coefficient of the pipe network in this search process.

The optimal results obtained according to the above steps are used as the pipe network resistance coefficients of the hydraulic model, making the hydraulic model more accurate and closer to the actual operation of the system.

2.2.3. Adam Optimization Algorithm for the LFD Model

"BP neural network" usually refers to multilayer feedforward neural networks trained with the error backpropagation (BP) algorithm. BP neural networks have been widely used in many engineering fields—such as pattern recognition, intelligent control, fault diagnosis, image recognition processing, and optimization computation—due to their nonlinear mapping capability, self-learning and self-adaptive capability, and generalization capability.

The Adam (adaptive moment estimation) optimization algorithm is an improved algorithm for traditional BP neural networks, which adopts independent adaptive learning rates for different parameters by calculating the first-order moment estimation and second-order moment estimation of the gradient during the training of the neural network [22]. It has been experimentally shown that neural networks based on the Adam optimization algorithm not only have faster training speed compared to other stochastic optimization methods, but also do not easily fall into local optima, and have excellent performance in practice. Therefore, this study uses the Adam optimization algorithm as the kernel for the LFD model of the ACWS.

This paper adopts the concept of hierarchy in the structure of the LFD model [23]. On the one hand, the ACWS undertakes the building's heat and cold load, and there are a variety of control methods, such as fixed-flow and variable-flow systems, variable-flow systems that contain the supply and return main-fixed temperature control systems, supply and return main-fixed differential pressure control systems, and the most unfavorable end-fixed differential pressure control systems. The control system is complex; on the other hand, the length of each pipe section in the ACWS is unevenly distributed, and the number of stages varies greatly—the long pipes may be hundreds of meters, while the short pipes may only be one or two meters. If all the pipes are diagnosed with leakage faults, this will affect the diagnosis time and reduce efficiency, while there is research showing that two-stage fault diagnosis, compared to single-fault diagnosis (a single diagnosis to identify the section of the leaky pipe and the leakage location), has higher diagnostic accuracy as well as relatively less training time [24], which not only reduces the complexity of fault diagnosis and the training time of diagnosis, but can also adjust the fault diagnosis process and improve the efficiency of fault diagnosis according to the user's needs in the actual application process. Therefore, this paper uses the two-stage LFD model for fault diagnosis of the ACWS pipe network.

The leakage conditions are simulated by the improved hydraulic model, and the sample datasets under different conditions are obtained. Then, the two-stage LFD model is trained by setting the parameters of the Adam optimization algorithm (such as the number of hidden layer nodes, the activation function of the hidden layer, and the regularization parameter). The training and testing process of the neural network was achieved in this study using the Python programming language.

2.2.4. LFD Model Performance Evaluation Indicators

In order to verify the application effect of the LFD model, it is also necessary to introduce indicators to measure the performance of the two-stage LFD model [25]. Since the two-stage LFD models solve different types of problems, different performance evaluation indicators need to be used, as shown in Table 1.

Table 1. Model performance evaluation indicators.

	Performance Evaluation Indicators	Definitions
	Accuracy	$Accuracy = \frac{TP+TN}{TP+TN+FP+FN}$
First-stage diagnosis model	P	$P = \frac{TP}{TP+FP}$
	R	$R = \frac{TP}{TP+FN}$
	F_1	$F_1 = \frac{2PR}{P+R}$
	MSE (mean squared error)	$MSE = \frac{1}{m} \sum_{i=1}^{m} (y_i - \hat{y}_i)^2$
Second-stage diagnosis model	*MAE* (mean absolute error)	$MAE = \frac{1}{m} \sum_{i=1}^{m} |y_i - \hat{y}_i|$
	R^2 (coefficient of determination)	$R^2 = 1 - \frac{\sum_{i=1}^{m}(\hat{y}_i - y_i)^2}{\sum_{i=1}^{m}(\overline{y}_i - y_i)^2}$

First-stage diagnosis is a typical multiple-classification task, so it uses the model performance evaluation indicators commonly used for classification tasks: precision, recall, and F_1 score. For the binary classification task, the samples can be classified into four cases according to the combination of true and predicted categories: *TP* (true positive), *FP* (false positive), *TN* (true negative), and *FN* (false negative).

Second-stage diagnosis is a typical regression task, so its performance needs to be evaluated using different evaluation indicators from first-stage leakage diagnosis. The commonly used performance evaluation indicators for regression tasks are shown in Table 1, where m denotes the total number of samples, y_i represents the true marker of x_i, $\hat{y}_i$ represents the prediction result of the learner for x_i, and $\overline{y}_i$ represents the average of the m sets of true values.

2.3. Case Study

Taking a Guangzhou (China) metro station's ACWS system pipe network project as an example, the ACWS form is a primary pump variable-flow system with fixed differential pressure control for the supply and return mains. The chilled water system has a supply and return water temperature of 10 and 17 °C, respectively. The system pressure point is set at the entrance of the circulating water pump, and the pressure is 32.3 kPa. Pressure sensors are set for the cold source and each terminal device's import and export, and flow sensors are set for the chilled water main pipe and each terminal device's return branch pipe.

In order to model the actual pipe network system on the Dymola platform, the system needs to be reasonably simplified, as follows: ① the chiller, the pump, and other equipment in the room are combined into one node (cold source); ② if two pipes are connected and there is no node in the middle, the two pipes are combined into the same pipe; ③ the local resistance of the fittings in the pipe is expressed using the length of the straight pipe section of the same diameter as the connected pipe (i.e., the local resistance's equivalent length). The supply and return water system, with pipe section numbers, is shown in Figure 5.

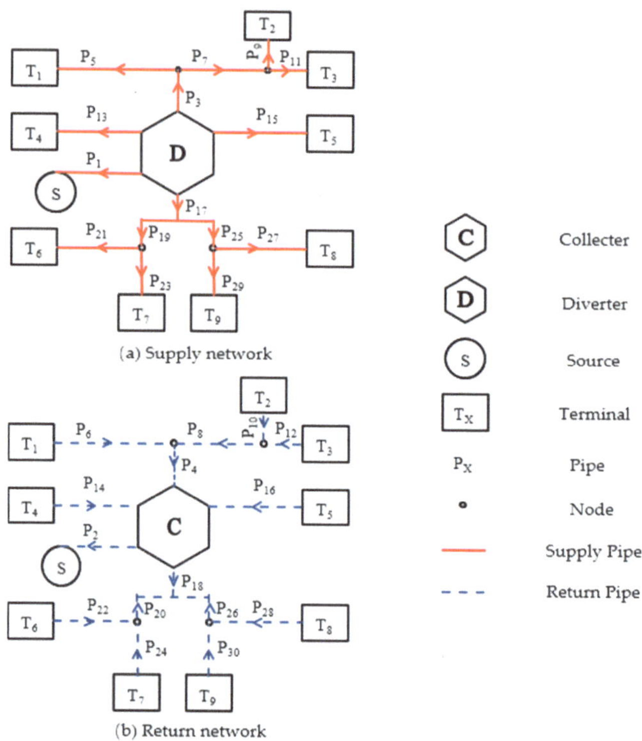

Figure 5. Supply and return of water in the ACWS network.

The case pipe network includes 1 cold source (S), 9 terminal devices (T1–T9), 30 pipes (pipe1–pipe30), and 31 sensor monitoring points (11 flow sensors and 20 pressure sensors), where the basic parameters of each pipe section are shown in Figure 6.

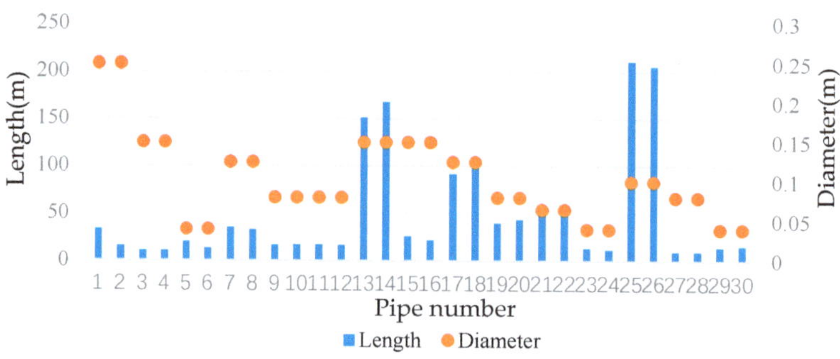

Figure 6. Basic parameters of the case pipe network.

The cuckoo search algorithm was used to identify the resistance characteristics of the case pipe network, and the parameters are shown in Table 2. The training curves of the optimization algorithms show that when the number of iterations reaches 100, the average and optimal fitness values of the algorithms no longer decrease significantly, and the algorithms can be considered to have reached convergence. In order to avoid the random error in the single identification process, the algorithm was operated independently 10 times, and the average value of the 10 identification results was taken as the final result of the identification of the resistance characteristics of the pipe network.

Table 2. Parameter settings of the cuckoo search algorithm used for the identification of resistance characteristics.

Parameter	Setting
Population size of each generation	50
Discovery probability	0.25
Maximum number of iterations	100

A sufficient number of training samples is a prerequisite for a satisfactory diagnostic performance of a pipe network LFD model, and the more comprehensive the leakage conditions contained, the more abundant the training data, and the better the performance of the final training diagnostic model. However, it is very difficult—almost impossible—to obtain a large and comprehensive set of fault condition data via experimental testing or historical data logging for the ACWS network in this case. Therefore, the Dymola model simulation was used to obtain the data samples required for the training of the LFD model. The simulated leakage conditions were as follows: for each pipe leak point setting in the case study, the ratio of the distance from the start of the pipe section to the total pipe length was selected as 0.05, 0.1, 0.15, 0.2, 0.25, 0.3, 0.35, 0.4, 0.45, 0.5, 0.55, 0.6, 0.65, 0.7, 0.75, 0.8, 0.85, 0.9, or 0.95, while the ratio of the possible leakage volume from the total circulating water volume was selected as 1%, 1.5%, 2%, 2.5%, 3%, 3.5%, 4%, 4.5%, or 5%.

According to the above setting conditions, for the 30 pipes in the case pipe network system, each pipe has 19 possible leakage points, and each leakage point can have 9 different degrees of leakage. For each leakage working condition, 5130 sets of simulated data samples can be obtained, one by one. At the same time, taking into account the random error of the sensor measurement process in the actual case, a certain amount of artificial noise is added to the original data generated by the simulation model, and the artificially added noise X follows a normal distribution with a mean of 0 and a standard deviation of σ. The accuracy level of the case pipe network sensor is 0.2%FS (full-scale, range), and according to the "3σ" criterion of normal distribution, σ is taken as 1/3 of 0.2% FS.

In the settings of the first-stage LFD model, all data samples are randomly divided into a training set and a test set at a ratio of 9:1. The random partitioning process takes the form of stratified sampling. The training set is used to train the BP neural network model, while the test set is used to replace the fault data monitored in the actual case. The neural network adopts a single-hidden-layer structure. The parameter settings of the Adam optimization BP neural network algorithm used for the first-stage LFD model are presented in Table 3.

Table 3. Parameter settings of the Adam optimization BP neural network algorithm used for the first-stage LFD model.

Parameter	Setting
Number of hidden layer nodes	31
Activation function of hidden layer	Identity function
Regularization parameter	0.0001
Maximum number of iterations	3000
Convergence precision	1×10^{-4}

In the settings of the second-stage LFD model, the length threshold $\theta_L = 50$ m is set as the basis for determining whether to perform secondary diagnosis of leaks according to the conditions of the case system. The output of the second-stage LFD model is the exact location of the leakage point in the ACWS, which is expressed as the distance of the leakage point from the beginning of the pipe section/the total length of the leakage pipe. Due to the fact that the second-stage LFD model is also based on the BP neural network model, its training and testing methods are essentially the same as those of the first-stage LFD model. Some of the settings are different, as follows: the training and testing sets are randomly divided at a ratio of 7:3, and the activation function is a ReLU function.

3. Results

3.1. Results of Identification of the Case Pipe Network's Characteristics

The flow and pressure parameters of four groups of case pipe networks under regular conditions were selected as the original data for solving the objective function of pipe network resistance characteristic identification, and another set of flow and pressure data were selected to verify the effect of identification of the pipe network resistance characteristics based on the cuckoo search algorithm. In order to avoid the accidental error of the single identification process, the algorithm was operated 10 times independently, and the average value was taken as the final result, as shown in Table 4.

As shown in Table 4, from the identification results of the pipe network's resistance characteristics, we found that the resistance characteristic coefficient of each pipe section was quite different; the smallest was 24 s^2/m^5, and the largest was 487,138.3 s^2/m^5. The main reason for this was the large difference in the pipes' length in the case system, and the internal situation of the pipe network was also different.

Due to the fact that the actual monitored values of the pipe network contain only two parameters of flow and pressure, it was necessary to substitute the results of the optimal pipe network resistance characteristic coefficients identified by the algorithm into the hydraulic model of the pipe network, and to evaluate the identification accuracy and precision of the algorithm by comparing the simulated values with the actual monitored values of flow and inlet/outlet pressure at each terminal, the results of which are shown in Figure 7.

Table 4. Results of identification of the case pipe network's characteristics.

Pipe Number	Coefficient of Pipe Resistance Characteristic (s^2/m^5)	Pipe Number	Coefficient of Pipe Resistance Characteristic (s^2/m^5)
1	53.3	16	588.0
2	24.0	17	6188.9
3	267.4	18	6749.5
4	277.5	19	32,186.2
5	487,138.3	20	32,118.5
6	344,413.0	21	119,729.4
7	2257.2	22	115,777.9
8	2249.1	23	426,030.6
9	11,666.6	24	383,965.0
10	11,909.8	25	52,034.0
11	12,438.9	26	43,475.3
12	12,068.8	27	7121.1
13	4295.8	28	7064.7
14	4421.0	29	347,589.0
15	718.9	30	414,020.7

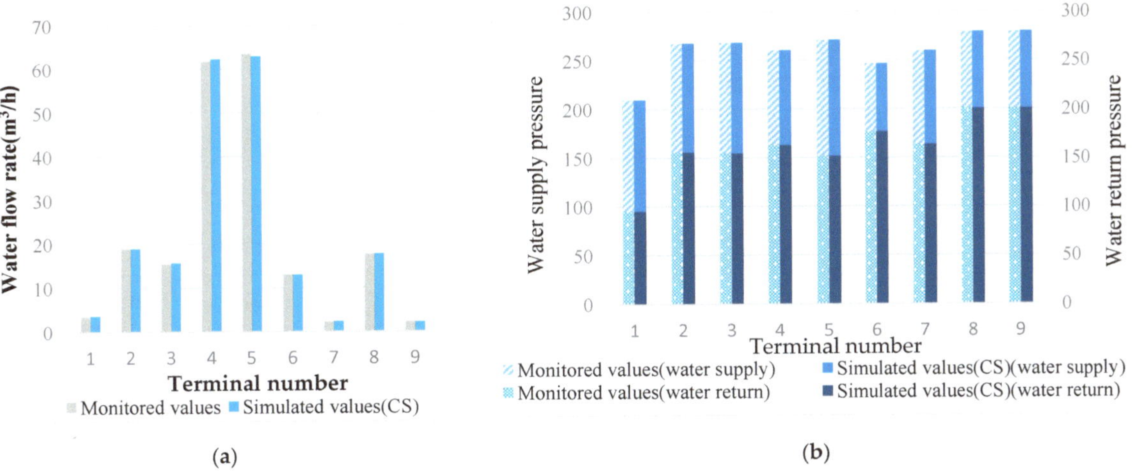

Figure 7. (**a**) Comparison of monitored and simulated flow values. (**b**) Comparison of monitored and simulated pressure values.

It was found that the optimal identification of the pipe resistance characteristic coefficients based on the cuckoo algorithm was largely consistent with the actual monitored values after being substituted into the original hydraulic model. As depicted in Figure 7a, the minimum difference in flow rate was 0.04 m^3/h, while the maximum difference was 0.63 m^3/h. As depicted in Figure 7b, the minimum difference in pressure was 0.02 kP$_a$, and the maximum difference was 0.3 kP$_a$. Further discussing the identification results of the pipe network's resistance characteristics, it can be observed that the average relative error between the monitored and simulated values of the flow rate at each terminal was 1.358%, the average relative error between the monitored and simulated values of the water supply pressure at each terminal was 0.057%, and the average relative error between the monitored and simulated values of the return water pressure at each terminal was 0.089%.

The above data show that the hydraulic model optimized by the cuckoo search algorithm to identify the resistance characteristics achieves an acceptable error range for the actual project. This verifies that this method can be used for the calibration and optimization of the ACWS hydraulic model, and also provides technical possibilities for application in other complex pipe networks, such as heating and municipal pipe networks. Due to the leakage condition dataset used to train the two-stage LFD model being generated from the hydraulic model simulation, the hydraulic model has a significant impact on the LFD model simulation results.

3.2. Results of the First-Stage LFD Model

Before testing the model, it is necessary to set the neural network's number of hidden layer nodes, because the number of nodes affects the performance of the LFD model. It has been shown in the literature [26] that if the number of hidden layer nodes is too small, the neural network will lack the necessary learning ability and information processing ability. If the number of hidden layer nodes is too large, it will not only greatly increase the complexity of the neural network structure, and make the neural network more likely to fall into local minima during the learning process, but also make the learning speed of the neural network very slow. The range of node numbers is determined with a commonly used empirical formula, and then tested and adjusted to obtain the optimal number of hidden layer nodes for the neural network.

According to the settings of the first-stage LFD model (detailed parameter settings are shown in Table 3), we used the Python programming language for the training and testing of the neural network. A total of 100 experiments were repeated to avoid the random error caused by a single experiment. Each time, the dataset was resampled to obtain a new training set and test set, and the average of 100 experiments was taken as the final result of the first-stage LFD model.

Table 5 presents each pipe's average precision, recall, and F_1 score from 100 experiments. The average precision was 87.26%, and the average recall was 86.96%. The first-stage LFD model's precision and recall were above 85%. However, there were still some pipe categories with poor diagnosis results such as pipe numbers 1, 2, 3, 4, 7, 8, 15, and 16 in the case network (discussed in Section 4.1).

At the same time, in order to further evaluate the performance of the first-stage LFD model, two evaluation indicators—*accuracy* and *macro-F_1*—were selected to evaluate the overall performance of the model in 100 test sets. Figure 8 shows the accuracy and *macro-F_1* of the BP-neural-network-based leak-stage diagnosis model for the ACWS network on the 100 test sets. The horizontal axis is the accuracy of the LFD model on the test set for each group of experiments, while the vertical axis is the macro-F1 of the LFD model on the test set.

Table 5. First-stage LFD model's average performance metrics.

Pipe Number	Precision (P)	Recall (R)	F_1	Pipe Number	Precision (P)	Recall (R)	F_1
1	62.40%	66.53%	0.6440	16	66.07%	66.65%	0.6636
2	64.85%	66.18%	0.6551	17	92.39%	85.90%	0.8902
3	51.19%	51.55%	0.5137	18	95.58%	90.14%	0.9278
4	57.51%	60.86%	0.5914	19	96.08%	95.91%	0.9599
5	99.47%	98.19%	0.9882	20	93.68%	95.21%	0.9444
6	99.64%	97.90%	0.9876	21	99.06%	98.54%	0.9880
7	72.19%	73.08%	0.7263	22	99.52%	97.13%	0.9831
8	84.09%	84.98%	0.8453	23	99.11%	98.24%	0.9868
9	93.16%	91.86%	0.9251	24	97.60%	97.49%	0.9755
10	95.38%	92.93%	0.9414	25	97.85%	92.82%	0.9527
11	93.41%	92.10%	0.9251	26	96.50%	91.98%	0.9418
12	94.61%	95.67%	0.9514	27	92.54%	98.01%	0.9520
13	89.31%	89.67%	0.8949	28	89.66%	96.37%	0.9289
14	91.69%	91.80%	0.9174	29	98.44%	96.48%	0.9745
15	57.31%	60.13%	0.5869	30	97.47%	94.62%	0.9602

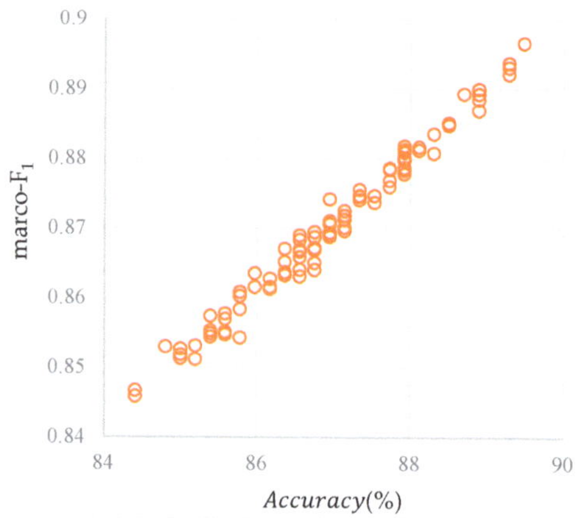

Figure 8. Performance metrics of the Adam-based LFD model on the test set.

As shown in Figure 8, the performance of the Adam-based LFD model was relatively stable in 100 sets of test experiments, and the classification *accuracy* of the LFD model for the case network was between 84.41% and 89.47%, while the *macro-F$_1$* was between 0.8458 and 0.8966. After the final calculation, the average *accuracy* of the BP-neural-network-based LFD model was 86.96%, and its average *macro-F$_1$* was 0.8709. The results show that the first-stage LFD model performs satisfactorily.

3.3. Results of the Second-Stage LFD Model

In this case, a total of eight pipes exceeded the threshold of 50 m. According to the settings of the second-stage LFD model (detailed parameter settings are shown in Table 3), the Python programming language was also used for the training and testing of the neural network.

Figure 9 presents the average *MAE*, *MSE*, and R^2 values for each of the eight pipes under 100 test experiments, with a stable overall performance. The minimum *MSE* was 0.00518 and the maximum was 0.01117, with the average value being 0.00708, which is less than 0.01; the minimum R^2 value was 0.85 and the maximum was 0.93, with the average value being 0.90, which is greater than 0.9.

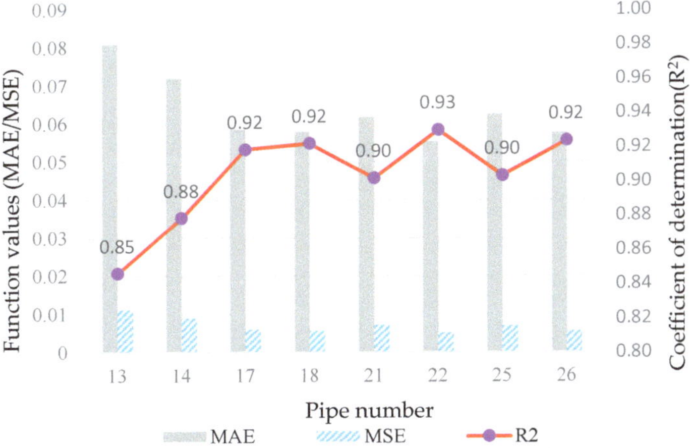

Figure 9. Second-stage LFD model performance in 100 sets of test experiments.

Taking the pipe13 leakage model as an example, the experimental results are presented for one randomly selected group out of 100 test experiments. Figure 10 shows the exact leakage locations and the model-predicted leakage locations in the test set. Except for a few outliers, the majority of the predicted values are close to the actual values, and it can be intuitively inferred that the BP-neural-network-based second-stage LFD model is able to predict the pipe leakage location accurately. Thus, the second-stage LFD model has good diagnostic efficacy.

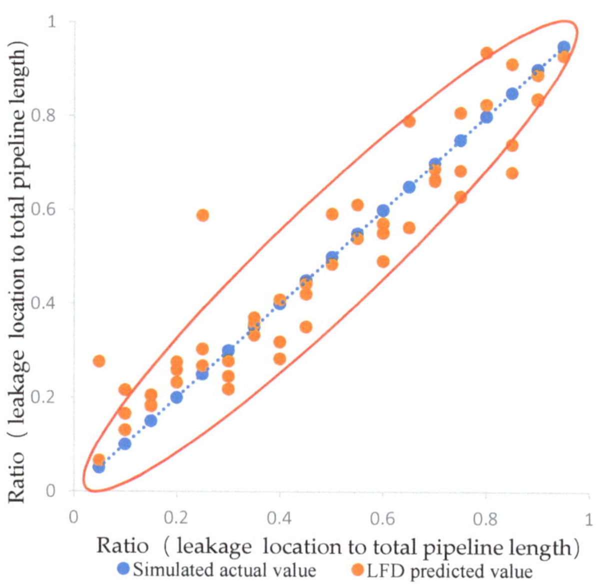

Figure 10. Distribution of simulated actual leakage locations and LFD-predicted locations for pipe13.

4. Discussions

4.1. Two-Stage LFD Model Discussion

The simulation results of the first-stage LFD model show that the diagnostic efficacy of some pipes is poor—such as pipe numbers 1, 2, 3, 4, 7, 8, 15, and 16—and the precision and recall rates are less than 85%. Therefore, the first-stage LFD model is prone to classification errors, resulting in incorrect diagnosis.

According to Table 4, the pipes with poor diagnostic efficacy of the first-level diagnosis model of pipe network leakage have the common characteristics of mutual connection and small resistance characteristic coefficients. Therefore, when these pipes have leakage failures, the parameter amplitude of the pipe network caused by leakage is limited—especially when the sensor reading itself has a certain random error—and the parameter change caused by the leakage of a pipe with a small resistance characteristic coefficient is easy to cover with noise, so it is difficult for the first-stage LFD model of the pipe network to achieve a diagnosis.

In response to the problem that the pipe network's LFD model is prone to misdiagnosis when leakage occurs in pipes with small resistance coefficients, this paper also seeks solutions from the following four perspectives:

(1) Improving the accuracy of sensors: Sensors with high accuracy should be installed and used as far as possible within the allowable range of conditions. The higher the accuracy level of the sensors, the less likely the reading error of the sensors to cover up the impact of pipe leakage, resulting in improved diagnostic accuracy of the LFD model.

(2) Cooperative diagnosis of multiple working conditions: In the event that it is difficult to make a diagnosis in a single working condition, the system's operating conditions can be switched to re-diagnose the leakage, and the fault diagnosis results of multiple working conditions can be used to collaboratively locate the leaky pipe section.

(3) Checking the second-ranking pipe based on probability distribution: Since the kernel of the LFD model is the probability distribution of the neural network algorithm, the second-ranking pipe in the probability ranking is checked, and if no fault occurs in this pipe, the pipes in the next ranking are checked by analogy.

(4) Combining pipe categories: Since the pipe categories prone to misclassification often have characteristics of low resistance and interconnection, it is useful to combine such pipes into one category and then carry out fault diagnosis, and when a fault is found in this category in practical application, all pipes within its combination should be inspected. Since the resistance characteristic coefficients of such pipes are small, and the pipe lengths are usually not too long, this does not theoretically increase the work difficulty of the maintenance personnel significantly.

In this case study, if pipes 1, 3, 7, and 15 and 2, 4, 8, and 16 are respectively combined into one category and other settings remain unchanged, the average accuracy of the BP-neural-network-based LFD model will be increased to 94.86%, which is 7.89% higher than the average accuracy before combining, It can be seen that the diagnostic effect of the first-stage LFD model can be improved by combining the pipe categories.

The simulation results of the second-stage LFD model show that the predicted value of the model is close to the simulated actual value under the leakage condition. The threshold set by the secondary leakage model in this paper is 50 m, and the average deviation between the predicted leakage point position and its actual position is 6.3% of the total length of the leaky pipe. This is acceptable for pipes over 50 m long. Applying the second-stage LFD model in engineering practice can help locate the exact leakage location of long pipes quickly, even if there are deviations, and when human initiative is exercised, maintenance workers can quickly locate the leakage point in the area near the resultant value, which plays an important role in reducing labor costs and improving time efficiency.

4.2. Limitations

In this paper, 11 flow sensors and 20 pressure sensors were installed in the ACWS network. The number of sensors in the case network is relatively complete, but in other projects, the number of sensors in many ACWSs is lower [27], which may reduce the performance of the LFD model, so the performance of the LFD model with different numbers of sensors should be explored in subsequent research.

At the same time, due to climate overheating [28], more passive design strategies are being added to buildings, which will lead to increased complexity and reduced reliability of air conditioning systems [29]. This directly affects the stability of ACWSs, so this factor should be considered in future research.

In addition, this paper is devoted to the study of pipeline leakage faults, which account for more than 80% of ACWS faults [30]. However, there are still blockage and junction problems in actual running water systems, which will be another direction for further research. Moreover, the ACWS pipe network leakage diagnosis method proposed in this paper only accounts for single-point pipe network leakage problems. Although the phenomenon of multiple simultaneous leakage failures is rare in the normal operation of running pipe network systems, there still exists a certain degree of possibility of this situation occurring. Therefore, the problem of multiple leakage faults occurring at the same time in ACWS pipe networks needs to be investigated in the future.

5. Conclusions

This paper proposes a leakage diagnosis method for ACWSs using an Adam optimization BP neural network algorithm, which is able to provide fault diagnosis. When a leak occurs in the actual pipe network system, the method locates the leaky pipe in the network and then locates the exact leakage location on the pipe. The main conclusions are summarized as follows:

(1) A hydraulic model that can simulate pipe network leakage was developed on the Dymola platform. In order to ensure the accuracy and reliability of the hydraulic model, a method of identifying the pipe network's resistance characteristics based on the cuckoo search algorithm was proposed, and the identification results of the algorithm were applied to the case of an ACWS pipe network. The average relative error of the flow rate at each terminal was 1.358%, the water supply pressure at each terminal was 0.057%, and the return

water pressure at each terminal was 0.089%. The results show that the average relative error between the simulated values and the actual monitored values obtained using the method was no more than 1.5%. The hydraulic model was consistent with the real system and it is practicable to generate the dataset under leakage conditions with this model.

(2) A two-stage LFD model based on the Adam-optimized BP neural network algorithm was proposed. The case study shows that the average accuracy of the first-stage LFD model for locating leaky pipes was 86.96%, and when the method of combining some pipe categories with smaller resistance characteristic coefficients was used, the average accuracy of the model increased to 94.86%. The second-stage LFD model of pipe network considers pipe's with a length of over 50 m in the ACWS system; the average R^2 of the second-stage LFD model was 0.9028, and the average error between the predicted location of the leakage point and its simulated actual location was 6.3% of the total length of the leakage pipe.

It should be noted that the hydraulic model of pipe network leakage built on the Dymola platform can be applied not only in ACWSs, but also in leakage studies of pipe networks of other engineering water systems (e.g., district heating pipe networks and municipal water supply networks). In addition, the proposed cuckoo search algorithm performs excellently in the identification of resistance characteristics of the pipe network, making the hydraulic model closer to the actual operating system, and providing a method to achieve an accurate hydraulic model. Meanwhile, the proposed Adam-optimized BP neural network algorithm for the two-stage fault diagnosis model shows accuracy in practical engineering applications, providing a promising and sustainable solution for actual ACWS leakage problems. More importantly, the design concept of hierarchical diagnosis opens up the possibility of complex water system diagnosis, and the operation and maintenance of water systems will be more intelligent and efficient in the future.

Author Contributions: Conceptualization, R.L. and Y.Z.; data curation, R.L.; writing—original draft preparation, R.L.; writing—review and editing, R.L.; visualization, R.L.; methodology, R.L. and Y.Z.; software, Y.Z.; validation, R.L., Y.Z. and Z.L.; formal analysis, R.L. and Y.Z.; investigation, Y.Z.; resources, Z.L.; supervision, Z.L.; project administration, R.L.; funding acquisition, R.L. All authors have read and agreed to the published version of the manuscript.

Funding: This research received no external funding.

Institutional Review Board Statement: Not applicable.

Informed Consent Statement: Not applicable.

Data Availability Statement: Data is contained within the article.

Conflicts of Interest: The authors declare no conflict of interest.

Appendix A

The theoretical calculation method is the most commonly used method for hydraulic calculation of pipe networks under leakage conditions. Based on the constant flow condition of the pipe, the import and export pressure difference of the pipe under leakage conditions can be calculated using the Darcy formula.

Example: In a straight pipe with a total length of 10 m, the nominal diameter of the pipe is DN65, the absolute roughness of the inner wall of the pipe is 0.2 mm and the distribution is uniform, the water temperature in the pipe is 10 °C, the inlet flow of the pipe is always maintained at 10 L/s, and the flow rate of the fluid in the pipe is 3.0 m/s without leakage.

In order to avoid the influence of contingency on the results of a single experiment, a total of 10 different sets of experiments were set up. Each experiment was set up with a different leakage volume and leakage location (the leakage location was expressed as the ratio of the distance of the leakage point from the starting point of the pipe to the total pipe length). The experimental results are shown in Table A1.

Table A1. Pressure difference between the inlet and outlet of a pipe, as determined by two methods.

Number	Leakage Volume (L/s)	Leakage Location	Pressure Difference between the Inlet and Outlet of the Pipe (kP$_a$)		Relative Error (%)
			Theoretical Value	Simulated Value	
1	0	0	19.011	18.962	−0.257
2	0.1	0.1	18.675	18.627	−0.260
3	0.5	0.1	17.365	17.318	−0.273
4	1.0	0.1	15.802	15.757	−0.289
5	0.1	0.5	18.825	18.756	−0.365
6	0.5	0.5	18.097	18.049	−0.266
7	1.0	0.5	17.288	17.181	−0.273
8	0.1	0.9	18.974	18.925	−0.258
9	0.5	0.9	18.828	18.780	−0.259
10	1.0	0.9	18.655	18.606	−0.261

The results show that under different leakage conditions, the simulation results of the inlet and outlet pressure difference of the experimental pipe are very close to the theoretical calculation results. The maximum absolute value of the relative error is 0.365%, and the average error is −0.276. The error is usually within the acceptable range of hydraulic calculation, so the developed leakage hydraulic model is suitable for simulating the actual pipe leakage.

References

1. Wang, J.; Huang, J.; Fu, Q.; Gao, E.; Chen, J. Metabolism-based ventilation monitoring and control method for COVID-19 risk mitigation in gymnasiums and alike places. *Sustain. Cities Soc.* **2022**, *80*, 103719. [CrossRef] [PubMed]
2. Li, H.; Li, H.; Pei, H.; Li, Z. Leakage detection of HVAC pipeline network based on pressure signal diagnosis. *Build. Simul.* **2019**, *12*, 617–628. [CrossRef]
3. Lu, Y. Practical Heating and Air Conditioning Design Manual. *Heat. Vent. Air Cond.* **2008**, *6*, 152.
4. Mulligan, S.; Hannon, L.; Ryan, P.; Nair, S.; Clifford, E. Development of a data driven FDD approach for building water networks: Water distribution system performance assessment rules. *J. Build. Eng.* **2021**, *34*, 101773. [CrossRef]
5. Puust, R.; Kapelan, Z.; Savic, D.A.; Koppel, T. A review of methods for leakage management in pipe networks. *Urban Water J.* **2010**, *7*, 25–45. [CrossRef]
6. Datta, S.; Sarkar, S. A review on different pipeline fault detection methods. *J. Loss Prev. Process Ind.* **2016**, *41*, 97–106. [CrossRef]
7. Zhou, Z.; Zhang, J.; Huang, X.; Zhang, J.; Guo, X. Trend of soil temperature during pipeline leakage of high-pressure natural gas: Experimental and numerical study. *Measurement* **2020**, *153*, 107440. [CrossRef]
8. Zhou, S.; O'neill, Z.; O'neill, C. A review of leakage detection methods for district heating networks. *Appl. Therm. Eng.* **2018**, *137*, 567–574. [CrossRef]
9. Abdulshaheed, A.; Mustapha, F.; Ghavamian, A. A pressure-based method for monitoring leaks in a pipe distribution system: A Review. *Renew. Sustain. Energy Rev.* **2017**, *69*, 902–911. [CrossRef]
10. Akkaya, A.E.; Talu, M.F. Extended kalman filter based IMU sensor fusion application for leakage position detection in water pipelines. *J. Fac. Eng. Archit. Gazi Univ.* **2017**, *32*, 1393–1404.
11. Abdulla, M.B.; Herzallah, R. Probabilistic multiple model neural network based leak detection system: Experimental study. *J. Loss Prev. Process Ind.* **2015**, *36*, 30–38. [CrossRef]
12. Shen, Y.; Chen, J.; Fu, Q.; Wu, H.; Wang, Y.; Lu, Y. Detection of District Heating Pipe Network Leakage Fault Using UCB Arm Selection Method. *Buildings* **2021**, *11*, 275. [CrossRef]
13. Zadkarami, M.; Shahbazian, M.; Salahshoor, K. Pipeline leakage detection and isolation: An integrated approach of statistical and wavelet feature extraction with multi-layer perceptron neural network (MLPNN). *J. Loss Prev. Process Ind.* **2016**, *43*, 479–487. [CrossRef]
14. Lei, C.; Zhou, P. Application of neural network in heating network leakage fault diagnosis. *J. Southeast Univ. (Engl. Ed.)* **2010**, *26*, 173–176.
15. Duan, P.; Duan, L.; Tian, Q. ANFIS in Leakage Fault Diagnosis of Heating Networks. *J. Zhengzhou Univ. (Eng. Sci.)* **2014**, *35*, 56–60.
16. Xue, P.; Jiang, Y.; Zhou, Z.; Chen, X.; Fang, X.; Liu, J. Machine learning-based leakage fault detection for district heating networks. *Energy Build.* **2020**, *223*, 110161. [CrossRef]
17. Banjara, N.K.; Sasmal, S.; Voggu, S. Machine learning supported acoustic emission technique for leakage detection in pipelines. *Int. J. Press. Vessel. Pip.* **2020**, *188*, 104243. [CrossRef]

18. Duan, H.F. Development of a TFR-Based Method for the Simultaneous Detection of Leakage and Partial Blockage in Water Supply Pipelines. *J. Hydraul. Eng.* **2020**, *146*, 04020051. [CrossRef]
19. Hamza, G.; Hammadi, M.; Barkallah, M.; Choley, J.Y.; Riviere, A.; Louati, J.; Haddar, M. Compact Analytical Models for Vibration Analysis in Modelica/Dymola: Application to the Wind Turbine Drive Train System. *J. Chin. Soc. Mech. Eng.* **2018**, *39*, 121–130.
20. Civicioglu, P.; Besdok, E. A conceptual comparison of the Cuckoo-search, particle swarm optimization, differential evolution and artificial bee colony algorithms. *Artif. Intell. Rev.* **2013**, *39*, 315–346. [CrossRef]
21. Iacca, G.; Dos Santos Junior, V.C.; De Melo, V.V. An improved Jaya optimization algorithm with Levy flight. *Expert Syst. Appl.* **2021**, *165*, 113902. [CrossRef]
22. Yi, D.; Ahn, J.; Ji, S. An Effective Optimization Method for Machine Learning Based on ADAM. *Appl. Sci.* **2020**, *10*, 1073. [CrossRef]
23. Fan, Q.; Guo, Y.; Wu, S.; Liu, X. Two-Level Diagnosis of Heating Pipe Network Leakage Based on Deep Belief Network. *IEEE Access* **2019**, *7*, 182983–182992. [CrossRef]
24. Wen, L.; Li, X.Y.; Gao, L. A New Two-Level Hierarchical Diagnosis Network Based on Convolutional Neural Network. *IEEE Trans. Instrum. Meas.* **2020**, *69*, 330–338. [CrossRef]
25. Fu, Q.; Li, K.; Chen, J.; Wang, J.; Lu, Y.; Wang, Y. Building Energy Consumption Prediction Using a Deep-Forest-Based DQN Method. *Buildings* **2022**, *12*, 131. [CrossRef]
26. Qin, F.W.; Bai, J.; Yuan, W.Q. Research on intelligent fault diagnosis of mechanical equipment based on sparse deep neural networks. *J. Vibroeng.* **2017**, *19*, 2439–2455. [CrossRef]
27. Wang, J.; Hou, J.; Chen, J.; Fu, Q.; Huang, G. Data mining approach for improving the optimal control of HVAC systems: An event-driven strategy. *J. Build. Eng.* **2021**, *39*, 102246. [CrossRef]
28. Ozarisoy, B. Energy effectiveness of passive cooling design strategies to reduce the impact of long-term heatwaves on occupants' thermal comfort in Europe: Climate change and mitigation. *J. Clean. Prod.* **2022**, *330*, 129675. [CrossRef]
29. Ozarisoy, B.; Altan, H. Regression forecasting of 'neutral' adaptive thermal comfort: A field study investigation in the south-eastern Mediterranean climate of Cyprus. *Build. Environ.* **2021**, *202*, 108013. [CrossRef]
30. Sun, J.L.; Wang, R.H.; Duan, H.F. Multiple-fault detection in water pipelines using transient-based time-frequency analysis. *J. Hydroinform.* **2016**, *18*, 975–989. [CrossRef]

buildings

Article

Optimal Control of Chilled Water System Based on Improved Sparrow Search Algorithm

Qixin Zhu [1,2,*], **Mengxiang Zhuang** [3], **Hongli Liu** [1] and **Yonghong Zhu** [4]

[1] School of Mechanical Engineering, Suzhou University of Science and Technology, Suzhou 215009, China; liuhl_sz@163.com
[2] Jiangsu Province Key Laboratory of Intelligent Building Energy Efficiency, Suzhou University of Science and Technology, Suzhou 215009, China
[3] School of Environmental Science and Engineering, Suzhou University of Science and Technology, Suzhou 215009, China; mxzhuang@mail.usts.edu.cn
[4] School of Mechanical and Electronic Engineering, Jingdezhen Ceramic Institute, Jingdezhen 333001, China; zyh_patrick@163.com
* Correspondence: qxzhu@mail.usts.edu.cn

Abstract: Chilled water systems have large time delays and large inertia, and the traditional PID controller has a poor control effect. In this paper, an improved sparrow search algorithm is proposed to optimize the control of chilled water systems. Firstly, the random walk strategy was used to randomly perturb the sparrows to improve the searching ability of the sparrows. Then, a Gauss mutation was added in the iteration process of sparrows to enhance the local search ability. Finally, the values of the PID parameters as obtained by the above methods were substituted into the controller for simulation. The simulation results show that the method proposed in this paper improves the search accuracy of the sparrow search algorithm and effectively solves the problems of large time delays and large inertia in the chilled water system. The method in this paper took the least amount of time for the system to reach the steady state at only 12.75 s. The control effect of the proposed method was also better than that of the improved ant colony optimization algorithm. The rise time was 2.713 s, and the adjustment time was 4.95 s.

Keywords: sparrow search algorithm; random walk strategy; Gauss mutation; PID parameter optimization; chilled water system

Citation: Zhu, Q.; Zhuang, M.; Liu, H.; Zhu, Y. Optimal Control of Chilled Water System Based on Improved Sparrow Search Algorithm. *Buildings* **2022**, *12*, 269. https://doi.org/10.3390/buildings12030269

Academic Editors: Shi-Jie Cao, Dahai Qi, Junqi Wang and Gwanggil Jeon

Received: 12 January 2022
Accepted: 21 February 2022
Published: 24 February 2022

Publisher's Note: MDPI stays neutral with regard to jurisdictional claims in published maps and institutional affiliations.

1. Introduction

A chilled water system is the main part of an HVAC system. It consumes a significant amount of energy and has an important influence on the refrigeration effect [1]. The energy consumption of a chilled water system comprises about 30% of the energy consumption of air conditioning [2,3]. In a chilled water system, energy efficiency can be achieved by changing the equipment, control strategy, pipeline layout [4] or temperature control [5] or using other methods. A chilled water system is time-varying and has a time delay [6]. A chilled water system with a better control effect would be conducive to creating energy-saving air conditioning systems [7]. Z.J. Ma et al. proposed an online control scheme. Compared with the conventional control method, the energy consumption of the chilled water pump was reduced by 11.99–24.86% [8]. Using the traditional PID controller to control the system will lead to the system to overshoot, resulting in slow responses and other problems. It does not work very well.

Swarm optimization algorithm has been applied in various fields, including ant colony optimization (ACO) [9], particle swarm optimization (PSO) [10], grey wolf optimization (GWO) [11] and so on. The above group optimization algorithms have the problem that they are easy to fall into the local optimum. Many people have improved the optimization algorithm. Y. Wang et al. used the improved ant colony algorithm to find the global optimal solution [12]. F. Zhang et al. used Gauss mutation and self-adjusting pheromones to improve the ant colony algorithm to optimize the controller in a chilled water system and improved the searching ability and convergence ability of the algorithm [13]. A. Dixit et al proposed a new differential evolution algorithm based on particle swarm optimization. The performance comparison showed that the algorithm could improve the convergence speed significantly and avoid premature closing [14]. J.C. Gu et al. added discrete heredity into the grey wolf optimization algorithm to further improve the search ability. The experimental data show that this algorithm was more effective than other algorithms [15]. J.K. Xue et al proposed the Sparrow Search Algorithm (SSA) in 2020, which had good convergence speed and stability [16]. X.C. Li et al. used the SSA, which can quickly and accurately search the required parameters [17]. X.L. Sun et al. proposed an improved sparrow algorithm and applied it to load forecasting [18,19].

In order to achieve a better control effect, this paper used the sparrow search algorithm to find the optimal PID control parameters. However, like other algorithms, the sparrow search algorithm is also prone to fall into local optimum and population diversity decreases with the increase in iteration times. Based on the above description, in order to solve the disadvantages of the sparrow search algorithm, this paper proposes an improved sparrow search algorithm and used it to optimize the control of a chilled water system. Firstly, the random walk strategy was used to perturb the position of the sparrows to enhance the global search ability of the sparrow. Secondly, Gauss mutation was applied to change the sparrows' position, so that individual sparrows could carry out a full search of their surroundings, and the local searching ability of the sparrow group was improved. Finally, the optimization algorithm was applied to a chilled water system and compared with the ant colony algorithm and particle swarm optimization algorithm. The simulation results show that the proposed method improved the search precision of the sparrow search algorithm and reduced the overshoot of the system. The control effect was also better than that of other optimization algorithms. The main contributions of the authors are as follows: (1) The sparrow search algorithm was improved using the random walk strategy and Gaussian variation in this paper. (2) The improved sparrow search algorithm was applied to a frozen water system.

The paper is divided into five sections including the introduction. Section 2 presents the related methods. Section 3 presents the main results of this paper. A discussion is given in Section 4. The conclusion is presented in Section 5.

2. Related Methods

2.1. Sparrow Search Algorithm Optimizes PID Parameters

The sparrow search algorithm optimizes the values of the three parameters of the PID controller, which are proportional, integral and differential, and its purpose is to find the corresponding PID parameters when the objective function value is the lowest, which can control the control system more effectively. In the sparrow swarm algorithm, the finders are responsible for finding the optimal region of the objective function values, and the participants will move close to them to form an orderly sparrow swarm [16]. Participants who are in very bad positions will go to other areas to search. The participants will constantly monitor the finders through the position update and the feedback of the objective function values. If the participants' positions are better than those of the finders at this time, they will become the finders.

We sorted the objective function values from smallest to largest and updated the positions of each sparrow in real time. Finders account for 20% of the sparrows, and the rest are participants. The position update of the finders is shown in Equation (1).

$$
x_{i,j}^{d+1} = \begin{cases} x_{i,j}^{d} \cdot e^{\left(\frac{-j}{\alpha \cdot d_{max}} \right)}, & \text{if} \quad R_2 < ST \\ x_{i,j}^{d} + Q \cdot L, & \text{if} \quad R_2 \geq ST \end{cases} \tag{1}
$$

where $x_{i,j}^d$ represents the position of the jth sparrow in the ith dimension in the dth iteration, $i = 1, 2, 3, j = 1, 2, 3, \ldots, n, d = 1, 2, 3, \ldots, d_{max}$; d_{max} represents the maximum number of iterations; α is a random number, and its value range is $(0, 1)$; Q is a random number subject to normal distribution; L represents a matrix of one row and three columns in which the elements are 1; R_2 represents the alert value, which is in the range of $(0, 1)$; ST represents the safe value, and the value range is $(0.5, 1)$; When $R_2 < ST$, this indicates that the sparrows have not found the predator, the population is in a safe state and the finders can perform a more extensive search, thus finding a better feeding area. When $R_2 \geq ST$, this indicates that some sparrows have spotted a predator and will alert others, who will immediately change their search strategy and move to a safe area.

The number of participants accounts for 80% of the sparrows, and the position update is shown in Equation (2).

$$x_{i,j}^{d+1} = \begin{cases} Q \cdot e^{\left(\frac{x_{worst}^d - x_{i,j}^d}{j^2}\right)}, & \text{if } j > \frac{n}{2} \\ x_p^d + \left| x_{i,j}^d - x_p^d \right| \cdot A^+ \cdot L, & \text{otherwise} \end{cases} \tag{2}$$

where x_{worst}^d represents the worst position in the group; x_p^t represents the best position occupied by the finders; A represents a matrix of one row and three columns in which the elements are randomly assigned to either -1 or 1, and $A^+ = A^T(AA^T)^{-1}$. When $j > n/2$, this indicates that the jth participant has a relatively low fitness value. It does not acquire food and is in a state of extreme hunger. It needs to fly somewhere else to find food. When $j \leq n/2$, this indicates that the jth participant will search for things around the best location found so far.

In the sparrows, the detection and early warning mechanism was also added, so that when the sparrow senses danger, and it will move to a safe place. In this paper, these sparrows account for 20% of the population, and the position update is shown in Equation (3).

$$x_{i,j}^{d+1} = \begin{cases} x_{best}^d + \beta \cdot \left| x_{i,j}^d - x_{best}^d \right|, & \text{if } f_j > f_g \\ x_{i,j}^d + K \cdot \left(\frac{\left| x_{i,j}^d - x_{worst}^d \right|}{(f_i - f_w) + \varepsilon} \right), & \text{if } f_j = f_g \end{cases} \tag{3}$$

where x_{best}^d is the best position in the group; β represents the step size control parameter and is a normal random number with a mean of 1 and a variance of 0. K is a random number with the value range of $[-1, 1]$. f_j is the fitness value of the jth sparrow; f_g and f_w represent the maximum and minimum fitness values in the population, respectively; and ε is a very small constant. It keeps the denominator from being 0, and it is 10^{-8}. When $f_j > f_g$, this indicates that sparrows are on the fringes of the population and are extremely vulnerable to predators. When $f_j \neq f_g$, this indicates that the sparrows in the middle of the group are aware of the danger and need to move closer to other sparrows to reduce the risk of predation.

Equations (1)–(3) are the updated formulae for the sparrow group position in the sparrow search algorithm. Parameters of the PID controller in the chilled water system are expressed in the form of a sparrow group, as shown in Equation (4).

$$X = \begin{bmatrix} x_{1,1}^d & x_{2,1}^d & x_{3,1}^d \\ x_{1,2}^d & x_{2,2}^d & x_{3,2}^d \\ \vdots & \vdots & \vdots \\ x_{1,n}^d & x_{2,n}^d & x_{3,n}^d \end{bmatrix} \tag{4}$$

where d represents the number of current iterations; and n represents the number of sparrows in the population.

The fitness function is shown in Equation (5).

$$f = \int_0^t e^2(t)\mathrm{d}t \tag{5}$$

where $e(t)$ is the difference between the input value and the output value.

In the group of sparrows, each column represents three parameters of PID. The fitness function value of each sparrow is shown in Equation (6).

$$F = \begin{bmatrix} f_1^d \\ f_2^d \\ \vdots \\ f_n^d \end{bmatrix} \tag{6}$$

2.2. The Main Improvement Steps of Sparrow Search Algorithm

2.2.1. Random Walk Strategy

The expression of the random walk strategy is shown in Equation (7).

$$X(t) = [0, cumsum(2r(rand(t,1)) - 1)] \tag{7}$$

where $X(t)$ is the set of steps of the random walk; *cumsum* is the formula for calculating the sum; and t is the number of steps of the random walk, and it is d_{max}. $r(t)$ is a random number, and its definition is shown in Equation (8).

$$r(t) = \begin{cases} 1, & rand(t,1) > 0.5 \\ 0, & rand(t,1) \le 0.5 \end{cases} \tag{8}$$

where $rand(t,1)$ is a matrix of t rows and 1 column, with the value range of $[0, 1]$.

After updating the positions of the sparrows, the boundary values of the variables are updated, following the rule that the larger the number of iterations is, the smaller the scope of search is. The form is shown in Equations (9) and (10).

$$lb_i^{d+1} = \frac{-lb_i^d}{I} + x_{best}^d \tag{9}$$

$$ub_i^{d+1} = \frac{ub_i^d}{I} + x_{best}^d \tag{10}$$

where lb_i^d represents the lower boundary value of the ith dimensional variable in the dth iteration; ub_i^d represents the upper boundary value of the ith dimensional variable in the dth iteration; and i represents the boundary reduction factor, as shown in Equation (11).

$$I = \begin{cases} 1 + 10^2 \cdot \frac{d}{d_{max}}, & \text{if} \quad d > d_{max} \cdot 0.2 \\ 1 + 10^3 \cdot \frac{d}{d_{max}}, & \text{if} \quad d > d_{max} \cdot 0.5 \\ 1 + 10^4 \cdot \frac{d}{d_{max}}, & \text{if} \quad d > d_{max} \cdot 0.7 \\ 1 + 10^5 \cdot \frac{d}{d_{max}}, & \text{if} \quad d > d_{max} \cdot 0.9 \\ 1 + 10^6 \cdot \frac{d}{d_{max}}, & \text{if} \quad d > d_{max} \cdot 0.95 \end{cases} \tag{11}$$

After the random walk, the sparrows' positions are updated, as shown in Equation (12).

$$x_i^d = \frac{(x_i^d - a_i) \cdot (ub_i^d - lb_i^d)}{(b_i - a_i)} + lb_i^d \tag{12}$$

where x_i^d represents the position of the ith dimension in the dth iteration of the sparrow; a_i is the minimum value of the random walk of the ith dimensional variable; b_i is the maximum value of the random walk of the ith dimensional variable.

2.2.2. Gauss Mutation

Gauss mutation uses the mutation factor of the genetic algorithm for reference. Gauss mutation was used to change the positions of sparrows in this paper. The mutation will produce a random number conforming to the normal distribution with a mean of μ and a standard deviation of σ. The fitness will be calculated according to the mutated value, and the original value will be chosen to be replaced. If the fitness value after the mutation is smaller than the value before the mutation, the original value will be replaced with the value after the mutation, and vice versa. The formula of the gauss mutation is shown in Equation (13).

$$x_i^d = x_i^d(1 + N(0,1)) \tag{13}$$

According to the characteristic of normal distribution, Gauss mutation has strong local search ability and can search the local area around sparrows adequately. The Gauss mutation improved the diversity of the sparrow search algorithm, which is conducive to finding the optimal position more quickly and accurately.

3. Results

3.1. Establishment of Optimization Model

The simulation of this paper was carried out in MATLAB 2019a. The mathematical model of the chilled water system is very complicated and belongs to a high-order system. Thus, a second-order model with time delay was used instead. The mathematical model of the chilled water system is shown in Equation (14).

$$G(s) = \frac{Ke^{-\tau s}}{(T_1 s + 1)(T_2 s + 1)} \tag{14}$$

where T_1 and T_2 are the inertial time constants; K is the amplification factor; and τ is the lag time parameter of the chilled water system.

A block diagram of the optimized chilled water system based on the improved sparrow search algorithm is shown in Figure 1.

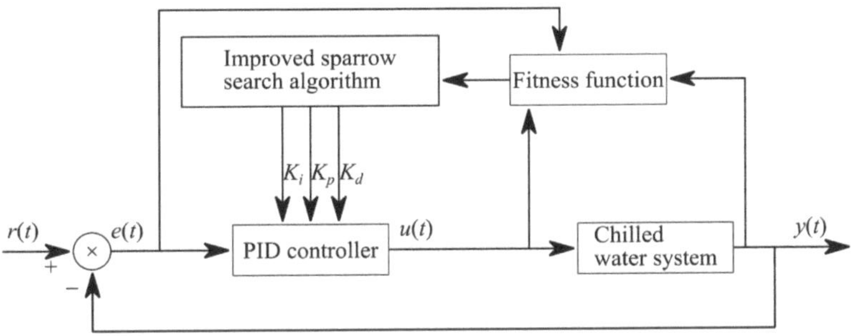

Figure 1. The block diagram of the chilled water system based on the algorithm.

3.2. Comparison of Simulation Effects

In order to verify the effectiveness of the proposed method, the proposed method was compared with the sparrow search algorithm, particle swarm optimization algorithm and ant colony optimization algorithm. The control method adopted was temperature difference control. The temperature difference between the supply and return water of the chilled water was set at 5 °C. The sampling time of the system was 0.5 s. The three parameters of the PID controller had an upper limit of 5 and a lower limit of 0. The group size of the swarm optimization algorithm in this paper was 50, and the maximum number of iterations was 100. Other parameters of the group algorithm were set as follows:

Particle swarm optimization algorithm: $w = 0.7$ (inertia factor) and $c1 = 2$, $c2 = 2$ (acceleration constant).

Ant colony optimization algorithm: *Rho* = 0.7 (Pheromone evaporation coefficient) *Q* = 1 (Pheromone intensity) and *Lam* = 0.2 (Crawling speed of ants).

The specific parameter values of the improved part and the sparrow search algorithm are detailed in Section 2.

The control object model adopted in this paper is shown in Equation (15). The optimal PID parameter values found by each algorithm are shown in Table 1. The simulation diagram of the system is shown in Figure 2.

Table 1. The optimization results of each algorithm.

Methods	K_p	K_i	K_d
This paper	3.6982	0.034355	3.3944
SSA	1.4747	0.027479	0.44203
PSO	4.2188	0.0326	5
ACO	3.8493	0.0339	3.7885

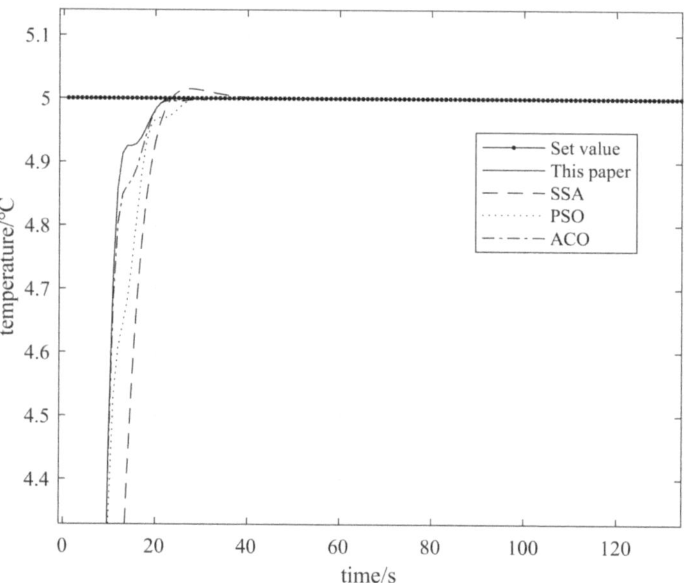

Figure 2. Simulation diagram of the system.

Since the application of the sparrow search algorithm in frozen water system has not been reported, in order to be more convincing, this paper also makes a comparison with literature. In Ref. [13], F. Zhang et al. used the improved ant colony algorithm to optimize the control of a chilled water system. In order to improve the searching ability and fast convergence, Gaussian variation and self-adjusting pheromones were introduced, and the sum of deviation squares was used as the objective function. The control method adopted was temperature difference control, which was controlled at 5 °C, and the outlet temperature of the chilled water was set at 7 °C. The purpose of the control was to keep the temperature of the return water at 12 °C. The model adopted by the control object is shown in Equation (15).

$$G(s) = \frac{12e^{-30s}}{(50s + 1)(s + 1)} \tag{15}$$

The sampling time of the system was 5 s, and the range of optimization parameters was $K_p \in [0, 0.6]$, $K_i \in [0, 0.5]$, $K_d \in [0, 1]$. The PID parameters optimized by the improved ant colony algorithm in [13] were 1.6561, 0.0325 and 0.8839. The PID parameters optimized by the method in this paper were 0.26399, 0.0052043 and 0.83927. The simulation results are shown in Figure 3. The control performances of the two methods are shown in Table 2.

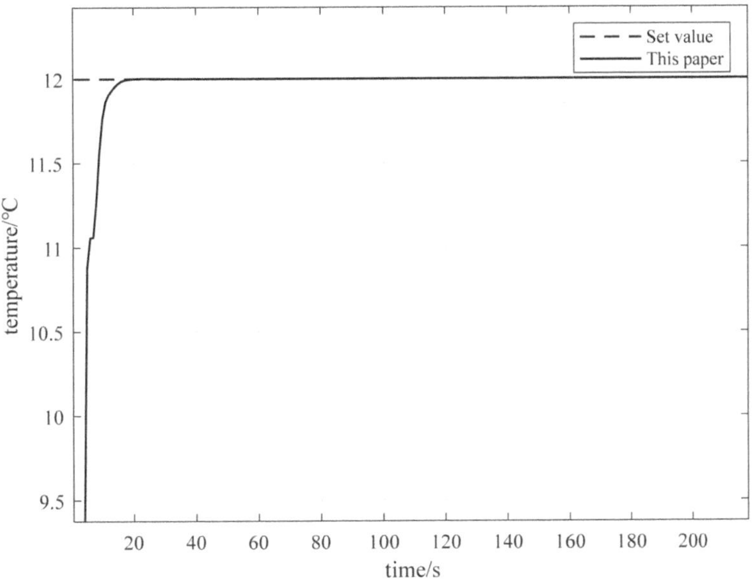

Figure 3. Simulation diagram of improved sparrow search algorithm.

Table 2. Comparison of control performance.

Methods	Overshoot	Rise Time	Adjustment Time
Literature [13]	0%	4.4 s	6.23 s
This paper	0%	2.713 s	4.95 s

4. Discussion

As can be seen from Figure 2, the simulation effect of the unimproved sparrow search algorithm was the worst, and even overshoot appeared. After the improvement in this paper, the overshoot was reduced. The method in this paper took the least amount of time for the system to reach the steady state at only 12.75 s. The time taken for the SSA, PSO and ACO to reach the steady state was 19.21 s, 17.45 s and 16.39 s, respectively. The proposed method has a better control effect than other methods.

From the perspective of control performance, the method proposed in this paper performed better than the control in [13] in both the rise time and adjustment time. The rise time was 2.713 s, and the adjustment time was 4.95 s. The control method proposed in this paper has a fast response speed, and its stability was improved accordingly. In view of the hysteresis of chilled water system, this method can improve the response speed of the controller.

5. Conclusions

In this paper, an improved sparrow search algorithm was proposed to optimize the control of chilled water systems. The random walk strategy was used to perturb the sparrows' positions, and Gauss mutation was added to improve the sparrows' search ability. Compared with the particle swarm optimization algorithm and ant colony optimization algorithm, the simulation results show that the improved method proposed in this paper not only improved the search ability of the sparrow search algorithm but also improved the control effect of the chilled water system. The method in this paper took the least amount of time for the system to reach the steady state at only 12.75 s. The control effect of the proposed method is also better than that of the improved ant colony optimization algorithm. The rise time was 2.713 s, and the adjustment time was 4.95 s. In future research, we will apply other methods to the swarm optimization algorithm to further improve the search ability of the swarm optimization algorithm.

Author Contributions: Methodology, Q.Z. and M.Z.; investigation, Q.Z.; writing—original draft preparation, M.Z.; writing—review and editing, Q.Z., M.Z. and H.L.; software, M.Z. and H.L.; validation, Q.Z. and H.L.; data curation, M.Z. and Y.Z.; supervision, Q.Z. and H.L. All authors have read and agreed to the published version of the manuscript.

Funding: This research was partially supported by the National Nature Science Foundation of China (Grant No.51875380 and 62063010).

Data Availability Statement: The data presented in this study are available on request from the corresponding author.

Acknowledgments: We acknowledge any support given which is not covered by the author contribution or funding sections.

Conflicts of Interest: We declare no conflict of interest.

References

1.　Deng, J.W.; He, S.; Wei, Q.P.; Liang, M.; Hao, Z.G.; Zhang, H. Research on systematic optimization methods for chilled water systems in a high-rise office building. *Energy Build.* **2020**, *209*, 109695. [CrossRef]

2.　Song, W.; Yang, J.M.; Ji, Y.B.; Zhang, C.X. Experimental study on characteristics of a dual temperature control valve in the chilled water system of an air-conditioning unit. *Energy Build.* **2019**, *202*, 109369. [CrossRef]

3.　Chang, C.; Jiang, Y.; Wei, Q.P. Evaluation of terminal coupling and its effect on the total delta-T of chilled water systems with fan coil units. *Energy Build.* **2014**, *72*, 390–397. [CrossRef]

4.　Gao, D.C.; Wang, S.W.; Sun, Y.J.; Xiao, F. Diagnosis of the low temperature difference syndrome in the chilled water system of a super high-rise building: A case study. *Appl. Energy* **2012**, *98*, 597–606. [CrossRef]

5.　Ma, X.W.; Gao, D.C.; Liang, D. Improved Control Strategy of Variable Speed Pumps in Complex Chilled Water Systems Involving Plate Heat Exchangers. *Procedia Eng.* **2017**, *205*, 2800–2806. [CrossRef]

6.　Li, J.W.; Ren, Q.C.; Long, H.; Feng, Z.X. Study on the Elman neural network operation control strategy of the central air conditioning chilled water system. *World J. Eng. Technol.* **2019**, *7*, 73–82. [CrossRef]

7.　Fang, X.; Jin, X.Q.; Du, Z.M.; Wang, Y.J.; Shi, W.T. Evaluation of the design of chilled water system based on the optimal operation performance of equipments. *Appl. Therm. Eng.* **2017**, *113*, 435–448. [CrossRef]

8.　Ma, Z.J.; Wang, S.W. An optimal control strategy for complex building central chilled water systems for practical and real-time applications. *Build. Environ.* **2009**, *44*, 1188–1198. [CrossRef]

9.　Wang, X.Q.; Li, H.Z.; Yang, J.; Yang, C.Y.; Gui, H.X. Optimal path selection for logistics transportation based on an improved ant colony algorithm. *Int. J. Embed. Syst.* **2020**, *13*, 200–208. [CrossRef]

10.　Fesharaki, J.J.; Madani, S.G.; Golabi, S. Best pattern for placement of piezoelectric actuators in classical plate to reduce stress concentration using PSO algorithm. *Mech. Adv. Mater. Struct.* **2018**, *27*, 141–151. [CrossRef]

11.　Natesan, G.; Chokkalingam, A. Task scheduling in heterogeneous cloud environment using mean grey wolf optimization algorithm. *ICT Express* **2019**, *5*, 110–114. [CrossRef]

12.　Wang, Y.; Yang, R.R.; Xu, Y.X.; Li, X.; Shi, J.L. Research on multi-agent task optimization and scheduling based on improved ant colony algorithm. In *IOP Conference Series: Materials Science and Engineering*; IOP Publishing: Bristol, UK, 2021. [CrossRef]

13.　Zhang, F.; Li, Z.J. Optimal control of central air-conditioning chilled water system by improved ant colony algorithm. *Comput. Eng. Des.* **2019**, *40*, 1311–1315. (In Chinese) [CrossRef]

14.　Dixit, A.; Mani, A.; Bansal, R. An adaptive mutation strategy for differential evolution algorithm based on particle swarm optimization. *Evol. Intell.* **2021**, *4*, 1–15. [CrossRef]

15.　Gu, J.C.; Jiang, T.H.; Zhu, H.Q.; Zhang, C. Low-Carbon Job Shop Scheduling Problem with Discrete Genetic-Grey Wolf Optimization Algorithm. *J. Adv. Manuf. Syst.* **2020**, *19*, 1–14. [CrossRef]

16.　Xue, J.K.; Shen, B. A novel swarm intelligence optimization approach: Sparrow search algorithm. *Syst. Sci. Control Eng.* **2020**, *8*, 22–34. [CrossRef]

17.　Li, X.C.; Ma, X.F.; Xiao, F.C.; Xiao, C.; Wang, F.; Zhang, S.C. Time-series production forecasting method based on the integration of Bidirectional Gated Recurrent Unit (Bi-GRU) network and Sparrow Search Algorithm (SSA). *J. Pet. Sci. Eng.* **2022**, *208*, 109309. [CrossRef]

48. Sun, X.L.; Li, S.X.; Liu, Q.Q.; Wang, K. Improved Sparrow Algorithm Based on Good Point Set and Inertia Weight. *Adv. Appl. Math.* **2021**, *10*, 3225–3232. [CrossRef]
49. Sun, X.L.; Li, S.X.; Liu, Q.Q.; Wang, K. Short Term Load Forecasting Based on Improved Sparrow Algorithm. *Comput. Sci. Appl.* **2021**, *11*, 2271–2279. [CrossRef]

Article

Building Energy Consumption Prediction Using a Deep-Forest-Based DQN Method

Qiming Fu [1,2,†], Ke Li [1,2,†], Jianping Chen [2,3,4,*], Junqi Wang [2], You Lu [1,2] and Yunzhe Wang [1,2]

1   School of Electronics and Information Engineering, Suzhou University of Science and Technology, Suzhou 215009, China; fqm_1@126.com (Q.F.); 1913041010@post.usts.edu.cn (K.L.); luyou@usts.edu.cn (Y.L.); yunzhe1991@mail.usts.edu.cn (Y.W.)

2   Jiangsu Province Key Laboratory of Intelligent Building Energy Efficiency, Suzhou University of Science and Technology, Suzhou 215009, China; junqi_wang@seu.edu.cn

3   School of Architecture and Urban Planning, Suzhou University of Science and Technology, Suzhou 215009, China

4   Chongqing Industrial Big Data Innovation Center Co., Ltd., Chongqing 400707, China

*   Correspondence: alanjpchen@aliyun.com

†   These authors contributed equally to this work.

Abstract: When deep reinforcement learning (DRL) methods are applied in energy consumption prediction, performance is usually improved at the cost of the increasing computation time. Specifically, the deep deterministic policy gradient (DDPG) method can achieve higher prediction accuracy than deep Q-network (DQN), but it requires more computing resources and computation time. In this paper, we proposed a deep-forest-based DQN (DF–DQN) method, which can obtain higher prediction accuracy than DDPG and take less computation time than DQN. Firstly, the original action space is replaced with the shrunken action space to efficiently find the optimal action. Secondly, deep forest (DF) is introduced to map the shrunken action space to a single sub-action space. This process can determine the specific meaning of each action in the shrunken action space to ensure the convergence of DF–DQN. Thirdly, state class probabilities obtained by DF are employed to construct new states by considering the probabilistic process of shrinking the original action space. The experimental results show that the DF–DQN method with 15 state classes outperforms other methods and takes less computation time than DRL methods. MAE, MAPE, and RMSE are decreased by 5.5%, 7.3%, and 8.9% respectively, and R^2 is increased by 0.3% compared to the DDPG method.

Keywords: energy consumption prediction; deep forest; deep Q-network; shrunken action space

Citation: Fu, Q.; Li, K.; Chen, J.; Wang, J.; Lu, Y.; Wang, Y. Building Energy Consumption Prediction Using a Deep-Forest-Based DQN Method. *Buildings* **2022**, *12*, 131. https://doi.org/10.3390/buildings12020131

Academic Editor: Geun Young Yun

Received: 22 December 2021
Accepted: 23 January 2022
Published: 27 January 2022

Publisher's Note: MDPI stays neutral with regard to jurisdictional claims in published maps and institutional affiliations.

1. Introduction

Global energy consumption increases drastically every year due to economic development and population growth. Building energy consumption is an integral part of the world's total energy consumption, accounting for 20.1% on average [1]. In many countries, this percentage is much higher; for example, it accounts for 21.7% and 38.9% of total energy consumption in China and America, respectively [2,3]. This increasing energy consumption exacerbates global warming and the scarcity of natural resources. Hence, improving building energy efficiency is crucial, as it can slow down global warming and promote the sustainable development.

Energy consumption prediction plays an important role in improving building energy efficiency, since it can facilitate the implementation of many building energy efficiency measures, namely demand response of buildings [4], urban energy planning [5], and fault detection [6]. It can also assist in assessing operation strategies of different systems, such as heating, ventilation, and air conditioning (HVAC) systems [7], and indirect evaporative cooling energy recovery systems [8] to save energy. Therefore, numerous studies have been concerned with energy consumption prediction, and many methods have been introduced to predict energy consumption.

1.1. Related Work

1.1.1. Energy Consumption Prediction

According to Ref. [9], all methods for energy consumption prediction can be roughly classified into engineering, statistical, and artificial intelligence methods. Table 1 shows the merits and demerits of these methods.

The engineering methods employ physics principles and thermodynamic formulas to calculate the energy consumption of each component of the building. Relationships between input and output variables are very clear, but these methods require detailed building and environmental information, which is often very difficult to obtain. Some researchers have tried to simplify engineering models to effectively predict energy consumption.

Yao et al. [10] proposed a simple method of formulating load profile (SMLP) to predict the daily breakdown energy demand of appliances, domestic hot water, and space heating. This method predicted energy demand for one season at a time since the average daily consumption for each component varied seasonally. Wang et al. [11] simplified the physical building model to predict cooling load. The parameters of the simplified models of the building envelopes were determined using easily available physical properties based on frequency response characteristic analysis. Moreover, they employed a thermal network of lumped thermal mass to represent a building's internal mass with parameters identified using monitored operation data. However, because they used simplified models, the prediction results may not have been completely accurate [12].

Statistical methods use mathematical formulas to correlate energy consumption data with influencing factors. Ma et al. [13] employed multiple linear regression (MLR) and self-regression methods to construct models based on the analysis of the relevant power energy consumption factors, such as specific population activities and weather conditions. They used the least square method to estimate parameters and predicted monthly power energy consumption for large-scale public buildings. Lam et al. [14] developed a new climatic index based on the principal component analysis (PCA) of three major climatic variables: dry-bulb temperature, wet-bulb temperature, and global solar radiation. Then, regression models were constructed to correlate the simulated daily cooling load with the corresponding daily new index. The calculation processes of these statistical methods are straightforward and fast, but they often cannot handle stochastic occupant behaviors and complex interactions between factors, so they are not flexible and often have poor prediction accuracy [15].

Artificial intelligence methods can learn from historical data, which are usually called data-driven methods, and they usually performance better than other methods [16]. In Ref. [17], all artificial intelligence methods were broadly divided into two categories: traditional artificial intelligence methods and deep learning methods. However, in practice, decision tree (DT), support vector machine (SVM), artificial neural network (ANN), and many other traditional machine learning methods can be regarded as traditional artificial intelligence methods. Azadeh et al. [18] proposed a method based on ANN and analysis of variance (ANOVA), which was used to predict annual electricity consumption. The method was more effective than the conventional regression model. Hou et al. [19] employed SVM to predict the cooling load of HVAC system, and the results indicated that the SVM method was better than auto-regressive integrated moving average (ARIMA) methods. In Ref. [20], DT, stepwise regression, and ANN methods were employed to predict electricity energy consumption in Hong Kong. The prediction results indicated that DT and ANN methods performed slightly better in the summer and winter phases, respectively. However, these traditional artificial intelligence methods adopt shallow structures for modeling, limiting models' prediction accuracy.

Deep learning (DL) methods may not always reflect physical behaviors, but they can learn more abstract features from raw inputs to construct better models [21]. Cai et al. [22] used recurrent neural network (RNN) and convolutional neural network (CNN) to forecast time-series building-level load in recursive and direct multi-step manners. The experimental results showed that the gated 24-h CNN method could improve prediction accuracy by 22.6%, compared with the seasonal auto-regressive integrated moving average with exogenous inputs (ARIMAX). Ozcan et al. [23] proposed dual-stage attention-based recurrent neural networks to predict electric load consumption. The method used the encoder and decoder for feature extraction as well as the attention mechanism. The experimental results indicated that the proposed method outperforms other methods.

Deep reinforcement learning (DRL) methods are another category of artificial intelligence methods that cannot be neglected. DRL methods combine the perception of DL with decision-making of reinforcement learning (RL) and have achieved many substantial results in many fields, such as games [24], robotics [25], and autonomous driving [26]. In the building field, DRL methods are mainly used to find the optimal control in HVAC systems [27,28]. Many researchers have also used DRL methods for energy consumption prediction and achieved satisfactory results. For instance, Liu et al. [29] explored the performance of DRL methods for energy consumption prediction, and the results showed that the deep deterministic policy gradient (DDPG) method achieved the highest prediction accuracy in single-step-ahead prediction. However, the potential of DRL methods has not been fully realized. One limitation of current works is that many researchers only focus on the DDPG method, but ignore the classical deep Q-network (DQN) method.

Table 1. Summary of merits and demerits of the prediction methods.

Method		Merits	Demerits
	Engineering [10,11]	Relationships between input and output variables are very clear	Detailed building information is required
	Statistical [13,14]	Straightforward and fast	Not flexible
Artificial intelligence	Traditional machine learning [18–20]	Learn from historical data	Adopt shallow structures for modeling
	Deep learning [21,22,29]	Extract more abstract features from raw inputs	May not always reflect the physical behaviors

1.1.2. Predictive Control

Predictive control is a multivariable control strategy based on prediction, and its aim is to minimize the cost function [30]. The predictive control strategy can reduce energy consumption and improve energy efficiency in the building field. Shan et al. [31] utilized chiller inlet guide vane openings as an indicator of chiller efficiency and cooling load to develop a robust chiller sequence control strategy. In the strategy, the opening or closing of a chiller was based on the measured cooling load and the predicted maximum cooling capacity, which was obtained by the MLR method. The experiment showed that the strategy could save 3% energy in a tested building compared with a typical strategy. In Ref. [32], predictive control was applied to space heating buildings. The heating demand of buildings was predicted using EnergyPlus software. Then, the predictive control strategy selected the appropriate operation schedule to minimize the electricity cost while meeting the heating demand of buildings. Finally, different buildings achieved cost savings of around 12–57% through a 7-day simulation. Imran et al. [33] proposed an IoT task management mechanism based on predictive optimization to minimize the energy cost and maximize thermal comfort. In this mechanism, the predictive optimization was based on the hybrid of prediction and optimization and used to optimize and control energy consumption. It was shown that predictive optimization-based energy management outperformed standalone prediction and optimization mechanisms in smart residential buildings.

In predictive control, methods of prediction and control are different. Prediction methods are usually employed to solve this regression problem, and control methods are used to find the optimal solution. DRL can also be used in predictive control as a model-free control method. For instance, Qin et al. [34] proposed a multi-discipline predictive intelligent control method for maintaining thermal comfort in an indoor environment. SVR was employed to construct an environmental prediction model, then the DRL method was applied to train an intelligent agent for intelligent control in the indoor environment. The results showed that the predictive control method could ensure thermal comfort and air quality in the indoor environment while minimizing energy consumption. Fu et al. [35] established a thermal dynamics model to predict the future trend of HVAC systems. Twin delayed deep deterministic policy gradient algorithm and model predictive control (TD3–MPC) were proposed to pre-adjust building temperatures at off-peak times. It was shown that TD3–MPC reduced energy consumption cost by 16% and thermal comfort RMSE by 0.4.

DRL methods can also be used to solve prediction problems. Some researchers have noted the power of the DDPG method, and use this method to predict HVAC system energy consumption [36]. However, the classical DQN method is often neglected and is rarely used for building energy consumption prediction.

1.2. The Purpose and Organization of This Paper

To date, the DDPG method has been investigated and developed more than other methods since it can process continuous action space problems; the DQN method with discrete action space is usually neglected. However, it cannot be ignored that the DDPG method often needs more computing resources and computation time to achieve high prediction accuracy. In contrast, the DQN method may not achieve such high prediction accuracy, but it can take less computation time than the DDPG method.

To obtain a higher prediction accuracy than the DDPG method and take less computation time than the DQN method, a deep-forest-based DQN (DF–DQN) method is proposed. The main contributions of this paper are as follows:

(1) The shrunken action space is proposed to replace the original action space, then DF–DQN method can quickly obtain the optimal action in the shrunken action space.

(2) State classes obtained by deep forest (DF) are used to determine the specific meaning of each action in the shrunken action space, and they can map the shrunken action space to a single sub-action space. Hence, the convergence of the DF–DQN method can be ensured.

(3) New states, composed of state class probabilities and historical energy consumption data, are constructed to improve the robustness of the DF–DQN method.

The remainder of this paper is structured as follows. In Section 2, theories of DQN and DF methods are simply described. Section 3 presents the overall framework of the DF–DQN method for energy consumption prediction. Then, the procedure of data pre-processing and MDP modeling are described in detail. Section 4 depicts experimental settings and adopted metrics of all methods, and experimental results are compared and analyzed. Some conclusions are given in Section 5.

2. Related Theories

2.1. Deep Reinforcement Learning

2.1.1. Reinforcement Learning

RL is an essential branch of machine learning; its final goal is to maximize the accumulative discount reward R_t [37], as shown in Equation (1):

$$R_t = \sum_{k=0}^{\infty} \gamma^k r_{t+k+1} \tag{1}$$

where γ is a discount factor and k represents different time steps. r_{t+k+1} represents the immediate reward in different time steps. Generally, RL problems can be modeled as a Markov decision process (MDP) to be solved. A MDP is a five-tuple (S, A, P, R, γ), where S is a set of states, A is a set of actions, P is a transition function, and R is a reward function. In the process of an agent interacting with the environment, the agent receives a state

s_t and executes an action a_t at time step t. Notably, the action a_t is selected by policy π, which represents a mapping from the state space S to the action space A. Then, the agent is transferred to the next state s_{t+1}, which is determined from the probability of state transition $P(s_{t+1}|s_t, a_t)$. Simultaneously, the immediate reward r_{t+1} is obtained by the environment.

In RL methods, the action value function Q represents the expectation of accumulative discount reward starting from state s and taking action a:

$$Q_\pi(s, a) = E_\pi \left[\sum_{k=0}^{\infty} \gamma^k R_{t+k+1}|s_t = s, a_t = a \right] \tag{2}$$

Policy π can be evaluated and improved by the action value function and optimal action value function, which can be denoted as:

$$Q_*(s, a) = \max_\pi Q_\pi(s, a) = E \left[R_{t+1} + \gamma \max_{a'} Q_* (s_{t+1}, a') \Big| s_t = s, a_t = a \right] \tag{3}$$

Finally, the optimal policy π_* is obtained, and the final goal R_t can be achieved by the optimal policy.

2.1.2. Deep Q-Network

Traditional RL methods, such as Q-learning and SARSA [38,39], can only tackle tasks with state spaces that are small and discrete. Recent methods have diverged from these restrictions by employing the deep neural network to approximate the action value function. However, these methods usually are not stable as they combine RL methods with function approximation, such as linear function or deep neural network [40]. This problem has recently been overcome by DQN with two specific techniques.

Firstly, the mechanism of experience replay is adopted to remove strong correlations between successive inputs, which means that experience tuples are stored in the replay memory and sampled randomly to train an agent. Here, experience tuples are generated by the interaction between the agent and the environment.

Secondly, to reduce correlations with targets, a separate network, namely the target Q-network, is constructed to generate targets and update the Q-network. Specifically, every J steps, the target Q-network is obtained by cloning the Q-network to calculate targets for the following J steps. The loss function at iteration i can be formulated by using two networks:

$$L(\theta_i) = E[(r + \gamma \max_{a'} Q(s', a'|\theta_i^-) - Q(s, a|\theta_i))^2] \tag{4}$$

where (s, a, r, s') is an experience tuple sampled in the replay memory, and a' is the selected action in state s'. θ_i^- and θ_i represent the parameters of the target Q-network and Q-network, respectively.

2.2. Deep Forest

DF is a novel decision tree ensemble method that can be applied for classification tasks [41]. Two techniques, namely multi-grained scanning and cascade forest structure, improve the performance of DF.

In the procedure of multi-grained scanning, all training samples are first transformed into instances using a sliding window. Then, all instances are employed to train a certain number of completely random tree forests and random forests, as well as the class vectors are generated. Finally, the transformed feature vectors are obtained by concatenating these class vectors.

The cascade forest structure is used to enhance the representational learning ability of DF. Each level receives feature information processed by its preceding level and outputs its processing result to the next level. The transformed feature vectors, which are the outputs of multi-grained scanning, are used to train the first level of the cascade forest. The final results are obtained at the last level and expressed as the maximum aggregated value.

3. DF–DQN Method for Energy Consumption Prediction

3.1. Overall Framework

Figure 1 depicts the overall framework of DF–DQN method for energy consumption prediction. The energy consumption data is divided into a training set and a test set according to the date. Then, the method, which can identify local outliers, is executed to detect outliers in the training set, and outliers are replaced considering date attribute and time factor. Feature extraction is then conducted to select h historical data as features, as well as samples and their corresponding labels can be constructed by these features. In addition, the features of each sample need to be normalized before they are transmitted into DF and DQN, which can enhance prediction accuracy.

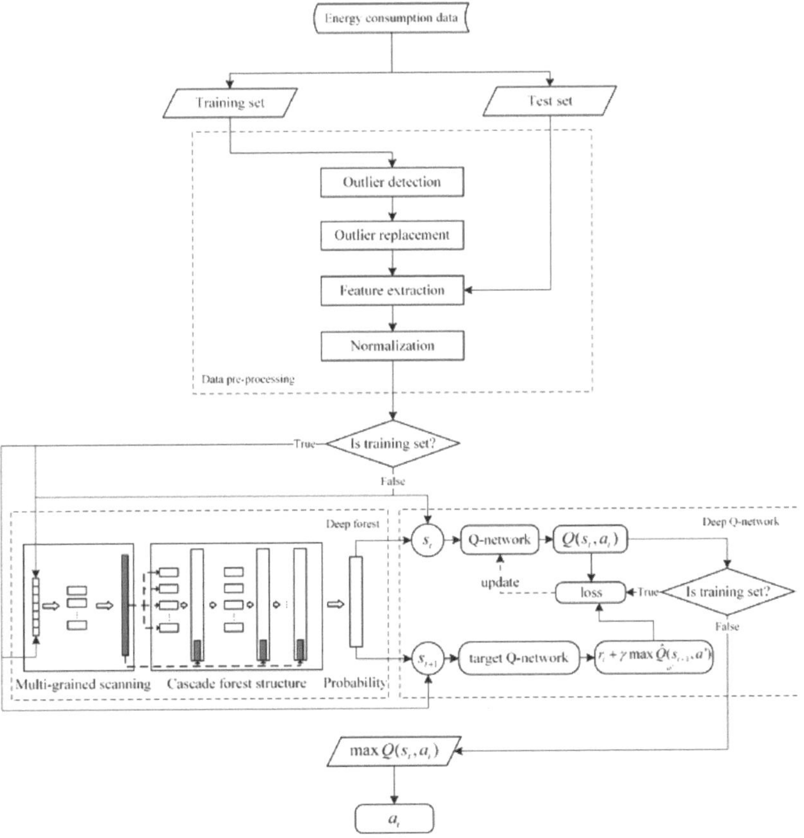

Figure 1. Overall framework of the DF–DQN method.

In the training process, a DF classifier is trained, firstly, by samples and labels, which are generated by the training set. Once the classifier training is complete, the normalized samples can be passed into a DF classifier as raw feature vectors. The transformed feature vectors are obtained in the procedure of multi-grained scanning. Then, the cascade forest structure takes the transformed feature vectors as inputs and outputs the probabilities of each class. The state at time step t (i.e., s_t) can be constructed, which is composed of the normalized historical data and the corresponding probabilities of each class. The Q-network takes s_t as input to calculate Q values for all actions, and an action with probability $1 - \varepsilon$

that has the maximum Q value or a random action with probability ε should be selected. The action selected at time step t (i.e., a_t) is the predicted energy consumption. Hence, the immediate reward can be obtained by the predicted energy consumption and the actual energy consumption at time step t. Similarly, the target Q-network can take the state at time step $t+1$ (i.e., s_{t+1}) as input to calculate target Q values for all actions. Finally, loss can be calculated by Q values and target Q values, and employed to update the Q-network.

In the test process, when the DF–DQN method receives h normalized historical data, the probabilities of each class are first obtained by DF. Then, a new state is constructed, and Q values for all actions are calculated. The action with the highest Q value is selected as the predicted energy consumption.

3.2. Data Pre-Processing

In this study, the data set concerned was the energy consumption data in an office building in Shanghai. The energy consumption data was collected every hour from 1 January 2015 to 31 December 2016. There were a total of 17,520 observation samples as the data from 29 February 2016 was not collected.

Since there may have been mixed use of electric meters, outliers were generated with high probabilities. We first detected outliers to improve the accuracy of energy consumption prediction. The local outlier factor (LOF) method is a density-based unsupervised method for identifying local outliers [42]. To find possible outliers, it can calculate local density deviation (i.e., LOF value) for each sample to their neighbors. If LOF values are high, samples have high probabilities of being treated as outliers. Similarly, samples with lower LOF values are more likely to be considered as normal data. Therefore, the LOF method can detect abnormal energy consumption data in the training set.

Outliers cannot be simply discarded after they are detected. The energy consumption data were collected at intervals of 1 h and had time-series periodicity. The absence of outliers would make the data series less accurate and make feature extraction and method execution more difficult. Therefore, outliers were replaced, which was more conducive to energy consumption prediction.

The replacement of outliers requires consideration of the time factor. In addition, the impact of holidays is not negligible for office buildings. These factors should be considered when replacing outliers. If the energy consumption data of workday is an outlier, it can be replaced by the average value, which is calculated by the sum of normal energy consumption data at the same time on the previous and following workday. The method used to replace holiday outliers is the same. This process can be formulated as below:

$$AE(d,t) = \begin{cases} NE(d-i,t) & condition1 \\ \frac{NE(d-i,t)+NE(d+j,t)}{2} & condition2 \\ NE(d+j,t) & condition3 \end{cases} \tag{5}$$

$$condition1.\, d-i \geq p,\, d+j > q,\, W(d-i) = W(d)$$
$$condition2.\, d-i \geq p,\, d+j \leq q,\, W(d-i) = W(d) = W(d+j)$$
$$condition3.\, d-i < p,\, d+j \leq q,\, W(d) = W(d+j)$$

where $i,j \in N$, AE, and NE denote the abnormal and normal energy consumption data. The date and time of energy consumption data are represented by d and t respectively. $W(d)$ can determine the date attribute of d (workday or holiday), and p and q represent the start and end date in the training set.

In this study, the date range of the training set was from 1 January 2015 to 31 October 2016. If one of the two normal energy consumption dates is outside the range, the outlier can be directly replaced with other normal energy consumption data. The dates of two normal energy consumption data used to replace outliers usually do not simultaneously fall outside the date range in the training set. The reason for this is that the date span is large. It should be mentioned that the above process of detecting and replacing outliers can only be performed in the training set, and the test data, which was unknown, cannot be used.

Feature extraction is the next step, selecting a certain number of historical energy consumption data as features. We set the number to h. If the energy consumption at time step t (denoted as E_t) needs to be predicted, h historical data from $t - h$ to $t - 1$ can be selected as features, which can be denoted as $(E_{t-h}, \ldots, E_{t-1})$. In other words, $(E_{t-h}, \ldots, E_{t-1})$ is used to predict the actual energy consumption at time step t. $(E_{t-h}, \ldots, E_{t-1})$ can be regarded as a sample, and E_t is the corresponding label. Therefore, when the total number of training data is M, $M - h$ samples and labels can be constructed.

In order to improve the accuracy of energy consumption prediction, each feature of the samples should be normalized as below:

$$\widetilde{X}_i^{(j)} = \frac{X_i^{(j)} - \mu^{(j)}}{\sigma^{(j)}} \tag{6}$$

where $X_i^{(j)}$ and $\widetilde{X}_i^{(j)}$ denote the previous and normalized value of j-th feature of the i-th sample, and $\mu^{(j)}$ and $\sigma^{(j)}$ denote the mean and standard deviation of the j-th feature, respectively.

3.3. MDP Modeling

When the DF–DQN method is employed to predict energy consumption, the prediction problem should be transformed into a control problem. This means that the process of energy consumption prediction should be modeled as an MDP, and the state, action and reward function should be defined.

The MDP constructed by the DF–DQN method improves settings in DQN method for energy consumption prediction. When the DQN method predicts energy consumption, all states are previous normalized samples. Specifically, h normalized historical energy consumption data, which is denoted as $\left(\widetilde{E}_{t-h}, \widetilde{E}_{t-h-1}, \ldots, \widetilde{E}_{t-1} \right)$, compose the state at time step t (i.e., s_t). In terms of the setting of actions, the range of historical energy consumption data determines the number of actions and action values, which is the predicted energy consumption. If the range of historical data is $[x, z]$, and the step size is g, the action selected at time step t (i.e., a_t) can be selected from $\{x, x + g, x + 2g, \ldots, z\}$, which represents the original action space, and the total number of actions is $(z - x)/g + 1$. By contrast, the DF–DQN method shrinks the original action space and introduces the DF classifier for energy consumption prediction.

3.3.1. Shrunken Action Space

This section describes the procedure of shrinking the original action space. Assuming that the historical data range is {10,59} and the step size is 1, the original action space X is generated as depicted in Figure 2. It should be mentioned that the action value is equal to the predicted energy consumption in the action space X. For example, the value of first action is 10, which means that the predicted energy consumption is 10. In practice, all action values in the action space X can be converted into other forms. Figure 2 illustrates this process, where action values of 10, … , 59 are transformed into 10 + 0, … , 50 + 9. Therefore, the original action space X can be replaced with the action space Y, and the action space size is not changed.

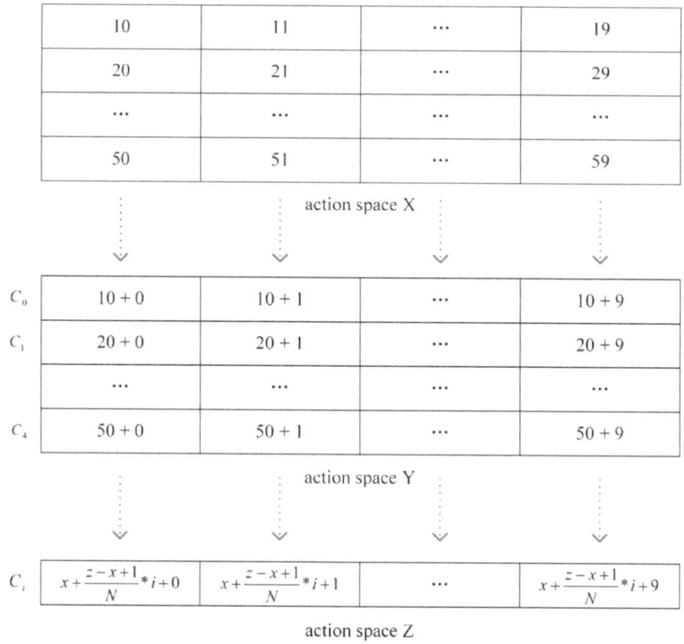

10	11	...	19
20	21	...	29
...	...	...	...
50	51	...	59

action space X

C_0 10 + 0	10 + 1	...	10 + 9
C_1 20 + 0	20 + 1	...	20 + 9
...	...	...	...
C_4 50 + 0	50 + 1	...	50 + 9

action space Y

C_i $x+\dfrac{z-x+1}{N}*i+0$	$x+\dfrac{z-x+1}{N}*i+1$	...	$x+\dfrac{z-x+1}{N}*i+9$

action space Z

Figure 2. An example of shrinking action space.

The action space Y can be equally divided into a certain number of sub-action spaces, and all states corresponding to each sub-action space can be classified into one class. Therefore, we assume that each row of actions in the action space Y composes a sub-action space, and its corresponding states are classified into one class. Finally, five sub-action spaces are generated, and all states that compose the state space can be classified into five classes (i.e., C_0, C_1, C_2, C_3, C_4). As shown in Figure 2, C_0 represents that all states that correspond to actions in the first sub-action space are classified into class 0. Similarly, the meaning of C_1, C_2, C_3, and C_4 can be obtained.

After the state classes are determined, the actions of the same sequence in all sub-action spaces can be represented uniformly. This is because the correlation of these actions can be established by using state classes. The exact formula is denoted below:

$$x + \frac{z - x + 1}{N} \times i + j \qquad i = 0, 1, 2, 3, 4. \qquad j = 1, \ldots, 9 \tag{7}$$

where x and z are equal to 10 and 59, which are the lower and upper bounds of the range of historical data. N and i denote the number of state classes and the i-th state class, respectively, and j represents the j-th action in the shrunken action space. The final result is shown at the bottom of Figure 2. Fifty actions in the action space Y are replaced with ten actions in the action space Z. At this point, Z can be regarded as the shrunken action space. Note that the step size is set to 1, and the number of actions in each sub-action space is 10 in the above example. In a more general sense, the step size is g, and the number of actions in each sub-action space is n. The actions of same sequence in all sub-action spaces can be represented as below:

$$x + \frac{\frac{z-x}{g} + 1}{N} \times i + j \qquad i = 0, 1, \ldots, N - 1. \qquad j = 0, g, 2g, \ldots, n \tag{8}$$

The result of applying this formula is that $N \times n$ actions in the original action space are replaced with n actions in the shrunken action space.

3.3.2. DF Classifier

Unlike the DQN method, the DF–DQN method needs a trained DF classifier. A DF classifier is chosen because it is not sensitive to hyperparameters and is easy to train. In Ref. [41], it was shown that DF could achieve excellent performance using almost the same settings of hyper-parameters when it is applied to different data across different domains. A DF classifier is introduced for two purposes. First, it can map the shrunken action space to a single sub-action space. Second, the state class probabilities obtained by the DF classifier are employed to construct new states.

The training of the DF classifier is based on the shrunken action space. In the process of shrinking the original action space, each sub-action space is generated by the range of energy consumption data. Further, different data ranges indicate different classes. As shown in Figure 3, {10–19} can be regarded as one sub-action space, and the data range {10–19} can represent the class C_0. Assuming that a sample obtained by feature extraction is $(E_{t-h}, E_{t-h+1}, \ldots, E_{t-1})$ and its corresponding label is E_t, if E_t is within the range, the extra class label of $(E_{t-h}, E_{t-h+1}, \ldots, E_{t-1})$ is C_0. Therefore, all samples obtained by feature extraction have additional class labels since their original labels must be within the energy consumption range of a sub-action space. These samples and class labels can compose the training set to train a DF classifier.

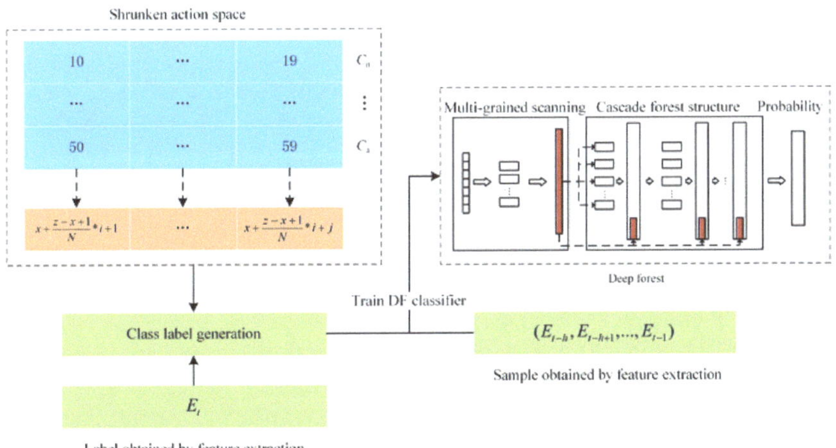

Figure 3. The training process of the DF classifier.

In the DF–DQN method, the input of the trained DF classifier is the original state composed of historical energy consumption data, and outputs are state class probabilities. These probabilities serve two purposes. Firstly, they can determine a single sub-action space. Secondly, they can be deployed to construct new states.

In the shrunken action space, each action has multiple meanings. For instance, the first action in the shrunken action space can represent the first action of each sub-action space. Similarly, in the Q neural network, one action has multiple meanings, which means that one neuron is used to represent multiple actions. The result is that the Q network cannot determine the specific meaning of the neuron at each time step, so it cannot converge effectively. State class probabilities, which are obtained by the DF classifier, can map the shrunken action space to a single sub-action space, then determine the specific meaning of each action. Assuming that the total number of classes is five, and the state class probabilities are (0.7, 0.05, 0.1, 0.1, 0.05), the corresponding state can be regarded as the first class. The shrunken action space is then equal to the first sub-action space, and it can be denoted as {10, 11, ... , 19}. This process is shown in Figure 4.

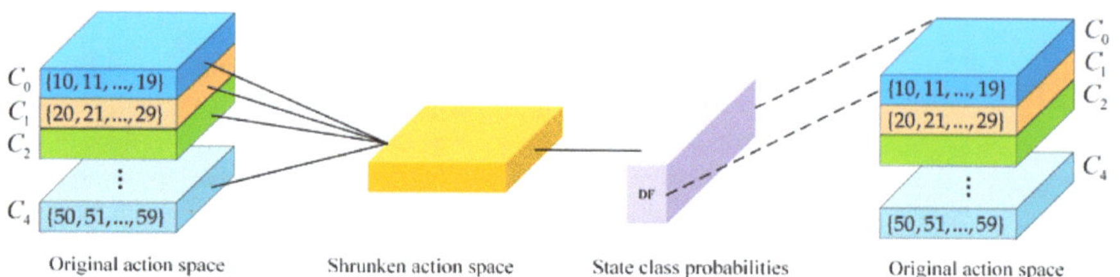

Figure 4. Mapping the shrunken action space to a single sub-action space.

Further, we constructed new states composed of the normalized historical data and its corresponding state class probabilities. This is because mapping the shrunken action space to a single sub-action space is a probability process. The Q-network should consider the probability factor when calculating the Q value, so new states are constructed to replace the original state. Moreover, new states can also be seen as integration. The performance of the DF classifier is integrated with the decision-making ability of DQN to enhance the prediction accuracy for energy consumption prediction.

Finally, the state of the DF–DQN method at time step t (i.e., s_t) is composed of normalized historical energy consumption data and corresponding state class probabilities, which is represented by a vector $(P_0, P_1, \ldots, P_{N-1}, \widetilde{E}_{t-h}, \widetilde{E}_{t-h-1}, \ldots, \widetilde{E}_{t-1})$. Here, P_i denotes the probability that $(\widetilde{E}_{t-h}, \widetilde{E}_{t-h-1}, \ldots, \widetilde{E}_{t-1})$ is determined to i-th class and $(\widetilde{E}_{t-h}, \widetilde{E}_{t-h-1}, \ldots, \widetilde{E}_{t-1})$ represents h normalized historical data. The action selected at time step t is a_t, which denotes that the predicted energy consumption is a_t at time step t. The immediate reward function can then be set as follows:

$$r_{t+1} = -|E_t - a_t| \tag{9}$$

where E_t represents the actual energy consumption at time step t. It should be mentioned that the closer the reward is to zero, the higher the prediction accuracy of the DF–DQN method.

3.4. DF–DQN Method

Once the problem of energy consumption prediction is modeled as an MDP, the DF–DQN method can be executed. Algorithm 1 depicts the main training process of the DF–DQN method for energy consumption prediction.

Algorithm 1 DF–DQN method for energy consumption prediction

---

(1)   Initialize state classes N

(2)   Initialize replay memory D

(3)   Initialize action-value function Q with random weights θ

(4)   Initialize target action-value function $\hat{Q}$ with weights $\theta^- = \theta$

(5)   Split the data set

(6)   Detect and replace outliers in the training set

(7)   Extract features to construct samples and labels

(8)   Train the deep forest classifier

(9)   **Repeat** (for each episode)

(10) Randomly select a sample

(11) Use DF classifier to obtain the possibility of each class

(12) Construct initial state (denoted as s_t)

(13) **Repeat** (for each step)

(14) Select a random action a_t with probability ε

(15) otherwise choose $a_t = \text{argmax}\, Q(s_t, a; \theta)$

(16) Execute action a_t and receive immediate reward r_t

(17) Construct state s_{t+1}

(18) Store transition (s_t, a_t, r_t, s_{t+1}) in D

(19) Sample a mini-batch $\left(s_j, a_j, r_j, s_{j+1}\right)$ from D

$$
(20)\ \text{Set } y_j = \begin{cases} r_j & \text{if episode terminates at step } j+1 \\ r_j + \gamma \max_{a'} \hat{Q}\left(s_{j+1}, a' \big| \theta^-\right) & \text{otherwise} \end{cases}
$$

(21) Update Q function using $\left(y_j - Q\left(s_j, a_j; \theta\right)\right)^2$

(22) Every J steps reset $\hat{Q} = Q$

(23) $s_t \leftarrow s_{t+1}$

(24) **Until** terminal state or maximum number of steps is reached

(25) **Until** maximum number of episodes is reached

---

4. Case Study

The MLR, SVR, DT, DQN, DF–DQN, and DDPG methods were used for energy consumption prediction. Sections 4.1 and 4.2 describe the experimental settings of all methods and four evaluation metrics of prediction accuracy. In Section 4.3, all methods are compared and analyzed from three perspectives, namely prediction accuracy, convergence rate, and computation time.

4.1. Experimental Settings

In the process of feature extraction, historical energy consumption data for the last 24 h were selected as features to predict future energy consumption. Hence, there were 24 neurons in the input layer of the DQN and DDPG methods. In contrast, the inputs of the DF–DQN method were composed of historical data and state class probabilities, so the number of neurons in the input layer was 24 + N (i.e., the number of state classes). In addition, the range of energy consumption was (129, 1063) in the training set, which means that all methods were performed in the continuous action space. Because the DQN and DF–DQN methods can only process discrete problems, the continuous action space was first converted into a discrete action space. We set the step size to 1 and the range of (129, 1063) was converted into 935 discrete points. The number of actions in the DQN method and the DF–DQN method were 935 and 935/N, respectively.

The hardware platform and package version used in the study are described in Tables 2 and 3, respectively. In addition, the hyper-parameters of all methods are summarized in Table 4. The DQN, DF–DQN, and DDPG methods used two hidden layers, and the number of neurons in each hidden layer was 32. In regard to the output layer, the number of neurons in the DQN and DF–DQN methods were consistent with the number of actions, while the DDPG method directly outputted the predicted energy consumption so that the number of neurons was 1. Further, the hyper-parameters of other methods obtained through extensive numerical experiments are also listed in Table 4.

Table 2. Hardware platform.

Hardware Platform	Configuration
Operating system	Windows 10
RAM	8 GB
CPU	Intel Core i5-9500
Programing language	Python
Programing software	PyCharm

Table 3. Package version.

Package	Version
TensorFlow	2.2.0
TensorLayer	2.2.3
NumPy	1.19.4
pandas	1.1.5
DeepForest	0.1.4

Table 4. Hyper-parameters of all methods.

Method	Parameters	Results
MLR	/	/
SVR	Kernel function	Linear
DT	Evaluation function	Mean squared error
	Maximum depth of the tree	16
DQN	Neurons	24,32,32,935
	Activation function	ReLu
	Learning rate	0.01
DF–DQN	Neurons	24+N,32,32,935/N
	Activation function	ReLu
	Learning rate	0.01
DDPG	Neurons (actor)	24,32,32,1
	Activation function (actor)	ReLu
	Learning rate (actor)	0.001
	Neurons (critic)	24,32,32,1
	Activation function (critic)	ReLu
	Learning rate (critic)	0.001

4.2. Evaluation Metrics

In order to compare the prediction accuracy of all methods, four evaluation metrics were adopted in this study, namely mean absolute error (MAE), mean absolute percentage error (MAPE), root mean square error (RMSE), and coefficient of determination (R^2). These evaluation metrics can be denoted as:

$$\text{MAE} = \frac{1}{m} \sum_{i=1}^{m} |y_i - y_i'| \tag{10}$$

$$\text{MAPE} = \frac{1}{m} \sum_{i=1}^{m} \left| \frac{y_i - y_i'}{y_i} \right| \tag{11}$$

$$\text{RMSE} = \sqrt{\frac{1}{m} \sum_{i=1}^{m} (y_i - y_i')^2} \tag{12}$$

$$R^2 = 1 - \frac{\sum_{i=1}^{m} (y_i - y_i')^2}{\sum_{i=1}^{m} (y_i - \bar{y})^2} \tag{13}$$

where m denotes the total number of samples, y_i and y_i' represent the predicted value and actual value of the i-th sample, respectively. $\bar{y}$ is the average of actual value.

4.3. Results and Analyses

In this section, each method was trained ten times under the settings of the hyperparameters presented in Table 4, and all experimental results were obtained as averages.

4.3.1. Prediction Accuracy

Figure 5 illustrates predicted results of the DF–DQN method with different state classes, where the horizontal axis reflects the predicted energy consumption and the vertical axis represents the actual energy consumption. In each sub-figure, the solid blue line indicates that the predicted energy consumption is equivalent to the actual energy consumption, and the blue dashed line denotes the 20% error. The shaded area represents that the predicted value differs less than 20% from the actual value. Therefore, the number of predicted points contained in the shaded area can reflect the prediction accuracy. The DF–DQN method with 15 and 19 state classes outperformed others, and the DF–DQN method with three state classes was the least effective. The predicted points show the trend of classification since the states were classified using the DF–DQN method. Specifically, the classification trend was noticeable in in the DF–DQN method with 5, 7, and 11 state classes.

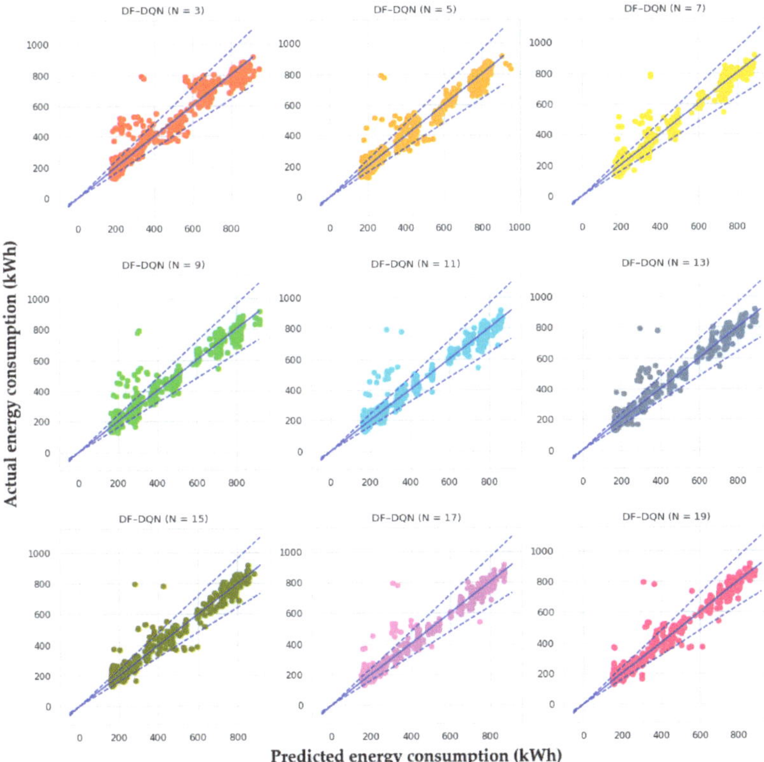

Figure 5. Predicted results of the DF–DQN method with different state classes.

Table 5 describes the prediction accuracy of the DF–DQN method with different state classes under four evaluation metrics, and the significant values are indicated in bold.

Notably, the DF–DQN method with 15 state classes had the lowest MAE, MAPE, and RMSE values, and the highest R^2 value. Therefore, the prediction accuracy of the DF–DQN method with 15 state classes was the highest. Table 5 also shows that the prediction accuracy increased roughly as state classes increased. However, some unexpected results were observed due to the classification accuracy and the influence of random factors in the training process. For example, the prediction errors of the DF–DQN method with two, four, and five state classes were lower than that of the DF–DQN method with six and seven state classes. Nonetheless, the trend that the increasing number of state classes improved the prediction accuracy was not affected. It also should be mentioned that excessive state classes decreased the prediction accuracy of the DF–DQN method due to the low classification accuracy. Therefore, the optimal number of state classes requires extensive experiments to obtain. In this experiment, the DF–DQN method with 15 state classes outperformed other numbers of state classes.

Table 5. Prediction accuracy of the DF–DQN method with different state classes.

N	Number of Actions	Accuracy of Classification	MAE	MAPE	RMSE	R^2
2	468	99.392%	23.333	7.960%	36.774	0.978
3	312	94.706%	29.630	9.664%	48.750	0.961
4	234	96.168%	22.950	7.828%	39.141	0.974
5	187	95.178%	23.357	7.866%	38.686	0.976
6	156	92.739%	27.439	9.476%	44.295	0.968
7	134	89.583%	27.512	9.655%	41.153	0.971
8	117	89.098%	23.810	8.074%	39.536	0.974
9	104	88.327%	23.152	7.886%	40.831	0.973
10	94	84.139%	24.456	8.370%	42.201	0.970
11	85	83.907%	23.643	8.246%	40.044	0.974
12	78	83.586%	22.254	7.845%	37.480	0.976
13	72	80.307%	21.921	7.379%	36.248	0.978
14	67	77.548%	20.912	7.231%	34.936	0.980
15	63	76.462%	**20.432**	**7.021%**	**34.057**	**0.981**
16	59	73.005%	20.975	7.390%	34.331	0.980
17	55	71.633%	20.971	7.545%	35.410	0.980
18	52	69.037%	20.590	7.315%	35.442	0.979
19	50	66.714%	20.596	7.367%	34.408	0.980
20	47	64.740%	20.623	7.272%	35.198	0.980

In addition, results using the DQN, DDPG, MLR, SVR, and DT methods were compared with those using the DF–DQN method. As shown in Figure 6, the DF–DQN method with 15 state classes and the DDPG method outperformed other methods. In order to further analyze the prediction accuracy of all methods, predicted results of a certain period were selected for display. Note that the energy consumption on workdays and holidays should only be compared with other results of their type, as the results from each category are quite different. Figure 7 depicts predicted results of all methods on workdays, where the horizontal axis reflects the time and the vertical axis represents the energy consumption. In each sub-figure, the blue line represents actual energy consumption, and the line of another color denotes the predicted energy consumption. It can be seen that all methods captured the energy consumption trend, and only the MLR method showed slight fluctuation. In contrast, the predicted results on holidays were inaccurate, as seen in Figure 8. The DQN, MLR, and SVR methods showed fluctuation, and the DT method had a noticeable error point. The DDPG method and the DF–DQN method with 15 state classes also captured part of the trend of actual energy consumption. A possible explanation for this might be that all methods were unable to learn the features of holiday energy consumption when only historical data were used as input.

Table 6 describes the prediction accuracy of all methods in detail. Notably, the DDPG method had advantages in energy consumption prediction, and its prediction accuracy was higher than the MLR, SVR, DT, and DQN methods. However, for the DF–DQN method with 15 state classes, MAE, MAPE, and RMSE decreased by 5.5%, 7.3%, and 8.9% respectively, and R^2 increased by 0.3% compared to the DDPG method. The results verify the superiority of DF–DQN, and demonstrate the potential of the collaboration between DF and DQN.

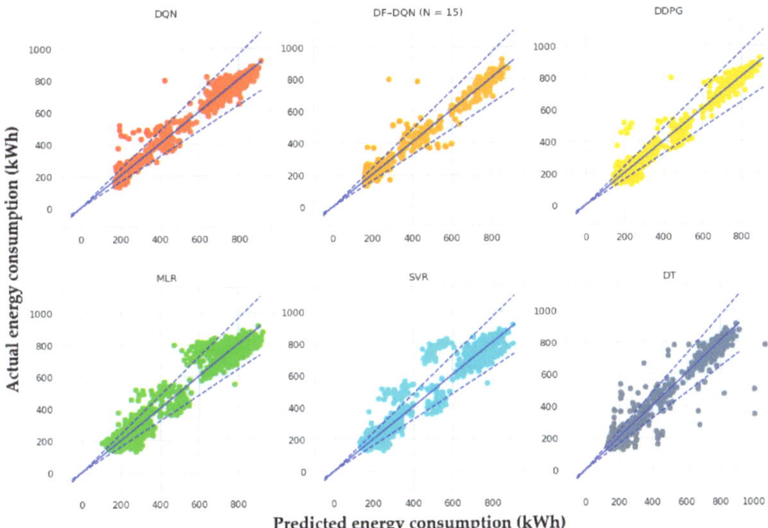

Figure 6. Predicted results of all methods.

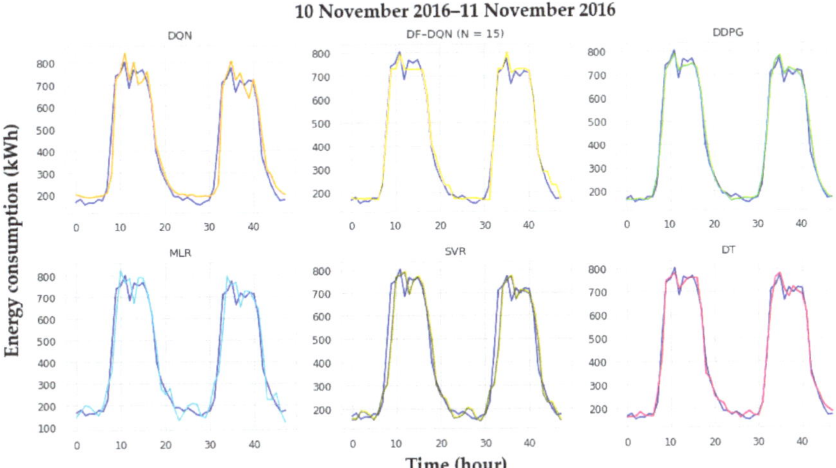

Figure 7. Predicted results of all methods on workdays.

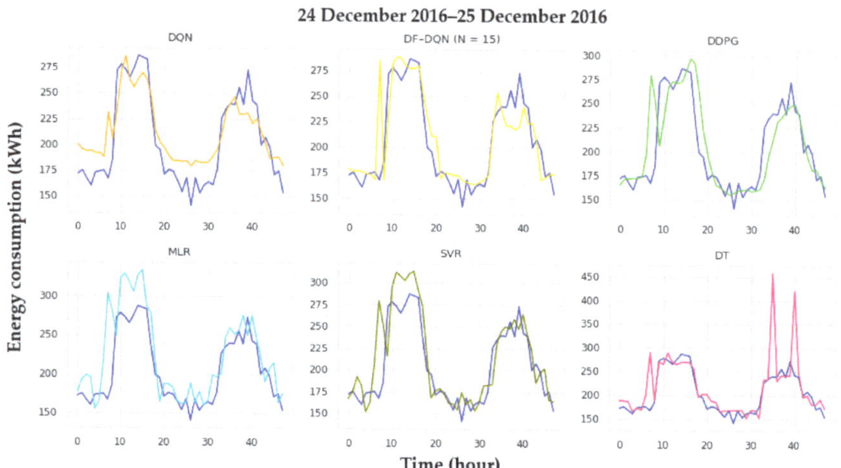

Figure 8. Predicted results of all methods on holidays.

Table 6. Prediction accuracy of all methods.

Method	MAE	MAPE	RMSE	R^2
MLR	41.069	12.869%	56.577	0.946
SVR	37.041	11.192%	63.150	0.930
DT	26.349	8.868%	50.470	0.959
DQN	27.942	9.362%	39.869	0.973
DDPG	21.619	7.573%	36.417	0.978
DF–DQN (N = 15)	**20.432**	**7.021%**	**34.057**	**0.981**

4.3.2. Convergence Rate and Computation Time

Figure 9 depicts MAE varying tendencies of DRL methods from the first episode, where the horizontal axis reflects iterations of the episode and the vertical axis represents MAE. It is evident that the convergence rate of the DQN method was the slowest, and the converged MAE was the highest among all DRL methods. However, the DDPG and DF–DQN methods could not be compared effectively since a very high MAE value was generated in the DDPG method. Therefore, a new figure from the fifth episode is shown to facilitate the analysis of the DDPG and DF–DQN methods. As shown in Figure 10, the convergence rate of the DDPG method was similar to that of the DF–DQN method with three state classes, and they converged near the 100th episode. The DF–DQN method with 15 state classes had the fastest convergence rate and roughly converged at the 75th episode. Further, the converged MAE value of the method was lower than other methods. It is also noteworthy that the larger the number of state classes, the faster the DF–DQN method converged, and the converged MAE value decreased as the total number of state classes increased. This is because the state classification lowered the initial value of MAE and accelerated the convergence rate of the DF–DQN method.

The convergence rates of the MLR, SVR, and DT methods are superior to DRL methods. This conclusion is presented in Table 7. The computation time of these methods is much lower than that of DRL methods. It is worth noting that the computation time of DRL methods was based on 200 episodes, as presented in Table 7. However, these DRL methods converged before the 200th episode. The computation time of the DQN method was about 625 s since the method converged at the 150th episode. Similarly, the DDPG and the DF–DQN methods with 15 state classes took 665 s and 262 s, respectively. Hence, of all DRL methods, the DF–DQN method with 15 state classes required the least computation time, and the DDPG method was slower than the DQN method. One important reason for this is that the DF–DQN method performs in the shrunken action space, and the number of actions was significantly lower than that of the DQN method. Regarding the DDPG method, two kinds of neural networks, actor and critic networks, should be trained. These networks take long computation times, even though the total number of parameters is lower than other methods.

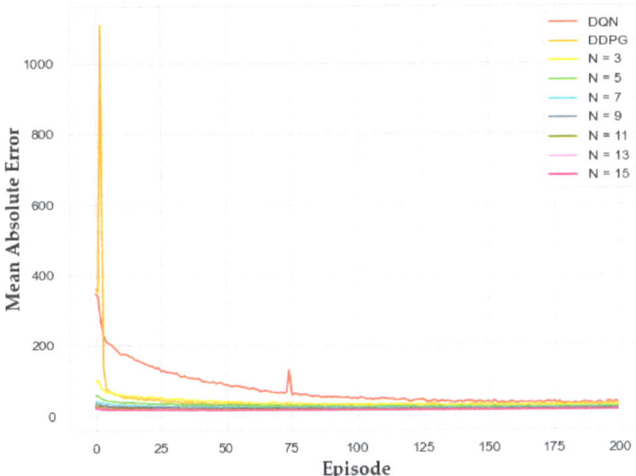

Figure 9. Variation tendency of MAE in DRL methods from first episode.

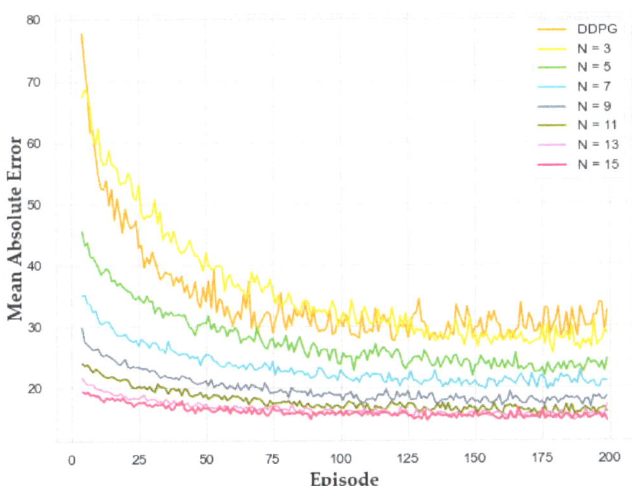

Figure 10. Variation tendency of MAE in DRL methods from the fifth episode.

Table 7. The computation time of all methods.

Method	Computation Time
MLR	0.07
SVR	7.734
DT	0.362
DQN	833.714
DDPG	1329.007
DF–DQN ($N = 15$)	699.529

5. Conclusions

This paper proposed a DF–DQN method to demonstrate the potential of DRL methods with discrete action space for energy consumption prediction. In the DF–DQN method, the original action space was divided into a certain number of sub-action spaces by the range of historical energy consumption data. Then, actions of the same sequence in all sub-action spaces were uniformly denoted to compose the shrunken action space. Compared with the original action space, the total number of actions in the shrunken action space was reduced greatly. Further, DF was introduced since each action has multiple meanings in the shrunken action space. State class probabilities obtained using DF can uniquely determine each action's specific meaning in the shrunken action space and map the shrunken action space to a single sub-action space, which ensures the convergence of the DF–DQN method. Moreover, the state class probabilities can also be employed to construct new states to improve the robustness of the DF–DQN method by taking into account the probabilistic process of shrinking the original action space. Based on the above operations, DF–DQN can find the optimal action quickly in the new shrunken action space.

The experimental results show that the prediction accuracy of the DF–DQN method with 15 state classes outperforms the MLR, SVR, and DT methods, even if it requires more computation time than these methods. In our experiments, for DRL methods, the DF–DQN method with 15 state classes had the highest prediction accuracy and fastest convergence rate, and required the least computation time. Specifically, compared to the DDPG method, the DF–DQN method with 15 state classes decreased MAE, MAPE, and RMSE by 5.5%, 7.3%, and 8.9%, respectively, and increased R^2 by 0.3%.

This study demonstrated that the DF–DQN method with discrete action space has great potential for predicting energy consumption. However, the study conducted in this paper may contain inaccuracies. The number of state classes is an additional hyperparameter and must be determined by extensive experiments. Future work will overcome the above deficiencies and explore the performance of the DF–DQN method for multi-step ahead prediction in recursive and direct multi-step manners.

Author Contributions: Conceptualization, K.L.; data curation, Q.F. and K.L.; formal analysis, Q.F. and K.L.; funding acquisition, J.C.; investigation, J.C., J.W., Y.L.; methodology, Q.F. and K.L.; project administration, J.W., Y.L. and Y.W.; software, Q.F. and K.L.; supervision, Q.F., J.C., J.W., Y.L. and Y.W.; validation, K.L.; writing—original draft, K.L.; writing—review & editing, Q.F. and K.L. All authors have read and agreed to the published version of the manuscript.

Funding: This work was financially supported by National Key R&D Program of China (No.2020YFC2 006602), National Natural Science Foundation of China (No.62072324, No.61876217, No.61876121, No.61772357), University Natural Science Foundation of Jiangsu Province (No.21KJA520005), Primary Research and Development Plan of Jiangsu Province (No.BE2020026), Natural Science Foundation of Jiangsu Province (No.BK20190942).

Institutional Review Board Statement: Not applicable.

Informed Consent Statement: Not applicable.

Data Availability Statement: The dataset is available at: https://github.com/gltzlike/DF-DQN-for-energy-consumption-prediction/tree/master/data (accessed on 25 January 2022).

References

1. Conti, J.; Holtberg, P.; Diefenderfer, J.; LaRose, A.; Turnure, J.T.; Westfall, L. *International Energy Outlook 2016 with Projections to 2040*; USDOE Energy Information Administration (EIA): Washington, DC, USA; Office of Energy Analysis: Washington, DC, USA, 2016.
2. China Building Energy Consumption Annual Report 2020. *Build. Energy Effic.* **2021**, *49*, 1–6.
3. Becerik-Gerber, B.; Siddiqui, M.K.; Brilakis, I.; El-Anwar, O.; El-Gohary, N.; Mahfouz, T.; Jog, G.M.; Li, S.; Kandil, A.A. Civil Engineering Grand Challenges: Opportunities for Data Sensing, Information Analysis, and Knowledge Discovery. *J. Comput. Civ. Eng.* **2014**, *28*, 04014013. [CrossRef]

1. Dawood, N. Short-term prediction of energy consumption in demand response for blocks of buildings: DR-BoB approach. *Buildings* **2019**, *9*, 221. [CrossRef]
5. Moghadam, S.T.; Delmastro, C.; Corgnati, S.P.; Lombardi, P. Urban energy planning procedure for sustainable development in the built environment: A review of available spatial approaches. *J. Clean. Prod.* **2017**, *165*, 811–827. [CrossRef]
6. Kim, J.; Frank, S.; Braun, J.E.; Goldwasser, D. Representing Small Commercial Building Faults in EnergyPlus, Part I: Model Development. *Buildings* **2019**, *9*, 233. [CrossRef]
7. Du, Y.; Zandi, H.; Kotevska, O.; Kurte, K.; Munk, J.; Amasyali, K.; Mckee, E.; Li, F. Intelligent multi-zone residential HVAC control strategy based on deep reinforcement learning. *Appl. Energy* **2021**, *281*, 116117. [CrossRef]
8. Min, Y.; Chen, Y.; Yang, H. A statistical modeling approach on the performance prediction of indirect evaporative cooling energy recovery systems. *Appl. Energy* **2019**, *255*, 113832. [CrossRef]
9. Zhao, H.-X.; Magoulès, F. A review on the prediction of building energy consumption. *Renew. Sustain. Energy Rev.* **2012**, *16*, 3586–3592. [CrossRef]
10. Yao, R.; Steemers, K. A method of formulating energy load profile for domestic buildings in the UK. *Energy Build.* **2005**, *37*, 663–671. [CrossRef]
11. Wang, S.; Xu, X. Simplified building model for transient thermal performance estimation using GA-based parameter identification. *Int. J. Therm. Sci.* **2006**, *45*, 419–432. [CrossRef]
12. Wei, Y.; Zhang, X.; Shi, Y.; Xia, L.; Pan, S.; Wu, J.; Han, M.; Zhao, X. A review of data-driven approaches for prediction and classification of building energy consumption. *Renew. Sustain. Energy Rev.* **2018**, *82*, 1027–1047. [CrossRef]
13. Ma, Y.; Yu, J.; Yang, C.; Wang, L. Study on power energy consumption model for large-scale public building. In Proceedings of the 2010 2nd International Workshop on Intelligent Systems and Applications, Wuhan, China, 22–23 May 2010; pp. 1–4.
14. Lam, J.C.; Wan, K.K.; Wong, S.; Lam, T.N. Principal component analysis and long-term building energy simulation correlation. *Energy Convers. Manag.* **2010**, *51*, 135–139. [CrossRef]
15. Ahmad, A.S.; Hassan, M.Y.; Abdullah, M.P.; Rahman, H.A.; Hussin, F.; Abdullah, H.; Saidur, R. A review on applications of ANN and SVM for building electrical energy consumption forecasting. *Renew. Sustain. Energy Rev.* **2014**, *33*, 102–109. [CrossRef]
16. Amasyali, K.; El-Gohary, N.M. A review of data-driven building energy consumption prediction studies. *Renew. Sustain. Energy Rev.* **2018**, *81*, 1192–1205. [CrossRef]
17. Tian, C.; Li, C.; Zhang, G.; Lv, Y. Data driven parallel prediction of building energy consumption using generative adversarial nets. *Energy Build.* **2019**, *186*, 230–243. [CrossRef]
18. Azadeh, A.; Ghaderi, S.; Sohrabkhani, S. Annual electricity consumption forecasting by neural network in high energy consuming industrial sectors. *Energy Convers. Manag.* **2008**, *49*, 2272–2278. [CrossRef]
19. Hou, Z.; Lian, Z. An application of support vector machines in cooling load prediction. In Proceedings of the 2009 International Workshop on Intelligent Systems and Applications, Wuhan, China, 23–24 May 2009; pp. 1–4.
20. Tso, G.K.; Yau, K.K. Predicting electricity energy consumption: A comparison of regression analysis, decision tree and neural networks. *Energy* **2007**, *32*, 1761–1768. [CrossRef]
21. Ozcan, A.; Catal, C.; Donmez, E.; Senturk, B. A hybrid DNN-LSTM model for detecting phishing URLs. *Neural Comput. Appl.* **2021**, *33*, 1–17. [CrossRef]
22. Cai, M.; Pipattanasomporn, M.; Rahman, S. Day-ahead building-level load forecasts using deep learning vs. traditional time-series techniques. *Appl. Energy* **2019**, *236*, 1078–1088. [CrossRef]
23. Ozcan, A.; Catal, C.; Kasif, A. Energy Load Forecasting Using a Dual-Stage Attention-Based Recurrent Neural Network. *Sensors* **2021**, *21*, 7115. [CrossRef]
24. Mnih, V.; Kavukcuoglu, K.; Silver, D.; Rusu, A.A.; Veness, J.; Bellemare, M.G.; Graves, A.; Riedmiller, M.; Fidjeland, A.K.; Ostrovski, G. Human-level control through deep reinforcement learning. *Nature* **2015**, *518*, 529–533. [CrossRef] [PubMed]
25. Levine, S.; Finn, C.; Darrell, T.; Abbeel, P. End-to-end training of deep visuomotor policies. *J. Mach. Learn. Res.* **2016**, *17*, 1334–1373.
26. Sallab, A.E.; Abdou, M.; Perot, E.; Yogamani, S. Deep reinforcement learning framework for autonomous driving. *Electron. Imaging* **2017**, *2017*, 70–76. [CrossRef]
27. Huang, Z.; Chen, J.; Fu, Q.; Wu, H.; Lu, Y.; Gao, Z. HVAC Optimal Control with the Multistep-Actor Critic Algorithm in Large Action Spaces. *Math. Probl. Eng.* **2020**, *2020*, 1386418. [CrossRef]
28. Wei, T.; Wang, Y.; Zhu, Q. Deep reinforcement learning for building HVAC control. In Proceedings of the 54th Annual Design Automation Conference 2017, Austin, TX, USA, 18–22 June 2017; pp. 1–6.
29. Liu, T.; Tan, Z.; Xu, C.; Chen, H.; Li, Z. Study on deep reinforcement learning techniques for building energy consumption forecasting. *Energy Build.* **2020**, *208*, 109675. [CrossRef]
30. Xue, J.; Kong, X.; Dong, B.; Xu, M. Multi-Agent Path Planning based on MPC and DDPG. *arXiv* **2021**, arXiv:2102.13283.
31. Shan, K.; Wang, S.; Gao, D.; Xiao, F. Development and validation of an effective and robust chiller sequence control strategy using data-driven models. *Autom. Constr.* **2016**, *65*, 78–85. [CrossRef]
32. Gholamibozanjani, G.; Tarragona, J.; De Gracia, A.; Fernandez, C.; Cabeza, L.F.; Farid, M.M. Model predictive control strategy applied to different types of building for space heating. *Appl. Energy* **2018**, *231*, 959–971. [CrossRef]
33. Imran; Iqbal, N.; Kim, D.H. IoT Task Management Mechanism Based on Predictive Optimization for Efficient Energy Consumption in Smart Residential Buildings. *Energy Build.* **2021**, *257*, 111762. [CrossRef]

34. Qin, H.; Wang, X. A multi-discipline predictive intelligent control method for maintaining the thermal comfort on indoor environment. *Appl. Soft Comput.* **2021**, *166*, 108299. [CrossRef]
35. Fu, C.; Zhang, Y. Research and Application of Predictive Control Method Based on Deep Reinforcement Learning for HVAC Systems. *IEEE Access* **2021**, *9*, 130845–130852. [CrossRef]
36. Liu, T.; Xu, C.; Guo, Y.; Chen, H. A novel deep reinforcement learning based methodology for short-term HVAC system energy consumption prediction. *Int. J. Refrig.* **2019**, *107*, 39–51. [CrossRef]
37. Sutton, R.S.; Barto, A.G. *Reinforcement Learning: An Introduction*; MIT Press: Cambridge, MA, USA, 2018.
38. Watkins, C.J.; Dayan, P. Q-learning. *Mach. Learn.* **1992**, *8*, 279–292. [CrossRef]
39. Rummery, G.A.; Niranjan, M. *On-Line Q-Learning Using Connectionist Systems*; Citeseer: State College, PA, USA, 1994; Volume 37
40. Tsitsiklis, J.N.; Van Roy, B. An analysis of temporal-difference learning with function approximation. *IEEE Trans. Autom. Control* **1997**, *42*, 674–690. [CrossRef]
41. Zhou, Z.-H.; Feng, J. Deep forest. *Natl. Sci. Rev.* **2019**, *6*, 74–86. [CrossRef]
42. Breunig, M.M.; Kriegel, H.-P.; Ng, R.T.; Sander, J. LOF: Identifying density-based local outliers. In Proceedings of the 2000 ACM SIGMOD International Conference on Management of Data, Dallas, TX, USA, 15–18 May 2000; pp. 93–104.

 *buildings*

Article

A New Configuration of Roof Photovoltaic System for Limited Area Applications—A Case Study in KSA

Ayman Al-Quraan [1,*], Mohammed Al-Mahmodi [2], Taha Al-Asemi [3], Abdulqader Bafleh [4], Mathhar Bdour [5], Hani Muhsen [6] and Ahmad Malkawi [7]

1. Electrical Power Engineering Department, Hijjawi Faculty for Engineering Technology, Yarmouk University, Irbid 21163, Jordan
2. Mechanical Engineering Department/Renewable Energy, The University of Jordan, Amman 11942, Jordan; mohammed.almahmodi@gju.edu.jo
3. Technology Utilities Trading Est, Riyadh 12274, Saudi Arabia; tahaalasemi@gmail.com
4. National Energy Works Company, Riyadh 11423, Saudi Arabia; Abdulq.bafleh@gmail.com
5. Department of Energy Engineering, School of Natural Resources Engineering and Management, German Jordanian University, Amman 11180, Jordan; madher.bdour@gju.edu.jo
6. Mechatronics Engineering Department, German Jordanian University, Amman 11180, Jordan; Hani.mohsen@gju.edu.jo
7. Mechatronics Engineering Department, The University of Jordan, Amman 11942, Jordan; ah.malkawi@ju.edu.jo
* Correspondence: aymanqran@yu.edu.jo

Citation: Al-Quraan, A.; Al-Mahmodi, M.; Al-Asemi, T.; Bafleh, A.; Bdour, M.; Muhsen, H.; Malkawi, A. A New Configuration of Roof Photovoltaic System for Limited Area Applications—A Case Study in KSA. *Buildings* 2022, 12, 92. https://doi.org/10.3390/buildings12020092

Academic Editors: Shi-Jie Cao, Dahai Qi, Junqi Wang and Gwanggil Jeon

Received: 13 December 2021
Accepted: 15 January 2022
Published: 19 January 2022

Publisher's Note: MDPI stays neutral with regard to jurisdictional claims in published maps and institutional affiliations.

Abstract: Increased world energy demand necessitates looking for appropriate alternatives to oil and fossil fuel. Countries encourage institutions and households to create their own photovoltaic (PV) systems to reduce spending money in electricity sectors and address environmental issues. Due to high solar radiation in the Kingdom of Saudi Arabia (KSA), the government urges people and institutions to establish PV systems as the best promising renewable energy resource in the country. This paper presents an optimal and complete design of a 300 kW PV system installed in a limited rooftop area to feed the needs of the Ministry of Electricity building, which has a high energy consumption. The design has been suggested for two scenarios in terms of adjusting the orientation angles. The available rooftop area allowed to be used is insufficient if a tilt angle of 22° is used, suggested by the designer, so the tilt angle has been adjusted from 22° to 15° to accommodate the available area and meet the required demand with a minimum shading effect. The authors of this paper propose a modified scenario "third scenario" which accommodates the available area and provides more energy than the installed "second scenario". The proposed panel distribution and the estimated energy for all scenarios are presented in the paper. The possibility of changing tilt angles and the extent of energy production variations are also discussed. Finally, a comparative study between measured and simulated energy is included. The results show that August has the lowest percentage error, with a value of 2.7%, while the highest percentage error was noticed in November.

Keywords: solar energy; PV module; power density; panel orientation; tilt angle

1. Introduction

In the last few decades, there has been a trend toward using clean energy sources in the world. Those intermittent energy sources are environmentally friendly and economical throughout the years. Developing countries, in particular, are inclined to utilize such resources to meet their energy demand. KSA is one of the highest energy-producing countries in the Middle East, with approximately 63 GW, most of which depends on oil [1–4]. This fact is considered a serious challenge for the electricity sector in the kingdom. Consequently, officials and specialists are keen to look for alternatives to address this problem adequately. Therefore, the National Renewable Energy Program in Saudi Arabia has been established to develop and maximize the utilization of renewable energy resources and to become one of the main sources rather than depending entirely on oil.

Solar energy resources are considered the breakthrough that can highly overcome such power generation problems in most Middle East countries. Therefore, KSA is going toward increasing the use of solar energy to reduce the dependence on oil in producing energy. The average radiation in KSA ranges between 5 and 7 kWh/m^2 [1,2,4–8]. This amount of solar radiation is relatively high compared to the neighboring countries [1,9–11]. To make the use of this free and clean energy possible, the KSA government issued legislation and laws that facilitate and assist the new installations and designs of solar energy systems on the roofs of residential and commercial buildings.

Designing a PV system requires knowledge about the solar radiation theory and its calculation. The determination of the peak sun hours, which refer to the amount of energy received by the panels or, in other words, how much energy is received during the day in a specific area, is one of the most important pieces of information needed to design a PV system [3,4]. Peak power or the kilowatt peak (kWp) of the power demand has to be calculated before determining the number of panels that have to be accommodated with the roof capacity [12].

Several research studies in the literature discussed the general assessment of solar radiation resources in Saudi Arabia and the vision of the kingdom for the upcoming decade. Network design, implementation, and data quality assurance are described in [1]. In addition, the authors analyze the first year of broadband solar resource measurements from a new monitoring network in Saudi Arabia developed by the King Abdullah City for Atomic and Renewable Energy (K.A.CARE). The analysis used 12 months (October 2013–September 2014) of data from 30 stations distributed across the country based on one-minute measurements of global horizontal irradiance (GHI), diffuse horizontal irradiance (DHI), direct normal irradiance (DNI), and related meteorological parameters. In [4], the authors provide a maiden attempt to investigate how much sustainability substance is in the 2030 Vision of Saudi Arabia. The Sustainable Society Index (SSI) has been employed to examine the 2030 Vision and understand the Kingdom's commitment to building resilient, inclusive, and sustainable societies. The Vision and National Transformation Program (NTP) texts were matched against five broad measures and 22 submeasures. Both the 2030 Vision and the NTP align with the SSI measures in some respect. The goals and objectives reflect the aspirations and context of Saudi Arabia. The carbon emission is expected to be zero in 2060 in KSA, so it is expected to be reduced by 60% in 2030 [5]. Several projects are carried out throughout the country, and different studies are presented to introduce the system regulations and policies. According to [6–10], certain policies that promote the renewable energy sector and regulate the relations with electricity and grid connection sectors are subsidized in KSA. The net metering and feed-in tariff policies are employed to exchange the surplus energy with the electricity network in KSA.

For rooftop PV systems, net-metering was tested in 2017 and found to be the most suitable policy for small-scale PV systems. The applicability of using feed-in tariff (FiT) policy in KSA is reviewed in [11], in comparison with that implemented in USA and Germany. It is concluded that KSA has the ability to adopt and use this technique.

Specific case studies in KSA have also been discussed in the literature to achieve the vision of the kingdom in the sector of solar energy. The authors in [12] examine the best tilt angle for a solar panel to maximize the collected amount of solar irradiation. A daily global and diffuse solar radiation measured on a horizontal surface is considered in this research. The optimal angle for each month allows capturing the most solar energy for the Madinah site, KSA. The authors concluded that the annual optimal tilt angle is almost equal to the location's latitude. In [13], a comprehensive analysis is presented in order to improve the solar energy performance of residential buildings in KSA by optimizing the building envelope elements. The elements included in energy cost and energy energy-saving analysis include the wall insulation, roof insulation, window area, window glazing, window shading, and thermal mass.

Figure 1 illustrates the anticipated progress towards increasing the energy produced by renewable energy resources. Overall, the targeted capacity by 2030 would cover 58.7 GW as stated by the kingdom vision [1–3]. The plan set an initial step of producing 9.5 GW of electrical energy, followed by producing 27 GW in the next five years. By maintaining efforts, the plan is expected to be realized, as previously presented, utilizing solar, wind, and CSP energy.

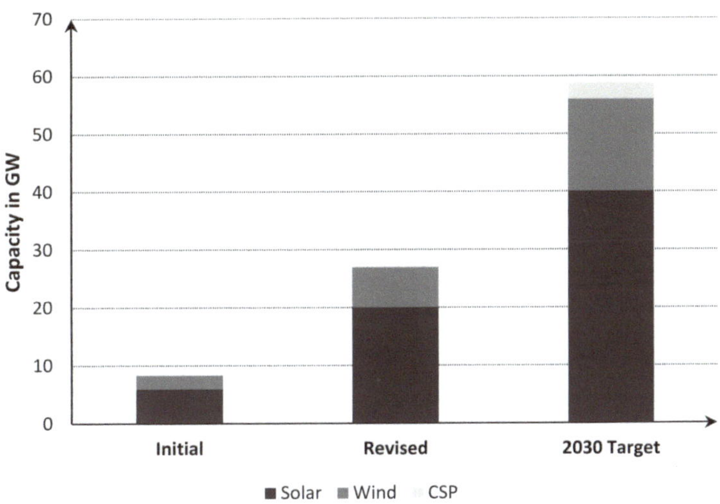

Figure 1. The vision of renewable energy in KSA for 2030.

When installing the PV system over buildings' roofs, one of the significant factors that should be considered is the sun path. The more solar panels exposed to the direct sun, the more energy is harvested. Each location in the world has a specific sun path and unique tilt, azimuth, and elevation angles [12]. Figure 2 represents the most significant angles that play an important role in PV system design.

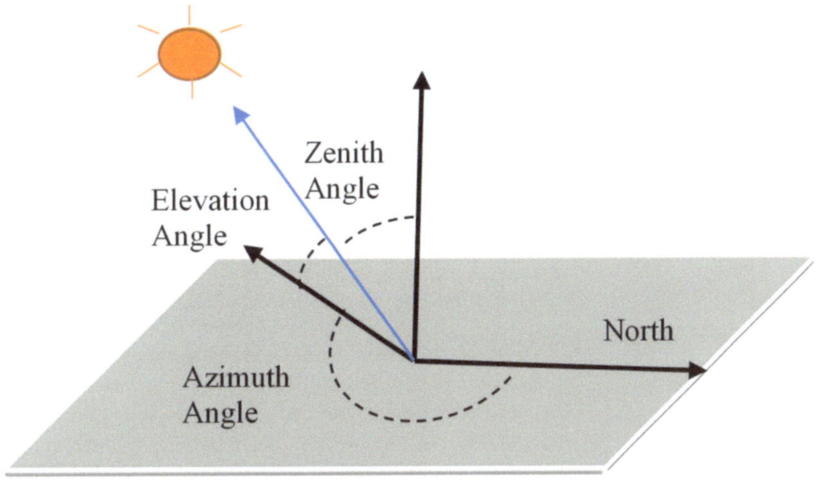

Figure 2. The main angles in PV system design.

The desired tilt angle varies according to the case study under test. The designers should ensure that the solar panels face the sunlight perpendicularly, as much as possible, since the solar panels produce the highest power when they face the sun directly. During the winter semester, the solar panels face the sun at a sharp angle. Therefore, the energy produced by the PV system is low compared to the energy produced during the summer semester. In the winter semester case, the operator should ensure harvesting large amounts of solar energy by choosing proper orientation angles. The optimum tilt angle in KSA is approximately equal to the site's latitude under test [13,14]. In addition, the elevation angle significantly affects the proper method of installing the panels with the most suitable spacing between rows. This must be determined to avoid the possibility of energy reduction that panels cause for each other. Figure 3 illustrates the relationship between elevation angle and spacing between rows in the PV system.

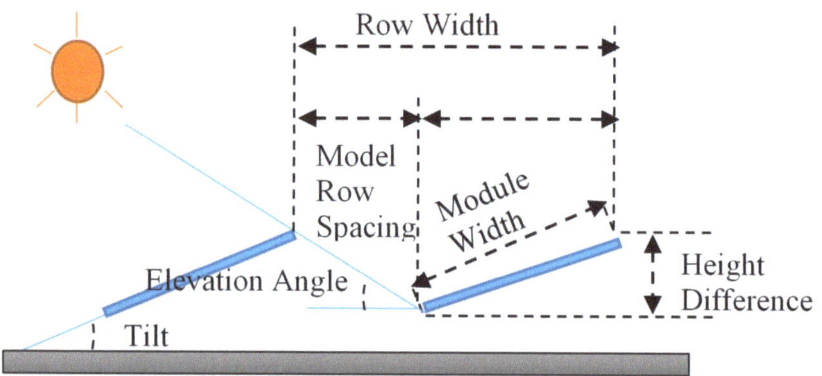

Figure 3. Relationship between elevation angle and spacing between rows.

Several studies consider the performance of rooftop solar panels by assessing different design and simulation methodologies. The effect of rooftop obstacles and shadow can be determined using photovoltaic software to identify the azimuth and tilt angles [15]. One study assesses the impact of a building's shading and analyzes the available rooftop area using hillshade analysis [16]. The study estimates the potential of a rooftop system when shaded area is excluded. PV systems in rooftops and facades and a shadow algorithm are developed in [17]. It is concluded that although facades receive lower solar radiation, they have a substantial impact on potential urban PV systems once they are utilized. Mangiante et al. in [18] carried out a comparison between rooftop solar arrays in terms of neighborhood orientation and tree shadows. These obstacles affect the solar energy potential. The height and age of the trees are found to have an effect as well.

Other research studies are conducted to investigate the available areas on the surface of the urban cities [19–23]. In [19], a review systematically presents the studies that consider estimating the rooftop area of the cities and the potential deployment of rooftop PV systems. A technical and geographical assessment of the rooftop area in urban cities using novel methodology is presented in [20]. A hierarchical methodology is employed for the estimation process, which encompasses three phases of estimation: physical, geographical, and technical potentials. Geometry calculations and irradiation analysis for tilted rooftop surfaces are performed in [21], including the use of image processing for shadow estimation algorithms in different sun positions. It has been found that in large urban cities, the appropriate analysis method needs three-dimensional data of the studied location.

Many articles focus on solar capacity assessment in rooftop systems in terms of sizing, installation angles, and the efficiency of photovoltaic systems [24–28]. The complete design of a grid-connected PV system is presented in [24,28]. Performance investigation and rooftop analysis are carried out in [26–32]. In [26], a financial and technical feasibility study is performed, which is applicable for only certain geographical locations and weather conditions. For evaluating system performance, the authors in [25] provide a photovoltaic plant design that operates with a seasonal tilt angle according to plant location. The tilt angle is set to be the same as the corresponding latitudinal value of the location to obtain the highest solar radiation. In [29], the rooftop area is assessed using the optimal tilt angle, and 60% of the area is found to be suitable for photovoltaic system implementation.

The research study in this paper aims to fill the gap by providing a comprehensive solar analysis to investigate the feasibility of establishing an on-grid PV system of 330 KW in KSA, exactly in Jeddah city, and introduce the proper building approach. The project is designed to feed the needs of the Ministry of Electricity building, which is a governmental complex with high energy-consuming appliances. The design aims to find the proper way to distribute the solar panels on the roof of the Ministry buildings. This task has been achieved by manipulating the tilt angle until obtaining the desired angle which enables the working staff to accommodate the available area with targeted production, considering that the shading must be avoided and reduced as much as possible.

The rest of the paper is organized as follows: Section 2 describes the design methodology of the PV system under test; two different scenarios are suggested in this section. The energy production for both scenarios is presented and compared. Panel orientation is discussed in Section 3, in which different orientation angles are defined and presented. The single-line diagram of the entire system is also presented in this section. PVsol simulator is used to estimate the performance ratio of this study. The size of the PV system for the second scenario is introduced in Section 4. Finally, the number of needed PV panels and inverters and the appropriate approach of connections are discussed in this section, prior to the conclusion in Section 5.

2. Design Methodology

The goal of establishing this project is to reduce the electricity bill by utilizing the empty spaces in the roof of the buildings and identifying the adequate distribution of solar panels. Table 1 shows the monthly and annual energy consumption of the building. The sizing of the project depends on the capacity of the available spaces. PVsol software was used to investigate the most appropriate orientation of PV panels. The targeted building needs 333 KW, which is the desired capacity. It is a serious challenge to accommodate such demand with a limited area. Undoubtedly, it is required for the designer to identify the proper orientation of solar panels. By adjusting the tilt angles, two scenarios were at disposal. The first scenario requires the utilization of three roofs of Ministry buildings while the second one requires only two roofs, which is the allowed case.

Table 1. Monthly and annual energy consumption of the building.

Month	Energy Consumption (kWh)
January	952,500
February	949,100
March	942,000
April	937,500
May	932,500
June	930,100
July	928,100
August	938,100
September	939,300
October	941,800
November	942,500
December	948,500
Annual energy (MWh)	11,282

2.1. First Scenario

As previously mentioned, the first scenario required the space of three roofs to install PV panels due to the proposed tilt angle. The first suggestion was to adjust the tilt angle to 22° to harvest solar energy as much as possible (see Figure 4). The advantage of this approach is that the system's efficiency is more significant than those with lower tilt angles The optimum tilt angle differs depending on the case study under consideration. Solar panels create the most power when they face the sun directly, so designers should make sure that the panels face the sun perpendicularly as much as feasible. In addition, large space is needed to meet the desired energy capacity at the optimum tilt angle of 22° (three complete roofs as previously mentioned), but only two complete roofs are available. Therefore, the second scenario is proposed by the designer to accommodate the limitations in the roof areas.

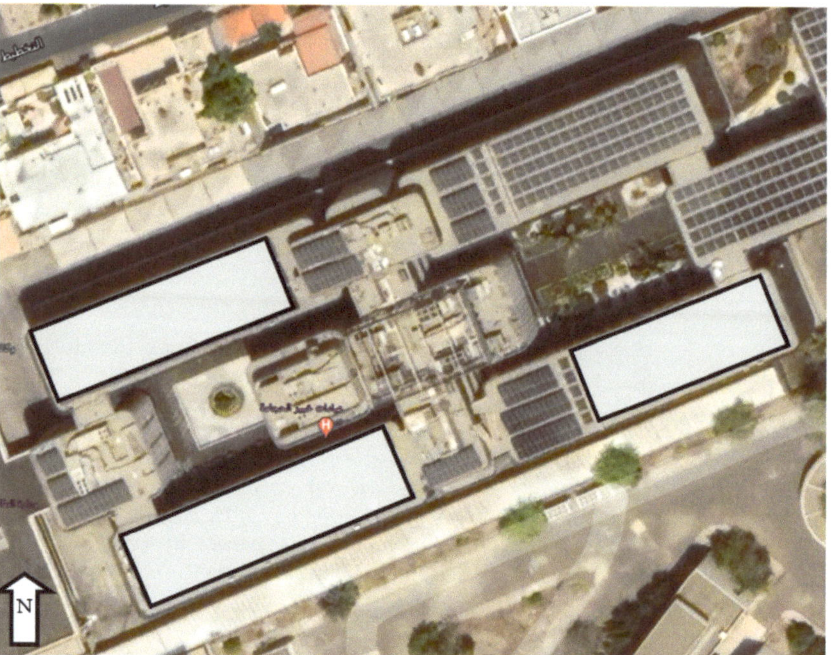

Figure 4. Distribution of the panels in the first scenario.

2.2. Second Scenario

To face the problem of being forbidden to use the entire area, one possible solution is to change the tilt angle to reduce shading and accommodate the area restriction with targeted capacity. This was done by decreasing the tilt angle to 15° rather than 22°. This will reduce the distance between the PV panels to prevent shading and consequently reduce the entire area of installation. Figure 5 presents the distribution of the PV panels on the available area using the second scenario in which the tilt angle is 15°. It is clear that the PV panels are installed over two complete roofs and three separate small, distributed roofs in which the total area is less than the first scenario area.

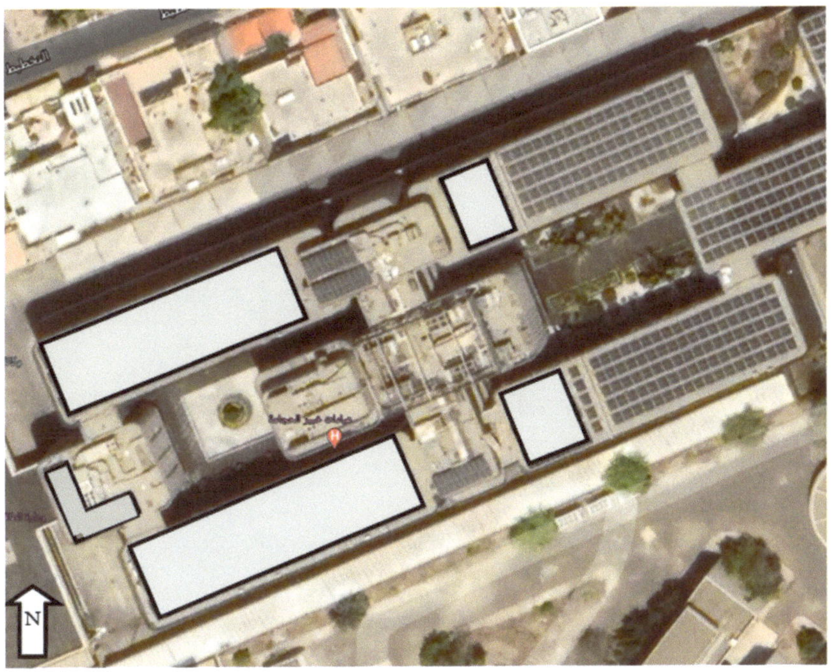

Figure 5. Distribution of the panels in the second scenario.

2.3. Third Scenario

An additional solution has been suggested by the authors to accommodate the limited area of the roof buildings. The solution is based on using the optimal tilt angle of 22° for the first string of the PV system and a tilt angle of 15° for the other strings. The advantage of using this configuration is that it provides more energy than the second scenario and accommodates the limited area of the roof building without shading. Figure 6 shows the authors' proposed configuration, which can be used over buildings' roofs where the area is limited. Figure 7 shows the relationship between elevation angle and spacing between rows for all scenarios. It is clear that the distance between rows is increased by increasing the tilt angle. Therefore, the first scenario has the maximum spacing between rows.

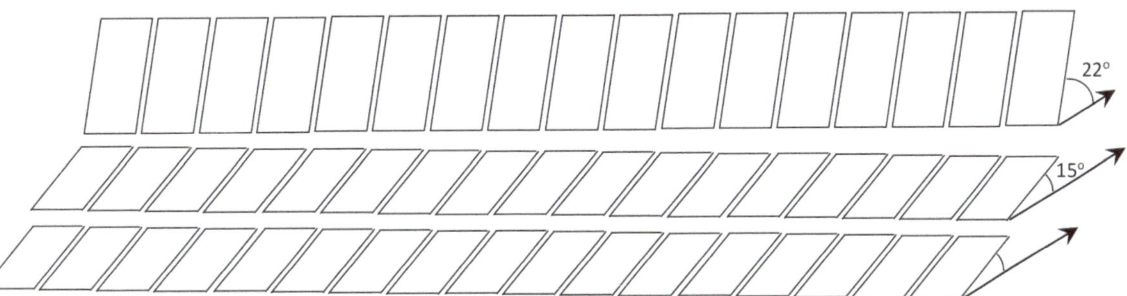

Figure 6. The configuration of the third scenario.

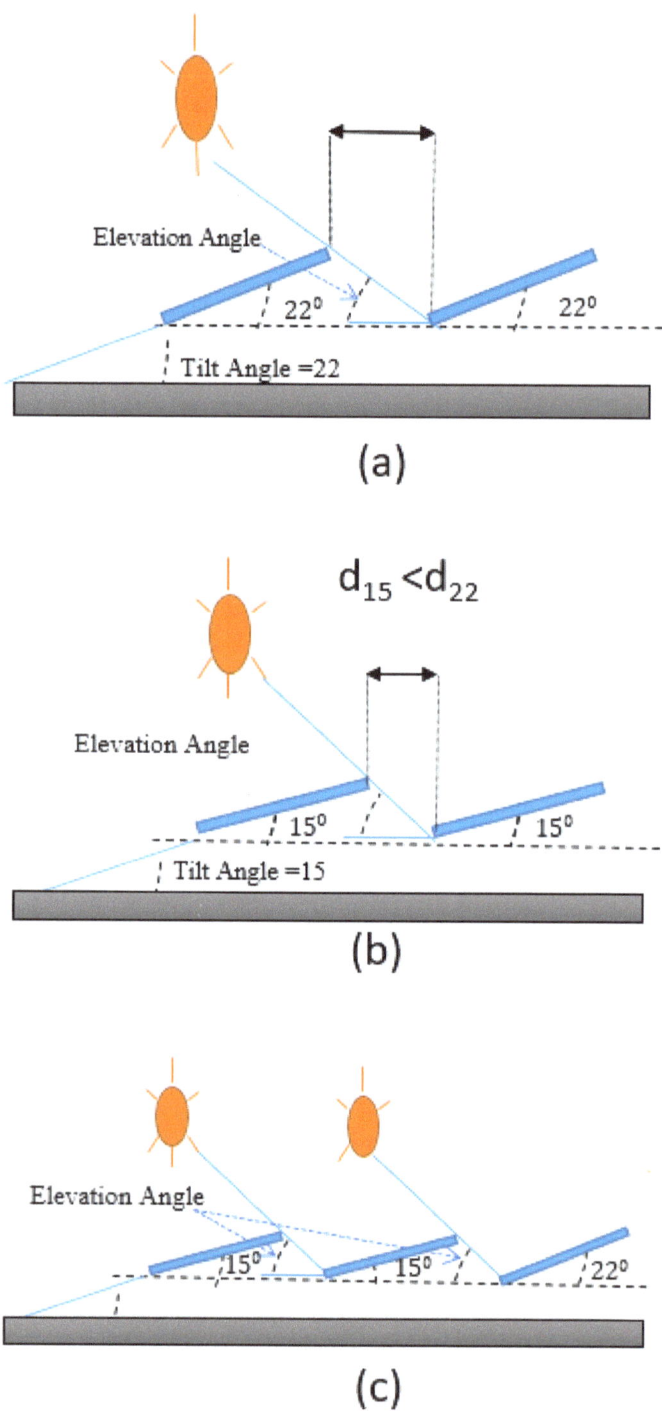

Figure 7. Relationship between elevation angle and spacing between rows: (**a**) Scenario 1, (**b**) Scenario 2, (**c**) Scenario 3.

2.4. Estimating of Energy Production

This section presents the method of estimating the energy production of the PV systems by varying the tilt angle. In addition, a comparative study between the simulated energy output of the PV system and the measured ones is also presented in this section.

The energy produced by the PV system (E) depends on several parameters as suggested by [32]:

$$E = A \times r \times G_R \times PR \tag{1}$$

where:

A—Surface area.
r—Solar panel efficiency.
G_R—Tilted surface mean solar radiation.
PR—Performance ratio.

The performance ratio used in this study is 0.798, according to the *PVsol* simulator for the system, while the average solar radiation for tilted surface, G_R, is given by [33,34]:

$$G_R = (G - D)R_B + DR_D + G_\rho R_R \tag{2}$$

where G and D are ground and diffuse solar radiation in kWh/m^2, respectively. The symbol ρ refers to the ground reflection, and the radiation coefficients R_P, R_D, and R_R are given by [35–39]:

$$R_P = \frac{\cos(L - \beta)\cos\delta\cos\omega_{ss} + \sin(L - \beta)\sin\delta}{\cos L \cos\delta\cos\omega_{ss} + \sin L \sin\delta} \tag{3}$$

where L, β, and ω_{ss} are location latitude, tilt angle, and hour angle, respectively. δ is the declination angle and can be given by:

$$\delta = 23.45 \ \sin\frac{360(284 + n)}{365} \tag{4}$$

where n is the number of days in the year. The reflected solar radiation R_R is calculated as:

$$R_R = \frac{1 - \cos\beta}{2} \tag{5}$$

$$R_D = \frac{1 + \cos\beta}{2} \tag{6}$$

By simulating and substituting the previous equations, the energy yield can be estimated for the intended location.

As mentioned before, the three scenarios would produce different amounts of energy even though the number of panels is the same. The energy production of the PV system varies during the year, in which the maximum energy production is in May and the lowest energy production is in December in both scenarios.

Figure 8 presents a comparison between the energy production for all scenarios. It can be noticed that the first scenario has the maximum energy production, followed by the third scenario. This is expected due to the higher tilt angle which exposes the panels to more solar energy than the second scenario. Furthermore, in the first and the third scenarios, the area of the panels exposed to the sunlight is greater than that in the second scenario. However, the second scenario has been installed by the designer due to the reasons mentioned before, while the third scenario is a new scenario suggested and presented in this paper.

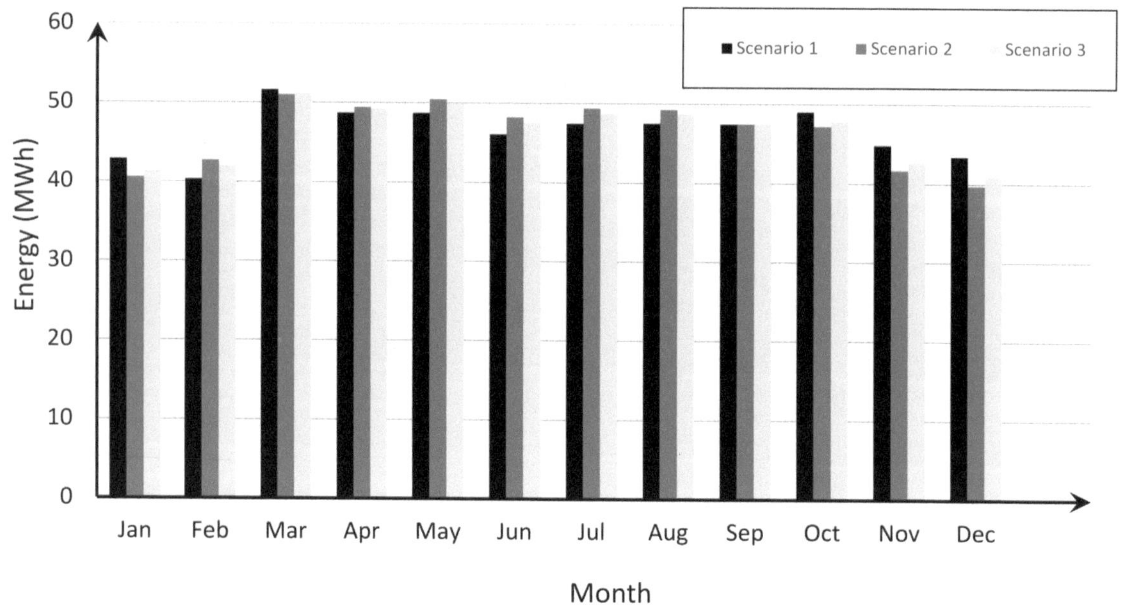

Figure 8. Energy production for all scenarios on a monthly basis.

3. System Configuration

The connection diagram between the panels and the inverters is shown in Figure 8. The panels are connected in four main groups: 2×17, 6×17, 2×15, and 7×17. Each group has a specific number of strings, in which panels are connected in series.

Table 2 gives details about the selected tilt and elevation angles, the distances between the PV panels, and the total roof areas for all scenarios. Table 3 shows solar irradiance in kWh/m^2 at different inclination angles.

Table 2. Selected angles, distances, and the total roof areas for all scenarios.

	Angles and Distances		
	Scenario 1	**Scenario 2**	**Scenario 3**
Tilt angle	22°	15°	22° (First String)/15° (Others)
Elevation angle	27°	19.2°	19.2°
Height difference (m)	1.3	1.04	1.3 (First String)/1.04 (Others)
Model row spacing (m)	1.73	1.42	1.42
Total roof area (m^2)	2267.8	1893.77	1893.77

In this section, the size of the PV system for the second scenario, which has been selected, is introduced. The number of needed PV panels and inverters and the appropriate approach of connections are discussed.

The required PV panels that met a demand of 300 kW are decided to be 790 modules of 380 W each. In addition, six inverters are needed to integrate the PV system with the grid [30,31]. The method of connecting the PV models with the inverters in the form of strings is shown in Figure 9. More details about the required system parameters and quantities are also shown in Table 4.

Table 3. Solar irradiance in (kWh/m^2) at different inclination angles.

Month	Inclination Angle (degree)	Solar irradiance (kWh/m^2)
January	60	164
February	60	147
March	30	189
April	30	205
May	10	241
June	10	254
July	10	262
August	10	245
September	30	229
October	30	215
November	60	183
December	60	156
Total energy		2490

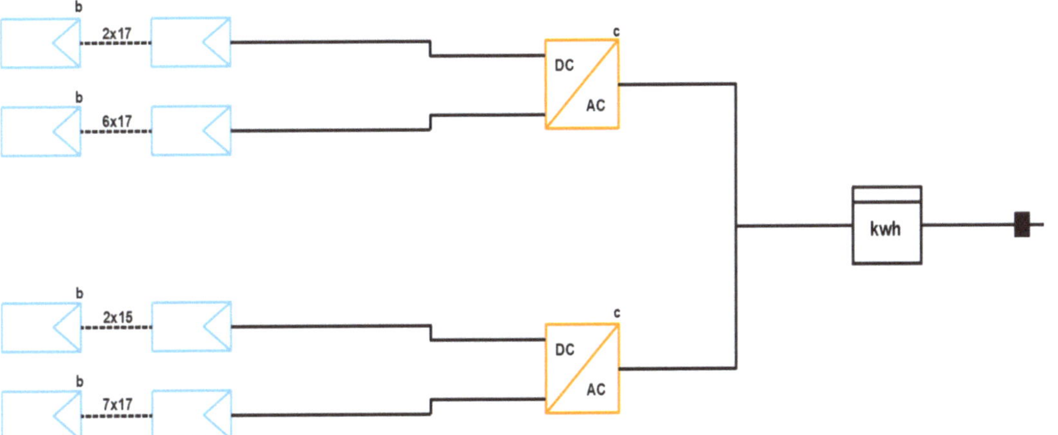

Figure 9. Connection diagram between the panels and the inverters.

Table 4. Equipment sizing details.

Case	Type or Quantity
PV pannel surface	2118 m^2
Dimensions of module	2000 mm $\times$ 991 mm $\times$ 40 mm
Brand of PV modules	JASOLAR
Model number	JAM72S03-380/PR
Rated maximum power of PV module (W)	380
Type of cell	Mono
Number of PV modules	790
Brand of inverters	Chint
Model number	CPS SCA50KTL-DO/400
Rated AC output power (kW)	50
Number of inverters	6
Installation type	Rooftop

The single-line diagram (SLD), shown in Figure 10, was drawn and sized according to the Ministry of Electricity standards of grid integration for the distribution system of low and medium voltages. Moving in detail to the SLD, the number of the connected solar panels is 790 panels distributed in an area of 2118 m^2. In addition, there are six inverters with a capacity of 50 kW each. The appropriate connection between these panels and inverters is to establish three groups of series panels, 15, 16, and 17, each group connected with one of the inverters' MPPTs. This distribution is determined by the inverters' voltage limits, which keep the voltage up to 950 volts.

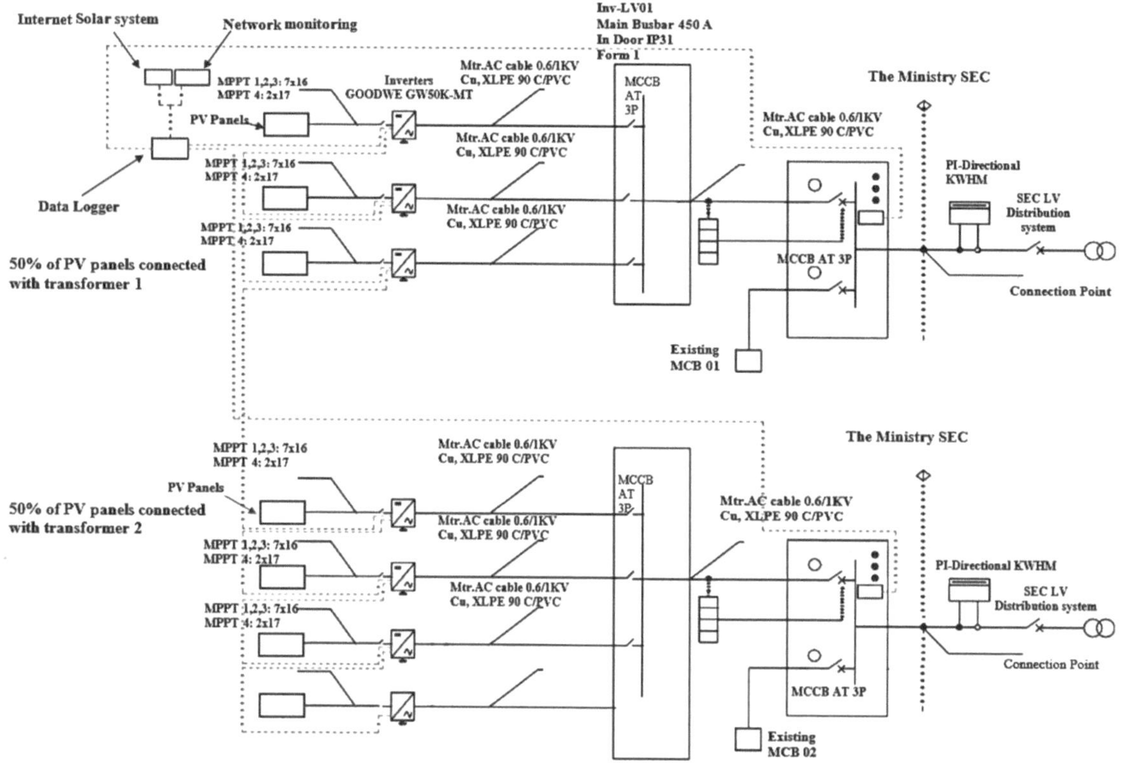

Figure 10. Single-line diagram of the entire system.

Moreover, the SLD represents the sizing of cables and their cross-sectional areas (CSAs) that connect the panels with the inverters. PV cables and AC cables that connect inverters with the grid are also presented in this diagram. Their cross-sectional areas are 6 mm^2 and 35 mm^2, respectively. From the main bus bar to the grid, cables of 185 mm^2 are used. A molded case circuit breaker (MCCB), with 122 A capacity, is connected between every inverter and the main bus bar. The main circuit breaker, which is established between the main bus bar and the low voltage grid, has a capacity of 500 A.

4. Comparison between Measured and Estimated Energy

Table 5 illustrates the estimated energy for the studied location considering the three different scenarios throughout the year. Nevertheless, the second scenario was established, and the measured energy for this scenario has been available since August 2019. Therefore, the comparison depends on the measured energy from August to December when the system was ready and successfully integrated with the grid. Table 6 gives information regarding the estimated average energy on a monthly basis. Generally, the estimated energy output is close to the measured output for the given period. The minimum value of the percentage error registered was in August, with a value of about 2.7%, while the maximum value appeared in November, about 18%.

Table 5. Monthly and annual estimated energy production of all scenarios in kWh.

Month	22° Tilt Angle	15° Tilt Angle	Configuration Proposed by the Author
January	42,863.72	40,586.62	41,346.98
February	43,120.12	42,701.92	41,905.32
March	51,605.42	50,982.12	51,189.88
April	48,698.62	49,404.42	49,169.15
May	48,710.22	50,453.52	49,872.42
June	46,055.82	48,218.22	47,497.42
July	47,434.52	49,388.12	48,736.92
August	47,504.22	49,234.62	48,657.95
September	47,435.52	47,469.221	47,458.12
October	49,030.32	47,195.12	47,806.98
November	44,819.02	41,655.71	42,710.34
December	43,419.12	39,732.51	40,961.58
Annual energy production	560,696.64	557,025.33	557,313.06

Table 6. Comparison between estimated and measured energy output of the PV system in kWh, with percentage error calculations.

	Estimated at 15° Tilt	Measured	Percentage Error %
August	49,234.62	47,909.40	2.69
September	47,469.221	44,873.24	5.46
October	47,195.12	40,545.59	14.10
November	41,655.71	33,842.27	18.75
December	39,732.51	34,447.50	13.30
January 2020	40,586.62	35,971.36	11.37

There are several circumstances for reducing the energy production of the PV system and therefore reducing the efficiency of the PV system. For instance, excessive heat, especially in KSA, and other nonanticipated weather conditions decrease the energy production of the system [40–46].

5. Conclusions

This paper proposed a new PV configuration over the buildings' roofs for limited-area applications. The proposed configuration is based on selecting an optimum tilt angle for the first string of the PV system that generates the largest amount of energy. The other strings of the PV system, in this configuration, can be selected to prevent shading in the limited-area roofs. The proposed new configuration was simulated and tested with other scenarios. The first scenario used the optimal tilt angle for the whole PV system, which generates the maximum amount of electric energy. However, this scenario can only be used if the roof's area is not limited, which is not a critical case. Knowing that, in the case of a roof area restriction, it is more efficient to change the tilt angle of the PV panels than to obtain the shading of the PV panel, as the amount of energy produced varies insignificantly. Therefore, the second scenario, which was applied to accommodate the limited roof area, produces lower energy than the proposed configuration.

The proposed configuration can be used routinely over buildings' roofs where the roof area is limited to produce a larger amount of energy. In addition, this configuration can be applied for small-capacity PV systems as well. Nevertheless, the area saved for such systems will be very small, since the saved area per module for each tilt angle is very small. The authors recommend using the new configuration where roof area is limited to increase the energy output of the whole PV system, although some regulations regarding PV system installations in some countries prevent the use of such a configuration over the roof of a building for aesthetic reasons.

Author Contributions: Conceptualization, A.A.-Q., M.A.-M., T.A.-A. and A.B.; methodology, A.A.-Q., M.A.-M., T.A.-A. and A.B.; software, T.A.-A. and A.B.; validation, A.A.-Q., M.A.-M.; formal analysis, A.A.-Q., M.A.-M., T.A.-A. and A.B.; investigation, A.A.-Q. and M.A.-M.; resources, T.A.-A. and A.B.; data curation, A.A.-Q. and M.A.-M.; writing—original draft preparation, M.A.-M.; writing—review and editing, A.A.-Q.; visualization, M.B., H.M. and A.M.; supervision, A.A.-Q., M.B., H.M. and A.M.; project administration, A.A.-Q.; funding acquisition, A.A.-Q., M.B., H.M. and A.M. All authors have read and agreed to the published version of the manuscript.

Funding: This research received no external funding.

Informed Consent Statement: Not applicable.

Data Availability Statement: The data presented in this study are available on request from the corresponding author.

Acknowledgments: The authors acknowledge Clean Environment Co. for implementing the project and providing the required information and data to prepare the theoretical study. They also acknowledge Yarmouk University, The University of Jordan, Trading Est Technology Utilities, National Energy Works Company and German Jordanian University for their support in this study.

Conflicts of Interest: The authors declare no conflict of interest.

References

1. Zell, E.; Gasim, S.; Wilcox, S.; Katamoura, S.; Stoffel, T.; Shibli, H.; Engel-Cox, J.; Al Subie, M. Assessment of solar radiation resources in Saudi Arabia. *Sol. Energy* **2015**, *119*, 422–438. [CrossRef]
2. Jurgenson, S.; Bayyari, F.M.; Parker, J. A comprehensive renewable energy program for Saudi Vision 2030. *Renew. Energy Focus* **2016**, *17*, 182–183. [CrossRef]
3. Yamada, M. *Vision 2030 and the Birth of Saudi Solar Energy*; Middle East Institute: Singapore, 2016; p. 13.
4. Alshuwaikhat, H.; Mohammed, I. Sustainability Matters in National Development Visions—Evidence from Saudi Arabia's Vision for 2030. *Sustainability* **2017**, *9*, 408. [CrossRef]
5. Heffron, R.J. Just Transitions Around the World. In *Achieving a Just Transition to a Low-Carbon Economy*; Palgrave Macmillan: Cham, Switzerland, 2021; pp. 87–124.
6. Delbeke, J.; Lamas, R. *Exploring Carbon Market Instruments for the Kingdom of Saudi Arabia (KSA)*; European University Institute: Fiesole, Italy, 2021.
7. Mosly, I.; Makki, A.A. Current status and willingness to adopt renewable energy technologies in Saudi Arabia. *Sustainability* **2018**, *10*, 4269. [CrossRef]
8. Mas'ud, A.A.; Wirba, A.V.; Alshammari, S.J.; Muhammad-Sukki, F.; Abdullahi, M.A.M.; Albarracín, R.; Hoq, M.Z. Solar energy potentials and benefits in the gulf cooperation council countries: A review of substantial issues. *Energies* **2018**, *11*, 372. [CrossRef]
9. Dehwah, A.H.; Asif, M. Assessment of net energy contribution to buildings by rooftop photovoltaic systems in hot-humid climates. *Renew. Energy* **2019**, *131*, 1288–1299. [CrossRef]
10. Ethraa Home Page. Net Metering and 4 Ways Saudi Arabia can implement it. Available online: https://www.ethraa-a.com/net-metering-and-4-ways-saudi-arabia-can-implement-it/ (accessed on 12 December 2021).
11. Ko, W.; Al-Ammar, E.; Almahmeed, M. Development of Feed-in Tariff for PV in the Kingdom of Saudi Arabia. *Energies* **2019**, *12*, 2898. [CrossRef]
12. Benghanem, M. Optimization of tilt angle for solar panel: Case study for Madinah, Saudi Arabia. *Appl. Energy* **2011**, *88*, 1427–1433. [CrossRef]
13. Alaidroos, A.; Krarti, M. Optimal design of residential building envelope systems in the Kingdom of Saudi Arabia. *Energy Build.* **2015**, *86*, 104–117. [CrossRef]
14. Gharakhani Siraki, A.; Pillay, P. Study of optimum tilt angles for solar panels in different latitudes for urban applications. *Sol. Energy* **2012**, *86*, 1920–1928. [CrossRef]
15. Nguyen, H.T.; Pearce, J.M. Automated quantification of solar photovoltaic potential in cities Overview: A new method to determine a city's solar electric potential by analysis of a distribution feeder given the solar exposure and orientation of rooftops. *Int. Rev. Spat. Plan. Sustain. Dev.* **2013**, *1*, 49–60. [CrossRef]
16. Hong, T.; Lee, M.; Koo, C.; Jeong, K.; Kim, J. Development of a method for estimating the rooftop solar photovoltaic (PV) potential by analyzing the available rooftop area using Hillshade analysis. *Appl. Energy* **2017**, *194*, 320–332. [CrossRef]
17. Redweik, P.; Catita, C.; Brito, M. Solar energy potential on roofs and facades in an urban landscape. *Sol. Energy* **2013**, *97*, 332–341. [CrossRef]
18. Mangiante, M.J.; Whung, P.Y.; Zhou, L.; Porter, R.; Cepada, A.; Campirano, E., Jr.; Licon, D., Jr.; Lawrence, R.; Torres, M. Economic and technical assessment of rooftop solar photovoltaic potential in Brownsville, Texas, USA. *Comput. Environ. Urban Syst.* **2020**, *80*, 101450. [CrossRef]

19. Byrne, J.; Taminiau, J.; Kurdgelashvili, L.; Kim, K.N. A review of the solar city concept and methods to assess rooftop solar electric potential, with an illustrative application to the city of Seoul. *Renew. Sustain. Energy Rev.* **2015**, *41*, 830–844. [CrossRef]
20. Izquierdo, S.; Rodrigues, M.; Fueyo, N. A method for estimating the geographical distribution of the available roof surface area for large-scale photovoltaic energy-potential evaluations. *Sol. Energy* **2008**, *82*, 929–939. [CrossRef]
21. Desthieux, G.; Carneiro, C.; Camponovo, R.; Ineichen, P.; Morello, E.; Boulmier, A.; Ellert, C. Solar energy potential assessment on rooftops and facades in large built environments based on lidar data, image processing, and cloud computing. methodological background, application, and validation in geneva (solar cadaster). *Front. Built Environ.* **2018**, *4*, 14. [CrossRef]
22. Mainzer, K.; Killinger, S.; McKenna, R.; Fichtner, W. Assessment of rooftop photovoltaic potentials at the urban level using publicly available geodata and image recognition techniques. *Sol. Energy* **2017**, *155*, 561–573. [CrossRef]
23. Singh, R.; Banerjee, R. Estimation of rooftop solar photovoltaic potential of a city. *Sol. Energy* **2015**, *115*, 589–602. [CrossRef]
24. Berwal, A.K.; Kumar, S.; Kumari, N.; Kumar, V.; Haleem, A. Design and analysis of rooftop grid tied 50 kW capacity Solar Photovoltaic (SPV) power plant. *Renew. Sustain. Energy Rev.* **2017**, *77*, 1288–1299. [CrossRef]
25. Kumar, B.S.; Sudhakar, K. Performance evaluation of 10 MW grid connected solar photovoltaic power plant in India. *Energy Rep.* **2015**, *1*, 184–192. [CrossRef]
26. Khatib, T.; Ibrahim, I.A.; Mohamed, A. A review on sizing methodologies of photovoltaic array and storage battery in a standalone photovoltaic system. *Energy Convers. Manag.* **2016**, *120*, 430–448. [CrossRef]
27. Kumar, A.; Andleeb, M.; Bakhsh, F.I. Design and Analysis of Solar PV Rooftop in Motihari. *J. Phys. Conf. Ser.* **2021**, *1817*, 12019. [CrossRef]
28. Kadyan, H.; Berwal, A.K. Design of a 12 kWp grid connected roof top solar photovoltaic power plant on school building in the Rohtak District of Haryana. *Int. J. Appl. Eng. Res.* **2018**, *13*, 11354–11361.
29. Mokhtara, C.; Negrou, B.; Settou, N.; Bouferrouk, A.; Yao, Y. Optimal design of grid-connected rooftop PV systems: An overview and a new approach with application to educational buildings in arid climates. *Sustain. Energy Technol. Assess.* **2021**, *47*, 101468. [CrossRef]
30. Khatib, T.; Mohamed, A.; Sopian, K.; Mahmoud, M. A New Approach for Optimal Sizing of Standalone Photovoltaic Systems. *Int. J. Photoenergy* **2012**, *2012*, 391213. [CrossRef]
31. Markvart, T.; Fragaki, A.; Ross, J.N. PV system sizing using observed time series of solar radiation. *Sol. Energy* **2006**, *80*, 46–50. [CrossRef]
32. Wissem, Z.; Gueorgui, K.; Hédi, K. Modeling and technical–economic optimization of an autonomous photovoltaic system. *Energy* **2012**, *37*, 263–272. [CrossRef]
33. Hay, J.E. Calculation of monthly mean solar radiation for horizontal and inclined surfaces. *Sol. Energy* **1979**, *23*, 301–307. [CrossRef]
34. Jaber, S.; Hawa, A.A. Optimal design of PV system in passive residential building in Mediterranean climate. *Jordan J. Mech. Ind. Eng.* **2016**, *10*, 39–49.
35. Duffie, J.A.; Beckman, W.A. *Solar Engineering of Thermal Processes*, 4th ed.; John Wiley & Sons: Hoboken, NJ, USA, 2013.
36. Radaideh, A.; Bodoor, M.; Al-Quraan, A. Active and Reactive Power Control for Wind Turbines Based DFIG Using LQR Controller with Optimal Gain-Scheduling. *J. Electr. Comput. Eng.* **2021**, *2021*, 1218236. [CrossRef]
37. Al-Mhairat, B.; Al-Quraan, A. Assessment of Wind Energy Resources in Jordan Using Different Optimization Techniques. *Processes* **2022**, *10*, 105. [CrossRef]
38. Al-Quraan, A.; Al-Qaisi, M. Modelling, Design and Control of a Standalone Hybrid PV-Wind Micro-Grid System. *Energies* **2021**, *14*, 4849. [CrossRef]
39. Yakup, M.A.b.H.M.; Malik, A.Q. Optimum tilt angle and orientation for solar collector in Brunei Durassalam. *Renew. Energy* **2001**, *24*, 223–234. [CrossRef]
40. Amelia, A.R.; Irwan, Y.M.; Leow, W.Z.; Irwanto, M.; Safwati, I.; Zhafarina, M. Investigation of the effect temperature on photovoltaic (PV) panel output performance. *Int. J. Adv. Sci. Eng. Inf. Technol.* **2016**, *6*, 682–688.
41. Dubey, S.; Sarvaiya, J.N.; Seshadri, B. Temperature Dependent Photovoltaic (PV) Efficiency and Its Effect on PV Production in the World—A Review. *Energy Procedia* **2013**, *33*, 311–321. [CrossRef]
42. Al-Quraan, A.; Stathopoulos, T.; Pillay, P. Comparison of Wind Tunnel and on Site Measurements for Urban Wind Energy Estimation of Potential Yields. *Elsevier J. Wind. Eng. Ind. Aerodyn.* **2016**, *158*, 1–10. [CrossRef]
43. Stathopoulos, T.; Alrawashdeh, H.; Al-Quraan, A.; Blocken, B.; Dilimulati, A.; Paraschivoiu, M.; Pillay, P. Urban Wind Energy: Some Views on Potential and Challenges. *Elsevier J. Wind Eng. Ind. Aerodyn.* **2018**, *179*, 146–157. [CrossRef]
44. Al-Quraan, A.; Pillay, P.; Stathopoulos, T. Use of a Wind Tunnel for Urban Wind Power Estimation. In Proceedings of the IEEE Power & Energy Society General Meeting, Washington, DC, USA, 27–31 July 2014.
45. Al-Quraan, A.; Alrawashdeh, H. Correlated Capacity Factor Strategy for Yield Maximization of Wind Turbine Energy. In Proceedings of the IEEE 5th International Conference on Renewable Energy Generation and Applications (ICREGA), Al-Ain, United Arab Emirates, 26–28 February 2018.
46. Al-Quraan, A.; Stathopoulos, T.; Pillay, P. Estimation of Urban Wind Energy-Equiterre Building Case in Montreal. In Proceedings of the International Civil Engineering for Sustainability and Resilience Conference (CESARE'14), Irbid, Jordan, 24–27 April 2014.

Article

Numerical Investigation on Energy Efficiency of Heat Pump with Tunnel Lining Ground Heat Exchangers under Building Cooling

Xiaohua Liu [1,2,*], Chenglin Li [3], Guozhu Zhang [3,*], Linfeng Zhang [3] and Bin Wei [2]

[1] College of Civil Engineering, Hunan University, Changsha 410082, China
[2] Shenzhen Transportation Design & Research Institute Co., Ltd., Shenzhen 518003, China; Weinbin@163.com
[3] Institute of Geotechnical Engineering, Southeast University, Nanjing 210096, China; lichenglinseu@163.com (C.L.); belzhang@seu.edu.cn (L.Z.)
* Correspondence: lxh.zj@163.com (X.L.); zhanggz@seu.edu.cn (G.Z.)

Abstract: For mountain tunnels, ground heat exchangers can be integrated into the tunnel lining to extract geothermal energy for building heating and cooling via a heat pump. In recent decades, many researchers only focused on the thermal performance of tunnel lining Ground Heat Exchangers (GHEs), ignoring the energy efficiency of the heat pump. A numerical model combining the tunnel lining GHEs and heat pump was established to investigate the energy efficiency of the heat pump. The inlet temperature of an absorber pipe was coupled with the cooling load of GHEs in the numerical model, and the numerical results were calibrated using the in situ test data. The energy efficiency ratio (EER) of the heat pump was calculated based on the correlation of the outlet temperature and EER. The heat pump energy efficiencies under different pipe layout types, pipe pitches and pipe lengths were evaluated. The coupling effect of ventilation and groundwater flow on the energy efficiency of heat pump was investigated. The results demonstrate that (i) the absorber pipes arranged along the axial direction of the tunnel have a greater EER than those arranged along the cross direction; (ii) the EER increases exponentially with increasing absorber pipe pitch and length (the influence of the pipe pitch and length on the growth rate of EER fades gradually as wind speed and groundwater flow rate increase); (iii) the influence of groundwater conditions on the energy efficiency of heat pumps is more obvious compared with ventilation conditions. Moreover, abundant groundwater may lead to a negative effect of ventilation on the heat pump energy efficiency. Hence, the coupling effect of ventilation and groundwater flow needs to be considered for the tunnel lining GHEs design.

Keywords: geothermal energy; tunnel lining GHEs; numerical model; energy efficiency; building cooling

Citation: Liu, X.; Li, C.; Zhang, G.; Zhang, L.; Wei, B. Numerical Investigation on Energy Efficiency of Heat Pump with Tunnel Lining Ground Heat Exchangers under Building Cooling. *Buildings* **2021**, *11*, 611. https://doi.org/10.3390/buildings11120611

Academic Editor: Fabrizio Ascione

Received: 6 November 2021
Accepted: 2 December 2021
Published: 4 December 2021

Publisher's Note: MDPI stays neutral with regard to jurisdictional claims in published maps and institutional affiliations.

1. Introduction

In China, the main source of building energy consumption is fossil energy, which amounts to 0.3 billion tonnes of coal equivalent per year [1]. The massive consumption of fossil energy has led to a series of problems, such as waste of resources and environmental degradation. Therefore, it is imperative to replace fossil energy with environment-friendly energy, especially in building cooling and heating, which is over 40% of building consumption [2]. Geothermal energy is a renewable and low-carbon energy source [3]. Ground-embedded structures, e.g., piles [4,5], underground diaphragm walls [6] and tunnel linings [7], can be coupled with ground source heat pump (GSHP) systems to extract geothermal energy for building heating and cooling. Many researchers pay attention to the tunnel lining Ground Heat Exchangers (GHEs) because of a large heat exchange area and no drilling, compared to traditional borehole GHEs. For the tunnel in a mountain environment, the tunnel lining GHEs can meet the energy requirement of adjacent users, such as the village away from the tunnel entrance of 1–2 km [8]. Moreover, the tunnel management center is usually only a few hundred meters from the tunnel entrance, which is a good option for using geothermal energy extracted by the tunnel lining GHEs. Figure 1 presents the diagram of the tunnel lining GHEs, the geothermal energy is transferred to tunnel lining GHEs by absorber pipes equipped between the primary and secondary linings, then absorbed geothermal energy is applied to heat and cool the building via the heat pump.

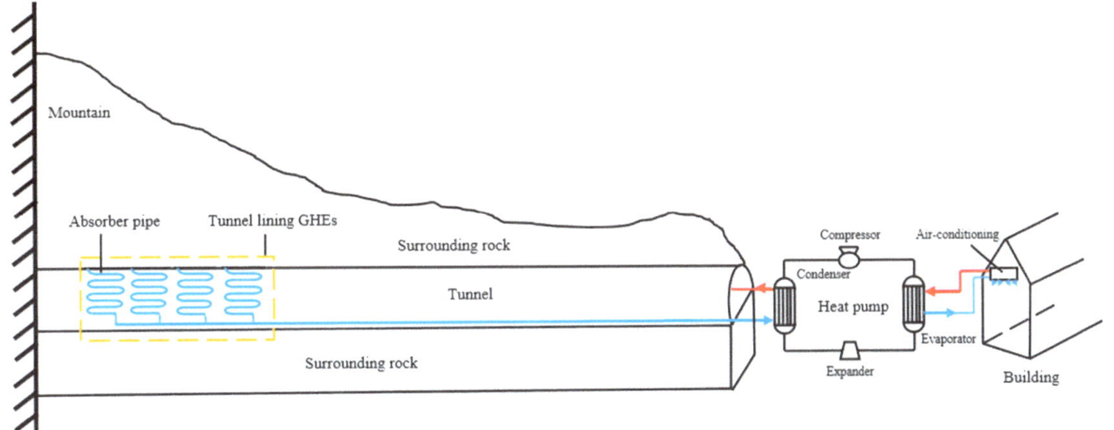

Figure 1. Diagram of tunnel lining GHEs.

The ground heat exchangers (GHEs) and heat pumps are the essential parts of the GSHP system. At present, there are many studies on the tunnel lining GHEs. Adam and Markiewicz [9] installed energy geotextiles into the tunnel lining GHEs in the experimental section of the Lainzer tunnel, which enhances efficiency of tunnel lining GHEs construction. Zhang et al. [10] assessed the thermal efficiency of tunnel lining GHEs in a mountain environment; the influencing factors on thermal performance of tunnel lining GHEs were further analyzed. Some research works showed that the groundwater conditions can improve thermal efficiency of tunnel lining GHEs, which help to recover ground temperature [11,12]. Di Donna and Barla [13] investigated the influence of hydraulic conductivity, thermal conductivity, groundwater temperature, and flow velocity on thermal performance of tunnel lining GHEs using a 3D numerical model of tunnel lining GHEs with a constant inlet temperature of absorber pipe. Ventilation conditions can improve the thermal performance of tunnel lining GHEs [14,15]. Li et al. [16] found that the effect of diurnal air temperature variation on the temperature of fluid inside the absorber pipe was slight and hysteretic. Dornberger et al. [17] proposed a design chart to summarize the effect of airflow characteristics in the tunnel on the geothermal energy potential of tunnel lining GHEs based on a 3D numerical model of tunnel lining GHEs. In this model, the convective heat transfer boundary was set on the tunnel internal surface to simulate the airflow inside the tunnel. Zhang et al. [18] conducted laboratory model tests to investigate the effect of ventilation and groundwater flow on the thermal efficiency of tunnel lining GHEs. The results showed that the temperature field of the surrounding rock was uneven under the influence of ventilation and groundwater flow, and the absorber pipes should be arranged upstream of the groundwater flow field. The geological conditions and ventilation conditions depend on the environment, which are the significant indicators to assess the heat exchange capacity of tunnel lining GHEs. Moreover, optimizations of tunnel lining GHEs design parameters and operation mode are also important [8]. Ogunleye et al. [19,20] investigated the effect of the design parameters and operation mode on tunnel thermal efficiency by the numerical methods. In this numerical model, the varying tunnel air temperature was imposed on the tunnel internal surface, the results showed that the absorber pipe length had the most significant impact, and an optimal intermittent ratio was necessary for improving thermal performance of tunnel lining GHEs.

The GSHP system changes not only the temperature of the building environment but also the temperature of the underground environment. The heat is injected into the underground by GHEs in cooling mode, which will induce a decrease in heat pump energy efficiency. However, most of the above studies only focused on the thermal performance of tunnel lining GHEs, ignoring heat pump energy efficiency. Hence, evaluation of the energy efficiency of heat pump under different absorber pipe layout types, pipe pitches and pipe lengths by a numerical model combining the tunnel lining GHEs and heat pump was necessary. In this model, the inlet temperature of the absorber pipe varied with time based on the cooling load. Meanwhile, the variation in the energy efficiency with pipe pitch and pipe length under different groundwater and ventilation conditions also were investigated. Based on the above numerical results, optimizations of the design parameters of the tunnel lining GHEs were discussed deeply, and some engineering suggestions for tunnel lining GHEs were given.

2. Methodology

2.1. Mathematical Formulation

The complex heat transfer process of the tunnel lining GHEs concerns convective heat transfer of liquid inside the absorber pipe, the convective heat transfer of tunnel air, and conduction heat transfer of a composite solid medium. To reduce computational costs, the assumptions were as follows: (1) the thermal properties of the solid medium are constant; (2) the thermal contact resistance of tunnel lining and surrounding rock is ignored [15,19]; (3) the heat transfer of absorber pipe wall follows a quasi-steady state [15].

The conduction heat transfer equation of tunnel secondary and primary lining can be as follows:

$$\rho_i C_{p,i} \frac{\partial T_i}{\partial t} = \nabla \cdot (k_i \cdot \nabla T_i) + Q_i (i = 1, 2) \tag{1}$$

where T_i denotes the temperature (°C); ρ_i denotes the density (kg/m^3); k_i denotes the thermal conductivity (W/m °C); $C_{p,i}$ denotes the specific heat capacity (J/(kg °C)); Q_i denotes the heat source (W/m^3); denotes the time (s); and $i = 1$, and 2 are the secondary lining and primary lining.

When the groundwater flow field was considered in the model, the rock was regarded as a porous medium. The heat transfer governing equations are as follows:

$$(\rho C_p)_{eff} \frac{\partial T_r}{\partial t} + (\rho C_p v_f)_{eff} \cdot \nabla T_r = \nabla \cdot \left(k_{eff} \cdot \nabla T_r \right) \tag{2}$$

$$(\rho C_p)_{eff} = (1 - \varepsilon_p)\rho_r C_{p,r} + \varepsilon_p \rho_f C_{p,f} \tag{3}$$

$$k_{eff} = (1 - \varepsilon_p)k_r + \varepsilon_p k_f \tag{4}$$

where ρ_f denotes the density of the water (kg/m^3); $C_{p,f}$ denotes the specific heat capacity of the water (J/(kg °C)); k_f denotes the thermal conductivity of the water (W/(m °C)); T_r denotes the temperature of the surrounding rock (°C); ρ_r denotes the density of the surrounding rock (kg/m^3); $C_{p,r}$ denotes the specific heat capacity of the surrounding rock (J/(kg °C)); k_r denotes the thermal conductivity of the surrounding rock (W/(m °C)); $(\rho C_p)_{eff}$ denotes the effective volumetric heat capacity (J/(m^3 °C)); k_{eff} denotes the effective thermal conductivity (J/(kg °C)); v_f denotes the groundwater flow rate (m/s); and ε_p denotes porosity.

In the groundwater seepage field, Darcy's law is usually employed to simulate the groundwater flow within a porous medium, and the governing equations are as follows:

$$\frac{\partial \varepsilon_p \rho_f}{\partial t} + \nabla (\rho_f v_f)_{eff} = 0 \tag{5}$$

$$v_f = -\frac{\kappa}{\mu_f} \cdot \left(\nabla p_f + \rho_f g \nabla D \right) \tag{6}$$

$$\kappa = -K_h \frac{\mu_w}{\rho_w g} \tag{7}$$

where κ denotes the permeability (m^2); μ_w denotes the dynamic viscosity of the water (Pa s); p_f denotes the pore pressure in the ground (Pa); g denotes the gravitational acceleration vector (m^2/s); D denotes the elevation along the direction of the vertical coordinate (m); and K_h denotes the hydraulic conductivity (m/s).

The transient convective heat transfer equations of liquid inside the absorber pipe are presented below.

The momentum equation is given by:

$$\rho_L \frac{\partial u_L}{\partial t} = -\nabla p_L - \frac{1}{2} f_D \frac{\rho_L}{d_h} |u_L| u_L \tag{8}$$

The continuity equation is defined as:

$$\frac{\partial \rho_L}{\partial t} + \nabla \cdot (\rho_L u_L) = 0 \tag{9}$$

The energy conservation equation is:

$$\rho_L A C_{p,L} \frac{\partial T_L}{\partial t} + \rho_L A C_p u_L \cdot \nabla T_L = \nabla \cdot (A k_L \nabla T_L) + \frac{1}{2} f_D \frac{\rho_L A}{d_h} |u_L|^3 + q_{wall} \tag{10}$$

$$q_{wall} = h_c (T_1 - T_L) \tag{11}$$

$$h_c = \frac{2\pi}{\frac{1}{d_{p,in} h_{in}} + \frac{1}{k_p} \ln\left(\frac{d_{p,out}}{d_{p,in}} \right)} \tag{12}$$

$$h_{in} = Nu \frac{k_L}{d_h} \tag{13}$$

where ρ_L denotes the liquid density (kg/m^3); $C_{p,L}$ denotes the liquid specific heat capacity (J/(kg $^\circ$C)); k_L denotes the thermal conductivity of the liquid (W/(m $^\circ$C)); u_L denotes liquid flow velocity (m/s); p_L denotes liquid pressure (Pa); f_D denotes the Darcy friction factor; d_h denotes the hydraulic diameter (m); $d_{p,out}$ and $d_{p,in}$ denote the outer and inner diameters of the absorber pipe (m), respectively; A denotes the inner cross-section of the pipe (m^2); k_p denotes the thermal conductivity of the absorber pipe (W/m $^\circ$C); q_{wall} denotes the heat flux (W/m); h_c denotes the equivalent convective heat transfer coefficient (W/m $^\circ$C); h_{in} denotes the convective heat transfer coefficient of the internal film of the pipe (W/m^2 $^\circ$C); and T_L denotes the liquid temperature ($^\circ$C).

2.2. Initial and Boundary Conditions

The initial ground temperature can be expressed as:

$$T_0(y,t) = T_M + A_S e^{-y\sqrt{\frac{\omega}{2a_r}}} \cos(\omega t - y\sqrt{\frac{\omega}{2a_r}}) \tag{14}$$

where T_M denotes the annual average temperature at the ground surface; A_S denotes the amplitude of the ground surface temperature variation; ω denotes the angular frequency of the annual temperature variation, $\omega = 2\pi/365$; and a_r is the rock thermal diffusivity.

The depth of ground temperature fields influenced by the air temperature is about 10–15 m [14]. The tunnel GHEs are located in the constant temperature layer with sufficient depth. The influence of air temperature on the temperature fields of surrounding rock can be neglected [14,15,21]. Hence, the constant temperature boundary conditions are imposed on the far-field boundaries. The adiabatic boundary conditions are imposed on the vertical surfaces crossing the tunnel axis due to the symmetry of the heat transfer model.

The dynamic air temperature is applied to interface of the tunnel and air as a convective heat transfer boundary to simulate airflow in the tunnel, which is defined as:

$$k_1 \nabla T_1 = h(T_1 - T_{air}(t)) \tag{15}$$

where $T_{air}(t)$ denotes the time-dependent air temperature (°C), and the h denotes convection heat transfer coefficient (CHTC) (W/m² °C).

The CHTC can be written as [22,23]:

$$h = aV^b + c \tag{16}$$

where V denotes wind speed (m/s); parameters a, b and c are 4.2, 1 and 6.2, respectively [15].

To simulate the groundwater flow, a hydraulic head difference is set on the upper and lower boundaries. The no-flow temperature boundary conditions are set on the lateral boundaries, which are shown in Figure 2.

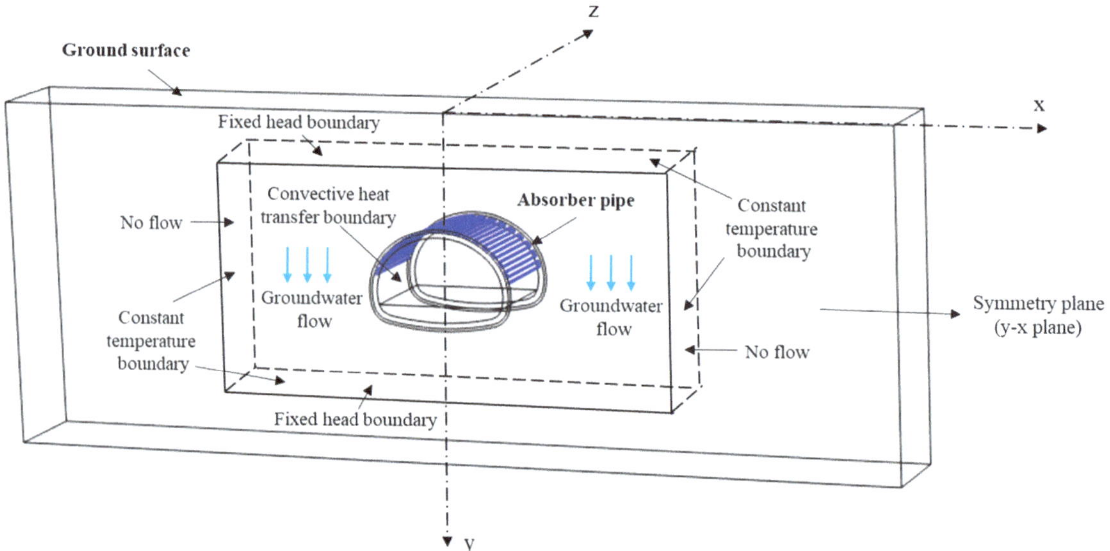

Figure 2. Schematic of heat transfer model.

2.3. Heat Pump Integration

The GHEs can be used in the GSHP system to meet the building load. The GHEs' load in the cooling modes are written as:

$$Q_{GHE}^{Cooling}(t) = Q_b^{Cooling}(t) + Q_{hp}^{Cooling}(t) \tag{17}$$

The energy efficiency ratio (EER) is applied to assess the heat pump energy efficiency in cooling modes. The EER is written as:

$$EER(t) = \frac{Q_b^{Cooling}(t)}{Q_{hp}^{Cooling}(t)} \tag{18}$$

where $Q_{hp}^{Cooling}(t)$ denotes the heat pump energy consumption (W); $Q_{b}^{Cooling}(t)$ denotes the load of the building; and $Q_{GHE}^{Cooling}(t)$ denotes the cooling load of the tunnel lining GHEs (W).

The EER can be calculated by a quadratic function of the outlet temperature [24,25]. The EER equation used in this study is given by:

$$EER(t) = M - NT_{out}(t) + ST_{out}(t)^2 \tag{19}$$

The parameters M, N and S of the heat pump model (Equation (19)) offered by the manufacturer [26] are 9.08, 0.179 and 0.00102, respectively, which are used in this study. The EER function could be easily extended to other different types of heat pumps.

As shown in Figure 3, the inlet temperature depends on the cooling load of the tunnel lining GHEs.

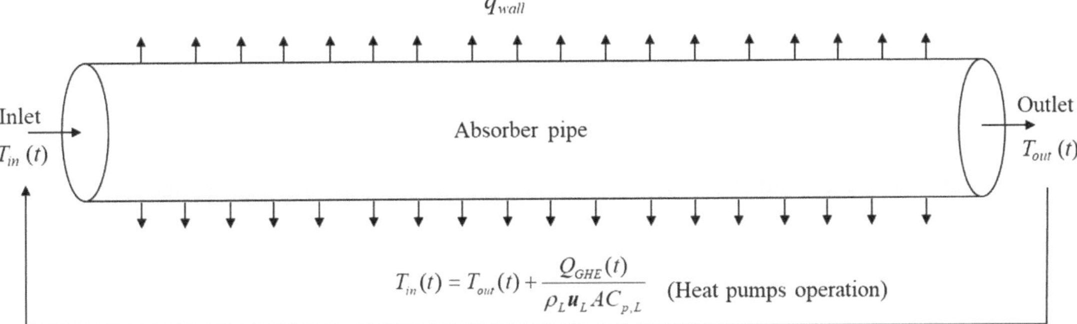

Figure 3. Schematic of inlet temperature variation.

This inlet temperature can be calculated by:

$$T_{in}(t) = T_{out}(t) + \frac{Q_{GHE}(t)}{\rho_L u_L A C_{p,L}} \tag{20}$$

where $T_{in}(t)$ denotes the inlet temperature (°C); $T_{out}(t)$ denotes the outlet temperature (°C); and $Q_{GHE}(t)$ denotes the cooling load of the tunnel lining GHEs (W).

3. Numerical Model

3.1. Model Validation

COMSOL Multiphysics software was employed to develop a 3D heat transfer numerical model of the tunnel lining GHEs.

The results of the field tests performed on real-scale tunnel lining GHEs section of Linchang Tunnel were reported in [10], which were used to verify the heat transfer model. The inner diameter of the tunnel was 5.7 m, and the secondary and primary linings had thicknesses of 35 cm and 17 cm. The other model parameters are presented in Table 1. A numerical model with a length of 100 m, a width of 60 m, and a height of 11 m was developed. Figure 4 presents the 3D view of the proposed heat transfer model.

Table 1. Parameters of model.

Material	Parameters	Unit	Value
Rock	Mass density (ρ_r)	kg/m^3	2530
	Thermal conductivity (k_r)	W/m °C	3.22
	Specific heat capacity ($C_{p,r}$)	J/kg °C	1670
Tunnel lining	Mass density (ρ_1, ρ_2)	kg/m^3	2400
	Thermal conductivity (k_1, k_2)	W/m °C	1.85
	Specific heat capacity ($C_{p,1}$, $C_{p,2}$)	J/kg °C	970
	Inner diameter ($d_{t,in}$)	m	5.7
	Primary lining thickness (δ_1)	m	0.17
	Secondary lining thickness (δ_2)	m	0.35
Absorber pipe	Thermal conductivity (k_p)	W/m °C	0.32
	Inner diameter ($d_{p,in}$)	mm	23
	Outer diameter ($d_{p,out}$)	mm	32
	Flow velocity (u_L)	m/s	0.6
	Pipe pitch (J)	m	0.5
	Pipe length (L)	m	70
Carrier liquid	Thermal conductivity (k_L)	W/m °C	0.56
	Specific heat capacity ($C_{p,L}$)	J/kg °C	4200
	Mass density (ρ_L)	kg/m^3	1000

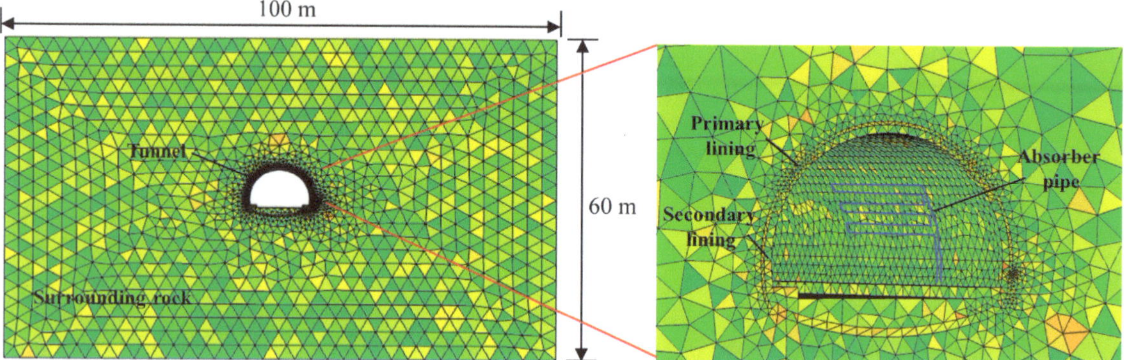

Figure 4. A 3D view of proposed heat transfer model.

In order to determine enough elements to achieve convergence, a grid-dependent numerical study was performed, and a finer grid was used around the absorber pipe. As shown in Table 2, the outlet temperature in continuous mode converged at 2,326,849 elements, achieving a grid-size independent solution. Figure 5 shows the experimental and numerical outlet temperatures of absorber pipe. The experimental outlet temperature of the absorber pipe has a large oscillation due to the periodic operation of the heat pump. The numerical results showed agreement with the experimental results. Therefore, we believe that the numerical model can accurately simulate the operation of tunnel lining GHEs.

Table 2. Grid study.

Elements Number	Temperature (°C)
1,362,852	8.46
1,662,152	8.42
1,971,699	8.38
2,326,849	8.36
2,666,951	8.36

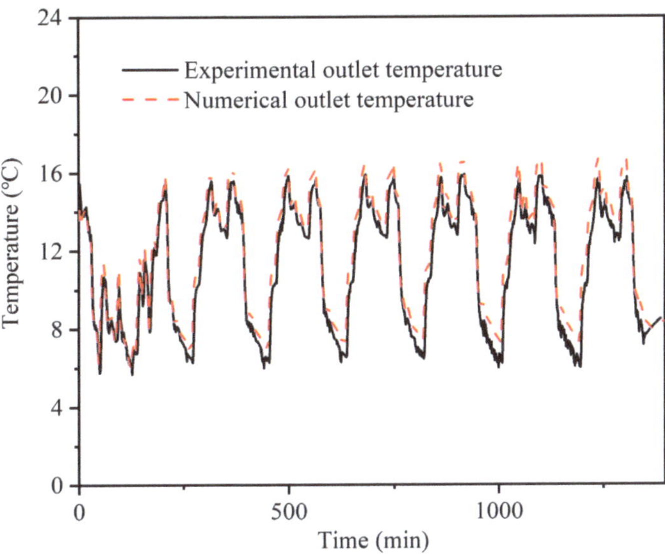

Figure 5. Numerical and experimental results.

3.2. Parametric Numerical Study

In the parametric numerical study, the operation of the tunnel lining GHEs was considered for three months of cooling. The model of the tunnel in Section 3.1 was used to perform the parametric numerical study. The measured air temperature was set on the boundary between the air and tunnel internal surface to simulate the effect of airflow. Different groups of tunnel lining GHEs could be arranged parallelly to meet the building cooling load. This study focused on one group of tunnel lining GHEs to investigate the effect of the tunnel lining GHEs parameters (absorber pipe layout types, pipe pitches and pipe lengths) on the energy efficiency of the heat pump. The temperature difference between the inlet and outlet was set to 5 °C according to the Design Code for Heating Ventilation and Air Conditioning of Civil Buildings (in China) [27], which corresponds to a tunnel lining GHEs cooling load of 5.235 kW. The maximum wind speed was determined to be 5 m/s according to the monitoring data of the wind speed range [15]. The porosity of the surrounding rock is assumed as 10% for simulating the groundwater flow. The parameters of parametric study are presented in Table 3.

Table 3. Parameters of parametric numerical study.

Characteristic	Unit	Value
Pipe pitch (J)	m	0.3, 0.4, 0.5 ₛ, 0.6
Pipe length (L)	m	250–400
Wind speed (V)	m/s	0.1 ₛ, 1, 3, 5
Groundwater flow rate (v_f)	m/s	0 ₛ, 10^{-6}, 10^{-5}, 10^{-4}, 10^{-3}

The parameters with subscripts are the standard values when the other parameters are investigated.

4. Energy Efficiency of Heat Pump with Tunnel Lining GHEs

4.1. Effect of GHEs Absorber Pipe Layout Types on Heat Pump Energy Efficiency

As shown in Figure 6, the absorber pipe layout types contain a type-1 absorber pipe arranged along the cross direction of the tunnel and a type-2 absorber pipe arranged along the axial direction of the tunnel. The area and length of the absorber pipe layout can be calculated by Equation (21).

$$\begin{cases} L_{total} = [(L_C/J) + 1] \times L_A + L_C \\ A = L_C \times L_A \end{cases} \tag{21}$$

where L_{total} denotes the absorber pipe length (m); L_C denotes the length along the tunnel cross direction (m); L_A denotes the length along the axial direction of the tunnel (m); J denotes the absorber pipe pitches (m); and A denotes the area of the absorber pipe layout (m^2).

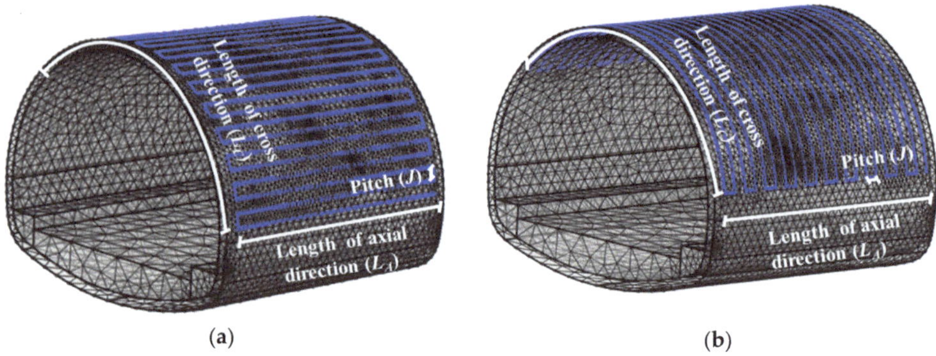

Figure 6. (**a**) Type-1 layout and (**b**) type-2 layout.

To facilitate a comparison between the effects of different absorber pipe layout types on the heat pump energy efficiency, the area and length of the type-1 absorber pipe are the same as those of the type-2 absorber pipe. The area and length of the two types of absorber pipe layouts are 100 m^2 and 220 m. As shown in Figure 7, the EERs of two types of absorber pipes layout decreased with the elapsed time for 90 days. The EER values of two types of absorber pipes layout reached 3.56 and 3.62 on the 90th day, respectively, and the EER values of the type-2 absorber pipe were greater than EER values of the type-1 absorber pipe, indicating that the type-2 absorber pipe layout was better than the type-1 absorber pipe layout.

To analyze the above results, the temperature field around the absorber pipe at the center of the tunnel model is presented, as shown in Figure 8.

From Figure 8, it is apparent that more heat accumulated around the type-1 absorber pipe compared with the type-2 absorber pipe, leading to a decrease in the EER. Figure 9 depicts the heat transfer schematic of different absorber pipe layout types at the tunnel surface. The heat transfer direction of type-1 absorber pipe was mainly along the axial direction of the tunnel, while the heat transfer direction of type-2 absorber pipe was mainly along the cross direction of the tunnel. Moreover, heat accumulated easily between the different groups of absorber pipes. Hence, there was more heat accumulation in the tunnel model with the type-1 layout compared with tunnel model with the type-2 layout.

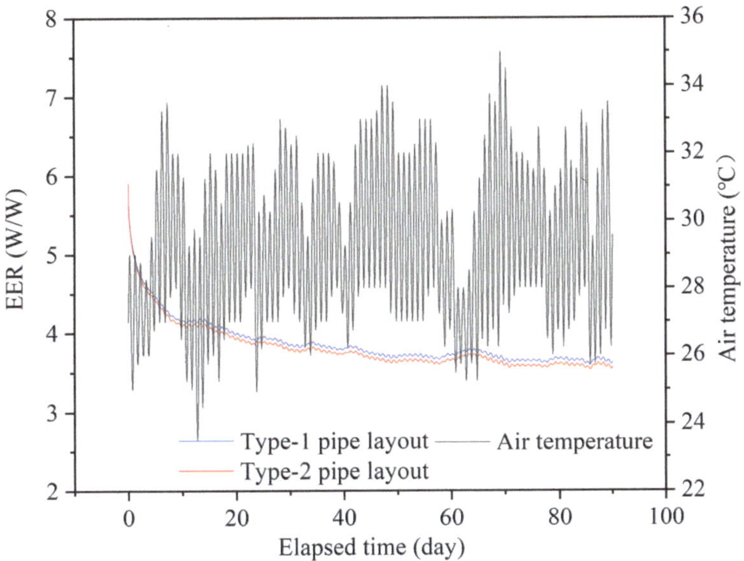

Figure 7. Variation in EER with the elapsed time.

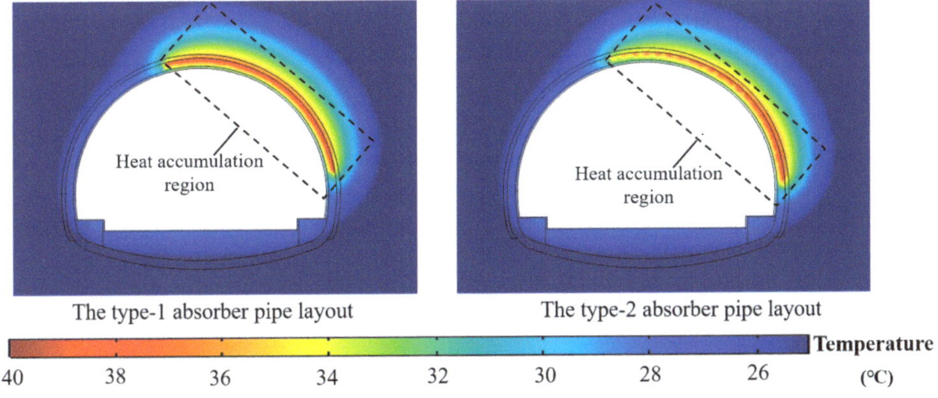

Figure 8. Ground temperature distributions of different absorber pipe layout types.

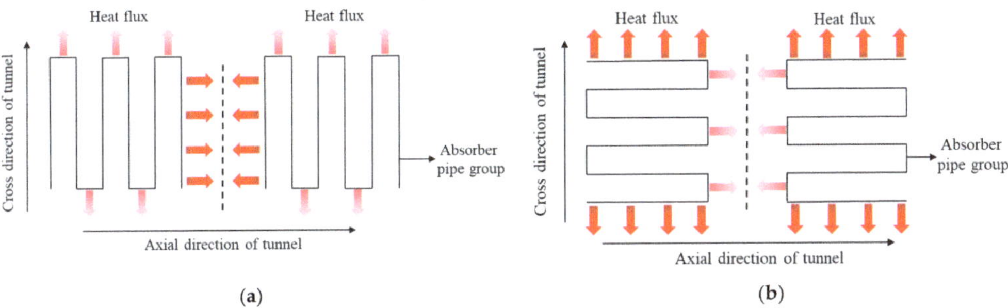

Figure 9. (**a**) Heat transfer schematic of type-1 absorber pipe layout at tunnel surface. (**b**) Heat transfer schematic of type-2 absorber pipe layout at tunnel surface.

Although the temperature change along the axial direction of tunnel and the buoyancy effect were not considered in this model, it was acceptable that the convective heat transfer boundary between the tunnel surface and air was used to simulate the convective heat transfer of tunnel air. Zhang et al. (2016) found that there was mainly forced convection along the axial direction of the tunnel in the mountain tunnel, and the maximum temperature difference between different temperature measurement points within the same cross section was only 0.18 °C, which meant that the temperature variation of the cross section could be neglected for the mountain tunnel. Peltier et al. (2019) found that constant values of the convection heat transfer coefficient could be used to describe the heat transfer performance in the tunnels driven by airflows when no disturbances of the thermal and velocity boundary layers were encountered with the longitudinal distance. This study focused on the mountain tunnel, a section of the mountain tunnel with a length of 11 m was used to investigate the effect of ventilation on the thermal performance of tunnel lining GHEs qualitatively. Hence, the variation in wind speed and air temperature could be neglected. In summary, compared with the type-1 layout, the type-2 layout exhibited a greater EER, recommending for the mountain tunnel lining GHEs design.

4.2. Effect of GHEs Absorber Pipe Pitch on Heat Pump Energy Efficiency

Based on the above results, the type-2 absorber pipe layout arranged along the axial direction of the tunnel had the higher heat pump energy efficiency. Hence, the type-2 absorber pipe layout was used in the investigation on the effect of GHEs absorber pipe pitch on the heat pump energy efficiency, and the absorber pipe length was fixed at 291.1 m.

As shown in Figure 10a, the EER increased exponentially with increasing wind speed for the same pitch. As shown in Figure 10b, the EER increased exponentially with an increase in the absorber pipe pitch. The influence of the pitch on the growth rates of EER had a diminishing trend as the wind speed increased. When the wind speed was 0.1 m/s, the growth rates of the EER corresponding to the absorber pipe pitches ranging from 0.3 to 0.4 m, 0.4 to 0.5 m, and 0.5 to 0.6 m were 5.88%, 3.29%, and 2.26%, respectively. When the wind speed was 5 m/s, the growth rates of the EER corresponding to the absorber pipe pitches ranging from 0.3 to 0.4 m, 0.4 to 0.5 m, and 0.5 to 0.6 m were 4.19%, 2.09%, and 1.58%, respectively.

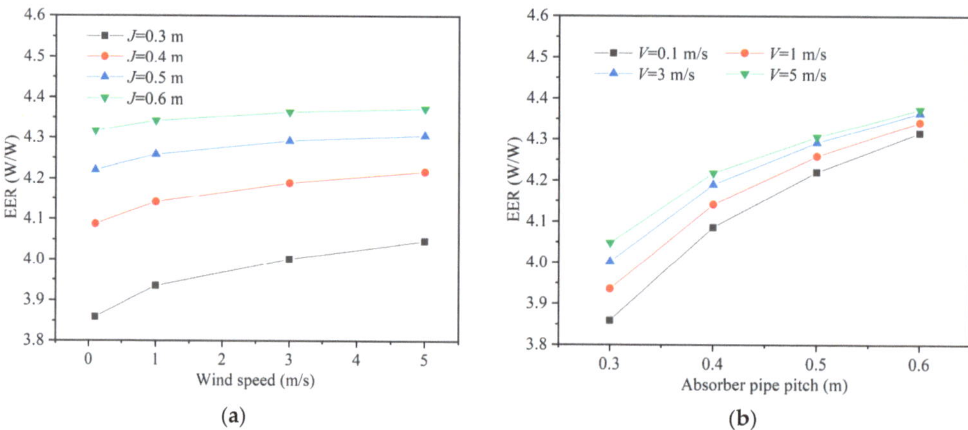

Figure 10. (**a**) Variation in EER with wind speed under different absorber pipe pitches. (**b**) Variation in EER with absorber pipe pitch under different wind speeds.

As shown in Figure 11a, the EER increased exponentially with increasing groundwater flow rate for the same pitch. The EER increased dramatically when the groundwater flow rates were from 0 m/s to 10^{-4} m/s. For instance, the EER under the pitch of 0.3 m increased from 3.86 to 4.91, resulting in a rate of increase of 27.20%, when the groundwater flow rates were from 0 m/s to 10^{-4} m/s. However, the growth rate of the EER was low when the groundwater flow rate was from 10^{-4} m/s to 10^{-3} m/s, resulting in a growth rate of 2.69%. As shown in Figure 11b, the influence of the pitch on the growth rates of EER had a diminishing trend as the groundwater flow rate increased. When the groundwater flow rate was 0 m/s, the growth rates of the EER corresponding to the absorber pipe pitches ranging from 0.3 to 0.4 m, 0.4 to 0.5 m, and 0.5 to 0.6 m were 5.88%, 3.29%, and 2.26%, respectively. When the groundwater flow rate was 10^{-3} m/s, the growth rates of the EER corresponding to the absorber pipe pitches ranging from 0.3 m to 0.4 m, 0.4 m to 0.5 m, and 0.5 m to 0.6 m were 2.26%, 1.38%, and 0.69%, respectively.

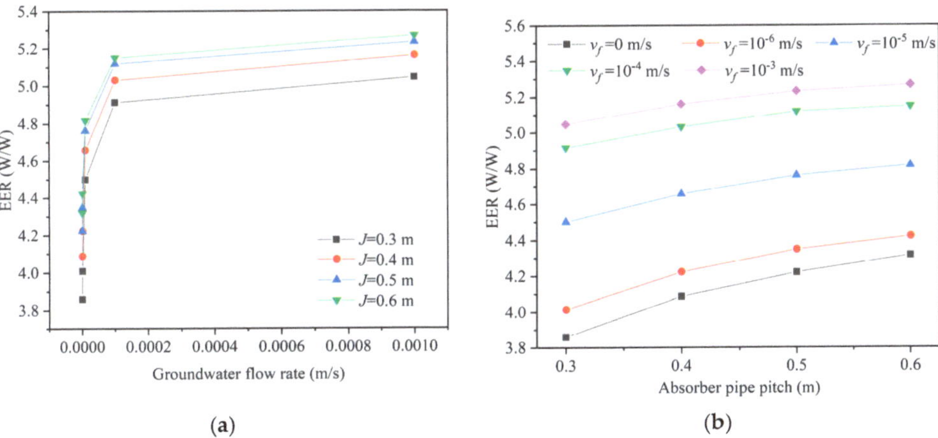

Figure 11. (**a**) Variation in EER with groundwater flow rate under different absorber pipe pitches. (**b**) Variation in EER with absorber pipe pitch under different groundwater flow rates.

4.3. Effect of GHEs Absorber Pipe Length on Heat Pump Energy Efficiency

The absorber pipe length was the most influential factor in the thermal performance of the tunnel GHEs [18]. To investigate the effect of GHEs absorber pipe length on heat pump energy efficiency, the variations in the EER with increasing the absorber pipe length under the same absorber pipe pitch of 0.5 m on the 90th day are plotted in Figures 12 and 13.

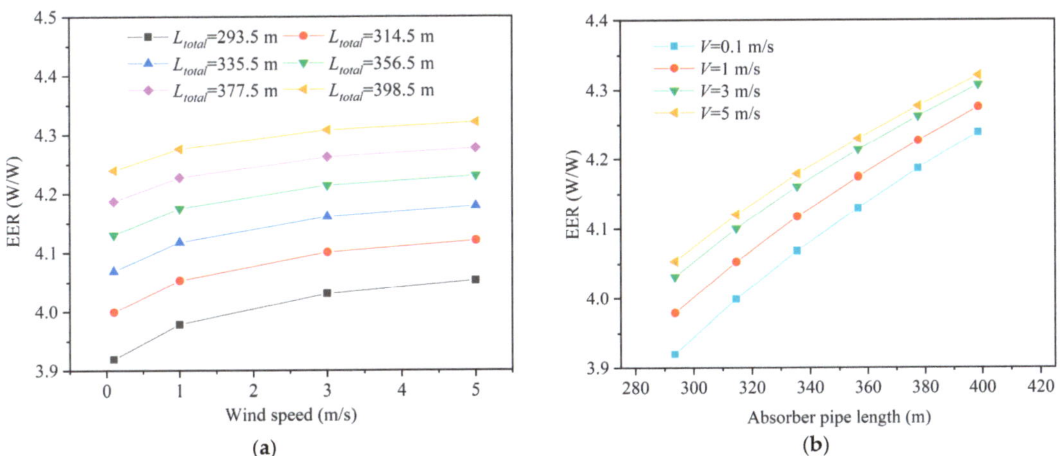

Figure 12. (**a**) Variation in EER with wind speed under different absorber pipe lengths. (**b**) Variation in EER with absorber pipe length under different wind speeds.

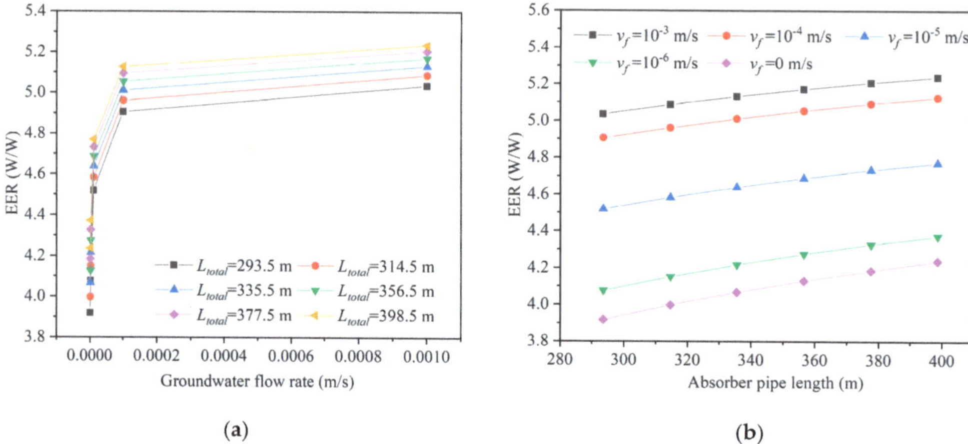

Figure 13. (**a**) Variation in EER with groundwater flow rate under different absorber pipe lengths. (**b**) Variation in EER with absorber pipe length under different groundwater flow rates.

As shown in Figure 12a, the EER increased exponentially with increasing wind speed for the same length. As shown in Figure 12b, the EER increased exponentially with an increase in the absorber pipe length. The influence of the length on the growth rates of EER had a diminishing trend as the wind speed increased. The absorber pipe length ranged from 293.5 m to 398.5 m. When the wind speed was 0.1 m/s, the EER increased from 3.92 to 4.24, resulting in a rate of increase of 8.16% in the EER. When the wind speed was 5 m/s, the EER increased from 4.05 to 4.32, resulting in a rate of increase of 6.67% in the EER.

As shown in Figure 13a, the EER increased exponentially as groundwater flow rate increased for the same length. Groundwater conditions could enhance the heat pump energy efficiency significantly. For instance, when the groundwater flow rate increased from 0 m/s to 10^{-5} m/s, the EER under the absorber pipe length of 293.5 m increased from 3.92 to 4.52, resulting in a rate of increase of 15.31%. As shown in Figure 13b, the influence of the length on the growth rate of EER had a diminishing trend with increasing groundwater flow rate. For instance, the absorber pipe length ranged from 293.5 m to 398.5 m. When the groundwater flow rate was 0 m/s, the EER increased from 3.92 to 4.23, resulting in a rate of increase of 7.91% in the EER. When the groundwater flow rate was 10^{-3} m/s, the EER increased from 5.04 to 5.24, resulting in a rate of decrease of 3.97% in the EER.

4.4. Discussion

Based on the above results, the effects of the mountain tunnel lining GHEs design parameters and coupling effect of the ventilation and groundwater flow on the heat pump energy efficiency were discussed deeply in this section.

According to Section 4.2, increasing absorber pipe pitch could enhance the heat pump energy efficiency. However, the pipe pitch could not increase freely when designing the tunnel lining GHEs. As shown in Figures 10 and 11, the growth rate of the EER decreased with increasing the pipe pitch. The growth rate of the EER would fall further when the wind speed and groundwater flow rate are increased. Moreover, increasing pipe pitch would lead to a larger layout area, which might be beyond the available layout area of the tunnel. Tinti et al. [8] believed that the design code of tunnel lining GHEs depended on the largest heat exchange surface area, the lowest pressure drops and investment cost. Hence, the pipe pitch design needs to consider the largest heat exchange area and shortest pipe length. The shortest pipe length can be determined based on the minimum EER [9], which is regarded as the critical pipe length. As shown in Figures 12 and 13, it can be seen that the good ventilation and groundwater conditions can decrease the critical pipe length, which is helpful to save the cost. Hence, cost savings can be accomplished by arranging the absorber pipes in the tunnel section with good ventilation and groundwater conditions.

To investigate the coupling effect of ventilation and groundwater flow on the heat pump energy efficiency, the absorber pipe pitch and length were fixed at 0.5 m and 293.5 m, respectively. The EERs of the heat pump under the different wind speeds and groundwater flow rates are presented in Figure 14. The effect of wind speed on the growth rate of EER reduced gradually with increasing groundwater flow rate, and the effect of groundwater flow rate on the growth rate of EER also decreased gradually as the wind speed increased. This is because the larger wind speed and groundwater flow rate are helpful to slow down the heat accumulation of the surrounding rock. Ventilation and groundwater conditions share responsibility for improvements in heat pump energy efficiency. It is worth noting that when the groundwater flow velocity reached 10^{-3} m/s, the ventilation did not enhance the EER. This is because the abundant groundwater reduced heat accumulation significantly, which led to a lower temperature of the absorber pipe than the air temperature. Hence, the coupling effect of ventilation and groundwater flow is significant for the tunnel lining GHEs design. Moreover, as shown in Figure 14, the groundwater conditions have a greater influence on the heat pump energy efficiency compared with the ventilation conditions.

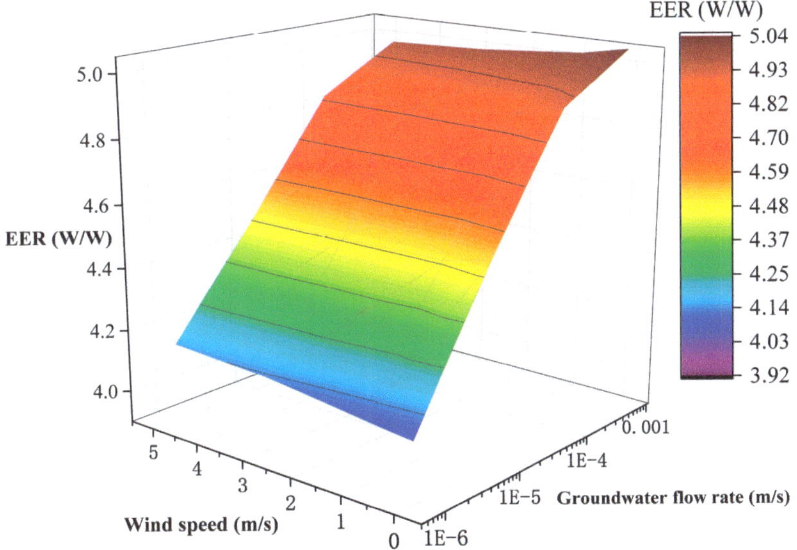

Figure 14. Coupling effect of ventilation and groundwater flow on energy efficiency.

5. Conclusions

In this paper, a numerical model of coupling the tunnel lining GHEs and heat pump was established to investigate the effects of the absorber pipe layout type, absorber pipe pitch, absorber pipe length, wind speed and groundwater flow rate on the energy efficiency of heat pump. The main conclusions can be summarized as follows:

(1) For the mountain tunnel, the absorber pipe arranged along the axial direction of the tunnel exhibits a higher energy efficiency for the heat pump with tunnel lining GHEs compared with the absorber pipe arranged along the cross direction of the tunnel.

(2) The EER increases exponentially with increasing absorber pipe pitch and length. The influences of pipe pitch and length on the growth rate of EER show a diminishing trend as the wind speed and groundwater flow rate increase.

(3) The influence of groundwater flow on the heat pump energy efficiency is more remarkable than that of tunnel ventilation. Moreover, abundant groundwater may lead to a negative effect of ventilation on the heat pump energy efficiency. Hence, the

coupling effect of ventilation and groundwater flow needs to be considered for the tunnel lining GHEs design.

Author Contributions: Conceptualization, G.Z. and L.Z; methodology, X.L. and C.L.; software, C.L.; validation, G.Z.; formal analysis, X.L. and C.L.; investigation, X.L. and C.L.; resources, X.L.; data curaion, X.L. and C.L.; writing—original draft preparation, X.L., C.L.; writing—review and editing, G.Z., L.Z.; supervision, G.Z. and L.Z.; project administration, B.W. All authors have read and agreed to the published version of the manuscript.

Funding: This research received no external funding.

Institutional Review Board Statement: Not applicable.

Informed Consent Statement: Not applicable.

Data Availability Statement: The data presented in this study are available on request from the corresponding author. The data are not publicly available due to privacy issues.

Conflicts of Interest: The authors declare no conflict of interest.

References

1. Building Energy Conservation Research Center of Tsinghua University. *Annual Report on China Building Energy Efficiency*, 2019 ed.; China Building Industry Press: Beijing, China, 2019.
2. Huang, S.F.; Xie, L.Y.; Lu, L.; Zhang, X.S. Design method for heating towers used in vapor-compression heat pump systems. *Build Environ.* **2021**, *189*, 107532. [CrossRef]
3. Lund, J.W.; Boyd, T.L. Direct utilization of geothermal energy 2015 worldwide review. *Geothermics* **2016**, *60*, 66–93. [CrossRef]
4. Jiang, G.; Lin, C.; Shao, D.; Huang, M.; Lu, H.; Chen, G.; Zong, C. Thermo-mechanical behavior of driven energy piles from full-scale load tests. *Energy Build.* **2021**, *233*, 110668. [CrossRef]
5. Yang, W.B.; Lu, P.F.; Chen, Y.P. Laboratory investigations of the thermal performance of an energy pile with spiral coil ground heat exchanger. *Energy Build.* **2016**, *128*, 491–502. [CrossRef]
6. Rui, Y.; Yin, M. Thermo-hydro-mechanical coupling analysis of a thermo-active diaphragm wall. *Can. Geotech. J.* **2018**, *55*, 720–735. [CrossRef]
7. Brandl, H. Energy foundations and other thermo-active ground structures. *Geotechnique* **2006**, *56*, 81–122. [CrossRef]
8. Tinti, F.; Boldini, D.; Ferrari, M.; Lanconelli, M.; Kasmaee, S.; Bruno, R.; Egger, H.; Voza, A.; Zurlo, R. Exploitation of geothermal energy using tunnel lining technology in a mountain environment. A feasibility study for the Brenner Base tunnel—BBT. *Tunn. Undergr. Space Technol.* **2017**, *70*, 182–203. [CrossRef]
9. Adam, D.; Markiewicz, R. Energy from earth-coupled structures, foundations, tunnels and sewers. *Geotechnique* **2009**, *59*, 229–236. [CrossRef]
10. Zhang, G.Z.; Xia, C.C.; Yang, Y.; Sun, M.; Zou, Y.C. Experimental study on the thermal performance of tunnel lining ground heat exchangers. *Energy Build.* **2014**, *77*, 149–157. [CrossRef]
11. Barla, M.; Di Donna, A.; Perino, A. Application of energy tunnels to an urban environment. *Geothermics* **2016**, *61*, 104–113. [CrossRef]
12. Insana, A.; Barla, M. Experimental and numerical investigations on the energy performance of a thermo-active tunnel. *Renew. Energy* **2020**, *152*, 781–792. [CrossRef]
13. Di Donna, A.; Barla, M. The role of ground conditions on energy tunnels' heat exchange. *Environ. Geotech.* **2016**, *3*, 214–224. [CrossRef]
14. Bidarmaghz, A.; Narsilio, G.A. Heat exchange mechanisms in energy tunnel systems. *Geomech. Energy Environ.* **2018**, *16*, 83–95. [CrossRef]
15. Zhang, G.Z.; Xia, C.C.; Zhao, X.; Zhou, S. Effect of ventilation on the thermal performance of tunnel lining GHEs. *Appl. Therm. Eng.* **2016**, *93*, 416–424. [CrossRef]
16. Li, C.; Zhang, G.; Xie, Y.; Liu, X.; Cao, S. Numerical investigation of energy tunnel lining ground heat exchangers in a mountain environment. *IOP Conf. Ser. Earth Environ. Sci.* **2021**, *861*, 072127. [CrossRef]
17. Dornberger, S.C.; Rotta Loria, A.F.; Zhang, M.; Bu, L.; Epard, J.-L.; Turberg, P. Heat exchange potential of energy tunnels for different internal airflow characteristics. *Geomech. Energy Environ.* **2020**, 100229. [CrossRef]
18. Zhang, G.Z.; Liu, S.Y.; Zhao, X.; Ye, M.; Chen, R.F.; Zhang, H.L.; Yang, J.D.; Chen, J.W. The coupling effect of ventilation and groundwater flow on the thermal performance of tunnel lining GHEs. *Appl. Therm. Eng.* **2017**, *112*, 595–605. [CrossRef]
19. Ogunleye, O.; Singh, R.M.; Cecinato, F.; Chan Choi, J. Effect of intermittent operation on the thermal efficiency of energy tunnels under varying tunnel air temperature. *Renew. Energy* **2020**, *146*, 2646–2658. [CrossRef]
20. Ogunleye, O.; Singh, R.M.; Cecinato, F. Assessing the thermal efficiency of energy tunnels using numerical methods and Taguchi statistical approach. *Appl. Therm. Eng.* **2021**, *185*, 116377. [CrossRef]

21. Ma, C.J.; Donna, A.D.; Dias, D.; Zhang, J.M. Numerical investigations of the tunnel environment effect on the performance of energy tunnels. *Renew. Energy* **2021**, *172*, 1279–1292. [CrossRef]

22. Defraeye, T.; Blocken, B.; Carmeliet, J. Convective heat transfer coefficients for exterior building surfaces: Existing correlations and CFD modelling. *Energy Convers. Manag.* **2011**, *52*, 512–522. [CrossRef]

23. Liu, J.; Heidarinejad, M.; Gracik, S.; Srebric, J. The impact of exterior surface convective heat transfer coefficients on the building energy consumption in urban neighborhoods with different plan area densities. *Energy Build.* **2015**, *86*, 449–463. [CrossRef]

24. Zhang, L.F.; Chen, J.Y.; Wang, J.Q.; Huang, G.S. Estimation of soil and grout thermal properties for ground-coupled heat pump systems: Development and application. *Appl. Therm. Eng.* **2018**, *143*, 112–122. [CrossRef]

25. Zhang, L.F.; Huang, G.S.; Zhang, Q.; Wang, J.G. An hourly simulation method for the energy performance of an office building served by a ground-coupled heat pump system. *Renew. Energy* **2018**, *126*, 495–508. [CrossRef]

26. Zhang, C.X.; Hu, S.T.; Liu, Y.F.; Wang, Q. Optimal design of borehole heat exchangers based on hourly load simulation. *Energy* **2016**, *116*, 1180–1190. [CrossRef]

27. Ministry of Housing and Urban-Rural Development of the People's Republic of China. *Design Code for Heating Ventilation and Air Conditioning of Civil Buildings*; GB50736-2012; China Architecture & Building Press: Beijing, China, 2012.

Article

The Framework of Technical Evaluation Indicators for Constructing Low-Carbon Communities in China

Yifei Bai [1], Weirong Zhang [1,*], Xiu Yang [2], Shen Wei [3] and Yang Yu [4]

[1] Key Laboratory of Green Built Environment and Energy Efficient Technology,
Beijing University of Technology, Beijing 100124, China; baiyifei@emails.bjut.edu.cn

[2] Institute of Climate Change and Sustainable Development, Tsinghua University, Beijing 100080, China;
yangxiuthu@tsinghua.edu.cn

[3] The Bartlett School of Construction and Project Management, University College London (UCL),
1-19 Torrington Place, London WC1E 7HB, UK; shen.wei@ucl.ac.uk

[4] School of Architecture, Xi'an University of Architecture and Technology, Xi'an 710055, China;
yuyang@xauat.edu.cn

* Correspondence: zhangwr@bjut.edu.cn

Abstract: In recent years, in order to promote the construction of low-carbon communities (LCCs) in China, many scholars have proposed an evaluation indicator system of LCC. The existing indicator systems are mostly established from the macro perspective of environmental impact and resource conservation, but few are from the micro technical perspective. Thus, the aim of this study is to construct a micro technical evaluation indicator system for LCCs. Firstly, the index system was divided into three categories: low-carbon building, low-carbon transportation, and low-carbon environment. Then, the technical indicators were selected through empirical analysis. The indicator weights were assigned by the improved analytic hierarchy process (AHP) and the multi-level fuzzy comprehensive evaluation method was used as the evaluation method of the indicators. Finally, in order to examine the practicality of the indicator system, two typical communities in Tianjin and Shanghai were selected as case studies. The results showed that the indicator system gave a reasonable low-carbon level for the two communities, which was in line with the actual low-carbon construction status of each community. In addition, the evaluation results pointed out that the low-carbon community (LCC) in Tianjin needs to further strengthen the construction of the low-carbon environment, including community compactness, rainwater collection and utilization, and waste recycling. For the LCC in Shanghai, it was pointed out that the construction of the low-carbon building and low-carbon transportation needs to be strengthened. The indicator system can be used as a tool for urban planning and construction personnel to evaluate the construction progress and low-carbon degree of LCC.

Keywords: low-carbon community; technical indicators; improved analytic hierarchy process; multi-level fuzzy comprehensive evaluation method

Citation: Bai, Y.; Zhang, W.; Yang, X.; Wei, S.; Yu, Y. The Framework of Technical Evaluation Indicators for Constructing Low-Carbon Communities in China. *Buildings* **2021**, *11*, 479. https://doi.org/10.3390/buildings11100479

Academic Editor: Audrius Banaitis

Received: 31 July 2021
Accepted: 11 October 2021
Published: 15 October 2021

Publisher's Note: MDPI stays neutral with regard to jurisdictional claims in published maps and institutional affiliations.

1. Introduction

In recent decades, with the rapid development of different countries, global carbon emissions have increased rapidly [1,2]. The increasing carbon emissions have increased the pressure on natural systems and resources, which directly leading to global climate change and the deterioration of the ecological environment [3]. The rapid urbanization process is one of the main factors leading to the increase in carbon emissions. Urban areas contribute more than 70% of the total energy demand and a corresponding proportion of the world's CO_2 emissions [4]. As the most basic unit in urban construction, the urban community is not only the main space carrier of human life, entertainment, and industrial production, but also the main carrier of urban carbon emissions [5,6]. Low carbon research at the community level is the foothold of the implementation of low-carbon urban planning strategies, and also plays a role in improving the low-carbon technology of single buildings [7,8]. Therefore, how to build an LCC has become a research hotspot in many countries.

As the largest developing country and the second largest economy in the world, China's rapid urbanization process consumes a lot of energy and produces a lot of carbon emissions, accounting for about a quarter of the world's total carbon emissions in the past five years [9]. The research on LCC in China is of great significance for both China and the world to reduce carbon emissions. At present, the research on LCC is mainly divided into quantitative and qualitative evaluation. In terms of quantitative evaluation of LCCs, song et al. [10] proposed an accounting framework for community carbon emissions based on the method of life cycle assessment (LCA), including direct fossil fuel combustion emissions, energy purchase (electricity, heat, and water) emissions, and supply chain emissions reflected in commodity consumption, which quantified the scale and mitigation potential of community carbon emissions. Yıldırım et al. [11] collected data on energy use, land demand, raw material consumption, and carbon emissions of communities, and quantified the environmental impact of different wastewater treatment options using LCA. Lin et al. [12] established a comprehensive accounting model based on the guidelines of the Intergovernmental Panel on Climate Change (IPCC) and the LCA method to more comprehensively and accurately quantify the carbon emissions and carbon sinks of communities. Although quantitative evaluation can give more intuitive results, its time and data requirements, complexity, and cost–benefit ratio have great uncertainty. There are still some difficulties in the application of quantitative evaluation in the actual LCC planning. The qualitative evaluation of LCC has the characteristics of simplicity, fewer data, and strong comprehensiveness, which is welcomed by scholars and urban planners. The United Kingdom has proposed a BREEAM community evaluation system for the construction of sustainable communities. The United States established the LEED-ND system to evaluate the sustainable development of community planning and construction. Japan has established the CASBEE-UD system to guide the ecological and green construction of communities [13,14]. Most of these evaluation index systems are built from the perspective of ecology, livability, and sustainability, and rarely from the perspective of low carbon. In terms of the construction of LCC evaluation indicators, the Chinese National Development and Reform Commission issued the Pilot Low-Carbon Communities Construction Guide, which defines the construction objectives, contents, and standards of China's LCCs. From the perspective of "carbon source control" and "carbon sink expansion", Wang et al. [15] established six evaluation indicators of LCCs. Based on the actual situation of urban LCCs in Guangdong Province, Xie et al. [16] established an evaluation system. Based on the theory and practice of LCC, Luo and Zhan [17] constructed the evaluation index of renewable energy utilization and green vegetation. Jiang and Guo [18] developed the evaluation index of LCC from the planning experience of LCC, such as energy conservation and creating a suitable ecological environment. Moghadam et al. [19] developed a new multicriteria spatial decision support system, which established the relationship between the energy of urban communities and the economic, social, technical, and environmental performance of transformation interventions, and provided meaningful community energy transformation schemes. However, most of the above LCC evaluation indicators are established from the macro perspective of reducing environmental impact and saving resources, and there are no evaluation indicators established from the micro technical perspective. Macroscopic evaluation indices can grasp the construction direction of LCCs, while technical evaluation indices are specific measures to reduce carbon emissions in each construction direction. Therefore, it is necessary to establish a technical evaluation index system of LCCs. The purpose of building the indicator system is to provide a useful tool for community planners to evaluate the low-carbon degree of a low-carbon community in the planning and construction stage or operation stage, and point out the low-carbon technologies that need to be further strengthened in a community.

In this study, firstly, combined with the previous construction contents of LCCs and the construction management departments of Chinese communities, the technical indicators of LCCs were divided into three categories: low-carbon building, low-carbon transportation, and low-carbon environment. Based on the examples of the LCC evaluation index system at home and abroad, and combined with the actual situation of China, a technical index pool was established, and the appropriate indicators were selected through empirical analysis. Then, the weight of the selected index was calculated by using the improved AHP, and the multi-level fuzzy comprehensive evaluation method was selected as the evaluation method of the index. Finally, two typical communities in Tianjin and Shanghai were selected to verify the practicality of the index system. The indicator system constructed in this study gave a reasonable low-carbon level for the two communities, which was in line with the actual low-carbon construction status of each community. The indicator system can not only evaluate the degree of low carbon in a community, but also indicate the aspects in which a community needs to strengthen the use of low-carbon technology, which can be used as a tool for community planning and construction personnel to evaluate the construction progress and degree of low carbon in LCCs.

2. Methodology

In the process of constructing an indicator system, the selection of the indicator and the assignment of the indicator weight are the two most important steps [20,21]. In terms of the selection of indicators, some scholars use frequency analysis; that is, the indicators with higher frequency are used preferentially [22,23]. However, in this method, the important indicators with lower frequency are often ignored. Lu et al. [24] established some sustainable indicator systems according to local regional geographical characteristics and the ecological environment, drawing lessons from the existing LCC evaluation indicator system, which are often not systematic. In this study, based on the current LCC evaluation system and famous LCC cases (including BedZED in the U.K., the Vauban District in Germany, the Hammarby community in Sweden, Beder in Denmark, and the Changxindian community in Beijing [25]), an indicator pool was established, and then the technical indicators were selected through empirical analysis to build a comprehensive technical indicator system for constructing LCCs [26]. For the assignment of indicator weight, the most commonly used method is the analytic hierarchy process (AHP) [27,28]. However, the traditional AHP is suitable for the comparative judgment of a small number of indicators, which is difficult to apply to the more complex evaluation indicators of LCCs. Additionally, the consistency test of the traditional AHP comparison matrix is complicated [29,30]. In this study, the improved AHP overcame the shortcomings of the traditional AHP, and the weight of each indicator could be calculated quickly and conveniently [31,32]. Then, for the index system, a suitable comprehensive evaluation method needs to be selected. Currently, the more commonly used comprehensive evaluation methods include the Grey Correlation Method, Artificial Neural Network Method, Technique for Order Preference by Similarity to an Ideal Solution (TOPSIS) Method, and Fuzzy Comprehensive Evaluation Method. The Grey Correlation Method is applicable to a large number of evaluation indicators, and some indicators have the characteristics of correlation or repetition [33,34]. The Artificial Neural Network Method is more suitable for the case of a large amount of data [35]. Because the TOPSIS Method has no definite method for the transformation of neutral indicators, the final result of comprehensive evaluation is not very accurate [36,37]. The Fuzzy Comprehensive Evaluation Method is a comprehensive evaluation method based on fuzzy mathematics. It has the characteristics of clear results and strong systematization, and it can better solve fuzzy and difficult-to-quantify problems [38]. The technical indicator system constructed in this study is a qualitative indicator that have the problem of being fuzzy and difficult to quantify in the evaluation. Based on the comprehensive analysis of the characteristics of different comprehensive evaluation methods, the Fuzzy Comprehensive Evaluation Method was selected. The technical route of this study is shown in Figure 1.

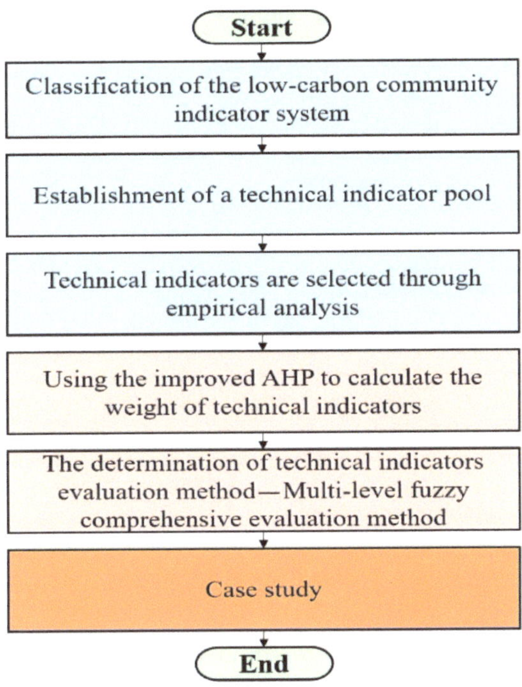

Figure 1. The technical route of the study.

2.1. Classification of the Technical Indicator System

Before constructing the evaluation technical indicator system for LCCs, it is necessary to clarify the classification of the indicator system for LCCs, which plays a very important role in the selection of the technical indicator and determines whether the technical indicator system is systematic [39,40]. At present, in China and the wider international context, there is the same definition of LCCs. It is generally believed that the construction of LCCs can reduce resource consumption and improve energy efficiency, thereby delaying global warming and ultimately achieving the development of a low-carbon economy [41,42]. Wang et al. [15] constructed an indicator system for LCCs based on six indicator categories: layout planning, transportation planning, architectural planning and design, environmental planning, municipal engineering planning, and construction management. Zhang et al. [43] analyzed the construction of indicators such as energy, transportation, waste management, and water management through cases in four communities. Jiang and Guo [18] summarized five aspects based on the successful experiences of two communities in terms of low-carbon planning including public participation, especially the important role of the government, making full use of energy, such as wind energy and solar energy, reasonable use of land, a reasonable layout of road traffic, and a convenient pedestrian transportation system.

To summarize, the indicator categories of LCCs mainly include building, transportation, land planning, and environmental life. In China, the construction of communities mainly involves three departments: building, transportation, and environment. In this study, the evaluation technical indicator system of LCCs was divided into three categories: low-carbon building, low-carbon transportation, and low-carbon environment. These three categories cover the main content of low-carbon community construction, and the indicators of each category can be managed by the corresponding departments. The following is an explanation of the three categories of indicators.

1. Low-carbon building: Buildings are the main sources of carbon emissions in communities. At present, there is no construction standard for low-carbon building in China. In this study, the detailed technical indicators for the construction of low-carbon building were established. The relevant evaluation technical indicators for the construction of LCCs that were established in this study were based on low-carbon buildings being used as a base point to radiate to low-carbon transportation and low-carbon environment.
2. Low-carbon transportation: Low-carbon transportation refers to optimizing a network structure, attaching importance to the construction of a slow-moving transportation system, and improving the convenience of a public transportation system. Every community needs to strengthen the management of motor vehicles and the application of advanced traffic management technology. At the same time, a community should pay attention to the promotion of low-carbon and environmentally friendly modes of transportation for residents.
3. Low-carbon environment: Low-carbon environment refers to scientific and reasonable community land layout planning, water environment planning, and household waste management. Low-carbon environment affects the choice of resident travel methods, which can create a good microclimate for a community and reduce environmental pollution.

2.2. Screening Method for Technical Indicators

2.2.1. Establishment of Technical Indicator Pool

In order to comprehensively construct the evaluation technical indicators of LCCs, in this research, a technical indicator pool was built by drawing on the concept of the topic pool in the "General Framework for Compiling Guidelines for Corporate Social Responsibility Reports in China". According to the research on the current development status of LCC assessments, all the main indicators related to LCCs, green ecological community, and sustainable community are defined in this framework.

China's indicator systems related to LCCs, which were mainly referred to in this study, include:

1. Technical announcement of the 11th Five Year Plan of the Ministry of Construction;
2. "Low carbon Housing Technology System Framework and Emission Reduction Indicators", issued by the China Real Estate Research Association and Housing Industry Development and Technology Committee;
3. "Key Points and Technical Guidelines for the Construction of Green Ecological Residential Areas", organized and compiled by the Housing Industrialization Promotion Center of the Ministry of Construction;
4. "Eco-residential Neighborhood Assessment Manual", published by the Industrialization Promotion Center Group, the Ministry of Construction; and
5. The Guide for Evaluation Technology for Low-Carbon Urban Areas and The Guide for Evaluation Technology for Low-Carbon Communities based on local Beijing standards.

The international common indicator system related to low-carbon communities that this study mainly referred to includes:

1. The LEED-ND formulated by the U.S. Green Building Council;
2. The BREEAM Communities formulated specifically for neighborhoods by the U.K.;
3. The CASBEE-UD created based on the CASBEE formulated by Japan, after taking into account urban areas and buildings.

2.2.2. Empirical Analysis

Empirical analysis is a method for comprehensively analyzing various factors based on researchers' professional knowledge and previous work experience. This method belongs to the category of qualitative analysis and has certain artificial subjectivity, but it is the most simple and feasible analysis method. In application, the method of increasing the number of researchers and synthesizing opinions can reduce subjectivity as much as possible and improve the accuracy of a conclusion [26]. As far as this study is concerned, there are many relevant evaluation technical indicators of LCCs, and there are overlapping concepts, inconsistent statistical calibers, and inconsistent data availability among some indicators. At the time of this study, the empirical analysis method could be used to compare and discriminate indicators to select the most appropriate evaluation technical indicators for LCCs. Experts judged each indicator according to the following standards: (1) each indicator is related to the community's carbon emissions; (2) each indicator should be logically related to each other; (3) the selected indicators should be typical and wraparound research field terms; (4) the indicators should be practical in a real situation and should be simple, accompanied by an explicit definition; (5) each indicator not only considers the present development condition of LCCs, but also includes the possibility of future development and changes. The results are denoted by "agreement, unsure, disagreement".

In the community, building carbon emissions, transportation carbon emissions, and living environment carbon emissions account for 54%, 40%, and 6%, respectively [44–46]. According to this feature, this study stipulated that the proportion of the number of experts in each field selected to the total number of experts should not be less than the proportion of carbon emissions in this field, so that more important indicators could be screened out. In addition, the experts selected by the research team have at least two professional backgrounds, which is to avoid the problem that experts with only one professional background often only choose indicators in their familiar field. A total of 24 experts were selected for this study. Experts in the field of building carbon emissions, transportation carbon emissions, and living environment carbon emissions accounted for 92%, 71%, and 63%, respectively, meeting the specified requirements. The results are shown in Table 1. In each row, the number of filled circles represents the number of professions that the expert is familiar with. The three filled circles represent experts with professional backgrounds in the three fields. The two filled circles represent experts with professional backgrounds in the two fields. Experts were asked to select indicators by filling out questionnaires. When experts selected indicators, only the ones agreed by more than two-thirds of the experts were selected into the system, which ensures the scientificity of indicator selection. Experts screened the indicators in the technical indicator pool according to the above criteria, eliminated indicators not related to LCC construction, and developed specific descriptions for each removed indicator.

Table 1. Professional background information of 24 experts.

Serial Numbers of the Experts	A	B	C
1	●	●	●
2	●	●	●
3	●		●
4	●		●
5	●	●	
6	●		●
7	●	●	
8	●	●	
9	●		●
10	●		●
11	●	●	
12	●		●
13	●		●
14		●	●
15	●	●	●
16	●	●	●
17	●	●	●
18	●	●	
19	●	●	●
20	●	●	

Table 1. *Cont.*

Serial Numbers of the Experts	A	B	C
21	•	•	
22	•	•	
23	•	•	
24		•	•

Note: "**A**" represents the field of building carbon emissions. "**B**" represents the field of transportation carbon emissions. "**C**" represents the field of living environment carbon emissions.

2.3. Weight Calculation Method for the Technical Indicators

2.3.1. Improved AHP

The technical evaluation indicators for low-carbon communities were characterized by strong systematization, wide coverage, and large quantities. The improved AHP was easy to operate, which overcame the difficulty of using fuzzy words such as "slightly" important, "relatively" important, and "extremely" important to express the relationship between the two elements accurately, and it did not require consistency tests to be conducted separately. The process was clear and simple.

1. After setting n indicators for a certain decision system A, i.e., $G_1, G_2 \ldots G_n$, the corresponding weights were $W_1, W_2 \ldots W_n$, and $W_1 + W_2 + \ldots + W_n = 1$. In order to construct the judgment matrix, a comparison matrix was established using the three-scale method.

$$C = \begin{pmatrix} c_{11} & c_{12} & \cdots & c_{1n} \\ c_{21} & c_{22} & \cdots & c_{2n} \\ \vdots & \vdots & \vdots & \vdots \\ c_{n1} & c_{n2} & \cdots & c_{nn} \end{pmatrix} = (c_{ij})_{n \times n} \tag{1}$$

where, if Gi is more important than G_j, c_{ij} is 1; if Gi is as important as G_j, c_{ij} is 0; if Gi is less important than G_j, c_{ij} is 1.

2. The comparison matrix C was used to calculate the optimal transfer matrix O through mathematical conversion.

$$O_{ij} = \frac{1}{n} \sum_{t=1}^{n} (c_{it} + c_{tj}) \tag{2}$$

$$O = \begin{pmatrix} O_{11} & O_{12} & \cdots & O_{1n} \\ O_{21} & O_{22} & \cdots & O_{2n} \\ \vdots & \vdots & \vdots & \vdots \\ O_{n1} & O_{n2} & \cdots & O_{nn} \end{pmatrix} = (O_{ij})_{n \times n} \tag{3}$$

3. The optimal transfer matrix O was transformed into the consistency matrix D, which was also called the judgment matrix of the indicator.

$$D_{ij} = \exp(O_{ij}) \tag{4}$$

$$D = \begin{pmatrix} D_{11} & D_{12} & \cdots & D_{1n} \\ D_{21} & D_{22} & \cdots & D_{2n} \\ \vdots & \vdots & \vdots & \vdots \\ D_{n1} & D_{n2} & \cdots & D_{nn} \end{pmatrix} = (D_{ij})_{n \times n} \tag{5}$$

4. The solution for the eigenvector W of D was determined.

The square root method was used to find $W = (W_1, W_2, \ldots, W_n)^T$, and the obtained eigenvector W_i could be used as the weight of each indicator [31,32].

2.3.2. Establishment of Tree Hierarchy Mode

Before calculating the weight of technical indicators, the hierarchical structure of an indicator model is established first. The tree hierarchy model uses a "directed tree" data structure to represent various entities and the relationships between entities. Each node in the tree represents a record type, and there is a clear structure and simple relationship between nodes [47]. This model takes the research object as a system and makes decisions according to the thinking mode of decomposition, comparative judgment, and synthesis. It has become an important tool of system analysis developed after mechanism analysis and statistical analysis. The establishment of the index model with tree structure is of great significance to the research and calculation of index weight.

2.4. Evaluation Method of Technical Indicators

2.4.1. Fuzzy Comprehensive Evaluation Method

The fuzzy comprehensive evaluation method is a comprehensive evaluation method based on fuzzy mathematics, which is based on the fuzzy set theory [48,49]. The following is the basic principle of the fuzzy comprehensive evaluation method.

The evaluation target is regarded as a fuzzy set composed of many factors, which is called the factor set. Then, the evaluation level that these factors can select is set. The fuzzy set that makes up the evaluation is called the evaluation set. The membership grade of each single factor for each evaluation level is called the fuzzy matrix. Then, according to the weight distribution of various factors in the evaluation target, the quantitative solution of the evaluation is obtained through calculation (called fuzzy matrix synthesis). The specific steps are as follows.

1. Determination of the factor set of the evaluation object:

$$U = \{u_1, u_2, \ldots, u_m\} \tag{6}$$

where U is the object of evaluation and u_m is m evaluation indices of the evaluation object.

2. Determination of the evaluation level set:

$$V = \{v_1, v_2, \ldots, v_n\} \tag{7}$$

For the evaluation target U, the evaluation results in n may be made, V represents the evaluation set of target U, and the specific level needs to be described in appropriate language according to the evaluation content. The ratings of the technical indicators screened out in this study were rated as "excellent", "good", "general", and "poor", as shown in Table 2.

Table 2. Ratings of the technical indicators.

Ratings	Specifications
Excellent	The application of the low-carbon technologies corresponding to the technical indicator is in full compliance with the local climate, resources, and other aspects. The specifications and installation position of the technical components are reasonable, the construction meets the requirements, and the operation is in good condition, which has a good effect on reducing carbon emissions.
Good	The application of the low-carbon technologies corresponding to the technical indicator is in line with the local climate, resources, and other aspects. The specification and installation position of the technical components are relatively reasonable, the construction essentially meets the requirements, and the operation condition meets the relevant standards, which has a certain effect on reducing carbon emissions.

Table 2. *Cont.*

Ratings	Specifications
General	The application of low-carbon technologies corresponding to the technical indicator did not fully meet the local conditions of climate and resources. The specifications and installation position of the technical components are not reasonable, the construction did not meet the corresponding requirements, and the emission reduction effect achieved is limited.
Poor	The application of low-carbon technologies corresponding to the technical indicator did not conform to the local climate, resources, and other aspects, and it had little effect on reducing carbon emissions, or there is no low-carbon technology corresponding to this technical indicator.

By analyzing the detailed planning diagram, construction description diagram, actual operation report, and other relevant documents of a low-carbon community, relevant professionals can evaluate the technical indicators constructed in this study according to the evaluation criteria.

3. Single factor evaluation and establishment of the fuzzy relationship matrix R:

$$R = \begin{pmatrix} R_{11} & R_{12} & \cdots & R_{1n} \\ R_{21} & R_{22} & \cdots & R_{2n} \\ \vdots & \vdots & \vdots & \vdots \\ R_{m1} & R_{m2} & \cdots & R_{mn} \end{pmatrix} \tag{8}$$

where R_{ij} $(i = 1, 2, \ldots, m; j = 1, 2, \ldots, n)$, indicating the membership degree of the evaluation target to the evaluation set v_j from the perspective of the factor u_i. In this step, experts or relevant personnel usually evaluate each factor of the evaluation target. R_{ij} refers to the ratio of the number of people whose evaluation result is v_j to the total number of evaluators for factor u_i.

4. According to the factor weight $W = \left(W_1, W_2, \ldots, W_n \right)^T$ obtained by the improved AHP, the evaluation result of F is calculated.

$$F = W \times R \tag{9}$$

2.4.2. Multi-Level Fuzzy Comprehensive Evaluation Method

The evaluation technical indicator system of the LCCs is multi-level, so the multi-level fuzzy comprehensive evaluation method is needed. The principle is the same as that of the first level fuzzy comprehensive evaluation method, but the factor set U is divided into S subsets according to the type of attribute, and the subsets are recorded as $U1, U2, \ldots, US$. For each subset Ui, the fuzzy comprehensive evaluation is carried out according to the first-level model. After the evaluation results are obtained, each Ui is taken as an element to continue to build the evaluation matrix, and so on [50]. The comprehensive scoring system of the LCCs in this study was divided into four grades according to the comprehensive scoring results of the low-carbon level, namely, grade I, grade II, grade III, and grade IV. The details are shown in Table 3.

Table 3. Comprehensive rating of low-carbon communities.

Level	I	II	III	IV
Total score	Excellent 85–100 (including 85)	Good 70–85 (including 70)	Genera 160–70 (including 60)	Not low carbon 0–60

3. Results and Discussion

3.1. Technical Evaluation Indicators of LCC

Through the establishment of the indicator pool and the empirical analysis of the indicator, the experts in the field of carbon emissions selected 34 indicators from the 73 indicators in the indicator pool. The selected indicators were highly systematic and scientific, and they fully represented the construction content of LCCs. The final determination results are shown in Table 4.

Table 4. Technical index and its specific explanation.

Indicators	Specifications
D1 Green planting system	Tree transplantation technology, artificial greening cultivation technology, and anti-seasonal planting technology
D2 Roof greening system	Light roof greening technology, thin substrate roof greening and vertical cultivation technology, and planting concrete planting roof technology
D3 Vertical greening system	Placement of suitable green plants on the exterior surface of a building
D4 Solar energy utilization technology	Passive solar energy utilization technology, solar power generation, solar heating, solar light utilization, solar thermal utilization, solar air conditioning, and refrigeration
D5 Geo-energy utilization technology	Geothermal power generation technology and geothermal heating technology
D6 Wind energy utilization technology	Passive wind energy utilization technology and wind power generation technology
D7 Biomass energy utilization technology	Straw gasification technology, and biogas application technology
D8 Wall	Wall insulation technology, coating insulation technology, and phase change wall materials
D9 Roof	Ventilation roofing, thermal insulation roofing, cold roof systems, and water storage roofing
D10 Door and window	Broken bridge energy-saving windows, composite energy-saving windows, and insulating glass doors and windows (inert gas insulating glass, low-E insulating glass)
D11 Shading technology	External shading systems, internal shading systems, body shading systems (hollow glass louver shading technology), and light-guided shading systems
D12 Building ground system	Floating floors, overhead floors, and phase change heat storage floors
D13 Water supply and drainage system	Water supply and drainage system optimization technology and water-saving appliances
D14 Heating system	Pipe insulation technology, central heating technology, decentralized heating technology, and heating supply end systems
D15 Ventilation system	Passive ventilation systems and high efficiency and energy-saving ventilation systems
D16 Lighting system	High efficiency and energy-saving lamp systems, light guide lighting systems, and light collection lighting systems
D17 Air conditioning system	Ice water storage air conditioning systems, air conditioning systems, variable air volume air conditioning systems, and air conditioning water systems
D18 Intelligent monitoring system	Intelligent lighting control systems, air conditioning, heating and ventilation equipment intelligent control systems, electrical equipment remote intelligent control systems
D19 Property management system	Waste disposal systems, building property intelligent management centers, and building property digital management control platforms
D20 Design technology for building outdoor environment	Reasonable building spacing, building plot ratios, and building densities

Table 4. *Cont.*

Indicators	Specifications
D21 Building design technology	Building shape coefficient controls, window wall ratio controls, and building orientation design
D22 High-quality public transportation system	The establishment of intelligent public transportation microcirculation networks, the establishment of one-stop transportation platforms, and the distance from the entrance and exit to the public transport station conforming to the national high quality public transport requirements
D23Slow traffic network	The establishment of a people-oriented slow traffic system (high-quality bicycle and pedestrian transportation network suitable for residents)
D24 Motor vehicle demand management	The implementation of the strict management of motor vehicle demand policies and encouraging the use of new energy vehicles
D25 Strict management standard for energy consumption and emission	Encouraging the use of clean energy vehicles and implementing strict energy consumption and emission management standards
D26 Advanced traffic management technology	Road traffic data collection, scientific transmission and processing information, and real-time release of traffic operation information
D27 Land utilization pattern of multi-functional mix	With comprehensive diversified functional space, the intensive use of land can be achieved, and a community functional network can be formed to meet the needs of diversified and multi-level activities
D28 Compact space pattern for low-carbon community	Advocating for rational population size and diversified land use, shortening the distance between activity spaces and families, reducing pollution and energy consumption
D29 Balanced layout of public service facilities	The establishment of balanced public service facilities to facilitate the life and travel of residents
D30 Rainwater collection technology	The roof use of rainwater and ground infiltration use of rainwater
D31Recycled water reuse technology	After the centralized treatment of domestic wastewater (bathing, washing, kitchen, and toilet), the wastewater can be reused for greening irrigation, vehicle washing, road washing, and household toilet flushing
D32 Permeable ground and constructed wetland technology	Increasing the proportion of permeable surfaces to rechargeable groundwater sources and establishing artificial wetlands
D33 Solid waste disposal and recycling technology	The waste disposal and recycling in the community to achieve the goal of the sustainable development of the community and to reduce the pollution to the environment
D34 Garbage classification and collection	The establishment of waste sorting collection devices and treatment and transportation systems

3.2. Weight Calculation for Technical Indicators

3.2.1. Tree Hierarchy Model of Technical Indicators

In this study, the hierarchical structure of the indicator model was divided into the target layer, primary indicator, secondary indicator, and tertiary indicator. First, the target layer was composed of LCCs. The division of the primary indicator was determined according to the three categories of the technical indicator system. The primary indicator could be divided into low-carbon building, low-carbon transportation, and low-carbon environment. The secondary indicator was determined according to the category of the screened indicator, and the tertiary indicator was the specific screened technical indicator. Indicators at all levels are shown in Table 5, and the hierarchy structure of the index model is shown in Figure 2.

3.2.2. Calculating the Weights of the Technical Indicators

In the process of calculating the index weight, 24 experts with two or more professional backgrounds in low-carbon fields were selected in this study. These experts comprehensively considered when comparing the weights of indicators, which avoided the problem that they tended to score a certain indicator higher due to cognitive limitations. The departments to which these experts belonged included universities, urban planning and design institutes, and government management departments. Experts in universities mainly considered problems from the level of basic theoretical knowledge, experts in urban planning and design institutes mainly considered problems from the aspect of practicability, and experts in government management departments mainly considered problems from the aspects of economic benefits and community management. Therefore, the calculated index weight was considered in different aspects, which increased the scientificity of index weight. In order to further eliminate the imbalance of index weight, the index weight given by all experts was arithmetically averaged. In this study, the Delphi method was used to issue the questionnaire, which made experts make a more objective and reasonable evaluation of the index weight. By programming the improved AHP with MATLAB and computing the collected questionnaire data, the weights of the indicators at all levels relative to the target layer were obtained, as shown in Figures 3–5.

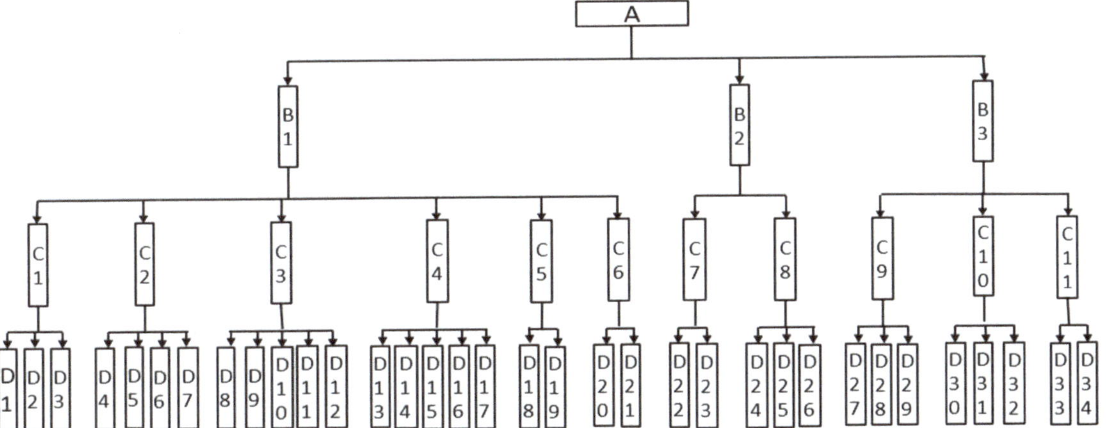

Figure 2. Hierarchical diagram of technical evaluation indicators for LCCs.

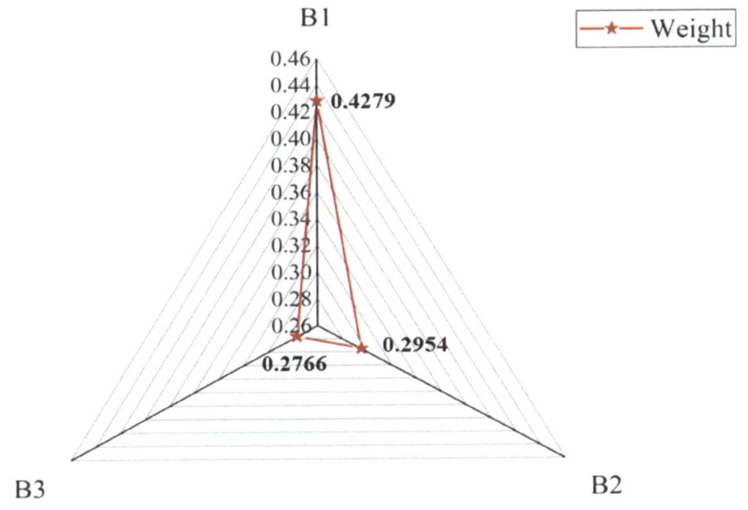

Figure 3. Radar for primary indicator weights.

Table 5. Indicators for all levels of low-carbon communities.

Target Layer	Primary Indicator	Secondary Indicator	Tertiary Indicator
A LCC	B1 Low-carbon building	C1 Low-carbon technology of greening system	D1 Green planting system, D2 Roof greening system, D3 Vertical greening system
		C2 Low-carbon technology for building energy supply	D4 Solar energy utilization technology, D5 Geo-energy utilization technology, D6 Wind energy utilization technology, D7 Biomass energy utilization technology
		C3 Low-carbon technology for building envelope	D8 Wall, D9 Roof, D10 Door and window, D11 Shading technology, D12 Building ground system
		C4 Low-carbon technology for building equipment	D13 Water supply and drainage system, D14 Heating system, D15 Ventilation system, D16 Lighting system, D17 Air conditioning system
		C5 Low-carbon technology for building operation management	D18 Intelligent monitoring system, D19 Property management system
		C6 Low-carbon technology of building design	D20 Design technology for building outdoor environment, D21 Building design technology
	B2 Low-carbon transportation	C7 Road planning	D22 High-quality public transportation system, D23 Slow traffic network
		C8 Traffic management	D24 Motor vehicle demand management, D25 Strict management standard for energy consumption and emission, D26 Advanced traffic management technology
	B3 Low-carbon environment	C9 Land layout planning	D27 Land utilization pattern of multi-functional mix, D28 Compact space pattern for low-carbon community, D29 Balanced layout of public service facilities
		C10 Water environment planning	D30 Rainwater collection technology, D31 Recycled water reuse technology, D32 Permeable ground and constructed wetland technology
		C11 Domestic waste management	D33 Solid waste disposal and recycling technology, D34 Garbage classification and collection

Among the primary indicators, B1 had the largest weight ratio, which was about 0.4279. Buildings are the largest carbon emission source in a community. Therefore, it is necessary to strengthen the construction of low-carbon building in the construction of low-carbon communities and make good use of various energy-saving and emission reduction measures. The weight of B2 and B3 accounted for about 0.28, and these two indicators had similar effects on low-carbon communities.

Among the secondary indicators, the top four were C7, C9, C8, and C2, accounting for 0.1733, 0.1271, 0.1221, and 0.0888, respectively. The construction of C7, C9, and C8 was mainly to reduce the travel by private cars, use public transportation as far as possible, and travel by walking in order to reduce the traffic carbon emissions. C2 low-carbon technology for building energy supply is a very important link in the construction of low-carbon buildings. Choosing appropriate renewable energy technology according to local natural resources is a very important way to reduce the carbon emissions of buildings. Among all the secondary indicators, the proportion of C1 was the lowest, which showed that compared with other indicators, the impact on the construction of LCCs was small.

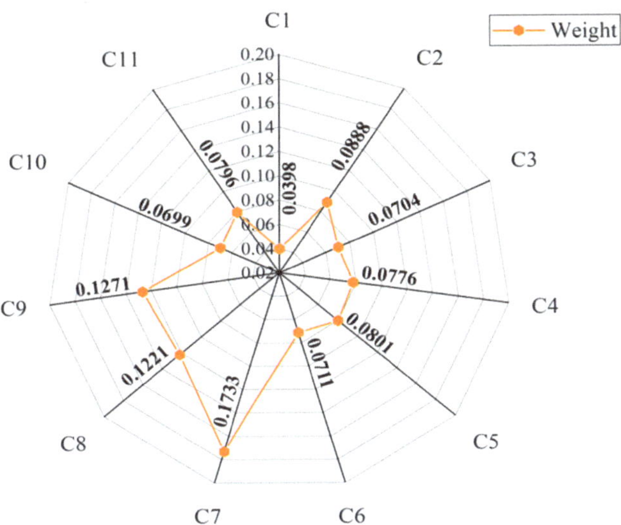

Figure 4. Radar for secondary indicator weights.

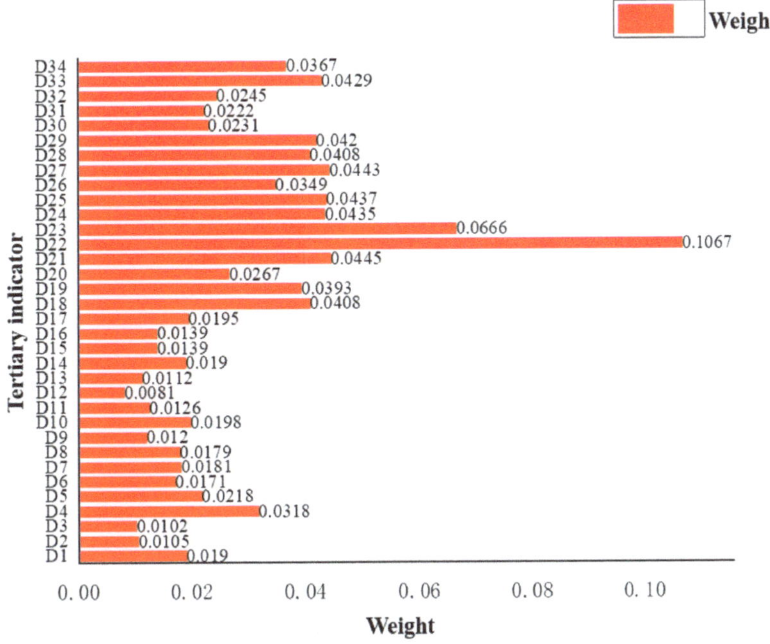

Figure 5. A bar chart for the tertiary indicator weights.

Among the tertiary indicators, according to the categories of B1, B2, and B3, the indicators at their respective levels were analyzed and discussed in order to be more meaningful for the construction of low-carbon communities. Within the scope of B1, the top-ranking indicators were D21, D18, D19, and D4. The scientific and reasonable building design is the premise of building utilization of natural resources. D18 and D19 play an important role in the operation of buildings and determine whether buildings can save energy efficiently. D4 is the most widely used and most mature technology in buildings [51,52]. The electricity and heat generated by solar energy reduce the dependence on traditional energy and play an important role in energy conservation and emission reduction. Within the scope of B2, the top-ranking indicators were D22, D23, D25, and D24. D22 and D23 are the most important indicators of low-carbon transportation planning, and they play an important role in reducing or even replacing private car travel. High-quality road planning can form a convenient system for public travel and pedestrian travel, and it can reduce a large number of traffic carbon emissions. In the B3 scope, the top-ranking indicators were D33, D29, and D28. The construction of D33 strengthens the utilization of resource cycle, and it is a landmark indicator for low-carbon communities to save resources and reduce energy consumption. D29 can facilitate the clothing, food, housing, and transportation of residents, reduce travel time, and promote the use of non-motor vehicles to travel for residents. D28 is conducive to the saving of land. The moderate increase in floor area ratios and building densities can effectively reduce road paving areas and the length of facilities and pipelines, and reduce the impact on ecological factors.

4. Case Study

In order to verify the practicality of the evaluation index system and the availability of index data, it was necessary to test the index system and apply it to the evaluation of actual communities. The study selected two communities as the application objects of the index system. The first community is the Tianjin Eco-city "Shimao New Town" community. The community is located on the north side of the Yongdingzhou Wetland Ecological Park and on the south side of the National Animation Industry Park, with a total area of 1.46 square kilometers. The "Shimao New Town" community was established in 2014. The community adopted a variety of energy-saving technologies. It is a typical green energy-saving community and has achieved suitable evaluation. The second community is the "Dongming" community under the jurisdiction of Shanghai's Pudong New Area, covering an area of 5.95 square kilometers. The community was established in 1999 and is an older community with relatively few applications of green energy-saving technologies.

By consulting the detailed planning documents and construction documents of the two communities, the low-carbon technologies used in community construction were sorted out. Ten experts from Beijing urban community planning and design institutes were invited. They have been engaged in low-carbon planning and design and actual construction of communities for a long time, and they are all professionals with senior professional titles. These experts have rich experience in the application of the index system. Each expert evaluated the low-carbon technologies adopted by the two communities and determined the rating of the tertiary indicators. Through the transformation of the multi-level fuzzy comprehensive evaluation method, the comprehensive scores of the two communities and the scores of primary and secondary indicators are shown in Tables 6 and 7. The score of the "Shimao New Town" community was 78, and it belonged to the II level. According to the comprehensive score of the LCCs, it could be concluded that the community was a better LCC, and some suggestions could be summarized from the scores of the indicators at all levels to guide the future construction of the community. It could be seen from the scores of the primary indicators that the low-carbon construction of B1 and B2 was better, and the low-carbon construction of B3 was poor. In terms of the construction of B1, the buildings in the "Shimao New Town" community adopted many green energy-saving technologies and had a good score. However, as can be seen from the scores of the secondary indicators, the construction of low-carbon technologies for a green system and building energy supply needed to be strengthened in the B1 field. The low-carbon construction of the community should adopt a variety of greening methods, such as the vertical greening of the building facade, the greening of the roof, and the greening of the area around the building. Buildings should make use of various forms of renewable energy technologies according to local conditions to reduce the use of traditional energy. The score of B2 was relatively high, and the low-carbon construction of road planning and traffic management was better. The "Shimao New Town" community focused on the construction of a slow traffic network and established a public transportation-oriented transportation system. Within the scope of B3, the scores of all indicators were relatively low, and the community needed to be further improved in terms of community compactness, rainwater collection and utilization, and waste recycling. The score of the "Dongming" community was 66, belonging to the III level. The community was a general low-carbon community. It could be seen from the scores of the primary indicators that the scores of B1 and B2 were relatively low, and the score of B3 was relatively high. The score of B1 was relatively low because the "Dongming" community was established earlier, and most buildings did not adopt suitable low-carbon and energy-saving technologies. The technical indicators contained in B1 needed to be further improved. The "Dongming" community did not pay attention to the planning and construction of low-carbon transportation in the construction stage, which led to the low score of B2. In the later period, the community implemented stricter traffic management policies to reduce traffic carbon emissions. Therefore, the secondary indicator C8 contained in B2 had a high score. The "Dongming" community had a better low-carbon construction in B3, because the community implemented energy-saving and emission-reduction policies in terms of low-carbon environment, including garbage classification, waste recycling, and rainwater recycling. It could be seen from the secondary indicators included in B3 that the community achieved relatively excellent scores in the construction of C10 and C11.

Table 6. Scores of the target layer and the indicator layers at all levels.

Target Layer A	Total Score of Target Layer	Primary Indicator B	Score of Each Primary Indicator	Secondary Indicator C	Score of Each Secondary Indicator
LCC	78	B1 Low-carbon building	78	C1 Low-carbon technology of greening system	68
				C2 Low-carbon technology for building energy supply	52
				C3 Low-carbon technology for building envelope	82
				C4 Low-carbon technology for building equipment	73
				C5 Low-carbon technology for building operation management	98
				C6 Low-carbon technology of building design	96
		B2 Low-carbon transportation	97	C7 Road planning	98
				C8 Traffic management	95
		B3 Low-carbon environment	58	C9 Land layout planning	64
				C10 Water environment planning	64
				C11 Domestic waste management	45

Table 7. Scores of the target layer and the indicator layers at all levels.

Target Layer A	Total Score of Target Layer	Primary Indicator B	Score of Each Primary Indicator	Secondary Indicator C	Score of Each Secondary Indicator
LCC	66	B1 Low-carbon building	58	C1 Low-carbon technology of greening system	69
				C2 Low-carbon technology for building energy supply	44
				C3 Low-carbon technology for building envelope	70
				C4 Low-carbon technology for building equipment	60
				C5 Low-carbon technology for building operation management	49
				C6 Low-carbon technology of building design	66
		B2 Low-carbon transportation	59	C7 Road planning	44
				C8 Traffic management	81
		B3 Low-carbon environment	84	C9 Land layout planning	80
				C10 Water environment planning	85
				C11 Domestic waste management	90

Figure 6 compared the scores of target layers and indicators at all levels of the two communities. As can be seen from Figure 6, the target layer score of the "Shimao New Town" community was higher than that of the "Dongming" community. The "Shimao New Town" community was well built in terms of low-carbon building and low-carbon transportation. In the field of low-carbon building, almost all technical indicators of the "Shimao New Town" community were better than those of the "Dongming" community, especially the construction of C5 and C6. In the field of low-carbon transportation, the "Shimao New Town" community was ahead of the "Dongming" community in the construction of C7 and C8. In terms of C7 construction, there was an obvious gap between the "Dongming" community and the "Shimao New Town" community. In the field of low-carbon environment, the "Dongming" community was better than the "Shimao New Town" community in various indicators. The two communities could learn from each other and further strengthen the application of low-carbon technologies for indicators with low scores.

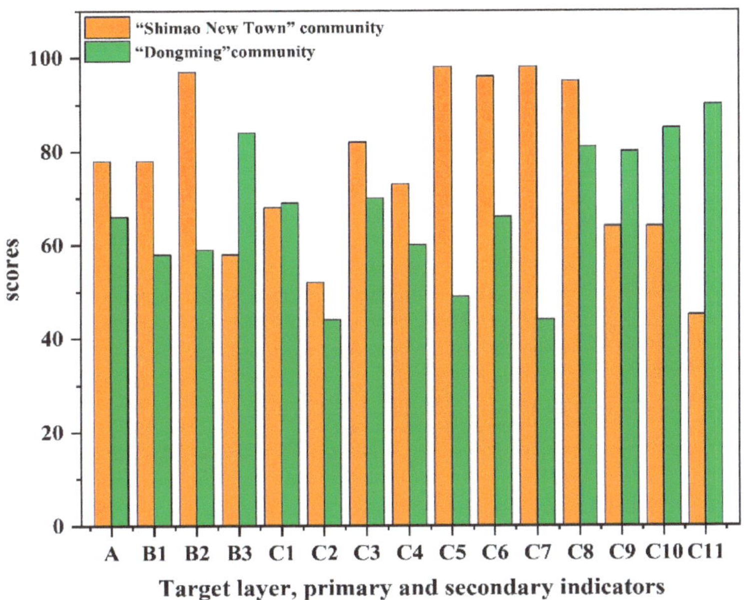

Figure 6. Scores of target levels and indicators at all levels of the two communities.

5. Conclusions

The study established a technical evaluation index system for low-carbon communities. According to the existing indicator for LCC construction, an indicator pool was established, and the relevant technical indicators of LCC construction were selected through empirical analysis. The improved AHP was selected to calculate the weight of the index, and the multi-level fuzzy comprehensive evaluation method was determined to be the evaluation method of the technical indicator.

On the above basis, the two typical communities in Tianjin and Shanghai were selected to verify the operability of the index system. The indicator system constructed in this study gave a reasonable low-carbon level for the two communities, which was in line with the actual low-carbon construction status of each community. In addition, for the low-carbon community in Tianjin with high scores, the evaluation results pointed out that the community had better construction in low-carbon building and low-carbon transportation indicators, and further pointed out that the community needed to strengthen the application of low-carbon technology of the greening system and building energy supply in the field of low-carbon building. In terms of the relatively poor low-carbon environmental indicators of the community, the evaluation results pointed out that the community needed to be further improved in terms of community compactness, rainwater collection and utilization, and waste recycling. For the low-carbon community in Shanghai with low scores, the evaluation results showed the advantages and disadvantages of the community in the construction of low-carbon building, low-carbon transportation, and low-carbon environment indicators. The community applied few low-carbon technologies in low-carbon building and needed to promote the application of various low-carbon technologies. Regarding the construction of low-carbon transportation, the evaluation results pointed out that the community needed to further strengthen the construction of road planning. The low-carbon environment construction of the community was relatively suitable, which could provide experience for the low-carbon construction of other communities. The indicator system can not only evaluate the degree of low carbon in a community, but also indicate the aspects in which a community needs to strengthen the use of low carbon technology, which can be used as a tool for community planning and construction personnel to evaluate the construction progress and degree of low carbon in LCCs. Furthermore, the method of constructing the technical index system in this study can provide a reference for future qualitative research. In the low-carbon construction of communities, planners should pay attention to the construction of low-carbon buildings, which is of great significance to the construction of low-carbon communities.

However, the study does have limitations. The empirical analysis method and improved AHP method used in this study are subjective methods, and the application scope of these methods needs to be further explored. In terms of the selection of the number of experts and the selection of indicators, more scientific methods need to be studied. The weight of indicators needs further analysis and discussion. Furthermore, in addition to establishing a qualitative evaluation index system for low-carbon communities, carbon measurement and quantitative evaluation of communities should also be considered. It is necessary to determine how to combine a qualitative technical indicator system with carbon dioxide emissions.

Author Contributions: Conceptualization, Y.B. and W.Z.; methodology, Y.B. and W.Z.; software, Y.B.; validation, Y.B. and W.Z.; formal analysis, Y.B. and W.Z.; investigation, W.Z. and X.Y.; resources, W.Z. and Y.Y.; data curation, Y.B., W.Z. and Y.Y.; writing—original draft preparation, Y.B. and W.Z.; writing—review and editing, W.Z., X.Y. and S.W.; visualization, Y.B., W.Z. and S.W.; supervision, W.Z., X.Y. and S.W.; project administration, W.Z. and Y.Y. All authors have read and agreed to the published version of the manuscript.

Funding: The research described in this paper was supported by the National Key R&D Program of China "Optimization Technology for Urban New District Planning and Design" (2018YFC0704600).

Institutional Review Board Statement: Not applicable.

Informed Consent Statement: Not applicable.

Data Availability Statement: The data presented in this study are available on request from the corresponding author.

Acknowledgments: The author would like to thank the experts in the field of carbon emission in the Institute for Sustainable Development Goals, Tsinghua University, and the technicians of Tianjin Eco-city.

Conflicts of Interest: We declare that we have no financial and personal relationships with other people or organizations that can inappropriately influence our work, there is no professional or other personal interest of any nature or kind in any product, service and/or company that could be construed as influencing the position presented in, or the review of, the manuscript entitled "The framework of technical evaluation indicators for constructing low-carbon communities in China".

References

1. Ssaa, B.; Aaac, D.; Ow, E.; Muee, F. The role of electricity consumption, globalization and economic growth in carbon dioxide emissions and its implications for environmental sustainability targets. *Sci. Total Environ.* **2020**, *708*, 134653.
2. Gu, G.; Wang, Z.; Wu, L. Carbon emission reductions under global low-carbon technology transfer and its policy mix with r&d improvement. *Energy* **2021**, *216*, 119300.
3. Cuffey, K.M.; Vimeux, F. Covariation of carbon dioxide and temperature from theVostok ice core after deuterium-excess correction. *Nature* **2001**, *412*, 523–527. [CrossRef]
4. Seto, K.C.; Satterthwaite, D. Interactions between urbanization and global environmental change. *Curr. Opin. Environ. Sustain.* **2010**, *2*, 127–128. [CrossRef]
5. Yang, J.; Zhan, Y.; Xiao, X.; Xia, J.C.; Sun, W.; Li, X. Investigating the diversityof land surface temperature characteristics in different scale cities based on local climate zones. *Urban Clim.* **2020**, *34*, 100700. [CrossRef]
6. Zhang, Y.; Liu, Y.; Wang, Y.; Liu, D.; Xia, C.; Wang, Z. Urban expansion simulation towards low-carbon development: A case study of Wuhan, China. *Sustain. Cities Soc.* **2020**, *63*, 102455. [CrossRef]
7. Lee, W.L.; Burnett, J. Benchmarking energy use assessment of hk-beam, breeam and leed. *Build. Environ.* **2008**, *43*, 1882–1891. [CrossRef]
8. Perez, M.; Rey, E. A multi-criteria approach to compare urban renewal scenarios for an existing neighborhood. case study in lausanne (switzerland). *Build. Environ.* **2013**, *65*, 58–70. [CrossRef]
9. Duo, C.; Deng, Z.; Liu, Z. China's non-fossil fuel CO2 emissions from industrial processes. *Appl. Energy* **2019**, *254*, 113537.
10. Song, D.; Su, M.; Yang, J.; Chen, B. Greenhouse gas emission accounting and management of low-carbon community. *Sci. World J.* **2012**, *2012*. [CrossRef]

11. Yıldırım, M.; Topkaya, B. Assessing environmental impacts of wastewater treatment alternatives for small-scale communities. *CLEAN–Soil Air Water* **2012**, *40*, 171–178. [CrossRef]
12. Lin, X.; Ren, J.; Xu, J.; Zheng, T.; Cheng, W.; Qiao, J.; Huang, J.; Li, G. Prediction of Life Cycle Carbon Emissions of Sponge City Projects: A Case Study in Shanghai, China. *Sustainability* **2018**, *10*, 3978. [CrossRef]
13. Sharifi, A.; Murayama, A. Neighborhood sustainability assessment in action: Cross-evaluation of three assessment systems and their cases from the US, the UK, and Japan. *Build. Environ.* **2014**, *72*, 243–258. [CrossRef]
14. Mcgranahan, G.; Balk, D.; Anderson, B. The rising tide: Assessing the risks of climate change and human settlements in low elevation coastal zones. *Environ. Urban.* **2007**, *19*, 17–37. [CrossRef]
15. Wang, X.; Zhao, G.; He, C.; Wang, X.; Peng, W. Low-carbon neighborhood planning technology and indicator system. *Renew. Sustain. Energy Rev.* **2016**, *57*, 1066–1076. [CrossRef]
16. Xie, Z.; Gao, X.; Feng, C.; He, J. Study on the evaluation system of urban low carbon communities in Guangdong province. *Ecol. Indic.* **2017**, *74*, 500–515. [CrossRef]
17. Luo, Q.L.; Zhan, Q.M. Best practice in low-carbon community planning. *Adv. Mater. Res.* **2012**, *450–451*, 1082–1085. [CrossRef]
18. Jiang, X.J.; Guo, Z.J. Foreign low carbon community planning comparative analysis. *Adv. Mater. Res.* **2011**, *233–235*, 1897–1900. [CrossRef]
19. Moghadam, S.T.; Lombardi, P. An interactive multi-criteria spatial decision support system for energy retrofitting of building stocks using CommuntiyVIZ to support urban energy planning. *Build. Environ.* **2019**, *163*, 106233. [CrossRef]
20. Wu, C. Indicator system construction and health assessment of wetland ecosystem—taking hongze lake wetland, china as an example. *Ecol. Indic.* **2020**, *112*, 106164. [CrossRef]
21. Wu, D.; Ding, H.; Chen, J.; Fan, Y. A delphi approach to develop an evaluation indicator system for the national food safety standards of china. *Food Control* **2020**, *121*, 107591. [CrossRef]
22. Yuan, W.; James, P.; Hodgson, K.; Hutchinson, S.M.; Shi, C. Development of sustainability indicators by communities in China: A case study of Chongming County, Shanghai. *J. Environ. Manag.* **2003**, *68*, 253–261. [CrossRef]
23. Spangenberg, J.H. Environmental space and the prism of sustainability: Frameworks for indicators measuring sustainable development. *Ecol. Indic.* **2002**, *2*, 295–309. [CrossRef]
24. Lu, Y.; Wang, R.; Zhang, Y.; Su, H.; Wang, P.; Jenkins, A. Ecosystem health towards sustainability. *Ecosyst. Health Sustain.* **2016**, *1*, 1–15. [CrossRef]
25. Marique, A.; Reiter, S. A simplified framework to assess the feasibility of zero-energy at the neighbourhood/community scale. *Energy Build.* **2014**, *82*, 114–122. [CrossRef]
26. Han, Y.; Dai, L.M.; Zhao, X.F.; Wu, Y. Construction and application of an assessment index system for evaluating the eco-community's sustainability. *J. For. Res.* **2008**, *19*, 154–158. [CrossRef]
27. Todd, J.A.; Crawley, D.B.; Geissler, S.; Lindsey, G. Comparative assessment of environmental performance tools and the role of the Green Building Challenge. *Build. Res. Inf.* **2001**, *29*, 324–335. [CrossRef]
28. Haapio, A.; Viitaniemi, P. A critical review of building environmental assessment tools. *Environ. Impact Assess. Rev.* **2008**, *28*, 469–482. [CrossRef]
29. Reed, M.; Fraser, E.D.; Morse, S.; Dougill, A.J. Integrating Methods for Developing Sustainability Indicators to Facilitate Learning and Action. *Ecol. Soc.* **2005**, *10*, 1–6. [CrossRef]
30. Alnaser, N.W.; Flanagan, R.; Alnaser, W.E. Model for calculating the sustainable building index (sbi) in the kingdom of bahrain. *Energy Build.* **2008**, *40*, 2037–2043. [CrossRef]
31. Englman, A.Y. A square-root method for the density matrix and its applications to lindblad operators. *Phys. A Stat. Mech. Its Appl.* **2006**, *371*, 368–386.
32. Ma, N.; Zhao, Z. Analysis of determining weight coefficient based on AHP and its improvement. *Water Conserv. Sci. Technol. Econ.* **2006**, *11*, 732–733.
33. Das, R.; Ball, A.K.; Roy, S.S. Optimization of E-jet Based Micro Manufacturing Process Using Grey Relation Analysis. *Mater. Today Proc.* **2018**, *5*, 200–206. [CrossRef]
34. Xia, X.; Sun, Y.; Wu, K.; Jiang, Q. Optimization of a straw ring-die briquetting process combined analytic hierarchy process and grey correlation analysis method. *Fuel Process. Technol.* **2016**, *152*, 303–309. [CrossRef]
35. Wang, X.; Peng, G.; Cheng, Y. Fuzzy synthetic evaluation of economic benefit of enterprise based on neural network. In Proceedings of the IEEE International Symposium on Computational Intelligence in Robotics & Automation, Kobe, Japan, 16–20 July 2003.
36. Wang, J.; Cheng, C.; Huang, K. Fuzzy hierarchical TOPSIS for supplier selection. *Appl. Soft Comput.* **2009**, *9*, 377–386. [CrossRef]
37. Jia, J.; Ying, F.; Guo, X. The low carbon development (lcd) levels' evaluation of the world's 47 countries (areas) by combining the fahp with the topsis method. *Expert Syst. Appl.* **2012**, *39*, 6628–6640. [CrossRef]
38. Zadeh, L.A. Similarity relations and fuzzy orderings. *Inf. Sci.* **1971**, *3*, 177–200. [CrossRef]
39. Lin, J.; Jacoby, J.; Cui, S.; Liu, Y.; Tao, L. A model for developing a target integrated low carbon city indicator system: The case of xiamen, china. *Ecol. Indic.* **2014**, *40*, 51–57. [CrossRef]
40. Lou, Y.; Jayantha, W.M.; Shen, L.; Liu, Z.; Shu, T. The application of low-carbon city (lcc) indicators—A comparison between academia and practice. *Sustain. Cities Soc.* **2019**, *51*, 101677. [CrossRef]

41. Middlemiss, L.; Parrish, B.D. Building capacity for low-carbon communities: The role of grassroots initiatives. *Energy Policy* **2010**, *38*, 7559–7566. [CrossRef]
42. Heiskanen, E.; Johnson, M.; Robinson, S.; Vadovics, E.; Saastamoinen, M. Low-carbon communities as a context for individual be-havioural change. *Energy Policy* **2010**, *38*, 7586–7595. [CrossRef]
43. Zhang, X.; Shen, G.Q.P.; Feng, J.; Wu, Y. Delivering a low-carbon community in china: Technology vs. strategy? *Habitat Int.* **2013**, *37*, 130–137. [CrossRef]
44. Zhao, R.; Huang, Y.; Yao, M.X.; Zhan, L.P.; Peng, D.P. Carbon emission assessment of an urban community. *Appl. Ecol. Environ. Res.* **2019**, *17*, 13673–13684. [CrossRef]
45. Akbarnezhad, A.; Xiao, J. Estimation and minimization of embodied carbon of buildings: A review. *Buildings* **2017**, *7*, 5. [CrossRef]
46. Bellucci, F.; Bogner, J.E.; Sturchio, N.C. Greenhouse gas emissions at the urban scale. *Elements* **2012**, *8*, 445–449. [CrossRef]
47. Zhan, G.; Yan, X.; Zhu, S.; Wang, Y. Using hierarchical tree-based regression model to examine university student travel frequency and mode choice patterns in China. *Transp. Policy* **2016**, *45*, 55–65. [CrossRef]
48. Chen, J.; Hsieh, H.; Do, Q.H. Evaluating teaching performance based on fuzzy AHP and comprehensive evaluation approach *Appl. Soft Comput.* **2016**, *28*, 100–108. [CrossRef]
49. Feng, S.; Xu, L.D. Decision support for fuzzy comprehensive evaluation of urban development. *Fuzzy Sets Syst.* **1999**, *1051*, 1–12 [CrossRef]
50. Wang, J.; Sun, Y. The Application of Multi-Level Fuzzy Comprehensive Evaluation Method in Technical and Economic Evaluation of Distribution Network. In Proceedings of the International Conference on Management and Service Science, Wuhan, China, 24–26 August 2010.
51. Esfahani, S.K.; Karrech, A.; Tenorio, R.; Elchalakani, M. Optimizing the solar energy capture of residential roof design in the southern hemisphere through evolutionary algorithm. *Energy Built Environ.* **2020**, *2*, 406–424. [CrossRef]
52. Qin, J.; Hu, E.; Li, X. Solar aided power generation: A review. *Energy Built Environ.* **2020**, *1*, 11–26. [CrossRef]

Article

Predicting Indoor Temperature Distribution Based on Contribution Ratio of Indoor Climate (CRI) and Mobile Sensors

Yanan Zhao [1], Zihan Zang [1], Weirong Zhang [1,*], Shen Wei [2] and Yingli Xuan [3]

[1]   Key Laboratory of Green Built Environment and Energy Efficient Technology, Beijing University of Technology, Beijing 100022, China; zhaoyn@emails.bjut.edu.cn (Y.Z.); CML9817@emails.bjut.edu.cn (Z.Z.)
[2]   The Bartlett School of Construction and Project Management, University College London (UCL), 1-19 Torrington Place, London WC1E 7HB, UK; shen.wei@ucl.ac.uk
[3]   Joint Usage/Research Center Wind Engineering Research Center Tokyo Polytechnic University, Tokyo Polytechnic University, Tokyo 164-8678, Japan; y.xuan@arch.t-kougei.ac.jp
*   Correspondence: zhangwr@bjut.edu.cn

Abstract: In practical building control, quickly obtaining detailed indoor temperature distribution is necessary for providing satisfying personal comfort and improving building energy efficiency. The aim of this study is to propose a fast prediction method for indoor temperature distribution without knowing the thermal boundary conditions in practical applications. In this method, the index of contribution ratio of indoor climate (CRI), which represents the independent contribution of each heat source to the temperature distribution, has been combined with the air temperature collected by one mobile sensor at the height of the working area. Based on a typical office model, the effectiveness of using mobile sensors was discussed, and the influence of its acquisition height and acquisition distance on the prediction accuracy was analyzed as well. The results showed that the proposed prediction method was effective. When the sensors fixed on the wall were used to predict the indoor temperature distribution, the maximum average relative error was 27.7%, whereas when the mobile sensor was used to replace the fixed sensors, the maximum average relative error was 4.8%. This indicates that using mobile sensors with flexible acquisition location can help promote both reliability and accuracy of temperature prediction. In the human activity area, data from a set of mobile sensors were used to predict the temperature distribution at four heights. The prediction accuracy was 2.1%, 2.1%, 2.3%, and 2.7%, respectively. However, the influence of acquisition distance of mobile sensors on prediction accuracy cannot be ignored. The distance should be large enough to disperse the distribution of the acquisition points. Due to the influence of airflow, some distance between the acquisition points and the room boundaries should be given.

Keywords: temperature distribution; prediction; CFD; contribution ratio of indoor climate (CRI); mobile sensors

Citation: Zhao, Y.; Zang, Z.; Zhang, W.; Wei, S.; Xuan, Y. Predicting Indoor Temperature Distribution Based on Contribution Ratio of Indoor Climate (CRI) and Mobile Sensors. *Buildings* **2021**, *11*, 458. https://doi.org/buildings11100458

Academic Editor: Fabrizio Ascione

Received: 28 July 2021
Accepted: 28 September 2021
Published: 4 October 2021

1. Introduction

As people spend about 90% of their time indoors [1], the indoor thermal environment becomes very important to their daily lives. Therefore, many researchers have made considerable efforts in creating comfortable indoor thermal environments to improve people's living conditions [2,3]. Meanwhile, the indoor thermal environment also has a significant impact on buildings' energy consumption, which is very important for sustainable development. Existing studies have shown that the energy consumption from buildings in China accounts for approximately 21% of the societal energy consumption, especially from urban buildings (75%) [4]. To reduce the energy consumption of buildings while ensuring thermal comfort, creating non-uniform indoor thermal environments has been considered. In this process, the buildings' ventilation mode gradually changes from traditional mixed ventilation (MV) to demand-oriented ventilation, such as displacement ventilation (DV) [5,6] and stratum ventilation (SV) [7,8]. The change of ventilation mode means that the indoor thermal environment cannot be considered as perfectly mixed, and obtaining its detailed temperature distribution becomes necessary. Computational fluid dynamics (CFD) has been used as an effective tool for studies of indoor thermal environments, such as airflow movement [9,10], heat transfer between indoor components [11,12] and pollutant dispersion [13,14]. Indoor temperature distribution, however, is affected by various heat sources, which are dynamically changing with time, hence difficult to obtain in practice. This means that accurate boundary conditions could not be identified in advance to support CFD simulation. Additionally, the requirements on both computational resources and time for CFD are high. Even if the required dynamic boundary conditions can be determined in advance, for example by supercomputers, the process still must set and calculate CFD repeatedly to achieve dynamic calculation of the indoor thermal environment, which is time-consuming. Considering the above, it is difficult to achieve real-time control of indoor thermal environments by this method.

In this regard, several new methods have been proposed to replace CFD for quick calculation of indoor temperature distribution. The CFD reduced-order method achieves the reduction of the order of large-scale simultaneous equations in CFD, using grid number as dimensions, and it can effectively increase the calculation efficiency with some sacrifice in accuracy. Sempey et al. simplified the CFD model by solving only the energy balance equation and reducing its order by proper orthogonal decomposition so that it can be applied to real-time control applications [15]. To improve the problems of CPU-unfriendly calculations as well as the complex meshing required, Cao et al. proposed applying the momentum source to the Navier-Stokes equations to simulate the motion of human bodiess/objects [16]. Fast fluid dynamics (FFD) is an intermediate approach between nodal models and CFD, which decouples pressure and velocity to achieve fast prediction of indoor airflow [17,18]. When integrating the models for HVAC systems with coupled multizone and CFD simulations for airflows, Tian et al. have incorporated the FFD model to improve computational efficiency [19]. Liu et al. have proposed an FFD-based joint method to accelerate the indoor-environment inverse design process and evaluated the effectiveness of four FFD models [20]. Contribution ratio of indoor climate (CRI) is an index extracted from CFD calculation results, and it quantitatively represents the independent contribution of each heat source on indoor temperature distribution. This index was first proposed by Kato et al. in 1994 [21], and it was derived from Sandberg's ventilation efficiency [22] and Kato's effectiveness of contamination exhaust [23]. Zhang et al. have extended both the concept and the calculation method of CRI to the natural convection airflow field [24]. Based on this, they further proposed a basic formula for calculating the temperature distribution and combined it with network models to improve the accuracy of long-term dynamic building performance simulation [25]. The accessibility, which is similar in concept to CRI, describes the independent effect of supply air and contaminant source on an arbitrary location within a finite time period [26]. Shao et al. have proposed a concise expression for fast prediction of indoor non-uniform temperature distribution using the accessibility and analyzed its reliability [27]. Additionally, Ma et al. further defined three transient accessibility indices to reveal the transient effects of supply air, contaminant source, and initial condition, respectively, and established a method predicting the concentration distribution in transient states [28]. The low-dimensional model (LM) uses the discrete method to greatly reduce the amount of high-resolution grid data in order to save computational time [29]. To verify its reliability [30], Ren and Cao elaborated the linear temperature model (LTM) to attain an LLTM (low-dimensional linear temperature model)-based CRI model for fast and reliable calculation of complicated temperature fields [31]. Although the methods noted above can accelerate the calculation of indoor temperature distribution to a certain extent, they still need significant calculation time in practical applications. For instance, although it has been reported that the calculation speed of FFD is 20–50 times faster than that of CFD simulation [32] and could potentially be accelerated with a graphic processing unit (GPU) and parallel computation [33,34], its calculation time is still much longer than that is required by BES. More importantly, these methods were all derived from CFD itself, so the requirement to have pre-determined indoor heat sources still exists, hence they are still difficult to use in practical applications.

To develop a temperature distribution prediction method suitable for practical control, Sasamoto et al. have developed a method that can quickly predict indoor temperature distribution using CRI and a finite number of air temperature measurements collected by fixed sensors [35]. However, due to the limitation of the fixed sensors in both installation location and installation number, its prediction accuracy needs to be improved. Therefore, this study has proposed to use one mobile sensor at the height of working areas, instead of fixed sensors, to collect air temperature, with the intention to get faster and more accurate prediction of temperature distribution. Through establishing a typical office model, the effectiveness of using mobile sensors instead of fixed sensors was explored and discussed. Meanwhile, through a comparison on prediction accuracy, the influences of both acquisition height and acquisition distance of this mobile sensor were further analyzed to guide practical applications.

2. Method Development

2.1. Definition of CRI

Based on an assumption that heat transfer is linear in a steady airflow field, the CRI index can be used to represent the independent contribution of each heat source to indoor temperature distribution. This concept has been extended to the response factor of heat sources in transient cases [36,37] and the evaluation of contaminant and moisture distribution [23,38]. Its specific definition in a forced convection airflow field is the ratio of temperature rise/drop at a location caused by one individual heat source to the absolute value of uniform temperature rise/drop caused by the same heat source. It indicates the range and degree of influence from each heat source within a steady airflow field, and its value is a relative intensity, in which the actual temperature rise/drop caused by each heat source is normalized by the absolute value of its own perfect mixing condition. For example, a CRI higher/lower than 1.0 at a location means that the influence of that heat source is higher/lower than that in the case of perfect mixing. The CRI of the heat source i at the location X_j is defined by Equation (1):

$$CRI_i(X_j) = \frac{\Delta\theta_i(X_j)}{\Delta\theta_{i,o}} = \frac{\theta_i(X_j) - \theta_n}{\theta_{i,o} - \theta_n} = \frac{\theta_i(X_j) - \theta_n}{q_i/C_p\rho F} \tag{1}$$

In one example, the heat emission from the heat source i is 200 W, resulting in a temperature rise of 0.8 °C at the location X_j. If heat diffuses uniformly through the whole space, the uniform temperature rise will be 1 °C. According to Equation (1), the CRI of the heat source i at the location X_j is 0.8, indicating that the heat source i has a smaller impact than the average indoor environment at the location X_j.

For more information about the basic premises, definitions, calculation methods, and mathematical meaning of CRI, please refer to Zhang et al., which gives a systematic and comprehensive introduction of this term [39].

2.2. Prediction Algorithm

As the airflow field can be considered as stable, the CRI of each heat source can be seen as constant. This means that if the heat emission or absorption from one heat source increases by a factor of 3, the temperature rise/drop within the field will increase by a factor of 3, regardless of location. Therefore, the CRI is an effective index for calculating dynamic temperature distribution without repeating CFD calculations. That is, when the heat emission or absorption from any heat source changes, the temperature change caused by this heat source at any location could be calculated by multiplying the heat by its CRI. A new temperature distribution can then be obtained by superimposing the effect of all heat sources. Therefore, when there are m heat sources in a room dominated by forced convection, the temperature rise/drop could be defined as $\Delta\theta_{i,o}$ and the CRI of heat source i to the location j as C_{ji}, when the heat emission or absorption from each heat source uniformly diffuses throughout the entire room. The temperature rise/drop from the neutral temperature at any location $\Delta\theta(X_j)$ could then be expressed by Equation (2):

$$\Delta\theta(X_j) = C_{j1} \cdot \Delta\theta_{1,o} + C_{j2} \cdot \Delta\theta_{2,o} + \ldots + C_{jm} \cdot \Delta\theta_{m,o} \tag{2}$$

Generally, the m heat sources include not only directly convective heat sources but also radiative heat sources, like walls. These radiative heat sources may not generate heat themselves, but they gain heat from other heat sources by radiation and then transfer heat to or absorb heat from indoor air by convection. Therefore, both directly convective heat source sand radiative heat sources were considered as contributing to the temperature distribution. For accurate prediction of temperature at any location, all CRI and the uniform temperature rise/drop caused by them need to be calculated.

By rewriting Equation (2) into matrix form, Equation (3) is obtained:

$$\Delta\theta(X_j) = \begin{bmatrix} C_{j1} & C_{j2} & \cdots & C_{jm} \end{bmatrix} \begin{bmatrix} \Delta\theta_{1,o} \\ \Delta\theta_{2,o} \\ \vdots \\ \Delta\theta_{m,o} \end{bmatrix} \tag{3}$$

In practice, the thermal boundary conditions of indoor thermal environments are generally dynamic and difficult to determine. Even if only steady-state calculations are considered, the intensity of each heat source still cannot be determined accurately. That is, the uniform temperature rise/drop caused by all heat sources cannot be determined accurately. Therefore, when calculating the uniform temperature rise/drop caused by each heat source, one mobile sensor was used to collect air temperature at n arbitrary locations in space (defined as $\Delta\theta_{si}$) and the measured temperature was then substituted into Equation (3), as shown in Equation (4):

$$\begin{bmatrix} \Delta\theta'_{s1} \\ \Delta\theta'_{s2} \\ \vdots \\ \Delta\theta'_{sn} \end{bmatrix} = \begin{bmatrix} C_{11} & C_{12} & \cdots & C_{1m} \\ C_{21} & C_{22} & \cdots & C_{2m} \\ \vdots & \vdots & \ddots & \vdots \\ C_{1m} & C_{2m} & \cdots & C_{nm} \end{bmatrix} \begin{bmatrix} \Delta\theta_{1,o} \\ \Delta\theta_{2,o} \\ \vdots \\ \Delta\theta_{m,o} \end{bmatrix} \tag{4}$$

According to Equation (4), in order to calculate $\Delta\theta_{i,o}$, it needs to multiply the inverse matrix C_{ij} on both sides of the equation. This requires that the number of air temperature measurements collected within a space should be equal to the number of heat sources, that is, $n = m$. The $\Delta\theta_{i,o}$ could then be obtained by Equation (5):

$$\begin{bmatrix} \Delta\theta_{1,o} \\ \Delta\theta_{2,o} \\ \vdots \\ \Delta\theta_{m,o} \end{bmatrix} = \begin{bmatrix} C_{11} & C_{12} & \cdots & C_{1m} \\ C_{21} & C_{22} & \cdots & C_{2m} \\ \vdots & \vdots & \ddots & \vdots \\ C_{1m} & C_{2m} & \cdots & C_{mm} \end{bmatrix}^{-1} \begin{bmatrix} \Delta\theta'_{s1} \\ \Delta\theta'_{s2} \\ \vdots \\ \Delta\theta'_{sn} \end{bmatrix} \tag{5}$$

By substituting Equation (5) into Equation (3), an expression for predicting temperature at any location can be obtained, as defined by Equation (6):

$$\Delta\theta(X_j) = \begin{bmatrix} C_{j1} & C_{j2} & \cdots & C_{jm} \end{bmatrix} \begin{bmatrix} C_{11} & C_{12} & \cdots & C_{1m} \\ C_{21} & C_{22} & \cdots & C_{2m} \\ \vdots & \vdots & \ddots & \vdots \\ C_{1m} & C_{2m} & \cdots & C_{mm} \end{bmatrix}^{-1} \begin{bmatrix} \Delta\theta'_{s1} \\ \Delta\theta'_{s2} \\ \vdots \\ \Delta\theta'_{sn} \end{bmatrix} \tag{6}$$

According to Equation (6), the temperature distribution can be predicted through calculating the CRI of each heat source in advance and collecting air temperature measurements equal to the number of heat sources.

2.3. Mobile Sensors

As noted earlier, Sasamoto et al. used some fixed sensors equal to the number of heat sources to collect air temperature measurements in a space and then combined them with CRI for temperature distribution prediction [35]. This method is suitable for practical applications without identified boundary conditions in advance. However, there are limitations in terms of both installation locations and installation numbers. For example, when the number of monitored data points increases, the number of fixed sensors also needs to increase, leading to higher expense and bigger installation space needs. Additionally, the installation locations of fixed sensors are not changeable and are usually far away from the target control area, such as near air outlets, at high positions on the wall, or where people cannot easily touch them. This means that the air temperature collected for Equation (6) is location-dependent, so it may not be representative of the real situation of each heat source.

With the development of mobile carriers in control engineering, mobile sensor technology has become a popular solution for controlling buildings' indoor environment, to overcome the limitations of fixed sensors. Xue and Zhai have developed a method using mobile sensors to capture both spatial and temporal distributions of pollutant concentrations in order to estimate the location of pollutant sources [40]. This method does not consider thermal factors, especially the influence of a change in the heat source on the airflow field, and it also adopts the probability method to explore the most probable location of pollutant sources, so it cannot achieve simple calculation of temperature at any location within a space. However, using mobile sensors to monitor environmental information provides a feasible and efficient idea for studies of indoor thermal environments.

Therefore, in this study, one mobile sensor (as shown in Figure 1) has been used to replace fixed sensors when collecting air temperatures within a space. Mobile sensors have the characteristics of variable acquisition height, diverse acquisition path, and flexible acquisition location. They are adjustable according to actual requirements or prediction demand to collect more suitable data with no restrictions, to achieve better temperature distribution prediction by Equation (6).

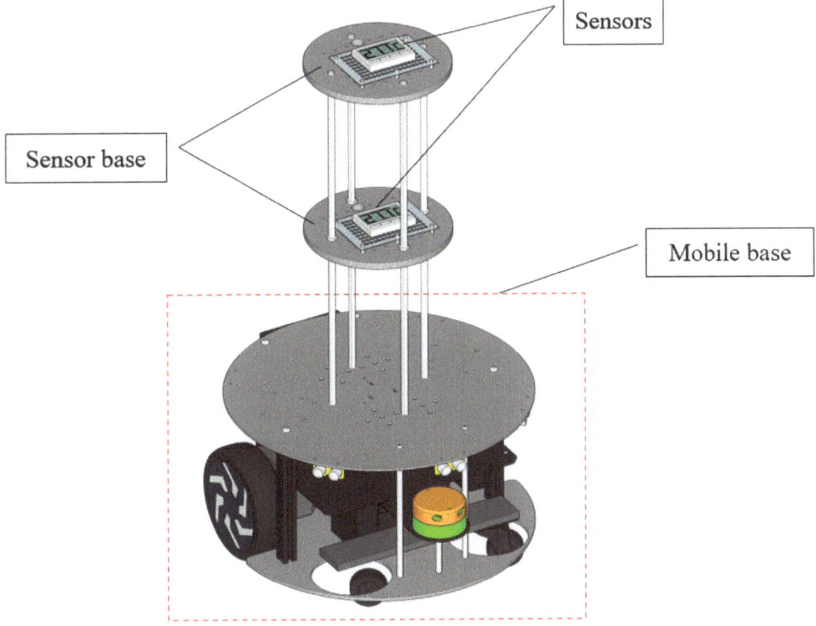

Figure 1. A mobile sensor.

3. Case Study

3.1. Verification of CFD Simulation

In this study, both indoor airflow field and temperature field were simulated using Ansys Fluent—a CFD tool that accounts for both indoor airflow and temperature distribution characteristics. The finite volume method was used to discretize the Reynolds averaged Navier-Stokes equations and the averaged energy and mass conservation equations. The standard $k - \varepsilon$ model and the discrete ordinates (DO) model were used to simulate both indoor turbulent flow and indoor radiation. The air materials adopted the Boussinesq model to consider the buoyancy term, which treats air density as a constant value in all solved equations, except for the buoyancy term in the momentum equation. The SIMPLE algorithm was adopted to solve the pressure-velocity coupling problem, and the standard scheme was applied for pressure discretization. The second-order upwind difference scheme was used for momentum and energy. A linear under-relaxation iteration was used to ensure convergence. The energy and other solutions were converged until the residuals of all cells in the simulation domain reached within 10^{-8} and 10^{-4}.

To validate the accuracy of simulation from ANSYS Fluent, a full-scale model was established and its prediction results were compared with data collected from the experiment done by Tian et al. [41], with a test chamber of 4.0 m (length) $\times$ 3.5 m (width) $\times$ 3.5 m (height), as shown in Figure 2, with thermally insulated walls, floor, and ceiling. There was a desk, a closet, a computer, and a thermal manikin in the test chamber, with three fluorescents installed on the ceiling. The manikin and the three fluorescents led to a total cooling load of 496 W, thus about 35 W/m^2. After a grid-independence test, 643,585 grids were adopted to balance accuracy and time. The air was supplied from a double grille diffuser located at the height of 1.36 m above the floor while the air was exhausted through a left-wall-based diffuser mounted at the height of 3.16 m above the floor. The ventilation rate of the chamber was 2.2 air changes per hour (ACH). Comparisons between simulation and measurement results along five sampling lines in the chamber are illustrated in Figure 3, showing good agreement between the two datasets (with an error of less than 1 °C in temperature and 0.1 m/s in velocity). Therefore, this simulation method was considered as accurate and usable in this study.

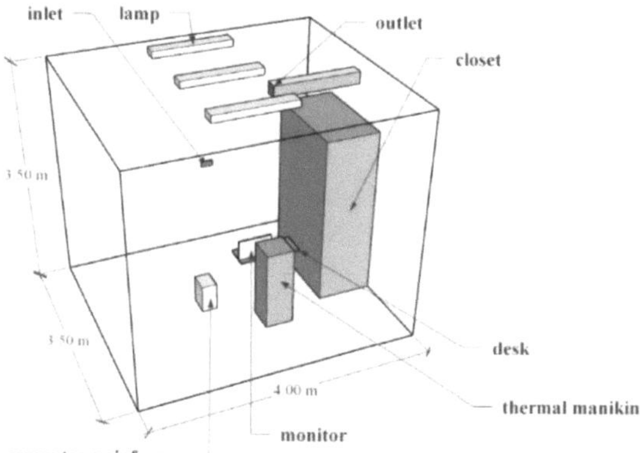

Figure 2. Configuration of the test chamber.

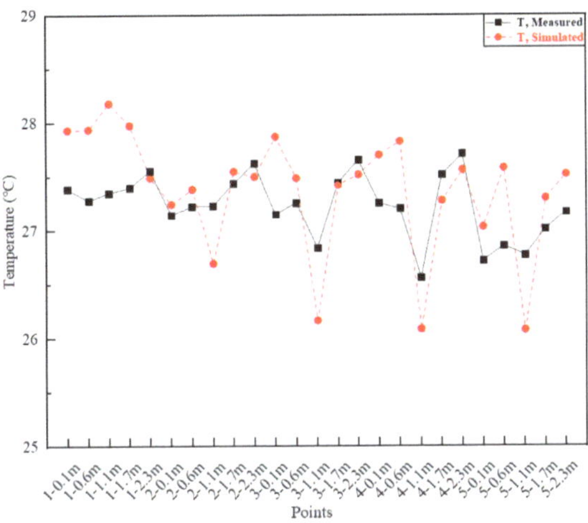

(**a**) Verification of temperature

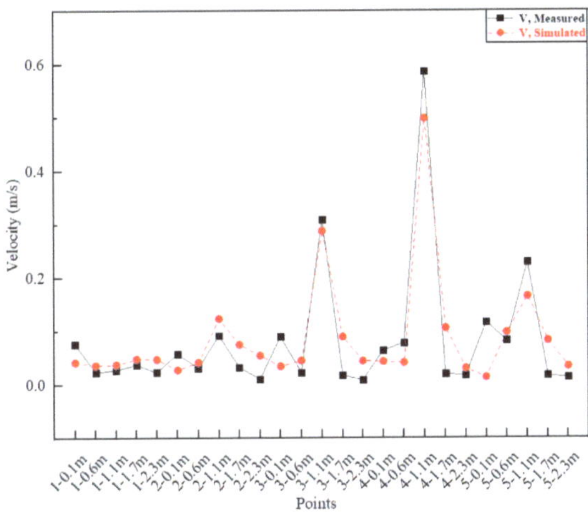

(**b**) Verification of velocity

Figure 3. Comparison between simulated and measured data at 5 heights at sampling lines 1–5 (remarks: measurement point X-Y m, where X is the number sampling line and Y is the heights of 0.1 m, 0.6 m, 1.1 m, 1.7 m and 2.3 m).

3.2. Description of the Model and Simulation

As shown in Figure 4, a typical office model has been established in this study, with dimensions of 14 m (length) × 10 m (width) × 4 m (height). There were four air supply inlets on the ceiling and four air exhaust outlets at the bottom of the two opposite walls. The walls, the ceiling, and the floor of this office were all thermally insulated. There were 6 lamps and 24 working positions (each with a person and a computer) in the office. To simplify the computational model, the radiative heat transfer of the walls and floor were considered as one heat source. Similarly, the radiative heat transfer of the ceiling and the heat emission from six lamps were considered as one heat source. This left the heat emission from four adjacent working positions, which were considered as one heat source. Therefore, there were a total of nine heat sources in this simulation work. The numerical method for calculating the temperature distribution of this office has been described above, and the specific boundary conditions are listed in Table 1. The neutral temperature in the office was assumed to be 25 °C. The office was discretized into 5,800,535, 7,784,596, and 10,746,488 hexahedral control volumes. After a grid-independence test, the middle definition was adopted to balance accuracy and time.

Table 1. Summary of numerical simulation conditions.

Surface	Boundary Condition
Walls/Ceiling/Floor	Wall; Adiabatic
Lamp	Wall; Heat flux: 150 W/m^2.
Person	Wall; Heat flux: 45 W/m^2
Computer	Wall; Heat flux: 70 W/m^2
Air supply	Velocity-inlet; Velocity: 1.0 m/s. Temperature: 21 °C
Air exhaust	Outflow

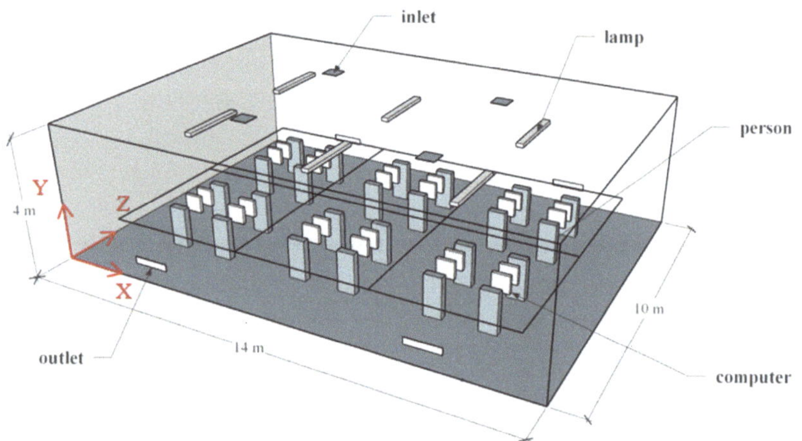

Figure 4. A model for the office.

According to the coupled calculation results of both convective and radiative heat transfers, the airflow field and the temperature field can be obtained. The cross sections *y = 1.0 m* and *z = 2.1 m* were taken as examples, with the velocity distribution and the temperature distribution are shown in Figures 5 and 6, respectively. The calculation results showed that the air was supplied vertically from the inlet toward the floor, with observable backflow along the floor and the walls. Simultaneously, the air supply directly reached the working area, which could effectively take away internal heat emissions and diffuse them to other places. Near the working positions, the air was warmed up significantly, so a heat plume around them could be observed. Due to the effect of backflow and buoyancy, the temperature distribution in the room was obviously stratified from bottom to top.

The sub-temperature distribution of each heat source was calculated with the airflow field described above. Taking one working position as an example, its CRI distribution obtained by Equation (1) is shown in Figure 7. The calculation results showed that the CRI of the heat emission from one working position (with four people and four computers) ranged between 0 and 4.67. It could be observed that near this heat source, the CRI was greater than 1.0. However, with the increase in distance, the heat diffusion was weakened. The CRI thus was less than 1.0 in most areas, meaning that the heat source had an obvious effect on the area around it, but a relatively small effect at larger distances.

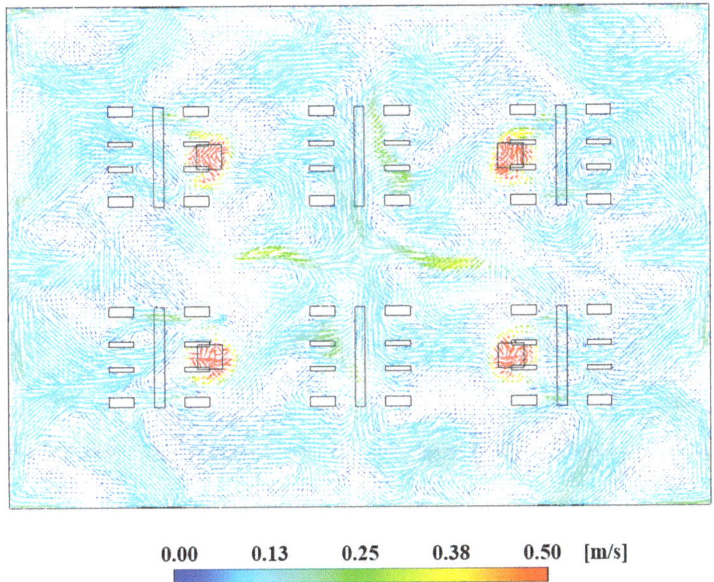

(**a**) Velocity distribution (*y = 1.0 m* cross section)

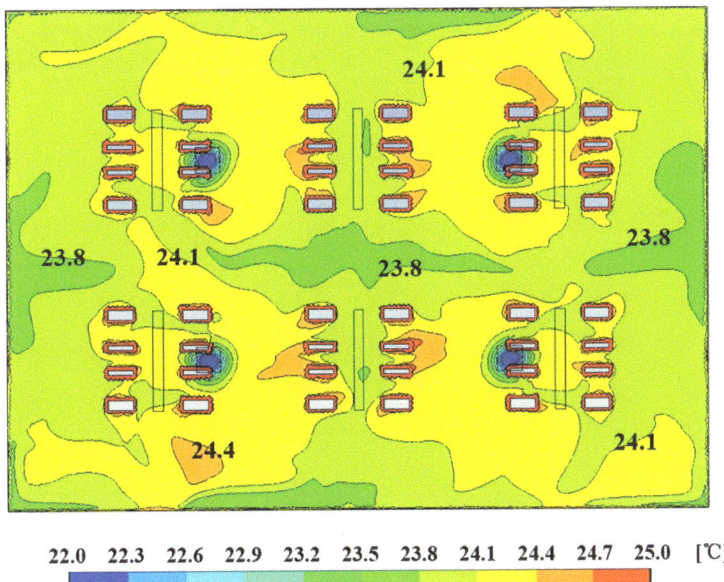

(**b**) Temperature distribution (*y = 1.0 m* cross section)

Figure 5. Velocity and temperature distributions (*y = 1.0 m* cross section).

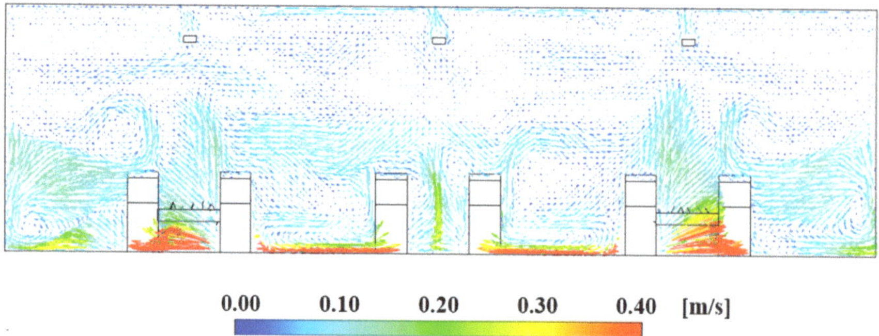

(**a**) Velocity distribution (*z = 2.1 m* cross section)

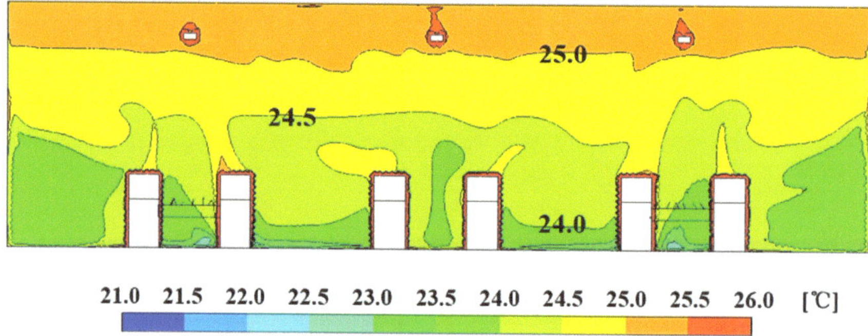

(**b**) Temperature distribution (*z = 2.1 m* cross section)

Figure 6. Velocity and temperature distributions (*z = 2.1 m* cross section).

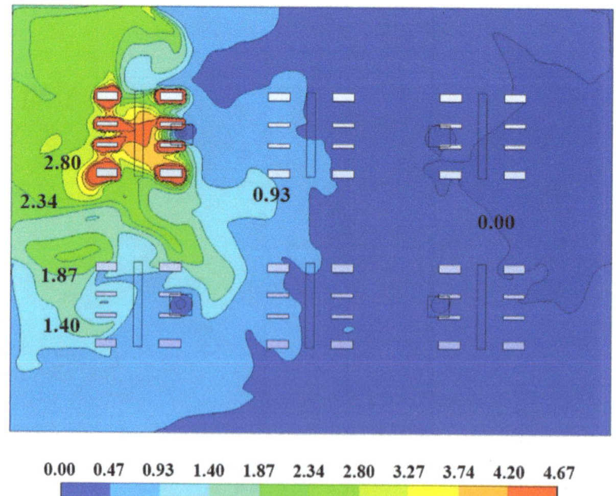

Figure 7. CRI distribution of the heat emission from the one working position (with four people and four computers) (*y = 1.0 m* cross section).

4. Results

4.1. Comparison on Prediction Accuracy between Fixed Sensors and Mobile Sensors

In the comparison, 10 temperature prediction points in the plane with the height of 1 m, 2 m and 3 m, respectively, were selected. They were named as L-1 to L-10, M-1 to M-10, and H-1 to H-10, respectively. It should be noted that in this study the locations of all prediction points were only different in height, but with identical plane coordinates. According to the requirements of the prediction algorithm, the number of collected air temperature measurements needs to be equal to the number of heat sources. Therefore, nine air temperatures were collected. The fixed sensor locations are shown in Figure 8. It is worth noting that the acquisition location of the sensors has a great impact on the prediction results. In other words, even if the number of sensors is the same, different results and prediction accuracy will be obtained by using the proposed algorithm with different acquisition locations. For example, in the study of Sasamoto et al. [35], the fixed sensors were installed near each heat source, and the prediction accuracy was acceptable. There are two reasons why fixed sensors were all installed on the walls in this study: (1) This is more suitable for the actual situation; (2) The purpose of this study is to verify the application disadvantages of fixed sensors. That is, taking the actual situation as a reference, the limitations of fixed sensors are analyzed and further the solutions are proposed. Meanwhile, a mobile sensor with the acquisition height of 1.2 m was used to collect the air temperature at several locations in the space, also shown in Figure 8. The above two groups of collected air temperatures were used to predict the temperature of 30 points at the three heights, with their prediction results shown in Figure 9.

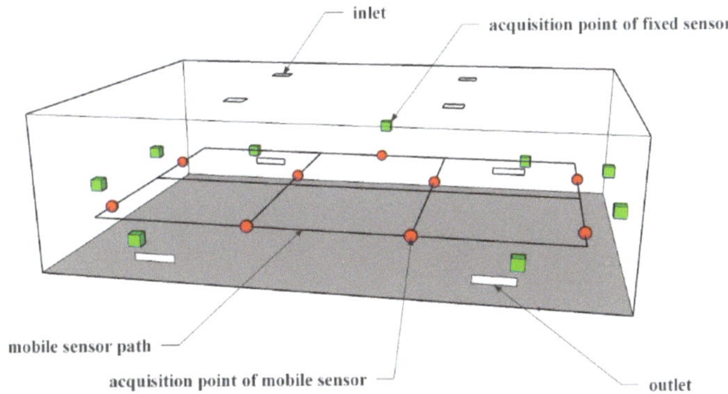

Figure 8. Acquisition points of fixed sensors and mobile sensors.

When the air temperature collected by fixed sensors was used for prediction, according to Table 2, the corresponding average relative errors were 5.7%, 10.8%, and 27.7% at the heights of h = 1.0 m, h = 2.0 m, and h = 3.0 m, respectively, which were relatively large. The reasons for the unsatisfactory results are considered as follows:

Table 2. Prediction of each point using fixed sensors and mobile sensors.

Average Relative Error	Fixed Sensors	Mobile Sensors
At the height of h = 1.0 m	5.7%	2.1%
At the height of h = 2.0 m	10.8%	3.3%
At the height of h = 3.0 m	27.7%	4.8%

First, the influence of the acquisition locations of the sensors on the prediction accuracy noted above cannot be ignored. Second, we suppose that it is mainly related to the influence of airflow distribution. The proposed prediction algorithm consists of two parts: the CRI matrix and the air temperature matrix. The assumption of a steady airflow field is the basic premise for calculating CRI. Specifically, when the airflow field is dominated by forced convection in the room, although the influences from other heat sources exist, the effect is smaller than that of forced convection, and thus can be ignored. That is, the airflow field can be considered as steady for a small variation range of supply air temperature and velocity. However, this is an ideal assumption, which has deviation in practical applications. For example, near the walls, air supply/exhausts, and local heat sources, due to the influence of boundary layer, air backflow and heat plume, the airflow distribution in these areas is complex and unstable, resulting in large errors in the calculation of CRI. In the proposed prediction algorithm, the application of sensors used to collect air temperature is combined with CRI to calculate the heat source intensity. Therefore, when using the fixed sensors on the walls with unstable airflow distribution to predict the temperature distribution, the inaccurate calculation of CRI at any sensor location was used repeatedly to calculate the intensity of each heat source. This leads to the influence of each CRI calculation error being superimposed and expanded, further affecting the prediction accuracy. This is the main reason for the lower prediction accuracy when using fixed sensors to predict temperature distribution.

Especially, Figure 9 showed that when using fixed sensors for temperature prediction, obvious abnormalities occurred at some locations, such as L-7, L-8, H-1, and H-10. In addition to the reasons noted above, we assume their locations also have an impact on the prediction accuracy. L-7 and L-8 were located at the lower part of the wall and close to the supply air, which was more vulnerable to the influence of air backflow after the supply air meets the floor and other surfaces. Therefore, it will affect the CRI calculation of each heat source here and further affect the prediction accuracy. Similarly, H-1 and H-10 were located at the higher part of the wall and close to the corner of the room. The airflow distribution here is affected by many factors, such as air backflow near the adjacent walls, heat plume above the heat source, etc. The prediction error thus also increases.

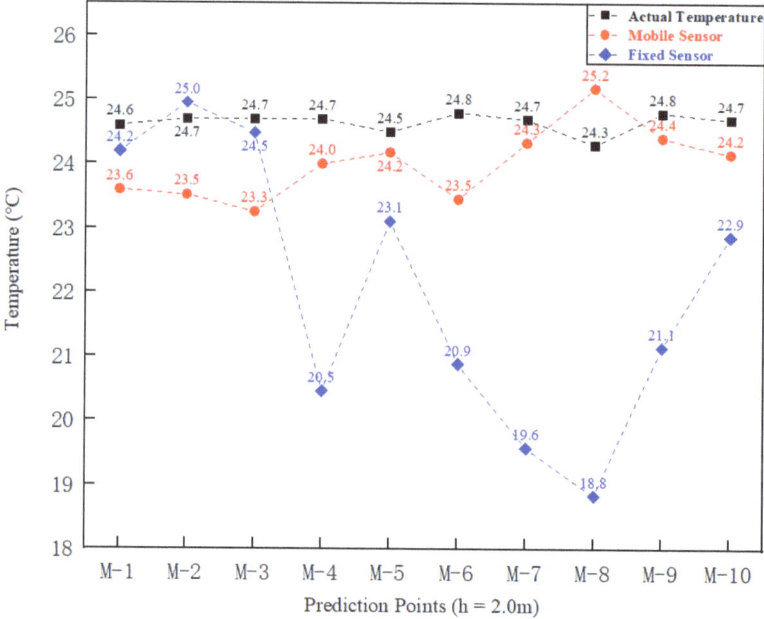

(**a**) Prediction at the height of 1.0 m using fixed sensors and mobile sensors

Figure 9. *Cont.*

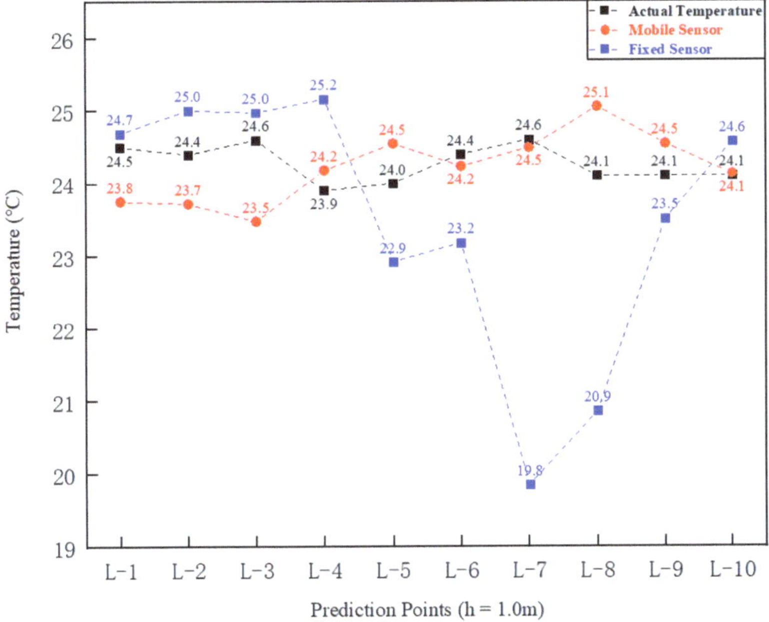

(**b**) Prediction at the height of 2.0 m using fixed sensors and mobile sensors

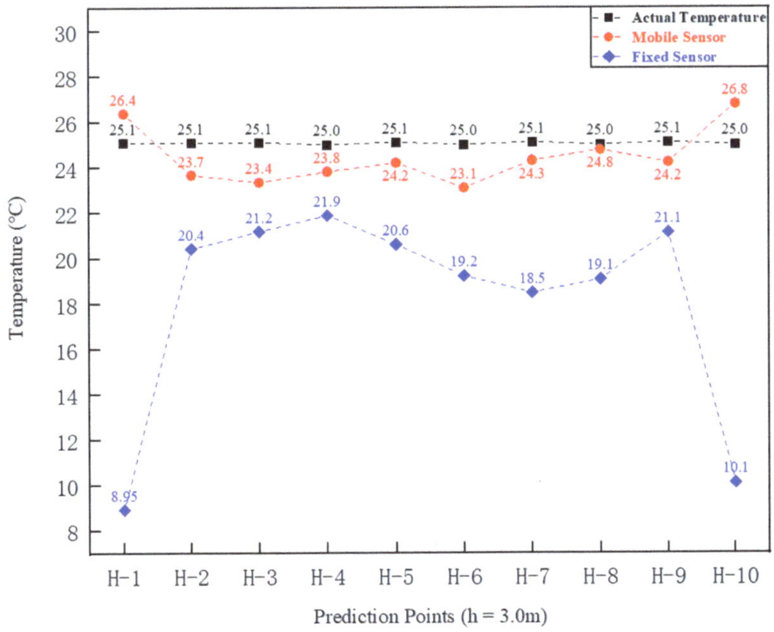

(**c**) Prediction at the height of 3.0 m using fixed sensors and mobile sensors

Figure 9. Prediction of different heights using fixed sensors and mobile sensors.

When using the mobile sensors with flexible acquisition locations instead of the fixed sensors for temperature prediction, the corresponding average relative errors were 2.1%, 3.3%, and 4.8%, respectively. This shows that the proposed temperature distribution prediction method based on CRI and finite air temperature is reliable, which is consistent with the research results of Sasamoto et al. [35]. Additionally, this indicates that due to some restrictions in practical applications, using mobile sensors instead of fixed sensors to predict the temperature distribution is appropriate. Besides, not in all cases, the prediction results obtained by using the mobile sensors are satisfactory.

4.2. Analysis on the Impact of Mobile Sensors Acquisition Height on Prediction Accuracy

The height of the mobile sensor can be adjusted in the vertical direction according to the actual situation and the prediction demand. Generally, the purpose of regulating the indoor thermal environment is twofold: reduce energy consumption and maintain comfortable conditions. Therefore, the air temperature within human activity areas is usually the main controlled variable. To figure out the influence of the acquisition height of mobile sensors on the prediction accuracy of this area, 10 prediction points at the heights of h = 0.7 m, h = 1.0 m, h = 1.2 m and h = 1.5 m, named 1-P1 to 1-P10, 2-P1 to 2-P10, 3-P1 to 3-P10, 4-P1 to 4-P10 and 5-P1 to 5-P10, respectively, were selected, with identical plane coordinates. Air temperatures obtained by the mobile sensor introduced in Section 4.1 was used to predict the air temperature at these points, with prediction results shown in Figure 10.

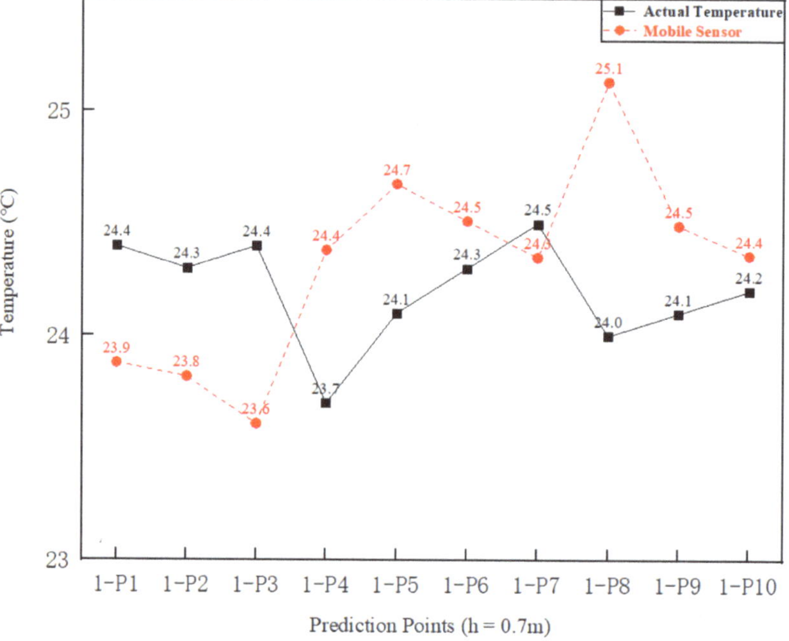

(**a**) Prediction at the height of 0.7 m using mobile sensors

Figure 10. *Cont.*

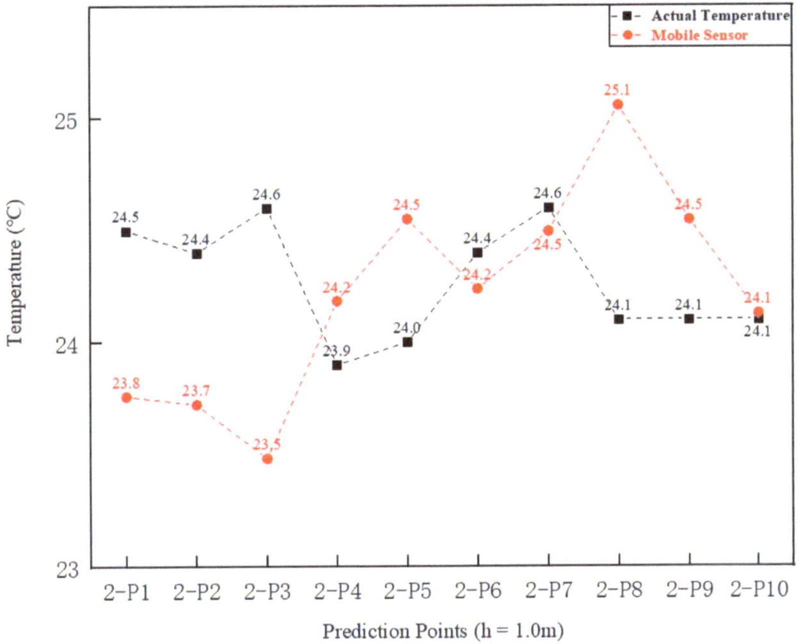

(**b**) Prediction at the height of 1.0 m using mobile sensors

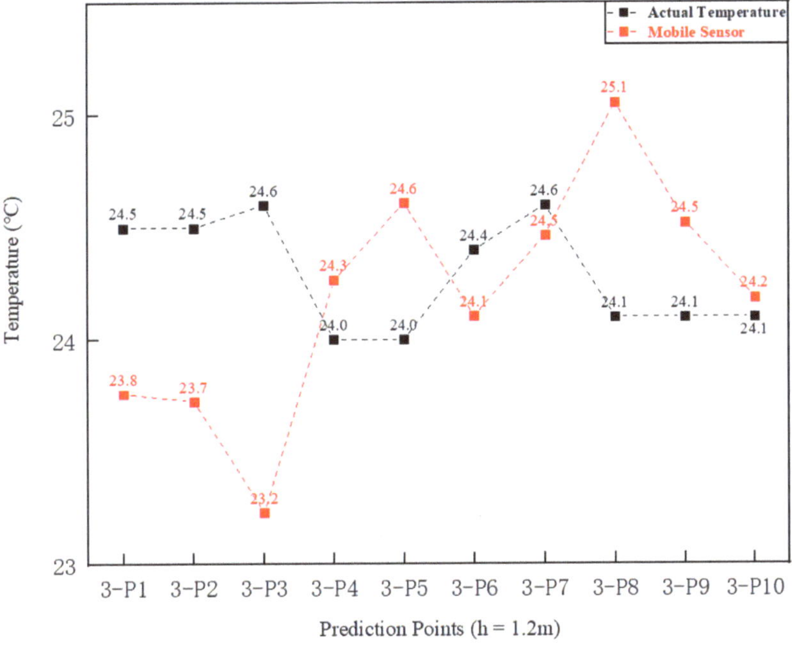

(**c**) Prediction at the height of 1.2 m using mobile sensors

Figure 10. *Cont.*

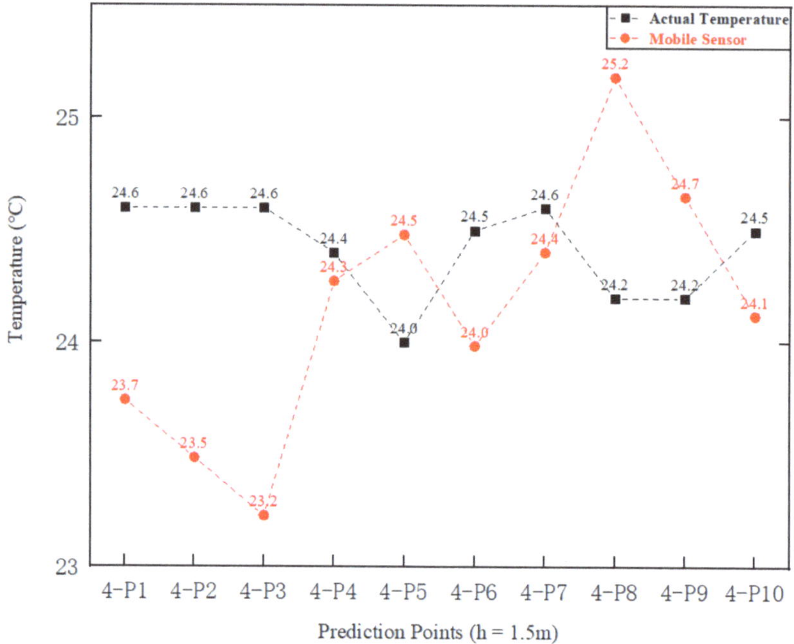

(**d**) Prediction at the height of 1.5 m using mobile sensors

Figure 10. Prediction of different heights using mobile sensors.

Similarly, according to Table 3, the corresponding average relative errors were 2.1%, 2.1%, 2.3%, and 2.7% at the heights of h = 0.7 m, h = 1.0 m, h = 1.2 m, and h = 1.5 m, respectively. Consequently, it can be concluded that in the human activity area, the acquisition height of mobile sensors has little influence on the prediction accuracy.

Table 3. Prediction of different heights using mobile sensors.

Average Relative Error	Mobile Sensors
At the height of h = 0.7 m	2.1%
At the height of h = 1.0 m	2.1%
At the height of h = 1.2 m	2.3%
At the height of h = 1.5 m	2.7%

4.3. Analysis on the Impact of Mobile Sensors Acquisition Distance on Prediction Accuracy

To better guide the application of mobile sensors in practical situations, it should not only consider how to set the acquisition height of mobile sensors, but also consider the acquisition distance of mobile sensors. To explore the impact from this aspect, given a mobile sensor acquisition path (the acquisition height was 1.2 m), several acquisition distances were designed, which were 1 m, 2 m, 3 m, 4 m and 5 m. In the case of each acquisition distance, 3, 3, 3, 3, and 2 acquisition point distributions were given, respectively, as shown in Figures 11–15. The selected prediction points were the same as those at the 1.2 m height discussed in Section 4.2, named 3-P1 to 3-P10.

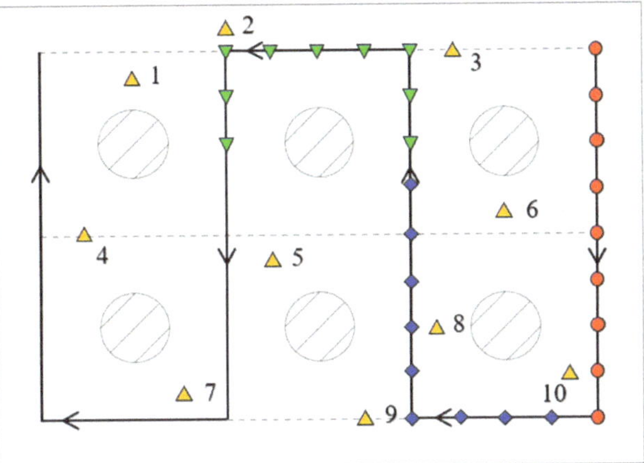

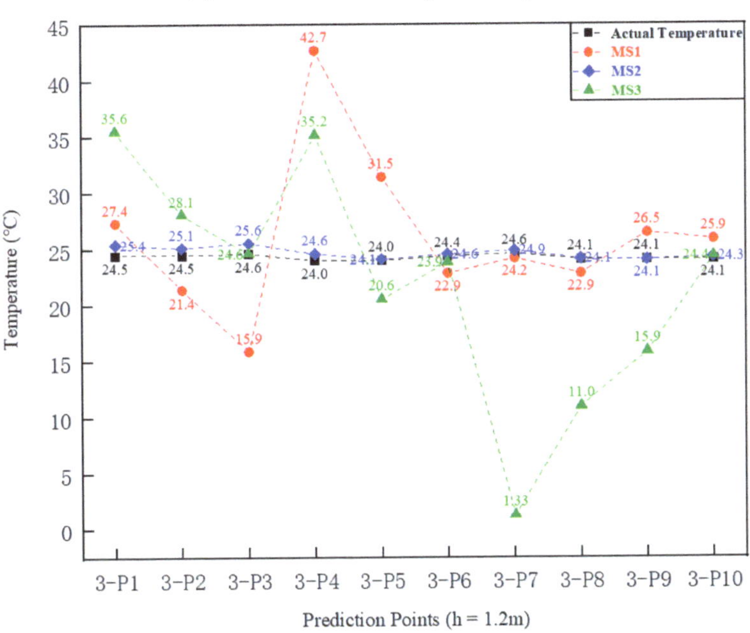

(**a**) Distributions of acquisition points

(**b**) Prediction results

Figure 11. Prediction of mobile sensors with acquisition distance of 1.0 m.

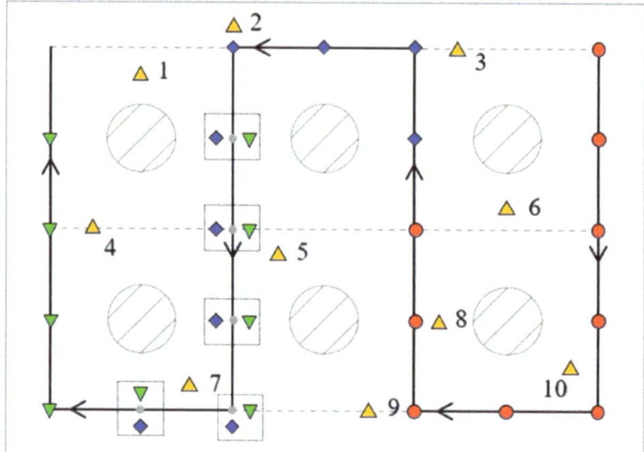

● - The 4th distribution of mobile sensors acquisition points (named **MS4**)

◆ - The 5th distribution of mobile sensors acquisition points (named **MS5**)

▽ - The 6th distribution of mobile sensors acquisition points (named **MS6**)

△ - Prediction points

(**a**) Distributions of acquisition points

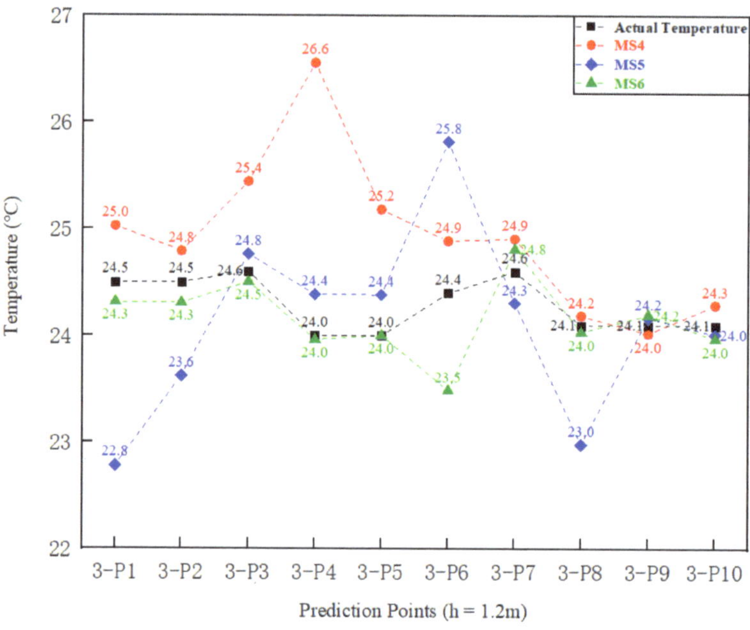

(**b**) Prediction results

Figure 12. Prediction of mobile sensors with acquisition distance of 2.0 m.

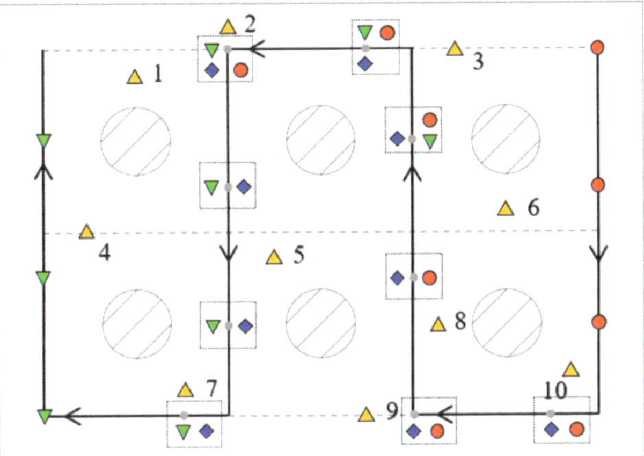

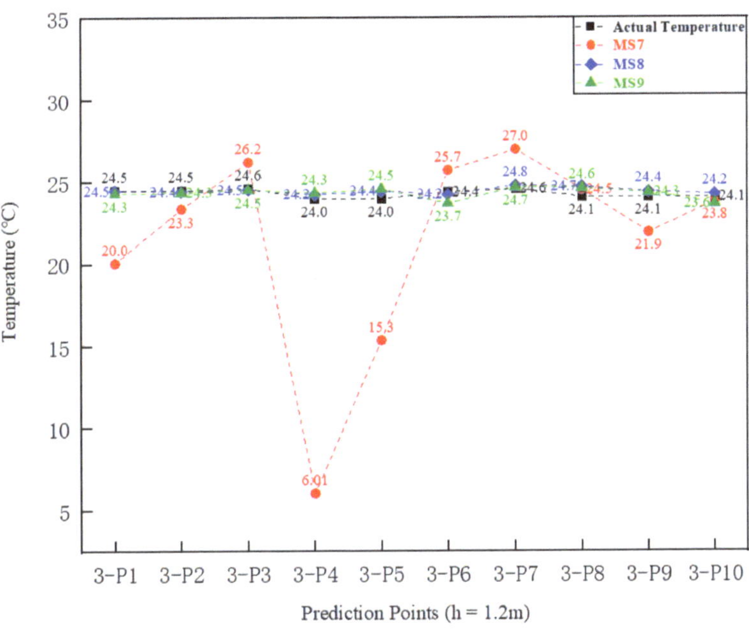

(**a**) Distributions of acquisition points

(**b**) Prediction results

Figure 13. Prediction of mobile sensors with acquisition distance of 3.0 m.

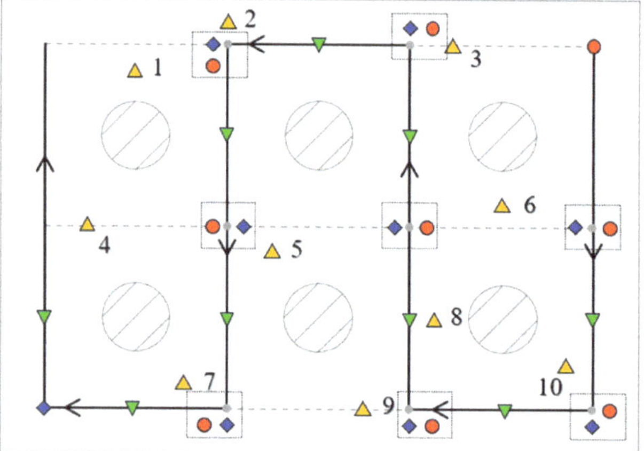

● - The 10th distribution of mobile sensors acquisition points (named **MS10**)

◆ - The 11th distribution of mobile sensors acquisition points (named **MS11**)

▽ - The 12th distribution of mobile sensors acquisition points (named **MS12**)

△ - Prediction points

(**a**) Distributions of acquisition points

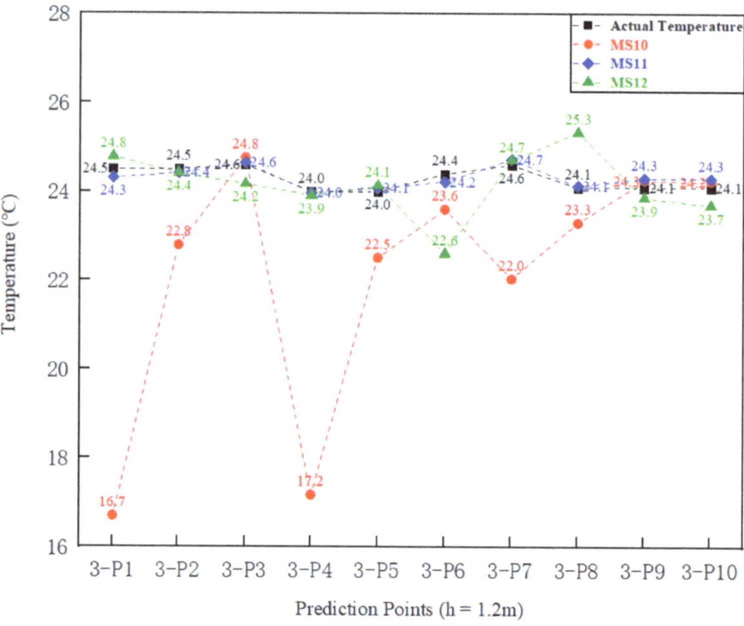

(**b**) Prediction results

Figure 14. Prediction of mobile sensors with acquisition distance of 4.0 m.

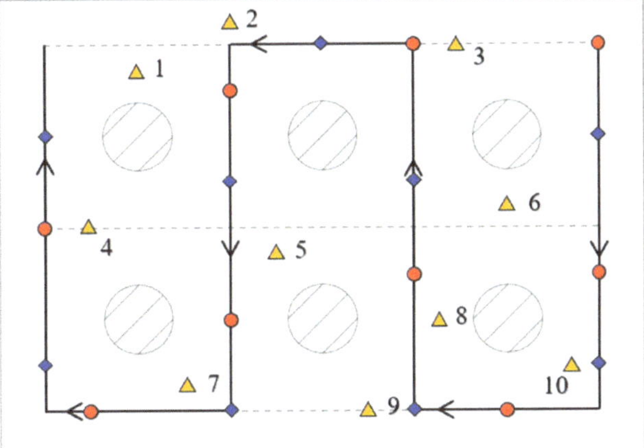

● - The 13th distribution of mobile sensors acquisition points (named **MS13**)

◆ - The 14th distribution of mobile sensors acquisition points (named **MS14**)

△ - Prediction points

(**a**) Distributions of acquisition points

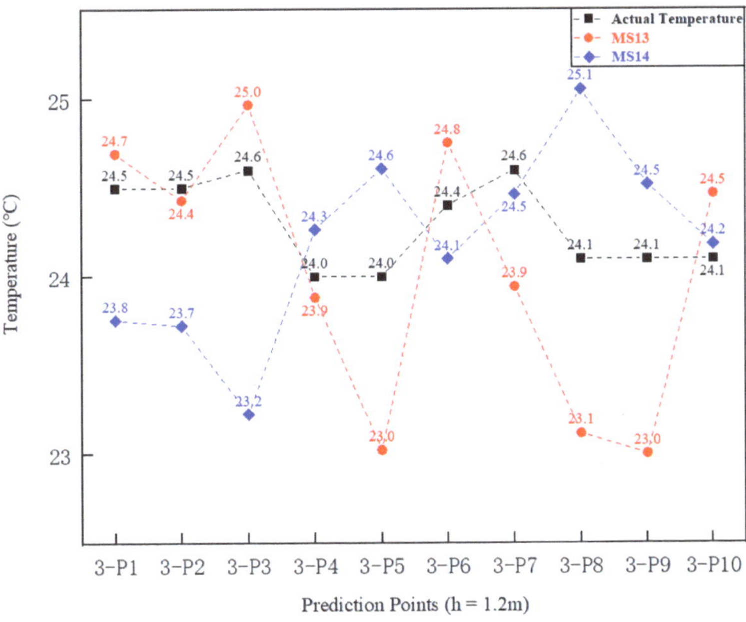

(**b**) Prediction results

Figure 15. Prediction of mobile sensors with acquisition distance of 5.0 m.

Figures 11–15 and Table 4 also show the prediction results and the corresponding average relative errors of each prediction point under the various distributions of acquisition points noted above. The prediction results showed that smaller acquisition distances would make the distribution of acquisition points more concentrated, hence leading to an obvious reduction in prediction accuracy. As shown in MS1 and MS3 in Figure 11, MS7 in

Figure 13, and MS10 in Figure 14, the corresponding average relative errors were 19.9%, 30.7%, 16.7%, and 9.2%, respectively. However, it also had the situation that the prediction accuracy was relatively high with small acquisition distances, for example, the average relative errors were 1.6% and 0.8% in the cases of MS2 in Figure 11 and MS6 in Figure 12, respectively. This indicates that the uncertainty of temperature distribution prediction using the air temperature collected by mobile sensors with smaller acquisition distance was larger, and the prediction accuracy at this time cannot be guaranteed. When controlling indoor thermal environments in practical applications, it must quickly and accurately obtain temperature distributions. Therefore, it is necessary to give a design criterion for selecting an acquisition distance of mobile sensors with lower uncertainty, higher accuracy, and wider application scope. Through a careful comparison of the prediction results under the distributions of acquisition points with various acquisition distances, as shown in Figures 11–15, a conclusion was drawn that the acquisition distance should be large enough to make the distribution of acquisition points more dispersed.

Table 4. Prediction of mobile sensors with different acquisition distances.

	MS1	MS2	MS3	MS4	MS5	MS6
Average relative error	19.9%	1.6%	30.7%	2.7%	2.7%	0.8%
	MS7	**MS8**	**MS9**	**MS10**	**MS11**	**MS12**
Average relative error	16.7%	0.9%	1.4%	9.2%	0.5%	2.0%
	MS13	**MS14**				
Average relative error	2.1%	2.3%				

5. Limitations

Although the application of mobile sensors instead of fixed sensors to collect air temperature in the area with stable airflow distribution can greatly improve the prediction accuracy, the influence of air distribution on the method proposed in this study cannot be easily ignored. An existing study has shown that the relationship between the sensor acquisition point and the airflow direction would affect the prediction accuracy [40].

Carefully observing the acquisition point distributions in the cases of MS1 and MS3 in Figure 11, MS7 in Figure 13, and MS10 in Figure 14, it can be seen that most of them were close to wall surfaces, where the airflow distribution is relatively complex. When the mobile sensor collects the air temperature near the boundary of the room, the prediction accuracy thus cannot be guaranteed. Whereas in the case with identical acquisition distances, with acquisition points that are located in the interior of the room and far away from the boundary of the room, the prediction accuracy was improved. For example, the average relative error of the acquisition point distribution shown in MS2 in Figure 11 is 1.6%, which is far less than the average relative error of the acquisition point distribution shown in MS1 and MS3 in Figure 11. Similarly, the comparison between the prediction results obtained under the distribution of each acquisition point shown in Figure 13 also verifies this conclusion. Therefore, to ensure the prediction accuracy, some distance between the acquisition points and the room boundaries should be given.

Additionally, the more detailed relationship between the acquisition location of mobile sensors and airflow distribution still needs to be further explored, which will be the focus of a future study. The optimization of acquisition distance and acquisition path of mobile sensors could be achieved to further reduce the time required for predicting temperature distribution in practical applications.

6. Conclusions

Because of the complexity and dynamic nature of indoor thermal environments and their impact on energy consumption, the control of indoor thermal environments has always been an important research focus. In this circumstance, it is necessary to obtain the indoor temperature distribution rapidly and in detail. In this study, a method predicting indoor temperature distribution has been proposed for the purpose of real-time prediction and precise control of indoor thermal environments in practical applications. In this method, the air temperature was collected by one mobile sensor at the working area height, and combined with the contribution ratio of indoor climate (CRI) to realize the rapid prediction of indoor temperature distribution. Through establishing a typical office room, the reliability and effectiveness of using mobile sensors instead of fixed sensors for air temperature collection for temperature distribution prediction has been verified through accuracy comparison. Furthermore, several acquisition heights and acquisition distances of mobile sensors were tested, and their impact on prediction accuracy was analyzed. The main findings from this study are summarized as follows:

(1) Due to some restrictions in practical applications, using mobile sensors instead of fixed sensors can realize the temperature distribution prediction of residential height by reducing the number of sensors. If there are no restrictions, the application of fixed sensors for prediction can also meet the requirements, but they are also limited by the acquisition height and acquisition path. Under this condition, it is possible that the combination of fixed sensors and mobile sensors can obtain higher prediction accuracy.

(2) The acquisition height of mobile sensors has shown little impact on prediction accuracy in human activity areas. By comparing the prediction accuracy of mobile sensors for temperature distribution at different heights, it was found that the difference between them was not significant. Therefore, when using mobile sensors to predict the temperature distribution in human activity areas, there is no need to specifically set the acquisition height.

(3) The acquisition distance should be large enough to make the distribution of acquisition points more dispersed. By comparing the prediction accuracy of mobile sensors with different acquisition distances, the results show that smaller acquisition distances made acquisition points more concentrated, hence reducing prediction accuracy. Considering the influence of airflow distribution, the acquisition points should be not very close to room boundaries.

From the above analysis, the method proposed in this study could be beneficial to the rapid prediction of non-uniform temperature distribution in the perspective of satisfying thermal comfort while improving energy efficiency. It will make outstanding contributions to the control strategy based on real-time response to the thermal environment.

Author Contributions: Conceptualization, W.Z. and Y.Z.; methodology, Y.Z., Z.Z. and W.Z.; validation, Z.Z.; formal analysis, Y.Z., Z.Z. and S.W.; investigation, Y.Z. and Z.Z.; resources, W.Z.; data curation, Y.Z. and Z.Z.; writing—original draft preparation, Y.Z.; writing—review and editing, Y.Z., Z.Z., W.Z., S.W. and Y.X.; visualization, Y.Z.; supervision, W.Z., S.W. and Y.X.; project administration, W.Z. and Y.X.; funding acquisition, W.Z. All authors have read and agreed to the published version of the manuscript.

Funding: This research was funded by Natural Science Foundation of China, grant number 5190080465; Special Fund of Beijing Key Laboratory of Indoor Air Quality Evaluation and Control, grant number BZ0344KF20-05; Joint Research Project of the Wind Engineering Research Center, Tokyo Polytechnic University, Mext (Japan) Promotion of Distinctive Joint Research Center Program Grant Number: JPMXP0619217840, JURC grant number 21212012.

Data Availability Statement: Not applicable.

Conflicts of Interest: The authors declare no conflict of interest.

Nomenclature

$u_j[\text{m/s}]$	air velocity
$\nu_t[\text{kg/(m}\cdot\text{s)}]$	turbulent viscosity
$t[\text{s}]$	time
$\text{Pr}_t[\text{J}\cdot\text{s}]$	turbulent Prandtl number
$C_p[\text{J/(kg}\cdot\text{K)}]$	specific heat of indoor air
$\rho[\text{kg/m}^3]$	air density
$q[\text{W}]$	heat emission and absorption of all heat sources
$q_i[\text{W}]$	heat emission or absorption of heat source i
$X_j[\text{m}]$	component of the spatial coordinates (j = 1,2,3)
$\theta[^\circ\text{C}]$	air temperature
$\theta_n[^\circ\text{C}]$	air neutral temperature, i.e., indoor initial air temperature
$\Delta\theta_i[^\circ\text{C}]$	temperature rise or drop caused by heat source i
$\theta_{i,o}[^\circ\text{C}]$	uniform air temperature caused by heat source i
$\Delta\theta_{i,o} = \theta_{i,o} - \theta_n[^\circ\text{C}]$	temperature rise or drop of the uniform air temperature caused by heat source i from θ_n
$\theta_i(X_j)[^\circ\text{C}]$	air temperature at the location X_j caused by heat source i
$\Delta\theta_i(X_j) = \theta_i(X_j) - \theta_n[^\circ\text{C}]$	temperature rise or drop at the location X_j caused by heat source i from θ_n
$F[\text{m}^3/\text{s}]$	volume of supply air
$V[\text{m}^3]$	room volume
m	number of heat sources
n	number of sensor points
C_{ji}	CRI of heat source i to location j
$\Delta\theta'_{si}$	temperature rise or drop collected by mobile sensors from θ_n

References

1. Klepeis, N.E.; Nelson, W.C.; Ott, W.R.; Robinson, J.P.; Tsang, A.M.; Switzer, P.; Behar, J.V.; Hern, S.C.; Engelmann, W.H. The National Human Activity Pattern Survey (NHAPS): A resource for assessing exposure to environmental pollutants. *J. Expo. Anal. Environ. Epidemiol.* **2000**, *11*, 231–252. [CrossRef]
2. Liu, S.M.; Cao, Q.; Zhao, X.W.; Lu, Z.C.; Deng, Z.P.; Dong, J.K.; Lin, X.R.; Qing, K.; Zhang, W.Z.; Chen, Q.Y. Improving indoor air quality and thermal comfort in residential kitchens with a new ventilation system. *Build. Environ.* **2020**, *180*, 107016. [CrossRef]
3. Park, D.Y.; Chang, S. Effects of combined central air conditioning diffusers and window-integrated ventilation system on indoor air quality and thermal comfort in an office. *Sustain. Cities Soc.* **2020**, *61*, 102292. [CrossRef]
4. Building Energy Conservation Research Centre, Tsinghua University. *Annual Report on China Building Energy Efficiency*; China Architecture & Building Press: Beijing, China, 2019. (In Chinese)
5. Gilani, S.; Montazeri, H.; Blocken, B. CFD simulation of stratified indoor environment in displacement ventilation: Validation and sensitivity analysis. *Build. Environ.* **2016**, *95*, 299–313. [CrossRef]
6. Chen, C.M.; Lai, D.Y.; Chen, Q.Y. Energy analysis of three ventilation systems for a large machining plant. *Energy Build* **2020**, *224*, 110272. [CrossRef]
7. Cheng, Y.; Zhang, S.; Huan, C.; Oladokun, M.O.; Lin, Z. Optimization on fresh outdoor air ratio of air conditioning system with stratum ventilation for both targeted indoor air quality and maximal energy saving. *Build. Environ.* **2019**, *147*, 11–22. [CrossRef]
8. Zhang, S.; Lin, Z.; Ai, Z.T.; Wang, F.H.; Cheng, Y.; Huang, C. Effects of operation parameters on performances of stratum ventilation for heating mode. *Build. Environ.* **2019**, *148*, 55–66. [CrossRef]
9. Zou, Y.; Zhao, X.W.; Chen, Q.Y. Comparison of STAR-CCM+ and ANSYS Fluent for simulating indoor airflows. *Build. Simul.* **2018**, *11*, 165–174. [CrossRef]
10. Taghinia, J.; Rahman, M.; Siikonen, T. Simulation of indoor airflow with RAST and SST-SAS models: A comparative study. *Build. Simul.* **2015**, *8*, 297–306. [CrossRef]
11. Lei, L.; Zheng, H.; Wu, B.; Xue, Y. Inverse determination of multiple heat sources' release history in indoor environments. *Build. Simul.* **2020**, *14*, 1263–1275. [CrossRef]
12. Yang, X.Q.; Wang, H.D.; Su, C.X.; Wang, X.; Wang, Y. Heat transfer between occupied and unoccupied zone in large space building with floor-level side wall air-supply system. *Build. Simul.* **2020**, *13*, 1221–1233. [CrossRef]
13. Choi, H.; Kim, H.; Kim, T. Long-term simulation for predicting indoor air pollutant concentration considering pollutant distribution based on concept of CRPS index. *Build. Simul.* **2019**, *12*, 1131–1140. [CrossRef]
14. Bazdidi-Tehrani, F.; Masoumi-Verki, S.; Gholamalipour, P.; Kiamansouri, M. Large eddy simulation of pollutant dispersion in a naturally cross-ventilated model building: Comparison between sub-grid scale models. *Build. Simul.* **2019**, *12*, 921–941. [CrossRef]

15. Sempey, A.; Inard, C.; Ghiaus, C.; Allery, C. Fast simulation of temperature distribution in air-conditioned rooms by using proper orthogonal decomposition. *Build. Environ.* **2009**, *44*, 280–289. [CrossRef]
16. Cao, S.J.; Cen, D.; Zhang, W.; Feng, Z. Study on the impacts of human walking on indoor particles dispersion using momentum theory method. *Build. Environ.* **2017**, *126*, 195–206. [CrossRef]
17. Liu, W.; Jin, M.G.; Chen, C.; You, R.Y.; Chen, Q.Y. Implementation of a fast fluid dynamics model in OpenFOAM for simulating indoor airflow. *Numer. Heat Transf. Part A Appl.* **2016**, *69*, 748–762. [CrossRef]
18. Liu, W.; Hooff, T.V.; An, Y.T.; Hu, S.; Chen, C. Modeling transient particle transport in transient indoor airflow by fast fluid dynamics with the Markov chain method. *Build. Environ.* **2020**, *186*, 107323. [CrossRef]
19. Tian, W.; Sevilla, T.A.; Zuo, W.D.; Sohn, M.D. Coupling fast fluid dynamics and multizone airflow models in Modelica Buildings library to simulate the dynamics of HVAC systems. *Build. Environ.* **2017**, *122*, 269–286. [CrossRef]
20. Liu, W.; You, R.Y.; Zhang, J.; Chen, Q.Y. Development of a fast fluid dynamics-based adjoint method for the inverse design of indoor environments. *J. Build. Perform. Simul.* **2017**, *10*, 326–343. [CrossRef]
21. Kato, S.; Murakami, S.; Kobayashi, H. New scales for assessing contribution of heat sources and sinks to temperature distributions in room by means of numerical simulation. In Proceedings of the ROOMVENT'94, Fourth International Conference on Air Distribution in Rooms, Krakow, Poland, 15–17 June 1994; pp. 539–557.
22. Sandberg, M. Ventilation effectiveness and purging flow rate—A review. In Proceedings of the International Symposium on Room Air Convection and Ventilation Effectiveness, Tokyo, Japan, 22–24 July 1992.
23. Kato, S.; Murakami, S. New ventilation efficiency scales based on spatial distribution of contaminant concentration aided by numerical simulation. *ASHRAE Trans.* **1988**, *94*, 309–330.
24. Zhang, W.R.; Kato, S.; Ishida, Y.; Hiyama, K. Calculation method of contribution ratio of indoor climate (CRI) by means of setting a uniform heat sink in natural convection air flow field. *J. Environ. Eng. (Trans. AIJ)* **2010**, *75*, 1027–1032. (In Japanese) [CrossRef]
25. Zhang, W.R.; Hiyama, K.; Kato, S.; Ishida, Y. Building energy simulation considering spatial temperature distribution for nonuniform indoor environment. *Build. Environ.* **2013**, *63*, 89–96. [CrossRef]
26. Li, X.T.; Zhao, B. Accessibility: A new concept to evaluate ventilation performance in a finite period of time. *Indoor Built Environ.* **2004**, *13*, 287–293. [CrossRef]
27. Shao, X.L.; Ma, X.J.; Li, X.T.; Liang, C. Fast prediction of non-uniform temperature distribution: A concise expression and reliability analysis. *Energy Build.* **2017**, *141*, 295–307. [CrossRef]
28. Ma, X.J.; Shao, X.L.; Li, X.T.; Lin, Y.W. An analytical expression for transient distribution of passive contaminant under steady flow field. *Build. Environ.* **2012**, *52*, 98–106. [CrossRef]
29. Cao, S.J.; Meyers, J. On the construction and use of linear low-dimensional ventilation models. *Indoor Air* **2012**, *22*, 427–441. [CrossRef]
30. Cao, S.J.; Ren, C. Ventilation control strategy using low-dimensional linear ventilation models and artificial neural network. *Build. Environ.* **2018**, *144*, 316–333. [CrossRef]
31. Ren, C.; Cao, S.J. Development and application of linear ventilation and temperature models for indoor environmental prediction and HVAC systems control. *Sustain. Cities Soc.* **2019**, *51*, 101673. [CrossRef]
32. Tian, W.; Han, X.; Zuo, W.D.; Sohn, M.D. Building energy simulation coupled with CFD for indoor environment: A critical review and recent applications. *Energy Build.* **2018**, *165*, 184–199. [CrossRef]
33. Zuo, W.D.; Chen, Q.Y. Fast and informative flow simulations in a building by using fast fluid dynamics model on graphics processing unit. *Build. Environ.* **2010**, *45*, 747–757. [CrossRef]
34. Tian, W.; Sevilla, T.A.; Zuo, W.D. A systematic evaluation of accelerating indoor airflow simulations using cross-platform parallel computing. *J. Build. Perform. Simul.* **2017**, *10*, 243–255. [CrossRef]
35. Sasamoto, T.; Kato, S.; Zhang, W.R. Control of indoor thermal environment based on concept of contribution ratio of indoor climate. *Build. Simul.* **2010**, *3*, 263–278. [CrossRef]
36. Hiyama, K.; Ishida, Y.; Kato, S. Thermal simulation: Response factor analysis using three-dimensional CFD in the simulation of air conditioning control. *Build. Simul.* **2010**, *3*, 195–203. [CrossRef]
37. Hiyama, K.; Kato, S. Integration of three-dimensional CFD results into energy simulations utilizing an advection-diffusion response factor. *Energy Build.* **2011**, *43*, 2752–2759. [CrossRef]
38. Huang, H.; Kato, S.; Hu, R.; Ishida, Y. Development of new indices to assess the contribution of moisture sources to indoor humidity and application to optimization design: Proposal of CRI(H) and a transient simulation for the prediction of indoor humidity. *Build. Environ.* **2011**, *46*, 1817–1826. [CrossRef]
39. Zhang, W.R.; Zhao, Y.N.; Xue, P.; Mizutani, K. Review and Development of the Contribution Ratio of Indoor Climate (CRI). *Energy Built Environ.* **2021**, in press. [CrossRef]
40. Xue, Y.; Zhai, Z.Q. Inverse identification of multiple outdoor pollutant sources with a mobile sensor. *Build. Simul.* **2017**, *10*, 255–263. [CrossRef]
41. Tian, X.; Li, B.Z.; Ma, Y.X.; Liu, D.; Li, Y.C.; Cheng, Y. Experimental study of local thermal comfort and ventilation performance for mixing, displacement and stratum ventilation in an office. *Sustain. Cities Soc.* **2019**, *50*, 101630. [CrossRef]

 *buildings*

Article

Improving Comfort and Health: Green Retrofit Designs for Sunken Courtyards during the Summer Period in a Subtropical Climate

Gang Han [1,2,†], Yueming Wen [1,2,†], Jiawei Leng [1,2,*] and Lijun Sun [1,2]

[1] School of Architecture, Southeast University, Nanjing 210096, China; han_gang007@163.com (G.H.); wenyueming66@163.com (Y.W.); s-lijun@163.com (L.S.)
[2] Future Underground Space Institute, Southeast University, Nanjing 210096, China
* Correspondence: jw_leng@seu.edu.cn
† Gang Han and Yueming Wen contributed equally to this work.

Abstract: The sunken courtyard has long been used in underground spaces and provides an important outdoor environment. It introduces natural elements to create a pleasant space for human activities. However, this study measured a typical sunken courtyard and found potential problems of excessive solar radiation and accumulated air pollutants in summer when at an acceptable outdoor temperature for human activities. To improve the comfort and health of a sunken courtyard, this research proposes some green retrofit designs. Firstly, compared with green wall, water and a tree, sunshade is a primary measure to improve thermal comfort. Combining sunshade, a green wall and water reduces the temperature by up to 5.6 °C in the activity zone during the hottest hour. Secondly, blocking/guiding wind walls can effectively improve the wind environment in a sunken courtyard, but only when the wind direction is close to the prevailing wind. A blocking wind wall was better at affecting velocity and uniformity, while the guiding wind wall was more efficient at discharging air pollutants. This study initially discusses the climate-adaptive design of underground spaces in terms of green, thermal comfort and natural ventilation. Designers should generally integrate above/underground and indoor/outdoor spaces using natural and artificial resources to improve comfort and health in underground spaces.

Keywords: comfort; health; green; sunken courtyard; retrofit design; climate-adaptive design

Citation: Han, G.; Wen, Y.; Leng, J.; Sun, L. Improving Comfort and Health: Green Retrofit Designs for Sunken Courtyards during the Summer Period in a Subtropical Climate. *Buildings* **2021**, *11*, 413. https://doi.org/10.3390/buildings11090413

Academic Editor: Cinzia Buratti

Received: 14 July 2021
Accepted: 13 September 2021
Published: 16 September 2021

Publisher's Note: MDPI stays neutral with regard to jurisdictional claims in published maps and institutional affiliations.

1. Introduction

Urban intensification and vertical development are driving underground spaces towards becoming important activity spaces, particularly in downtown and railway areas. The sunken courtyard (or sunken plaza), which is a long-standing spatial form, plays a significant role in enclosing underground spaces in terms of connecting to the outdoors, improving thermal comfort, introducing natural elements and reducing traffic noise and energy consumption [1,2]. However, greening is disappearing from current sunken courtyards and being replaced by artificial materials, such as tiles, metals and plastic. Vernacular architecture and scientific studies have shown that soil, water and plants effectively modify sunlight, temperature, humidity and air quality [3,4]. In the summer, artificial materials with high albedo increase the radiant temperature, causing courtyards to be unsuitable for human activities. This situation also causes potential health risks, thereby preventing people from moving around outdoors during epidemics [5] and can cause health hazards, such as the release of total volatile organic compounds (TVOCs) and particulate matters (PMs) [6]. Therefore, there are research requirements and practical needs for monitoring environmental parameters and analyzing potential deficiencies in artificial sunken courtyards. Moreover, green retrofit designs can be developed purposefully and by scientifically combining microclimate simulations.

Existing research has indicated that location, orientation, form, geometry and construction affect the physical performances of sunken courtyards, including lighting, ventilation, energy and thermal comfort [1,4,7–9]. However, these factors are determined during the early design and construction phase. Subsequently, changing them would be extremely costly or even difficult because the construction of underground spaces is highly irreversible, particularly for those with other buildings or facilities aboveground. This study uses site measurements and environmental conditions as a basis for focusing on retrofit designs to improve the thermal comfort and air quality of existing sunken courtyards, including internal layout, building surface, greening and ventilation potential. In the summer weather, excessive radiation and air pollutants have been identified by analyzing the variation and correlation of air temperature, black-bulb temperature, relative humidity, wind velocity, CO_2, TVOCs and PMs, as the two most important problems. It has been identified that unshaded areas and building materials lead to high radiant temperatures. Physiological equivalent temperature (PET) is adopted by comparing several thermal comfort indicators. Furthermore, excessive TVOCs and PMs indicate that airflow in the courtyard is unable to discharge air pollutants. A wind vector is adopted to evaluate the ventilation potential driven by ground winds.

The simulation results supported the hypothesis that there is excessive solar radiation and accumulated air pollutants. The thermal effects of a sunshade, a green wall, water and the presence of a tree, as well as the natural ventilation effects of ground winds were simulated using ENVI-met Version 4 and Cradle scSTREAM Version 14. On the one hand, sunshades are a primary measure to improve thermal comfort in the summer, reducing the physiological equivalent temperature (PET) by up to 3 °C in the activity zone during the hottest hours. Green walls and shallow water reduce PET by up to 3 °C and 0.8 °C, respectively, when in specific positions. Combining a sunshade, a green wall and water reduce PET by up to 5.6 °C in the activity zone where the measured maximum of 39.5 °C drops to below 35 °C (which is hot for PET). On the other hand, a guiding wind wall introduced ground winds and created suitable and uniform wind fields in the activity zone and discharged air pollutants. A guiding wind wall can also be combined with a green wall and controllable louvers to provide controllable and comfortable ventilation. Actual measurements and retrofit designs indicate that underground spaces with good thermal performance often disregard climate-adaptive designs in the early stages. This study initially discusses the climate-adaptive design of underground spaces in terms of green, thermal comfort and natural ventilation. Introducing additional natural elements into sunken courtyards improves physiological comfort and also visual and psychological delight. This study is a direct reference for retrofit designs and newly built courtyards, providing numerous comfortable and healthy spaces for urban activities.

2. Methodology

2.1. Field Measurements

The measured sunken courtyard used in this research is located at an underground shopping mall in the center of Nanjing, China (Figure 1). Nanjing has a subtropical monsoon climate with an annual average temperature of 15.4 °C, an annual high temperature of 39.7 °C, and it belongs to the hot summer and cold winter climate zone in China. Nanjing is known as a hot summer city in China and is experiencing significant warming. Summer in Nanjing is also becoming longer and arrives in mid-May [10,11]. However, dates where the temperatures were above expected levels were not studied because, from observations during the previous summer, people were generally reluctant to stay in the sunken courtyard being measured for long periods when it exceeded 33 °C. Furthermore, China sets 35 °C as the threshold for high temperature warnings and outdoor work allowances. In the previous 140-day summer in Nanjing, only 19 days exceeded 33 °C. Three adjacent weekends, 29–30 May and 5 June 2021 (the weather was cloudless and sunny with a high temperature of 32 °C and a Beaufort Wind Force of 2–3), were selected in advance based on weather forecasts. The study period began at 10:00 (mall opening) and ended at 18:00 (dinner) when solar radiation decreased significantly to a steady phase. Three measurements showed highly consistent patterns and characteristics, which were sufficient to support validity and representativeness. Data from 30 May were chosen because there were only a few unforeseen interferences and numerous evident interactions of environmental parameters.

Figure 1. The measured sunken courtyard in Nanjing, China.

The measurement instrument was placed on a table and its probe was approximately 1.1-m high, which was also the height of the consumers' faces when sitting for a long time in the courtyard. Table 1 summarizes the specifications of the JT integrated monitor of thermal environment and air quality. The sampling interval was set at 10 s to sensitively determine the environmental effects of changing winds. A total of 2881 data sets were analyzed over an eight-hour period (i.e., 10:00–18:00). Meanwhile, environmental parameters required by ENVI-met were measured. Air temperatures, relative humidity (RH) and building surface temperatures were measured at a height of 2 m in the center of the courtyard. Air temperatures and RH were measured using a TANDD TR-72wf thermo recorder with a precision of 0.5 °C and 5% RH. Surface temperatures were measured using a Fluke 59 mini-infrared thermometer with a temperature precision of 2 °C. Wind velocities and directions were measured at a height of 10 m from the courtyard floor. Wind velocities were measured using Testo 405i hot-wire anemometer with a velocity precision of 0.1 m/s and temperature precision of 0.5 °C.

Table 1. Specifications of JT integrated monitor of thermal environment and air quality.

Parameters	Range	Precision
Air temperature	−40–85 °C	±0.3 °C
Globe temperature	20–85 °C	±0.3 °C (20−40 °C)
Wet-bulb temperature	5–40 °C	±0.5 °C
Relative humidity	0–100% RH	±2%RH
Wind velocity	0.05–2 m/s	±(0.03 m/s + 2% reading)
$PM_{2.5}$	0–999 ug/m^3	±10% reading
CO_2	0–5000 ppm	±30 ppm
TVOCs	0.1–0.6 ppm	±10% reading

2.2. Thermal Comfort Evaluation

Outdoor thermal comfort indices can be divided into three categories: (1) cold and hot thermal stress indices based on regression analysis, such as Wet-bulb globe temperature (WGBT) and predicted mean vote amended by Jendritzky (PMV*) [12]; (2) indices based on steady-state heat transfer models, such as OUT Standard Effective Temperature (OUT-

SET*) [13] and PET [14]; and (3) indices based on unsteady heat transfer models, such as the Universal Thermal Climate Index (UTCI) [15]. Table 2 summarizes the factors involved in some commonly used indicators. Based on the literature review, PET and UTCI have been recommended and widely used in recent years. Compared with UTCI, PET is considerably more sensitive to wind velocity and solar radiation [16,17]. Moreover, PET requires wind velocity to be measured at 1.1 m, whereas UTCI requires wind velocity to be measured at 10 m. Therefore, PET is better adapted to the requirements of the current study. At any given place, outdoors or indoors, PET is equivalent to the air temperature at which, in a typical indoor setting, the heat balance of the human body (work metabolism 80 W of light activity was added to basic metabolism; heat resistance of clothing 0.9 clo) is maintained with core and skin temperatures equal to those under the conditions being assessed [14]. Table 3 shows PET ranges corresponding to different thermal sensations. In the current study, clothing insulation (0.6 clo) and metabolic rate (1.0 met or 60 W/m^2) were predicted according to field observation and *ASHRAE Standard 55-2017: Thermal Environmental Conditions for Human Occupancy* [18].

Table 2. Commonly used outdoor thermal comfort indices.

Index	Physical Factor (Climatic Conditions)				Physiological Factor (Physiological Regulation)				
	Air Temperature	Humidity	Air Velocity	Radiation	Skin Temperature	Skin Wettedness	Core Wettedness	Clothing Insulation	Metabolic Rate
WGBT	+	+	+	+					
PMV*	+	+	+	+	+	+		+	+
OUT-SET*	+	+	+	+	+	+		+	+
PET	+	+	+	+	+	+	+	+	+
UTCI	+	+	+	+	+	+	+	+	+

Table 3. Thermal sensations and corresponding PET ranges.

Thermal Sensation	Very Cold	Cold	Cool	Slightly Cool	Neutral	Slightly Warm	Warm	Hot	Very Hot
PET range (°C)	<4	4–8	8–13	13–18	18–23	23–29	29–35	35–41	>41

The RayMan Version 1.2 model was applied to calculate PET from the measurement data. The RayMan model is a diagnostic micro-scale radiation model developed by the department of meteorology and climatology at the Albert Ludwigs University of Freiburg. This model is designed to calculate radiation fluxes in simple and complex environments [19]. Mean radiation temperature (T_{mrt}) is a key index for calculating PET and can be approximately calculated from the SkyHelios model and Sky View Factor in RayMan. However, the current study adopted a considerably accurate calculation by measuring the globe temperature which is explained in *ISO 7726:1998-Ergonomics of the thermal environment-Instruments for measuring physical quantities* [20]. It is obtained using the following equation:

$$T_{mrt} = \left[\left(T_g + 273.15\right)^4 + \frac{1.1 \times 10^8 v^{0.6}}{\varepsilon D^{0.4}} \times \left(T_g - T_a\right) \right]^{1/4} - 273$$

where T_g = globe temperature (°C), v = wind velocity (m/s), ε = globe emissivity (0.95), D = globe diameter (mm) and T_a = air temperature (°C).

2.3. Simulation of Retrofit Designs

The microclimate effects of four retrofit designs were modeled using Rhinoceros software and simulated in ENVI-met software, including a water pond, a tree, wall greening and a sunshade. ENVI-met is widely used and validated for predicting microclimates for buildings and green areas. The courtyard in this study is 15 m × 23 m and has a calculation domain of 45 m × 69 m; the simulation results shown in similar medium-sized courtyards were satisfactory [21–24]. Furthermore, the ground environment of the measured courtyard is a green garden. No large buildings are within three times the scale of the courtyard and as such do not affect the microenvironment [25].

Table 4 shows the environmental parameters at the heights required by the ENVI-met simulation. These measured parameters were added to the epw.weather file. The epw. file that is recommended for ENVI-met provided geographical location, solar angle and radiation in this study; this was downloaded from https://www.ladybug.tools/epwmap/ (accessed on 12 September 2021) supported by the US Department of Energy. This study compared the simulation results of three grid sizes (i.e., 2 m, 1 m and 0.5 m, which are the smallest supported by ENVI-met) and found that a grid size of 1 m significantly balanced simulation precision and computation time. Figure 2 shows the sizes and materials of the original and modified courtyards.

Table 4. Environmental parameters required by ENVI-met simulation.

	Time	2-m Height (Courtyard Center)		10-m Height (Ground Surface)	
		Air Temperature (°C)	RH (%)	Wind Velocity (m/s)	Wind Direction (°)
	09:00	31.6	58.2	2.5	202.5
	10:00	32.2	52.8	1.6	247.5
	11:00	32.8	48.5	2.3	202.5
Hourly averaged	12:00	33.5	38.6	1.6	270.0
environmental	13:00	34.9	35.8	2.7	270.0
parameters	14:00	35.8	35.3	1.6	292.5
	15:00	37.0	31.7	2.4	225.0
	16:00	33.8	36.6	1.3	135.0
	17:00	32.7	39.1	2.1	292.5
	18:00	31.7	42.2	1.9	112.5
	19:00	31.2	48.6	2.7	112.5
Initial building temperature (09:00)			30.2 °C		

Note: 0-m height of the sunken courtyard floor.

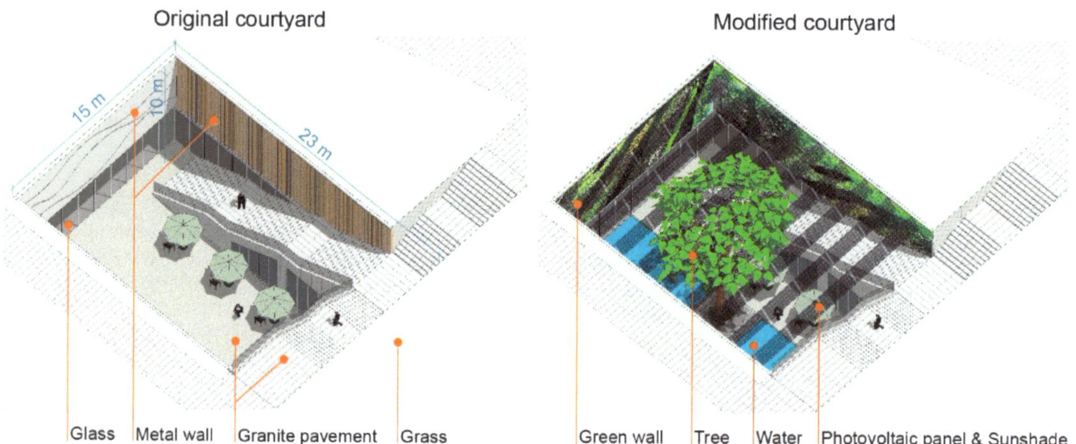

Figure 2. Models of original and modified courtyards.

Numerical models and settings were validated by comparing the simulated and measured temperatures and RH. However, the measurements of the hourly average wind velocities were 3–8 times that of those simulated (0.06 m/s–0.28 m/s), which seriously affects the calculation of PET. This result supported our assumption that the sunken courtyard was minimally affected by ground winds. Therefore, the measured wind velocity was used to modify the simulated PET. The comparative results showed that temporal trends are more consistent with errors of T_a (0.9–4.0%), RH (6.8–19.7%) and PET (6.0–15.6%)

(Figure 3). ENVI-met calculated high solar radiation and low humidity, which were also found in previous studies, but were at acceptable levels [22,26,27]. The effect of green renovation was mainly assessed by comparing the differences of the simulated PET.

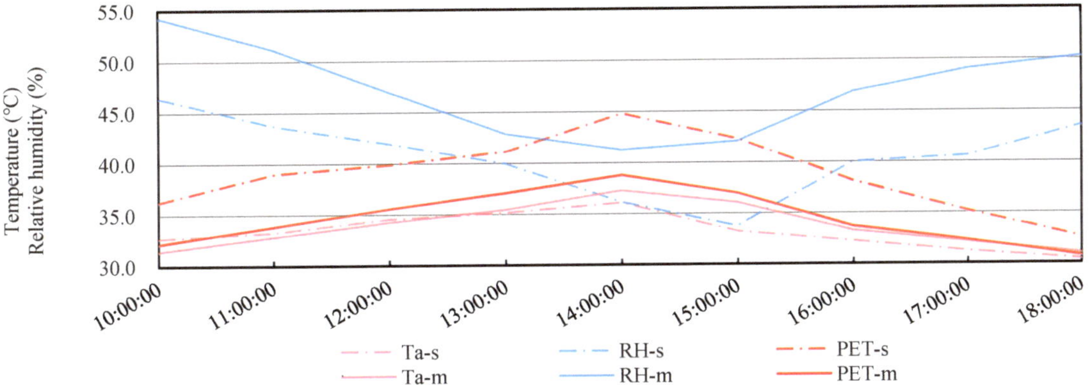

Figure 3. Model validation (s: simulation, m: measurement).

Given the shortcomings of ENVI-met in fine-scale computerized fluid dynamics (CFD) simulations, this study adopted Cradle scSTREAM Version 14, which is a specialized computational fluid dynamics software, to accurately simulate the wind environment in the sunken courtyard. Referring to Chinese standards, "GB 50736-2012, Design code for heating ventilation and air conditioning of civil buildings" [28], Table 5 summarizes the computational settings that meet the requirements of outdoor wind environment analysis.

Table 5. Computational settings of CFD simulations.

Computational Domain	
Material	Air (incompressible)
Analysis type	Turbulence flow, heat, solar radiation, steady state
Turbulence model	Renormalization group (RNG) k-ε model
Ambient temperature	31.6 °C
Initial solid temperature	30.2 °C
Boundary Condition	
Basic type	External flow (winds blowing through buildings)
Prevailing wind	2.6 m/s (10 m height), SSE
Roughness category	Urban area formed by medium-rise buildings (4–9 story) mainly
Flow boundary	Power law (inlet), static pressure (outlet)
Wall boundary	Noslip (power law or smooth)
Thermal boundary	Adiabatic (outer), heat transfer (fluid-solid), conduction (solids)
Solar Radiation	
Location	Nanjing, 32°00′ (Latitude), 118°48′ (Longitude)
Date and time	30 May, 9:00 a.m.
Solid absorptance	Architectural material

3. Results

3.1. Analysis of Measurement Data

3.1.1. Correlation Analysis

A correlation analysis was conducted using SPSS (Table 6). Firstly, PET was strongly positively correlated with T_a and T_g and strongly negatively correlated with RH. Secondly, TVOCs were strongly positively correlated with air temperature, which also led to a strong negative correlation with humidity. Thirdly, wind velocity has a limited effect on the concentration of air pollutants, which is contrary to common sense in ventilation. A positive correlation amongst CO_2, PM2.5 and TVOCs initially proved that there was air stagnation in the sunken courtyard.

Table 6. Pearson correlation analysis of environmental parameters.

	PET	T_{mrt}	T_a	T_g	RH	v	CO_2	PM2.5	TVOC
PET	1	0.982 **	0.986 **	0.988 **	−0.848 **	−0.227 **	0.374 **	0.290 **	0.495 **
T_{mrt}		1	0.941 **	0.999 **	−0.782 **	−0.108 **	0.275 **	0.273 **	0.396 **
T_a			1	0.953 **	−0.897 **	−0.262 **	0.462 **	0.300 **	0.579 **
T_g				1	−0.797 **	−0.134 **	0.295 **	0.277 **	0.416 **
RH					1	0.224 **	−0.495 **	−0.193 **	−0.596 **
v						1	−0.266 **	−0.092 **	−0.264 **
CO_2							1	0.449 **	0.854 **
PM2.5								1	0.364 **
TVOC									1

Absolute	Correlation
0.0–0.1	Not
0.1–0.3	Weak
0.3–0.5	Moderate
0.5–1.0	Strong

**. Correlation significant at 0.01 level (two-tailed).

3.1.2. Excessive Solar Radiation

Temporal trends of the measured data showed a significant concordance amongst PET, T_{mrt}, T_a and T_g (Figure 4). Humidity showed a significant negative correlation with temperature in terms of overall trends and local changes. Given the effect of solar radiation on ambient temperature and humidity, solar radiation can be deduced as the decisive factor influencing the thermal environment in the sunken courtyard. Local changes in wind velocity had evident cooling effects but did not affect the overall trend of PET (Figure 5). From 10:00 to 15:00, solar radiation led to $T_{mrt} > T_g > $ PET $ > T_a$. Before 15:00, the four temperatures steadily increased and started to exceed 35 °C (the hot threshold for PET) at approximately 11:30. After reaching a maximum of 39.5 °C (PET) at 15:00, these temperatures decreased rapidly because of self-shading of the sunken courtyard and a reduction of direct solar radiation. Additionally, T_{mrt} and T_g were occasionally lower than PET, which may be related to increasing humidity. These temperatures decreased gradually and converged after 17:00. Therefore, controlling solar radiation is the primary means of improving thermal comfort, with relative potential for increasing humidity.

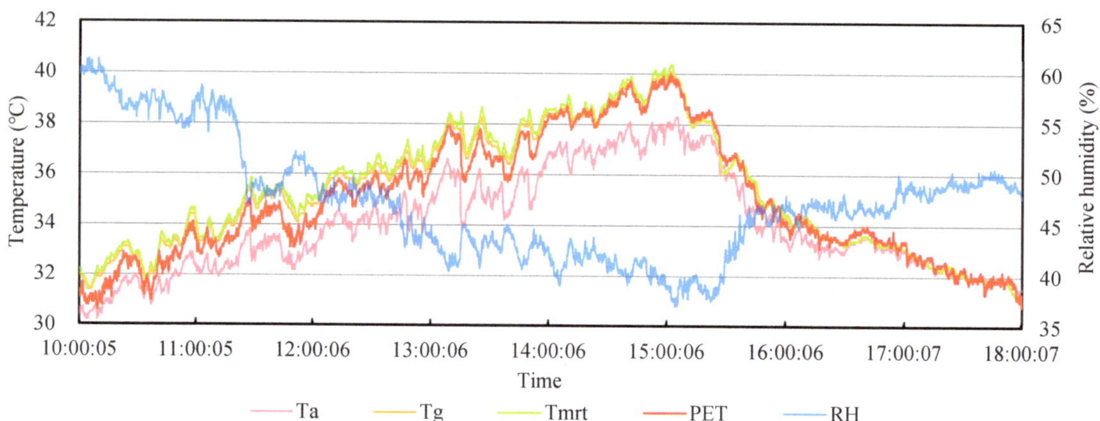

Figure 4. Temporal trend of strongly correlated thermal comfort parameters.

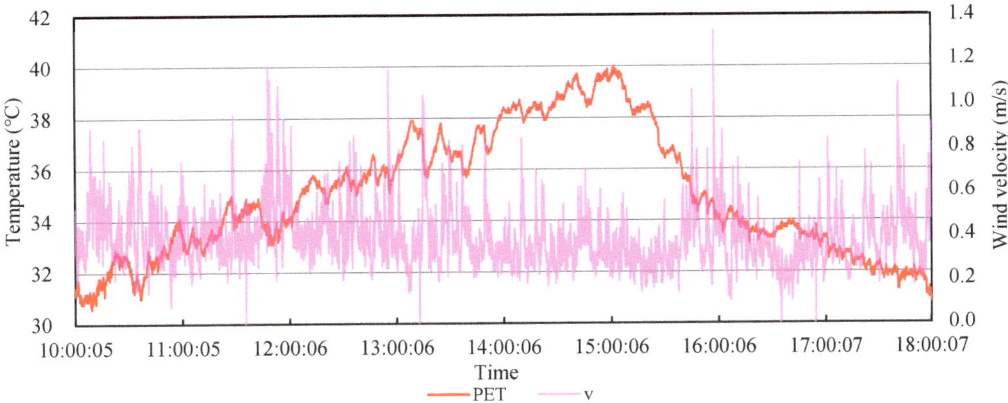

Figure 5. Temporal trend of weakly correlated thermal comfort parameters.

3.1.3. Air Stagnation and Pollutant Accumulation

Concentrations of PM2.5, CO_2 and TVOCs were consistently stable, even with evident winds (Figure 6). However, there was an unforeseen scenario: at approximately 15:30, a consumer stayed close to the instrument for 20 min smoking and eating, thereby leading to a sudden increase in these concentrations. Table 7 shows the healthy building standards in China and the US, in which PM2.5 (average 36.4 $\mu g/m^3$) and TVOC (0.65 ppm) were consistently exceeded, whilst CO_2 (522.8 ppm) was below the threshold. Specifically, TVOCs were nearly four times over the thresholds, which may be derived from the interior air in the mall and decoration materials in the courtyard. Although the sunken courtyard is an open outdoor space, it is not connected to the ground wind environment, thereby preventing the discharge of air pollutants. Additionally, fresh air from the above ground garden cannot enter the sunken courtyard. Occasional there were breezes in the measurements that were mainly caused by air convection between open doors. Therefore, the sunken courtyard should be modified to introduce substantial natural winds.

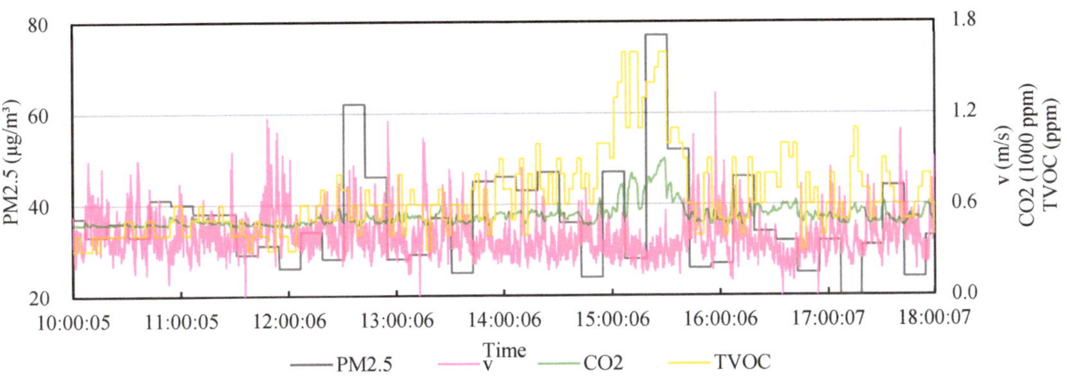

Figure 6. Temporal trends of wind velocity and air pollutants.

Table 7. Air pollutant thresholds in healthy building standards.

Air Pollutant	T/ASC 02-2016: Assessment Standard for Healthy Building	The WELL Building Standard-V2	Remark
PM2.5 ($\mu g/m^3$)	35	35	24-h mean
CO_2 (ppm)	1000	800	24-h mean
TVOC (ppm)	0.15	0.125	8-h mean

3.2. Simulation of Green Retrofit Designs

3.2.1. Retrofit Design for Thermal Comfort

Figure 7 shows the PET reduction when using a sunshade, a green wall, a tree, water and their combination in the middle section of the courtyard. Table 8 summarizes ENVI-met parameters of green retrofit measures. Firstly, the sunshade reduces PET throughout the courtyard, most noticeably immediately below the sunshade by 2–4 °C. In the activity zone (0–2 m high), it reduces the temperature by 1–3 °C. Secondly, the green wall increases the surrounding PET by 0–2 °C, which is caused by sunlight reflection and heat storage in the leaves. However, it reduces the temperature by 2–4 °C in the central activity zone. Thirdly, contrary to common experience, a tree with a dense canopy increases the PET in most of the courtyard, but reduces the temperature by 2–3 °C in the central activity zone. A comparison of 15 m and 20 m tall trees shows that the canopy should preferably be above ground level to avoid blocking surface winds. Fourthly, shallow water has a very limited cooling effect, only cooling its surroundings by less than 0.8 °C. Fifthly, combining the four designs reduces the temperature by up to 4.9 °C in the activity zone. Finally, considering the counteraction of the trees, the combination of a sunshade, a green wall and water reduces the temperature by up to 5.6 °C in the activity zone. This design integrates the cooling effects of three designs to reduce the measured maximum temperature of 39.5 °C to below 35 °C (which is hot for PET).

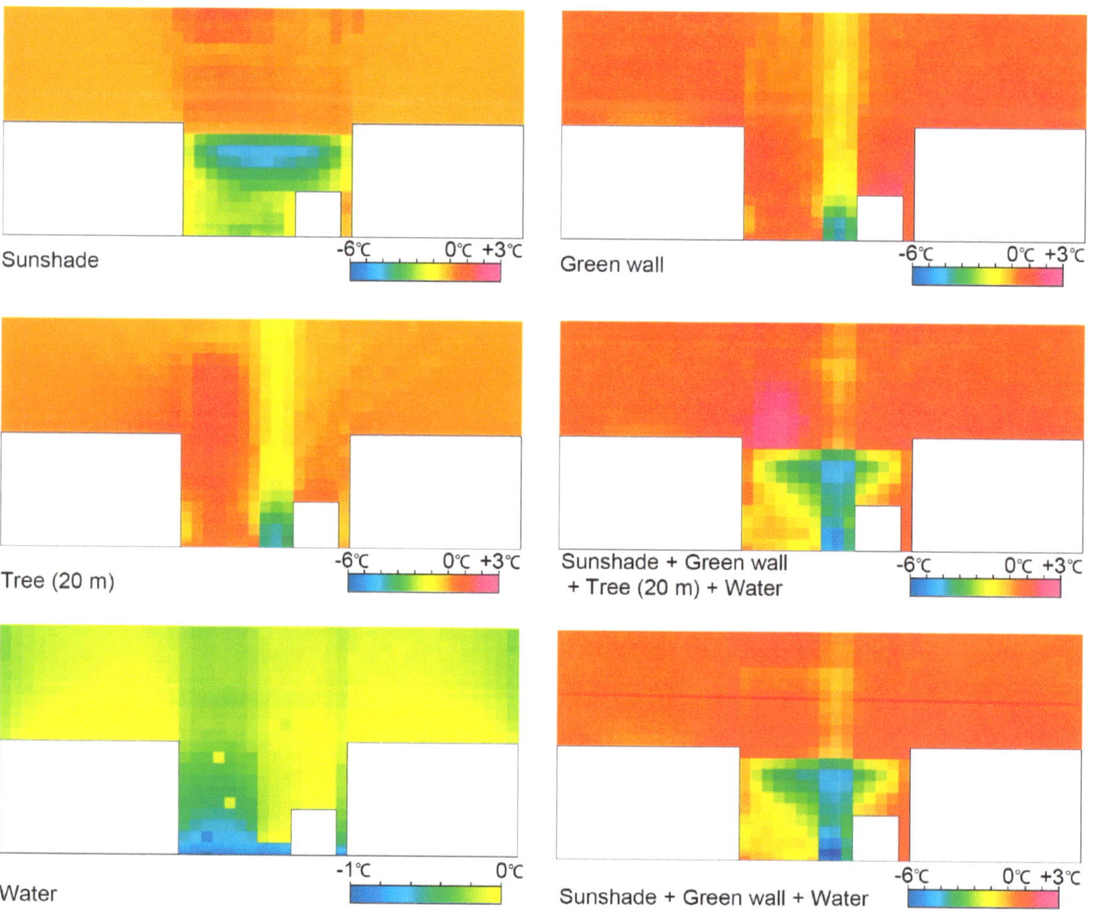

Figure 7. ENVI-met simulation results of improved designs for PET.

Table 8. ENVI-met parameters of green retrofit measures.

Green Retrofit Measure	Number	Parameter
Sunshade	000001	Single wall, SunSail, material (PV), thickness (0.2)
Green wall	01AGDS	Greenings with air gap, green and mixed substrate
Tree	0000BS	20 m height, dense, distinct crown layer, albedo (0.2), transmittance (0.3)
Water	0000WW	default

3.2.2. Retrofit Design for Natural Ventilation

Pressure difference is fundamental for organizing natural ventilation. This study aims to promote wind-pressure ventilation to introduce ground wind into the sunken courtyard and exhaust accumulated air pollutants. The two retrofit designs and original courtyard were compared by CFD simulation (Figure 8) in a prevailing SSE wind.

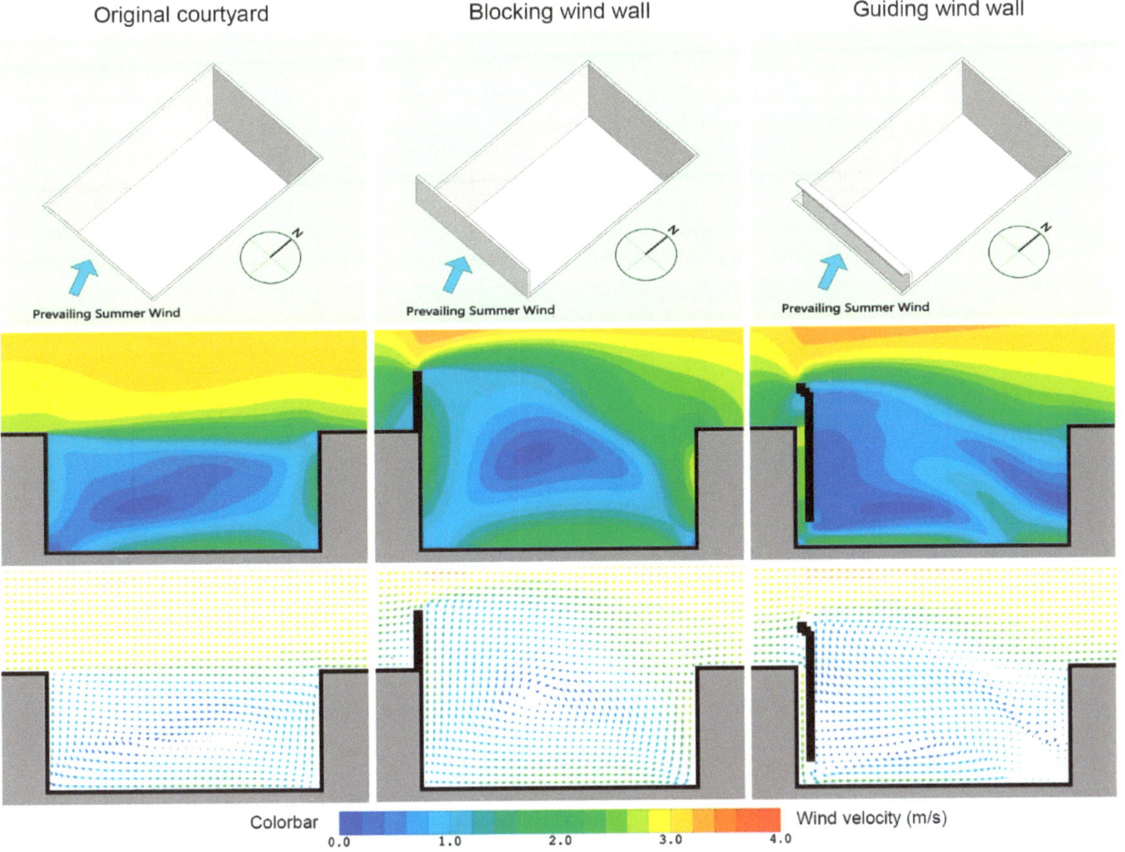

Figure 8. CFD simulation results of improved designs for natural ventilation.

First, the original courtyard had a wind velocity of 0.2–1.4 m/s in the activity zone, which was just perceptible but has a limited cooling effect in the summer. The vector diagram showed airflows circulating in the courtyard without being noticeably carried by ground winds. Therefore, the sunken courtyard becomes an outdoor ventilation "dead zone" and accumulates air pollutants, thereby confirming the uncertainties of previous measurements and ENVI-met simulation.

Thereafter, the 5-m high wall blocked ground winds to create a negative pressure zone in the courtyard and promote ground winds that sucked out underground airflows. The airflow field was effectively increased to 0.8–2.2 m/s in the activity zone. This condition was inspired by the upward wind in the surrounding area induced by the windward side of buildings in an urban wind environment. A blocking wall on the south side of the courtyard improved wind velocity, airflow exchange and the ventilation area better than the north side according to a comparison simulation. Furthermore, airflows were steadily circulated from the sunken courtyard to the ground, although many polluted airflows enter the courtyard again.

Lastly, guiding surface winds into the courtyard to create a positive pressure zone enables effective and consistent ventilation. This condition was inspired by widely used wind towers and wind deflectors in both aboveground and underground buildings. The 3-m high guiding wind wall increased the airflow field to 0.8–1.8 m/s in the activity zone. The guiding wind wall was installed 1 m from the original wall and did not encroach on a significant area. Although the guiding wind wall did not create a more uniform and faster wind than the blocking wind wall, it directed ground winds into the sunken courtyard more directly and with less circulation of polluted airflows.

In addition, both wind walls can be combined with greening or water cooling to reduce airflow temperature and upward buoyancy in the summer. Guiding wind walls can also be combined with controlled louvers in the top opening to avoid winter wind and undesirable airflow. Figure 9 shows ventilation performances of guiding wind walls for north, south and west (similar to east) ground winds with the same velocity as the prevailing wind (2.6 m/s). For wind direction close to the prevailing wind, there was a better performance on increasing wind velocity and discharging air pollutants. For wind direction contrary to the prevailing wind, it was no better than the effect of a slight discharge passage. For wind direction perpendicular to the prevailing wind, it was ineffective but had no negative effect.

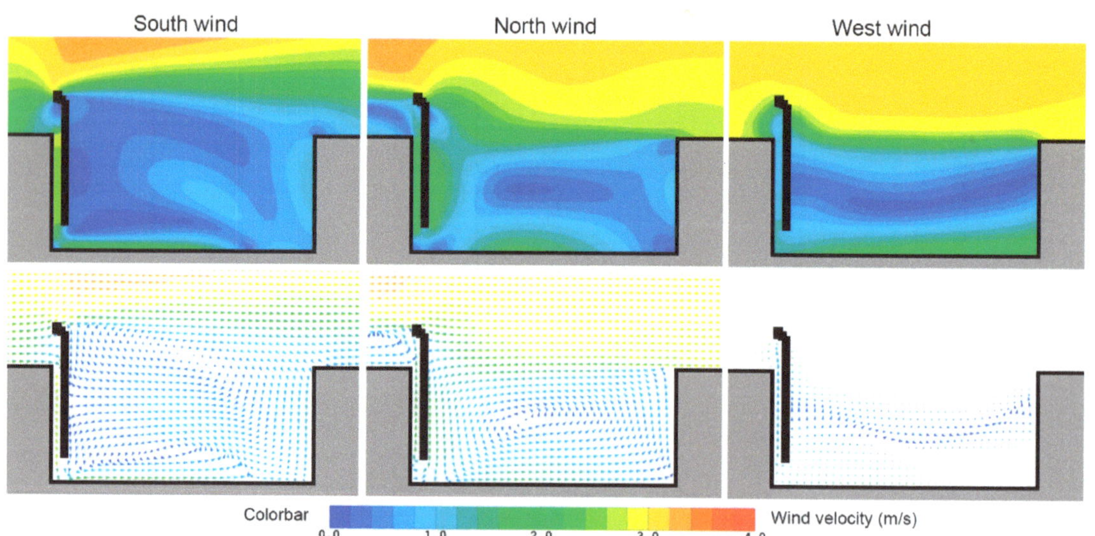

Figure 9. Ventilation performance of guiding wind wall in different wind directions.

Both wind walls can effectively improve the wind environment in the sunken courtyard, but only in wind directions close to the prevailing wind. A blocking wind wall was better at improving the wind environment for velocity and uniformity, while the guiding wind wall was more efficient at discharging air pollutants. When controlled louvers are closed, the blocking wind wall became a prevailing wind wall as well.

3.3. Summary

The main microclimatic problems in the measured sunken courtyard, that is, excessive solar radiation and accumulated air pollutants, were identified through analyzing field measurement and software simulations. First, compared to a green wall, water and a tree, sunshade was the primary measure to improve thermal comfort. When applying the findings to other sunken courtyards, position, size and angle of a sunshade should be adapted to local climate in order to balance summer shading with winter heating. Deciduous liane and photovoltaic panels can be combined with a sunshade to improve shading effects and ecological benefits. Secondly, large trees that provide partial shade for courtyards not only cause solar radiation to accumulate under the crown layer, but also weaken natural ventilation above and underground. Therefore, the traditional and empirical design concept of "more trees is better" may not apply. Appropriate plants should be chosen for different architectural interfaces to address corresponding environmental problems. Lastly, this study only applied the default setting of water in ENVI-met and found little effect on the courtyard microclimate. This requires further research on water of different depths, sizes and volumes. The design implications described above are further extended in Section 4.

4. Discussion

From this study of the thermal environment and air quality of the sunken courtyard, a classic underground space, it is clear that many empirically well and comfortably built environments conceal comfort and health problems, such as bad thermal conditions, accumulated pollutants and a lack of natural elements. In the early stages of building design, the designer should incorporate modeling and simulation to predict environmental quality. In turn, building design and environmental control should be adapted to the climatic conditions and usage requirements. This facilitates energy savings, natural access and reduces potential risks of the equipment used in the spaces. To counter global warming and the health crises, underground spaces cannot simply be artificially designed, due to the difficulties of construction and management, but should, as much as possible, be naturalized and healthy. The following section provides extended discussions on "greening underground spaces" and "adaptive designs for thermal environments".

4.1. Greening Underground Spaces

4.1.1. Multiple Benefits

Greening underground spaces can enhance vitality and positive impressions. It can also help to regulate the microclimate in order to improve thermal comfort and air quality and promote healthy urban and social environments.

Firstly, greening revitalizes underground spaces into pleasant, comfortable and lively public spaces. The public has negative impressions of underground spaces, mostly from traditional cultures or unpleasant experiences, such as gloomy, damp, stuffy or lost environments. The sunken courtyard connects to the ground, bringing in natural elements, such as light, wind and water, which naturalize the underground environment and show the weather and time of day. This condition enhances multi-sensory experiences and also eliminates the unsafe and unhealthy psychological cues of underground spaces [29]. Furthermore, sunken courtyards enhance a sense of place and identity by providing external views and attractive images, which also echoes vernacular architecture and traditional context [30].

Secondly, plants and water effectively regulate the microclimate to enhance environmental quality whilst reducing energy consumption and carbon emissions. Plants enhance thermal comfort through transpiration, the absorption of carbon dioxide and the release of oxygen through photosynthesis. Plants can also absorb air pollutants (e.g., Vanda against nicotine and Tortoise against formaldehyde). Water acts as an efficient heat sink owing to its high thermal capacity and enhances absolute humidity and thermal comfort [31]. Meanwhile, these measures reduce operational loads for environmental control equipment, thereby reducing energy consumption and carbon emissions.

Thirdly, planted spaces in urban areas are beneficial to healthy environments in numerous ways [32]. Green spaces directly contribute to public health, notably because they enable citizens to engage in various activities that reduce obesity and prevent diseases, such as cardiovascular and lung disease. Additionally, outdoor activities are promoted to reduce the risk of indoor infections during epidemics.

4.1.2. Greening Design

The use of limited underground space to create a green atmosphere requires an elaborate design. Floors, walls and ceilings are available for plant design, creating a green underground space for a surrounding, immersive experience. Firstly, plants on the floor are easily accessible to people and provide spatial orientation. Fixed planting beds provide the soil depth required for plant growth. However, it inevitably takes up the already insufficient underground space and affects space availability. Removable containers can be arranged flexibly to meet different spatial requirements and can also be changed for seasonal scenes and various experiences. Secondly, a green wall does not encroach on horizontal spaces and provides an evident greening effect and recognizability. Modular green walls can easily be removed and replaced, providing a green underground space throughout the year. A green wall is also a widely used measure for regulating the microclimate and has been studied for its effectiveness in reducing radiation, enhancing thermal comfort and air quality and blocking noise [4,33,34]. Thirdly, a green ceiling contributes to the surrounding natural atmosphere, does not encroach on spaces available for human activities and provides coherent orientations. Additionally, a green ceiling can be combined with an outdoor pergola or sunshade to create a markedly pleasant environment in sunken courtyards. Lastly, water can be used in three spatial interfaces and with plants and has superior effects on thermal comfort. Note that water can provide animated landscapes and pleasant soundscapes. Water curtain walls, fountains and water spray [35,36] increase the thermal comfort, liveliness and vitality of underground spaces, giving occupants multi-sensory experiences.

Sunlight, air and water limit plant selection in underground spaces. Sunlight determines the photosynthesis and transpiration of plants. Outdoors, natural light is introduced as much as possible whilst balancing thermal temperatures and daylight [37]. Indoors, natural light can be supplied through direct, reflected or light-guided techniques, as well as by lighting that simulates the natural spectrum. Air movement affects temperature and humidity, which affect the root health and bacterial growth of plants. Poor air movement increases indoor carbon dioxide during plant respiration and affects human health. Water can be flexibly regulated through rainwater collection, grey water and irrigation systems. Therefore, drought-loving, shade-loving and cryptogamous plants are suited to underground greening, particularly in environments lacking in natural elements. Table 9 lists some plants suitable for underground spaces.

Table 9. Some plants suitable for underground spaces.

Green Methods	Plant Types	Plants
	Arbor	Royal palm, hemp palm, ficus lyrata
Green floor	Shrub	Octophylla, monstera, rohdea
	Herbage, liane, pteridophyta	Scindapsus, begonia cathayana hemsl, ophiopogon japonicus, moss
Green wall	Herbage, liane	Euphorbia humifusa, ivy, wisteria
Green ceiling	Herbage, liane	Scindapsus, ivy, chlorophytum comosum
Green water	Herbage	Eichhornia crassipes, iris hexagonus, lotus

4.2. Adaptive Designs for Thermal Environments

Thermal environments cannot simply be designed to meet certain environmental standards. A healthy and energy-efficient thermal environment requires an adaptive design and flexible regulation. On the one hand, standard thermal comfort indicators should be modified according to the characteristics of underground spaces and the target population. On the other hand, energy-efficient-oriented thermal design should integrate indoor and outdoor environments. Moreover, energy-oriented thermal regulation should integrate indoor and outdoor environments to fully use natural and artificial cooling and heat sources. Traditional designs focus on the interior and only consider using outdoor wind and light to reduce energy consumption of indoor environmental controls. Air and light between indoor and outdoor environments can be organized to maximize energy efficiency. For example, excessive cold air can penetrate from the indoors to the courtyard through open doors rather than mechanical exhaust systems, thereby reducing the energy consumption of equipment and also enhancing the thermal comfort of a sunken courtyard. Therefore, after analyzing the thermal demands generated by the environment and people, designers should integrate the surrounding environment and resources to design buildings, facilities, greening and operations with adaptive approaches.

4.2.1. Demand Analysis for Thermal Environment and Ventilation

The two common methods of thermal comfort are the predicted mean vote (PMV) and percentage of people dissatisfied (PPD) methods developed by Fanger [38], and the adaptive model proposed by Nicol and Humphreys [39]. Adaptive thermal comfort models combine field studies and linear regression to obtain specific temperatures adapted to location, climate and population [40]. Li et al. conducted a long-term and comprehensive study of underground thermal environments in various climatic zones in China. Firstly, four climatic zones for underground engineering in China were classified. Secondly, thermal comfort models and recommended temperature ranges for the different climate zones are proposed by regression analysis through field measurement and thermal sensation survey [41,42]. The current study classified Nanjing in the humid climate zone and suggested an indoor neutral temperature (25.43 °C) and an indoor acceptable temperature (23.80–27.65 °C). Lastly, the effects of temperature, relative humidity, wind velocity and the length of time that people dwell in the space, on thermal comfort in an underground mall were analyzed. The high temperature of transitional spaces between indoors and outdoors positively affected thermal comfort and energy consumption [43]. The research has been instructive for sunken courtyards as transitional spaces. The acceptable temperatures indoors are substantially lower than outdoor temperatures. Cool indoor air can naturally exhaust from the courtyard to enhance thermal comfort during hot times. Furthermore, an adaptive temperature gradient can provide dynamic thermal pleasure from the ground, to a courtyard, to the interior [44]. This condition also reduces cardiovascular and respiratory diseases caused by sudden exposure to hot and cold.

High thermal inertia of the surrounding soil provides a stable and delayed thermal environment to reduce the effects of outdoor climate fluctuations. The air-conditioning load of underground buildings is less than that of aboveground. However, excessive humidity in the summer, which causes condensation and insufficient sunlight, increases the dehumidification load of air-conditioning systems [45,46]. Humid air indoors also moderates dry environments in the sunken courtyard on summer afternoons.

Typical indicators of indoor air quality in underground buildings include concentrations of formaldehyde, TVOCs, CO_2 and radon, which are the main risk factors for sick building syndrome (SBS) and building-related illness (BRI) [45]. The most significant air pollutants in underground malls are TVOCs and formaldehyde from indoor decorations, catering and merchandise. These pollutant concentrations are significantly correlated with wind velocity and significantly positively correlated with air temperature and relative humidity [6]. Additionally, radon is a common problem for underground building air quality. Ventilation, coating level, decorating materials and geological formation are the main influencing factors [47]. Air pollutants infiltrate and accumulate in sunken courtyards, making them an environment which has a greater health risk than indoor environments with mechanical exhaust systems. Therefore, promoting ventilation using natural or artificial resources is key to the environmental health of courtyards.

4.2.2. Adaptive Design for Sunken Courtyard Microclimates

Site layout, building form, building configuration, building surfaces, greening and amenities of the built environment design significantly affect the thermal environment and energy efficiency. Vernacular experience and passive design of traditional buildings can be adapted to contemporary architectural design using innovation [48–50]. Even the built environment determines the long-term problems faced by microenvironments. Common problems in underground spaces include high humidity, lack of sunlight and views, poor air circulation and air quality and difficulty diffusing out noise. Considerable equipment, energy and space are required to compensate for the health and comfort deficiencies caused by the built environment [51–53]. The sunken courtyard is a significant solution that ideally connects and regulates indoor and outdoor microclimates and introduces the two most important natural elements: sunlight and wind.

Several studies have extensively explored adaptive solar, wind and thermal designs. Orientation, form, geometry and movable sunshade can be designed to provide substantial shadow in the summer and receive extensive sunlight in the winter. Reducing solar radiation in the summer should be balanced with heating in the winter, because sunken courtyards may not provide sufficient sunlight and acceptable thermal comfort in the winter [2]. Li studied, measured and simulated the ventilation and lighting performance of a sunken courtyard and found that its height and width are positively correlated with a ventilation effect. Furthermore, the height–lighting relationship suggested a north–south orientation and 5-m height adapted to the research field [54]. For additional daylighting and energy saving, Omrani et al. studied how the depth (D), width (W) and length (L) of sunken courtyards affect daylighting performance inside rooms and showed a Well Index (WI) = 0.5 that refers to D/W for a square plan or D(W + L)/2 WL for a rectangular plan [9]. Additionally, some studies on building surfaces and greening have been instructive. Ghaffarianhoseini et al. evaluated the ability of unshaded courtyards to provide thermally comfortable outdoor spaces according to different design configurations and scenarios, including orientation, height and wall albedo and vegetation [55]. Similarly, high albedo surfaces, a water pond and vegetation were suggested to mitigate heat loads and moderate the microclimate of courtyards [23,56].

Natural ventilation can effectively improve thermal comfort and air quality in daily life and during epidemics and has the immense potential to control efficiency and save energy [57–59]. However, adequate and stable natural ventilation in a separate underground space is difficult to obtain. In shallow underground spaces and hot-humid climates, the generation of buoyancy by the vertical temperature difference between underground and aboveground is not evident [60,61]. Thermal pressure ventilation needs stable temperature differences, which can be obtained by active and passive designs. Firstly, solar chimneys and photovoltaic–thermal collectors can collect solar energy to create temperature differences and promote stable ventilation [62]. Secondly, earth–air–heat exchanger systems use the constant temperature of the subsoil to heat or cool outdoor air [63]. Underground water cooling and spray cooling can be applied adaptively to reduce the length of cooling ducts. Thirdly, temperature differences created by connected courtyards or atria, which are widely used in traditional group buildings, can also create stable natural ventilation. Lastly, the previously discussed wind pressure ventilation can be further enhanced by the design of wind towers, wind ducts and wind catchers [8,58]. Figure 10 presents some concepts for active and passive designs to promote natural ventilation in underground spaces.

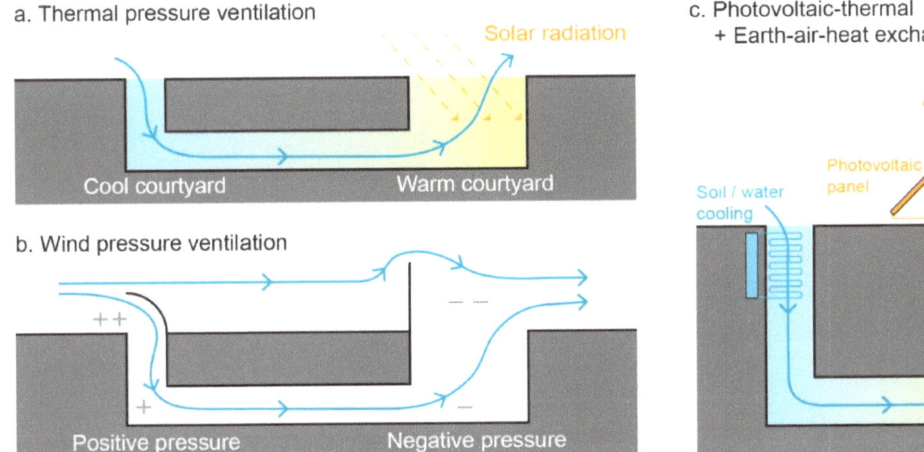

Figure 10. Active and passive techniques for natural ventilation in underground spaces.

5. Conclusions

Sunken courtyards, as major outdoor spaces underground, can provide comfortable, healthy and natural environments. Sunlight, wind and greening are the key factors to improving thermal comfort, air quality and usage experience. This study identified the existing problems of excessive summer solar radiation and accumulated air pollutants in a sunken courtyard through field measurements and simulations.

1. Measurements showed that excessive solar radiation caused PET to peak at 39.5 °C in the sunken courtyard when the weather forecast was below 33 °C. Increasing humidity and wind velocities can reduce PET to a limited extent.
2. Ground wind conditions hardly affected the sunken courtyard causing poor thermal comfort and accumulated pollutants. PMs and TVOCs consistently exceeded health standards and were likely to originate from artificial building materials, catering and merchandise.
3. In the activity zone, the sunshade most effectively reduced PET by 1–3 °C. The green wall reduced the temperature by 2–4 °C in the central zone. The shallow water only cooled its surroundings by less than 0.8 °C. Contrary to common experience, the tree with a dense canopy increased the PET in most of the courtyard but reduced the temperature by 2–3 °C in the central zone. Combining a sunshade, a green wall and water reduced the temperature by up to 5.6 °C and reduced the maximum temperature of 39.5 °C to below 35 °C (which is hot for PET).
4. Blocking/guiding wind walls can effectively improve the wind environment in the sunken courtyard, but only in wind directions close to the prevailing wind. A blocking wind wall was better at velocity and uniformity, while the guiding wind wall was more efficient at discharging air pollutants.
5. Climate-adaptive designs have immense potential and demand in underground spaces. Green-adaptive design considers the growth characteristics and environmental effects of plants to avoid a negative impact on an underground microclimate. Thermal-adaptive design balances daylight and heating to increase building self-shading in the summer and provide additional light in the winter. Wind-adaptive design integrates indoor and outdoor spaces to improve thermal comfort and passive ventilation through the use of natural and artificial sources.

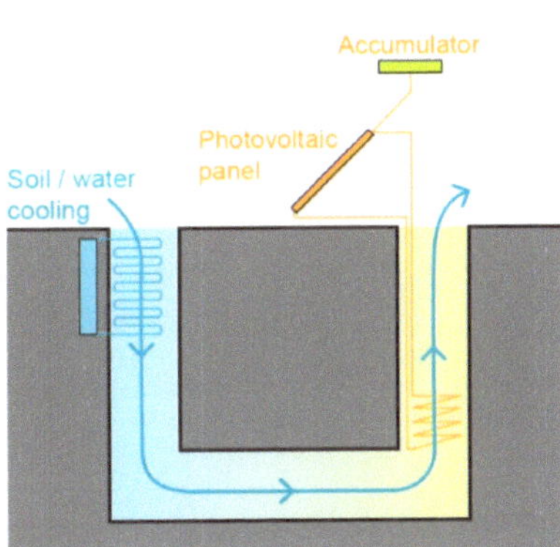

The limitations of this study are the simulation errors of solar radiation and humidity in ENVI-met, and its grid sizes and greening functions. The potential of wind pressure ventilation in underground courtyards was only discussed initially and lacked realistic simulations. Future research and engineering should combine significantly accurate environmental simulations with comprehensive adaptive design to improve the comfort and health of underground spaces.

Author Contributions: Conceptualization, J.L. and G.H.; methodology, G.H. and Y.W.; software, G.H. and Y.W.; validation, J.L. and L.S.; formal analysis, J.L.; investigation, G.H., Y.W. and L.S.; resources, J.L.; data curation, Y.W.; writing—original draft preparation, G.H. and Y.W.; writing—review and editing, J.L.; visualization, G.H., Y.W. and L.S.; supervision, J.L.; project administration, J.L.; funding acquisition, J.L. and Y.W. All authors have read and agreed to the published version of the manuscript.

Funding: This research was funded by the National Natural Science Foundation of China, grant number 52178009, the Postgraduate Research and Practice Innovation Program of Jiangsu Province, grant number KYCX20_0112, and the open fund for Jiangsu Province Key Laboratory of Intelligent Building Energy Efficiency, grant number BEE201902.

Conflicts of Interest: The authors declare no conflict of interest.

References

1. Al-Mumin, A.A. Suitability of sunken courtyards in the desert climate of Kuwait. *Energy Build.* **2001**, *33*, 103–111. [CrossRef]
2. Amirbeiki Tafti, F.; Rezaeian, M.; Emadian Razavi, S.Z. Sunken courtyards as educational environments: Occupant's perception and environmental satisfaction. *Tunn. Undergr. Space Technol.* **2018**, *78*, 124–134. [CrossRef]
3. Soflaei, F.; Shokouhian, M.; Mofidi Shemirani, S.M. Traditional Iranian courtyards as microclimate modifiers by considering orientation, dimensions, and proportions. *Front. Archit. Res.* **2016**, *5*, 225–238. [CrossRef]
4. Fatemeh, V.; Siavash, R.; Reza, A. Investigating the cooling effect of living walls in the sunken courtyards of traditional houses in Yazd. *Eur. J. Sustain. Dev.* **2016**, *5*, 27. [CrossRef]
5. Venter, Z.S.; Barton, D.N.; Gundersen, V.; Figari, H.; Nowell, M. Urban nature in a time of crisis: Recreational use of green space increases during the COVID-19 outbreak in Oslo, Norway. *Environ. Res. Lett.* **2020**, *15*, 104075. [CrossRef]
6. Tao, H.; Fan, Y.; Li, X.; Zhang, Z.; Hou, W. Investigation of formaldehyde and TVOC in underground malls in Xi'an, China: Concentrations, sources, and affecting factors. *Build. Environ.* **2015**, *85*, 85–93. [CrossRef]
7. Guo, H.; Deng, M.; Li, Y. Effect of sunken plaza on ventilation performance of underground commercial buildings. *J. South China Univ. Technol.* **2014**, *42*, 114–120.
8. Wen, Y.; Leng, J. A spatial prototype of natural ventilation in underground public spaces combining courtyard with wind tower. *IOP Conf. Ser. Earth Environ. Sci.* **2021**, *703*, 12003. [CrossRef]
9. Omrani, M.; Lian, Z.; Xuan, H. Effects of the courtyard's geometry in dig pit underground dwellings on the room's daylighting performance. *Build. Simul.* **2019**, *12*, 653–663. [CrossRef]
10. Xu, C. Nanjing Official Declares Summer Lasts 140 Days Longest in This Century and Enters Autumn. Available online: http://jsnews.jschina.com.cn/hxms/202010/t20201010_2641923.shtml (accessed on 4 September 2021).
11. Wu, L. Simulation capacity assessment of the start date and length of the four seasons in Nanjing. *Technol. Innov. Appl.* **2021**, *11*, 52–54.
12. Jendritzky, G.; Bucher, K.; Laschewski, G.; Walther, H. Atmospheric heat exchange of the human being, bioclimate assessments, mortality and thermal stress. *Int. J. Circumpolar Health* **2000**, *59*, 222–227. [PubMed]
13. de Dear, R.; Pickup, J. (Eds.) An Outdoor Thermal Comfort Index (OUT-SET*): Part I-The Model and Its Assumptions. In Proceedings of the 5th International Congress of Biometeorology and International Conference on Urban Climatology, Sydney, NSW, Australia, 1 January 1999; Macquarie University: Macquarie Park, NSW, Australia, 1999.
14. Höppe, P. The physiological equivalent temperature-a universal index for the biometeorological assessment of the thermal environment. *Int. J. Biometeorol.* **1999**, *43*, 71–75. [CrossRef] [PubMed]
15. ISB Commission 6. Universal Thermal Climate Index. Available online: http://www.utci.org/ (accessed on 8 July 2021).
16. Fröhlich, D.; Gangwisch, M.; Matzarakis, A. Effect of radiation and wind on thermal comfort in urban environments-Application of the RayMan and SkyHelios model. *Urban Clim.* **2019**, *27*, 1–7. [CrossRef]
17. Fang, Z.; Lin, Z.; Mak, C.M.; Niu, J.; Tse, K.-T. Investigation into sensitivities of factors in outdoor thermal comfort indices. *Build. Environ.* **2018**, *128*, 129–142. [CrossRef]
18. ASHRAE Standard 55-2017. *Thermal Environmental Conditions for Human Occupancy*; American Society of Heating, Refrigerating and Air-Conditioning Engineers: Atlanta, GA, USA, 2017.
19. Andres, M.; Frohlich, D. RayMan. Available online: https://www.urbanclimate.net/rayman/ (accessed on 8 July 2021).
20. ISO 7726. *Ergonomics of the Thermal Environment: Instrments for Measuring Physical Quantities*, 2nd ed.; International Organization for Standardization: Geneva, Switzerland, 1998.
21. Nasrollahi, N.; Hatami, M.; Khastar, S.R.; Taleghani, M. Numerical evaluation of thermal comfort in traditional courtyards to develop new microclimate design in a hot and dry climate. *Sustain. Cities Soc.* **2017**, *35*, 449–467. [CrossRef]
22. Forouzandeh, A. Numerical modeling validation for the microclimate thermal condition of semi-closed courtyard spaces between buildings. *Sustain. Cities Soc.* **2018**, *36*, 327–345. [CrossRef]

23. Taleghani, M.; Tenpierik, M.; van den Dobbelsteen, A.; Sailor, D.J. Heat in courtyards: A validated and calibrated parametric study of heat mitigation strategies for urban courtyards in the Netherlands. *Sol. Energy* **2014**, *103*, 108–124. [CrossRef]

24. Soares, R.; Corvacho, H.; Alves, F. Summer Thermal conditions in outdoor public spaces: A Case study in a Mediterranean climate. *Sustainability* **2021**, *13*, 5348. [CrossRef]

25. Liu, S.; Pan, W.; Zhao, X.; Zhang, H.; Cheng, X.; Long, Z.; Chen, Q. Influence of surrounding buildings on wind flow around a building predicted by CFD simulations. *Build. Environ.* **2018**, *140*, 1–10. [CrossRef]

26. Acero, J.A.; Herranz-Pascual, K. A comparison of thermal comfort conditions in four urban spaces by means of measurements and modelling techniques. *Build. Environ.* **2015**, *93*, 245–257. [CrossRef]

27. Salata, F.; Golasi, I.; de Lieto Vollaro, R.; de Lieto Vollaro, A. Urban microclimate and outdoor thermal comfort. A proper procedure to fit ENVI-met simulation outputs to experimental data. *Sustain. Cities Soc.* **2016**, *26*, 318–343. [CrossRef]

28. GB 50736-2012. *Design Code for Heating Ventilation and Air Conditioning of Civil Buildings*; Ministry of Housing and Urban-Rural Development of the People's Republic of China: Beijing, China, 2012. Available online: http://www.jianbiaoku.com/webarbs/book/16582/1663584.shtml (accessed on 12 September 2021).

29. Breton, E. Underground vegetal scenography: Benefits and installation conditions. *Procedia Eng.* **2016**, *165*, 369–378. [CrossRef]

30. Sun, L.; Wang, Y.; Leng, J. A study of museum courtyard space in eastern China. *J. Asian Archit. Build. Eng.* **2019**, *18*, 28–42. [CrossRef]

31. Imam Syafii, N.; Ichinose, M.; Kumakura, E.; Jusuf, S.K.; Chigusa, K.; Wong, N.H. Thermal environment assessment around bodies of water in urban canyons: A scale model study. *Sustain. Cities Soc.* **2017**, *34*, 79–89. [CrossRef]

32. Elliott, H.; Eon, C.; Breadsell, J.K. Improving City vitality through urban heat reduction with green infrastructure and design solutions: A systematic literature review. *Buildings* **2020**, *10*, 219. [CrossRef]

33. Tan, C.L.; Wong, N.H.; Jusuf, S.K. Effects of vertical greenery on mean radiant temperature in the tropical urban environment. *Landsc. Urban Plan.* **2014**, *127*, 52–64. [CrossRef]

34. Lau, S.-K.; Zhu, X.-F.; Lu, Z. Enhancing the acoustic absorption of vegetation with embedded periodic metamaterials. *Appl. Acoust.* **2021**, *171*, 107576. [CrossRef]

35. Farnham, C.; Emura, K.; Mizuno, T. Evaluation of cooling effects: Outdoor water mist fan. *Build. Res. Inf.* **2015**, *43*, 334–345. [CrossRef]

36. Wai, K.-M.; Xiao, L.; Tan, T.Z. Improvement of the outdoor thermal comfort by water spraying in a high-density urban environment under the influence of a future (2050) climate. *Sustainability* **2021**, *13*, 7811. [CrossRef]

37. Yu, F.; Wennersten, R.; Leng, J. A state-of-art review on concepts, criteria, methods and factors for reaching 'thermal-daylighting balance'. *Build. Environ.* **2020**, *186*, 107330. [CrossRef]

38. Fanger, P.-O. *Thermal Comfort: Analysis and Applications in Environmental Engineering*; Danish Technical Press: Copenhagen, Denmark, 1970.

39. Nicol, J.F.; Humphreys, M.A. Adaptive thermal comfort and sustainable thermal standards for buildings. *Energy Build.* **2002**, *34*, 563–572. [CrossRef]

40. Toe, D.H.C.; Kubota, T. Development of an adaptive thermal comfort equation for naturally ventilated buildings in hot–humid climates using ASHRAE RP-884 database. *Front. Archit. Res.* **2013**, *2*, 278–291. [CrossRef]

41. Li, Y.; Geng, S.; Yuan, Y.; Wang, J.; Zhang, X. Evaluation of climatic zones and field study on thermal comfort for underground engineering in China during summer. *Sustain. Cities Soc.* **2018**, *43*, 421–431. [CrossRef]

42. Li, Y.; Geng, S.; Zhang, X.; Zhang, H. Study of thermal comfort in underground construction based on field measurements and questionnaires in China. *Build. Environ.* **2017**, *116*, 45–54. [CrossRef]

43. Li, Y.; Geng, S.; Chen, F.; Li, C.; Zhang, X.; Dong, X. Evaluation of thermal sensation among customers: Results from field investigations in underground malls during summer in Nanjing, China. *Build. Environ.* **2018**, *136*, 28–37. [CrossRef]

44. Liu, S.; Nazarian, N.; Hart, M.A.; Niu, J.; Xie, Y.; de Dear, R. Dynamic thermal pleasure in outdoor environments-temporal alliesthesia. *Sci. Total. Environ.* **2021**, *771*, 144910. [CrossRef]

45. Yu, J.; Kang, Y.; Zhai, Z. Advances in research for underground buildings: Energy, thermal comfort and indoor air quality. *Energy Build.* **2020**, *215*, 109916. [CrossRef]

46. Dong, X.; Wu, Y.; Chen, X.; Li, H.; Cao, B.; Zhang, X.; Yan, X.; Li, Z.; Long, Y.; Li, X. Effect of thermal, acoustic, and lighting environment in underground space on human comfort and work efficiency: A review. *Sci. Total. Environ.* **2021**, *786*, 147537. [CrossRef]

47. Li, X.; Zheng, B.; Wang, Y.; Wang, X. A survey of radon level in underground buildings in China. *Environ. Int.* **2006**, *32*, 600–605. [CrossRef]

48. Paramita, B.; Fukuda, H.; Perdana Khidmat, R.; Matzarakis, A. Building Configuration of low-cost apartments in bandung—Its contribution to the microclimate and outdoor thermal comfort. *Buildings* **2018**, *8*, 123. [CrossRef]

49. Sun, F. Chinese Climate and vernacular dwellings. *Buildings* **2013**, *3*, 143–172. [CrossRef]

50. Santy; Matsumoto, H.; Tsuzuki, K.; Susanti, L. Bioclimatic Analysis in pre-design stage of passive house in Indonesia. *Buildings* **2017**, *7*, 24. [CrossRef]

51. Wen, Y.; Leng, J.; Yu, F.; Yu, C.W. Integrated design for underground space environment control of subway stations with atriums using piston ventilation. *Indoor Built Environ.* **2020**, *29*, 1300–1315. [CrossRef]

52. Wen, Y.; Leng, J.; Shen, X.; Han, G.; Sun, L.; Yu, F. Environmental and health effects of ventilation in subway stations: A Literature review. *Int. J. Environ. Res. Public Health* **2020**, *17*, 1084. [CrossRef] [PubMed]

53. Leng, J.; Wen, Y. Environmental standards for healthy ventilation in metros: Status, problems and prospects. *Energy Build.* **2021**, *245*, 111068. [CrossRef]

54. Li, K. Research and Optimization on Sunken Courtyard Building Natural Ventilation and Lighting in Xi'an. Master's Thesis, Xi'an University of Architecture and Technology, Xi'an, China, 2013.

55. Ghaffarianhoseini, A.; Berardi, U.; Ghaffarianhoseini, A. Thermal performance characteristics of unshaded courtyards in hot and humid climates. *Build. Environ.* **2015**, *87*, 154–168. [CrossRef]

56. Taleghani, M.; Sailor, D.J.; Tenpierik, M.; van den Dobbelsteen, A. Thermal assessment of heat mitigation strategies: The case of Portland State University, Oregon, USA. *Build. Environ.* **2014**, *73*, 138–150. [CrossRef]

57. Chenari, B.; Dias Carrilho, J.; Gameiro da Silva, M. Towards sustainable, energy-efficient and healthy ventilation strategies in buildings: A review. *Renew. Sustain. Energy Rev.* **2016**, *59*, 1426–1447. [CrossRef]

58. Mukhtar, A.; Yusoff, M.Z.; Ng, K.C. The potential influence of building optimization and passive design strategies on natural ventilation systems in underground buildings: The state of the art. *Tunn. Undergr. Space Technol.* **2019**, *92*, 103065. [CrossRef]

59. Leng, J.; Wang, Q.; Liu, K. Sustainable design of courtyard environment: From the perspectives of airborne diseases control and human health. *Sustain. Cities Soc.* **2020**, *62*, 102405. [CrossRef]

60. Khan, N.; Su, Y.; Riffat, S.B. A review on wind driven ventilation techniques. *Energy Build.* **2008**, *40*, 1586–1604. [CrossRef]

61. Liu, P.-C.; Lin, H.-T.; Chou, J.-H. Evaluation of buoyancy-driven ventilation in atrium buildings using computational fluid dynamics and reduced-scale air model. *Build. Environ.* **2009**, *44*, 1970–1979. [CrossRef]

62. Xiang, B.; Yu, T.; Yuan, Y.; Sun, L.; Cao, X. Study on influencing factor of solar chimney effect in underground space based on photovoltaic-thermal. *Trans. Chin. Soc. Agric. Eng.* **2017**, *33*, 141–147.

63. Agrawal, K.K.; Misra, R.; Agrawal, G.D.; Bhardwaj, M.; Jamuwa, D.K. The state of art on the applications, technology integration, and latest research trends of earth-air-heat exchanger system. *Geothermics* **2019**, *82*, 34–50. [CrossRef]

 *buildings*

Article

Detection of District Heating Pipe Network Leakage Fault Using UCB Arm Selection Method

Yachen Shen [1,2], **Jianping Chen** [2,3,*], **Qiming Fu** [1,2,*], **Hongjie Wu** [1,2], **Yunzhe Wang** [1,2] **and You Lu** [1,2]

1   School of Electronics and Information Engineering, Suzhou University of Science and Technology, Suzhou 215009, China; 1913041027@post.usts.edu.cn (Y.S.); hongjiewu@usts.edu.cn (H.W.); yunzhe1991@mail.usts.edu.cn (Y.W.); luyou@usts.edu.cn (Y.L.)
2   Jiangsu Province Key Laboratory of Intelligent Building Energy Efficiency, Suzhou University of Science and Technology, Suzhou 215009, China
3   School of Architecture and Urban Planning, Suzhou University of Science and Technology, Suzhou 215009, China
*   Correspondence: alan@usts.edu.cn (J.C.); fqm_1@usts.edu.cn (Q.F.)

Abstract: District heating networks make up an important public energy service, in which leakage is the main problem affecting the safety of pipeline network operation. This paper proposes a Leakage Fault Detection (LFD) method based on the Linear Upper Confidence Bound (LinUCB) which is used for arm selection in the Contextual Bandit (CB) algorithm. With data collected from end-users' pressure and flow information in the simulation model, the LinUCB method is adopted to locate the leakage faults. Firstly, we use a hydraulic simulation model to simulate all failure conditions that can occur in the network, and these change rate vectors of observed data form a dataset. Secondly, the LinUCB method is used to train an agent for the arm selection, and the outcome of arm selection is the leaking pipe label. Thirdly, the experiment results show that this method can detect the leaking pipe accurately and effectively. Furthermore, it allows operators to evaluate the system performance, supports troubleshooting of decision mechanisms, and provides guidance in the arrangement of maintenance.

Keywords: contextual bandit; linear upper confidence bound; reinforcement learning; district heating pipe network; fault detection

Citation: Shen, Y.; Chen, J.; Fu, Q.; Wu, H.; Wang, Y.; Lu, Y. Detection of District Heating Pipe Network Leakage Fault Using UCB Arm Selection Method. *Buildings* **2021**, *11*, 275. https://doi.org/10.3390/buildings11070275

Academic Editor: Brent Stephens

Received: 10 May 2021
Accepted: 18 June 2021
Published: 27 June 2021

Publisher's Note: MDPI stays neutral with regard to jurisdictional claims in published maps and institutional affiliations.

1. Introduction

Intelligent fault detection is a very important part of future city digital development [1]. District Heating (DH) is an indispensable public energy service that transfers heat from heat sources to satisfy users who live in buildings [2]. A District Heating System (DHS) is shown in Figure 1 [3]. A DHS [4] is made up of three main components: heat sources, district heating networks, and substations. The temperatures of supply water and return water of the district heating networks (DHNs) are approximately 75–90 °C and 40–50 °C, respectively [5]. District heating networks distribute heat for residential and commercial heating purposes and domestic hot water in buildings. It is necessary to create a comfortable and pleasant indoor climate and guarantee the productive and domestic water [6]. Although DHSs can bring convenience to our lives, they will malfunction for several reasons. Even heat cessation may occur in severe cases. Heat cessation will cause severe harm to social activities and inhabitants' lives. Accordingly, a reliable and online fault detection method should be applied to detect real-time faults.

Several problems may occur in the operation of DHNs as time goes on. Heat transfer causes temperature reduction. Friction of hot water against the pipe shell causes pressure losses. Both of these can lead to heat loss in the system. Moreover, pipe corrosion, insulation layer damage or fall off, leakage and other reasons may lead to pipe network malfunction. Among them, the phenomenon of hot water leaking from a damaged insulation layer or pipe shell cracking is common. Unfortunately, in existing DHNs, the observational data of leakage faults are relatively rare and cannot cover all leakage cases [7]. In order to obtain more data and realize online fault detection, it is necessary to simulate a district heating network, which can not only adapt to temperature fluctuations and user needs, but also anticipate component or entire system failures through fault detection and diagnosis (FDD). This will ultimately reduce costs for both utility companies and end-users.

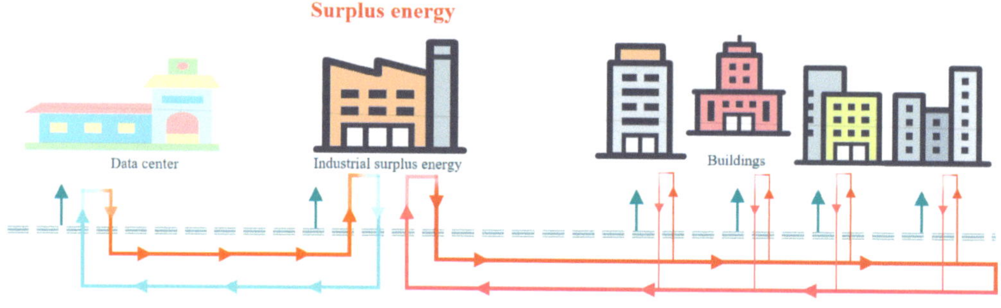

Figure 1. A District Heating System (DHS).

In general, traditional FDD methods can be divided into: (1) signal processing-based methods; (2) analytical model-based methods; and (3) knowledge-based methods. These methods can achieve certain detection accuracies and basically detect these leakage faults, but they need large modeling efforts and lack accuracy and flexibility. Furthermore, with the development of artificial intelligence technology, a hybrid detection system combined with a variety of different intelligent technologies is the development trend of intelligent fault detection [8]. In the building pipe network, the sensors of pressure and flow are typically installed at each heat source, substation, and user terminal. In order to support the operation and maintenance of district heating systems, Supervisory Control and Data Acquisition (SCADA) systems can monitor and record running data in real time. Specifically, the leakage fault of DHSs will cause slight changes in the flow and pressure parameters compared with normal circumstances, which inspires researchers to locate leakage faults through these subtle changes. Based on this point, several leakage fault detection (LFD) methods have been implemented to locate leakage points. Zhao et al. [9] studied the leakage detection and location of natural gas pipelines based on negative pressure and combined the negative pressure wave method with the signal theory to propose a solid part method to find the singularity. In order to locate the leakage, the gas velocity in the Romberg and the Dichotomy Searching methods are considered in the location formula. Jia et al. [10] provided a new pipeline leakage location method that combined the advanced FBG circumfluence strain sensor with an effective classification algorithm based on a BP neural network. Xue et al. [11] proposed a machine learning-based detection method for heating pipe network leakage by establishing a hydraulic simulation system to obtain a leakage dataset, adding a strong integrated algorithm, XGBoost, to the model, which finally outputs the leaking pipe label. Lei et al. [12] used a BP neural network to detect leakage faults both in a branch-shaped heating network and loop-shaped heating network. At the same time, he also used an SVM to make improvements. Morteza et al. [13] proposed a leakage detection method based on Artificial Neural Networks (ANNs). Berg et al. [14] proposed using a thermal image enhancement analysis method to reduce the number of false alarms in the leakage of heating networks. Most of the pipe network LFD methods discussed above focus on wave detection or supervised learning. The DHN is a closed circulation network consisting of an equal number of supply and return pipes. However, due to the cost problem, in most cases, there are not enough sensors to monitor all pipes' situations. Thus, more efficient LFD methods are necessary. A reliable LFD method for DHNs ought to have three features: high accuracy, low investment, and online and real-time detection capabilities.

Reinforcement learning is the closest to the human learning style in machine learning, which provides an alternative solution for the fault detection of a smart city energy system. Reinforcement learning is a powerful unsupervised learning method in which the environment gives agent feedback and the agent selects the optimal action with the goal of obtaining the maximum expected cumulative reward [15]. Based on the idea of "only using the current state to obtain the optimal action" in reinforcement learning, this paper proposes a method for the rapid online detection of pipe network leakage faults based on Contextual Bandit [16,17]. In this paper, reinforcement learning is used to carry out some exploratory research in the field of pipe leakage fault detection. The results show that the fault detection accuracy is improved, and our method has a high adaptability for different pipe networks. Moreover, the proposed method does not depend on the model of the problem. Based on the collected sensor data, it can perform the online training automatically. Thus, it also features low investment and online real-time detection capabilities. Three main components of this research are summarized as follows.

1. A reinforcement learning-based approach needs a large number of samples associated with all possible leakage fault situations. Unfortunately, in existing district heating networks, the observational data of leakage faults are relatively rare and cannot cover all leakage cases. Therefore, the hydraulic simulation model established by Xue [11] is used to obtain a leakage dataset [18]. In order to ensure the accuracy of the results, an impedance identification method was also used;

2. When a malfunction occurs, the overall DHN make-up water will often change greatly, which will trigger the alarm. In order to enhance system robustness, a delayed alarm triggering algorithm is applied to check the make-up flow rate regularly to indicate whether a leakage has occurred;

3. The core of the leakage fault detection model is Contextual Bandit (CB). It mainly includes model parameter synchronization, model prediction, an exploitation–exploration mechanism, real-time feature recording and storage, etc. The model uses the observed data as states to indicate agent arm selection which is a leaking pipe label.

2. Theoretical Background

2.1. Contextual Bandit

In probability theory and machine learning, the multi-armed bandit problem (also called the K- or N-armed bandit problem) is a problem to which a fixed set of finite resources should be allocated among different choices to maximize the cumulative expected payoff. This is a typical reinforcement learning problem, which reflects the exploration–exploitation tradeoff dilemma. The gambler must decide which machines to play, how many times to play each machine and in which order to play them, and whether to continue with the current machine or try a different machine. In this problem, each machine provides a random reward based on a probability distribution specific to that machine. The gambler goal is to maximize payoff through a series of lever pulls.

Figure 2 compares the relationship between the state and the action in different bandit algorithms. In the top subfigure, as a multi-armed bandit problem, the reward is only affected by the action. In the middle one, the contextual bandit problem, both states and actions can affect the reward. Additionally, in the bottom one, a full RL problem, the next state will be affected by the action, and the reward will be affected by both states and actions and it will also be delayed at the same time [19].

In a multi-armed bandit problem, the agent picks a pull from multiple arms of that bandit, and a payoff corresponding to the value between 0 and 1 is obtained. The problem is considered solved when the agent always chooses the arm that can return a relatively large payoff. In this case, the agent completely ignores the state of the environment, as there is only a single unchanging state [20].

In a contextual bandit problem, at each iteration, based on a state and the rewards of the arms played in the past, which is often represented as a d-dimensional eigenvector (contextual vector), an agent can choose which arm to play with. In the learning process, the agent has to try to collect more and more information, which is about the relationship between the state and the reward. In this way, it can choose the best arm to pull according to the current state [21].

LinUCB is an online linear method of Contextual Bandit. The basic idea is to assume a linear relation between the expected reward of an action and its contextual state, and a set of linear predictors is also used to model the representation space [22].

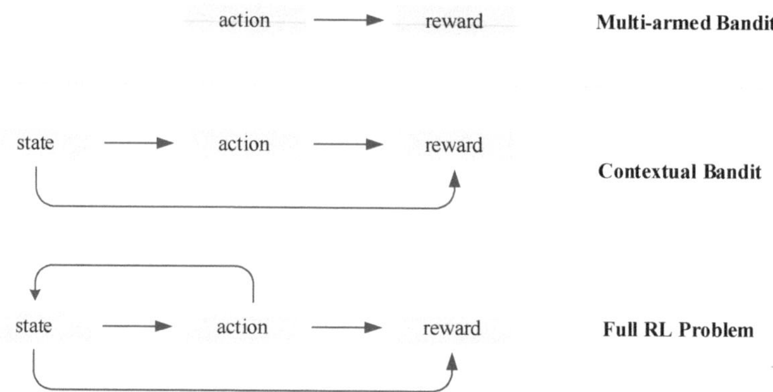

Figure 2. The relationship between state and action in different bandit algorithms.

2.2. Upper Confidence Bound (UCB)

Rather than performing the exploration by simply selecting an arbitrary action, it is better to define a heuristic information formula for the arm selection. The UCB algorithm uses uncertainty in the action-value estimations for balancing exploration and exploitation. With UCB, A_t, the action selected at time step t, is:

$$A_t = \underset{a}{\operatorname{argmax}} \left[Q_t(a) + c \sqrt{\frac{\ln t}{N_t(a)}} \right] \tag{1}$$

where t denotes the total operational numbers of each arm currently; $N_t(a)$ denotes the number of times action a has been selected before time t, and c is a confidence value that controls the level of exploration. If $N_t(a) = 0$, a is considered as the most likely action to be chosen.

Equation (1) can be thought of as being formed from two distinct parts. $Q_t(a)$ represents the exploitation part. UCB is based on the principle of "optimism in the fact of uncertainty", which basically means if you do not know which action is best, then select the one that currently seems to be the best—that is, the action with the highest estimated reward will be selected.

The second half of the equation represents the exploration, where the degree of exploration is controlled by hyper-parameter c. Effectively, this part of the equation provides a measure of the uncertainty for the action's reward estimation. If an action has not been selected frequently, or has not been selected at all, then $N_t(a)$ will be very small. Therefore, the uncertainty term will be large, which will make this action more likely to be selected. Every time an action is taken, the agent become more confident about its estimation. In this case, $N_t(a)$ increases, and so the uncertainty term decreases, which will make it less likely to be selected as exploration (although it may still be selected as the action with the highest value, mainly due to the exploitation term). When an action is not being selected, the uncertainty term will grow slowly, due to the *ln* function, whereas every time that the action is selected, the uncertainty will decrease rapidly due to the increase in $N_t(a)$. Gradually, the exploration part decreases (since $N_t(a)$ goes to infinity, the square root term goes to zero), and eventually actions are selected based only on the exploitation part [23].

3. LFD Method Based on Reinforcement Learning

3.1. Delayed Alarm Triggering Algorithm

The amount of make-up water is used to measure whether a leakage has occurred. Nevertheless, due to the influence of measurement error and environmental noise, an instantaneous peak value will inevitably appear [24]. Inspired by electric power systems, this paper uses a delayed alarm triggering algorithm to reduce the effects of these interferences.

It is not recommended to trigger the alarm signal immediately when the amount of make-up water just exceeds the threshold value G_m^* (typically set to 1% of the total circulating flow rate G_m). The maximum tolerance M (typically set to $\frac{1}{2}N_0$) acts as a buffer. When the buffer is full, the alarm will be triggered. For each check, the maximum observed value N_0 can be set according to the sampling interval. The simulation systems often set the sampling intervals to less than 10 min. Thus, waiting for several successive observations can reduce the disturbance of measurement errors and noise, which makes the algorithm more robust.

3.2. CB-Based Leakage Fault Detection

3.2.1. Fault Detection Process

Figure 3 shows the leakage fault detection process using the Contextual Bandit algorithm. Firstly, the establishment of a small DHN pipe network is used for simulating all leakage faults that can occur in the networks, which can be used to construct a dataset. Then, the simulated leakage data and real leakage data are used to train a CB model. Secondly, when the amount of the overall network make-up water exceeds the threshold, the alarm system will not be triggered until the buffer is full. It can effectively mitigate the interference of measurement errors and noise. Finally, when the leakage occurs, the observed data are sent to the CB model for the best arm selection, which is the leaking pipe label [25].

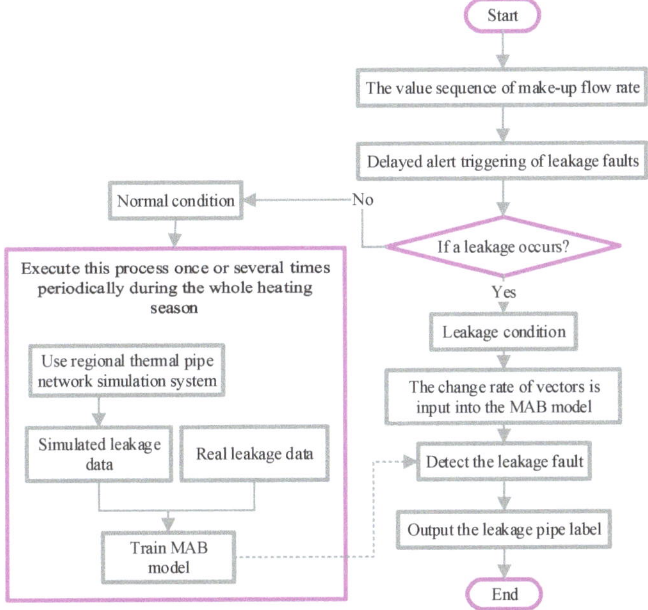

Figure 3. Flowchart of leakage fault detection for Contextual Bandit.

3.2.2. LinUCB for Disjoint Linear Model

This method solves context-independence problem in a traditional MAB and considers the influence of the state on arm selection.

We assume that the expected payoff of an arm a is linear in the d-dimensional feature $x_{t,a}$, with some unknown coefficients vector θ_a^*—namely, for all t:

$$E[r_{t,a}|x_{t,a}] = x_{t,a}^T \theta_a^* \tag{2}$$

where $x_{t,a}$ is the contextual information, i.e., the information about the eigenvectors of a pipe network. The parameters of the model are not shared among different arms. Each arm has a set of weights with a weighted relationship to the d-dimensional features to obtain the expected payoff. Considering the total loss function of multiple experiments on a single arm, we define the square loss function as follows:

$$loss(\theta) = \|c_a - D_a\theta_a\|^2 + \|I_d\theta_a\|^2 \tag{3}$$

We use the L2 regularization $\|I_d\theta_a\|^2$ to prevent overfitting, where I_d is the $d \times d$ identity matrix. By making the derivative of θ_a in Equation (3) equal to zero, we obtain:

$$\frac{\partial loss(\theta)}{\partial\theta} = 2D_a^T(D_a\theta_a - c_a) + 2I_d^T I_d\theta_a = 0 \tag{4}$$

$$\hat{\theta}_a = \left(D_a^T D_a + I_d\right)^{-1} D_a^T c_a \tag{5}$$

Let Da be a $m \times d$ matrix at trail t, where the rows correspond to m training inputs, and $c_a \in R^m$ is the corresponding reward vector. Since it is an extension of the UCB method, in addition to obtaining the expected value, we also need a confidence upper bound. Fortunately, an upper bound has been found that is at least $1 - \delta$ [26].

$$P\left\{\left|x_{t,a}^T\hat{\theta}_a - E[r_{t,a}|x_{t,a}]\right| \le \alpha\sqrt{x_{t,a}^T(D_a^T D_a + I_d)^{-1}x_{t,a}}\right\} \le 1 - \delta \tag{6}$$

where $\alpha = 1 + \sqrt{\ln(2/\delta)/2}$ is a constant, for any $\delta > 0$ as well as $x_{t,a} \in R^d$. The UCB arm selection strategy can be obtained from the inequality above. At each trial t, choose:

$$a_t \overset{def}{=} \text{argmax}_{a\in A_t}\left(x_{t,a}^T\hat{\theta}_a + \alpha\sqrt{x_{t,a}^T A_a^{-1}x_{t,a}}\right) \tag{7}$$

where $A_t \overset{def}{=} D_a^T D_a + I_d, b = D_a^T c_a$.

Ridge regression can also be seen as a Bayesian point estimate, where the posterior distribution of the coefficient vector, denoted as $p(\theta_a)$, is a Gaussian with mean $\hat{\theta}_a$ and covariance A_a^{-1}. The predicted variance of the expected payoff $x_{t,a}^T\theta_a^*$ is evaluated as $x_{t,a}^T A_a^{-1}x_{t,a}$, and then $\sqrt{x_{t,a}^T A_a^{-1}x_{t,a}}$ becomes the standard deviation. Moreover, in the information theory, the differential entropy of $p(\theta_a)$ is defined as $-\frac{1}{2}\ln\left((2\pi)^d\det A_a\right)$. The entropy of $p(\theta_a)$ is updated with the addition of the new point $x_{t,a}$. Then, it becomes $-\frac{1}{2}\ln\left((2\pi)^d\det(A_a + x_{t,a}x_{t,a}^T)\right)$. The entropy reduction in the model posterior is $\frac{1}{2}\ln(1 + x_{t,a}^T A_a^{-1}x_{t,a})$. The contribution from $x_{t,a}$ is evaluated by this quantity for model improvement. Therefore, the arm selection criterion in Equation (7) can also be seen as a tradeoff between the payoff estimation and reduction in the uncertainty in the model [27].

3.2.3. Algorithm Design

Firstly, the datasets measured by the sensors are processed by splicing into matrixes, which are regarded as different state spaces Da in CB. There are n flow sensor data $D_f^1 = \{d_1, d_2, d_3, \ldots, d_n\}_{n\times d}$, and $m-n$ pressure sensor data $D_p^2 = \{d_1, d_2, d_3, \ldots, d_{m-n}\}_{(m-n)\times d}$, which are combined to form $D_a = \left\{ d_f^1, d_f^2, d_f^3, \ldots, d_f^m, d_p^1, d_p^2, \ldots, d_p^{m-n} \right\}_{m \times d}$, modeled as states in RL. a pipes can be modeled as actions in RL. The arm selection in CB is just the action selection, which also means locating the leakage pipe in DHS, $a_t = \mathrm{argmax}_{a \in A_t} \left(x_{t,a}^T \hat{\theta}_a + \alpha \sqrt{x_{t,a}^T A_a^{-1} x_{t,a}} \right)$. The reward function is set to $c_a = Bandit(a)$, where $Bandit(a)$ corresponds to a normal distribution function between 0 and 1. Additionally, the leaking pipe corresponds to the maximum value of $Bandit(a)$. Iteratively updating the A and b values is carried out to update the weights θ. The overall algorithm is shown in Algorithm 1.

Algorithm 1. Leakage fault detection algorithm based on Contextual Bandit

---

Input : $D_a = \left\{ d_f^1, d_f^2, d_f^3, \ldots, d_f^m, d_p^1, d_p^2, \ldots, d_p^{m-n} \right\}_{m \times d}$ flow and pressure sensor data.

$\quad\quad\quad G_m$, total mass flow of replenished water

$\quad\quad\quad G_m^*$, flow threshold, set to 10% of G_m

$\quad\quad\quad N_0$, maximum number of observations in one inspection

Output: a, selected action (select a leaky pipe)

(a) loop

(b)   initialize α, $m^{(0)}$, s = false, M = 0.5 N_0

(c)   for t = 1,2, $\ldots$, N_0 do:

(d)   if s = false then:

(e)   if $G_m^{(t)} > G_m^*$: $m^{(t)} = m^{(t-1)} + 1$

(f)   if $m^{(t)} \geq M$: s = true

(g)   break

(h)   else: $m^{(t)} = m^{(t-1)} - 1$

(i)   else for t = 1,2,3, $\ldots$:

(j)   get the current contextual association vector for all arms

(k)   for all a:

(l)   if a is new:

(m)   set A_a to d-dimensional unit matrix

(n)   set b_a to d-dimensional zero vector

(o)   calculate $\hat{\theta} = A_a^{-1} b_a$

(p)   calculate arm selection probability $a_t = \mathrm{argmax}_{a \in A_t} \left(x_{t,a}^T \hat{\theta}_a + \alpha \sqrt{\left(x_{t,a}^T A_a^{-1} x_{t,a} \right)} \right)$

(q)   update $A_{a_t} = A_{a_t} + x_{t,a_t} x_{t,a_t}^T$

(r)   update $b_{a_t} = b_{a_t} + r_t x_{t,a_t}$

---

4. Experimental Analysis

4.1. Model Parameters

There are 16 users in our simulation model. The flow parameters of each pipe in the simulation model are given in Table 1.

Table 1. Heat source and user flow design information.

User-ID	Pipe Name	Mass	User-ID	Pipe Name	Mass
U0	n1	2196.4	U8	n30	458.9
U1	n23	75.7	U9	n31	118.7
U2	n24	172.7	U10	n32	49.6
U3	n25	214.4	U11	n33	183.2
U4	n26	116.2	U12	n34	187.4
U5	n27	148.3	U13	n35	67.2
U6	n28	25.3	U14	n36	143.4
U7	n29	16.9	U15	n37	218.5

We used the stratified sampling method to divide the leakage dataset into a training set and a test set. In total, 70% of the whole leakage dataset was used as the training set and the rest were used as the test set. Table 2 shows the design information and data quantity of the pipe network.

Table 2. Pipe network design information and data quantity.

Parameter	Number
Number of main pipes (supply water and return water)	78
Number of flow sensors	16
Number of pressure sensors	31
Number of data collected per pipeline leakage	100–400
Number of training sets	10,609
Number of test sets	4506

The supply water network is shown in Figure 4, and the return water network flows in the opposite direction to the supply water network, with pipe sections numbered n′ [28].

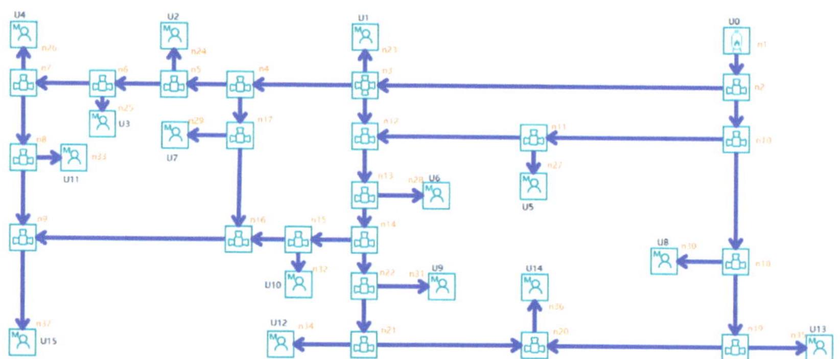

Figure 4. Supply water network in a DHS (diagram).

4.2. Evaluation Criteria

In order to implement the LinUCB algorithm for the given dataset, we first parsed each line of the input text file in the following way:

1. Strip every line of new line character;
2. Iterate over each line of input, which act as individual time steps, and split the line based on a single space. This gives us a list of 48 elements;
3. Pop the head of the list and assign it as the arm for the current step;
4. Take the remaining 47 elements and assign them to the context array for the current step.

This gives us all the parameters required to perform the online reward prediction of the arms [29].

Then, with all the required parameters, we calculated the coefficient, payout and standard deviation for each arm at every step and chose the arm with the highest payoff (i.e., upper confidence bound) as our selection. This prediction was followed by an update of matrixes "A" and "b" for the predicted arm. This was repeated for all time steps.

In order to evaluate the accuracy of our algorithm, we used the cumulative take-rate replay which at time T is defined as:

$$C(T) = \frac{\Sigma_{t=1}^{T} y_t \times 1[\pi_{t-1}(x_t) = a_t]}{\Sigma_{t=1}^{T} 1[\pi_{t-1}(x_t) = a_t]} \tag{8}$$

Whenever the selected arm is equal to the current arm, the identity function evaluates to 1 and the CTR is updated for that time stamp [30].

4.3. Analysis of Experimental Results

4.3.1. Comparison with Other Methods

At present, supervised learning methods are mainly used for pipe network fault leakage detection, such as XGBoost, forward neural networks, and support vector machines, etc. XGBoost is an optimized version of gradient tree promotion, which has had a good effect on multi-classification tasks. In the application scenario of this paper, the classification accuracy of XGBoost can reach 86.55% [11], the traditional BP network and SVM only reach 85% [12], and the accuracy of improved support vector machine can reach 92% [13].

Specifically, we consider a dynamic environment and apply the learned model to each new leakage pipe situation. It can perform experiments in the environment, obtain samples online, extract experience from the experiments, and modify the weights θ according to the tendency of past pipe damage. Our method, compared with other supervised learning methods (1) can acquire samples online without manual labeling and (2) enables online learning and has greater adaptability to new changes.

In the training phase, since reinforcement learning searches in a large space, the convergence speed is slower than that of neural networks. In practice, for example, the online learning characteristics of reinforcement learning make the speed of convergence depend on online sample acquisition. After the model stabilized, the Contextual Bandit algorithm supports the addition and deletion of dynamical candidate pipes. When a new pipe is added, it will be initialized in real time, added to the arm selections, given a certain exploration rate. In contrast, the neural network-based multi-classification approach has to add an input to the input layer, retrain the neural network, and correct the weights when a new pipe is added.

A comparison of accuracy rates of the different research methods is shown in Table 3.

Table 3. Accuracy comparison of different research methods.

Research Methods	Accuracy
XGBoost (Loop network fault) [11]	86.55%
BP (Secondary leakage fault) [12]	80%
SVM (Secondary leakage fault) [12]	85%
HKLS-SVM [13]	92%
CB (Loop network fault)	95.08%

Although the fault detection algorithm proposed in this paper has a slower convergence time than other supervised learning methods. However, our method can realize online learning, support the addition and deletion of new pipes, and improve the accuracy at the same time. This is extremely helpful for DHN companies and end-users.

4.3.2. Arm Selection Analysis

As shown in Figure 5, the UCB method is compared with the random selection method, ε–*greedy* method, and the Boltzmann method. After comparing these four methods, we found that the randomly selection method has the worst performance and the other three methods have a small difference in cumulative reward. However, the UCB strategy fluctuates less and is very stable, which not only guarantees the accumulation of rewards, but also an accurate estimate of the real rewards of each arm.

Figure 6 shows the situation of arm selection when a leakage occurs in the 15th and 21st pipes, respectively. In this case, the pipe with the largest UCB value has the maximum likelihood of being selected, followed by the pipe with the second-largest average UCB value. It validates the fact that LinUCB selects the pipe with the highest upper confidence bound. Additionally, it shows that the algorithm is correct and feasible.

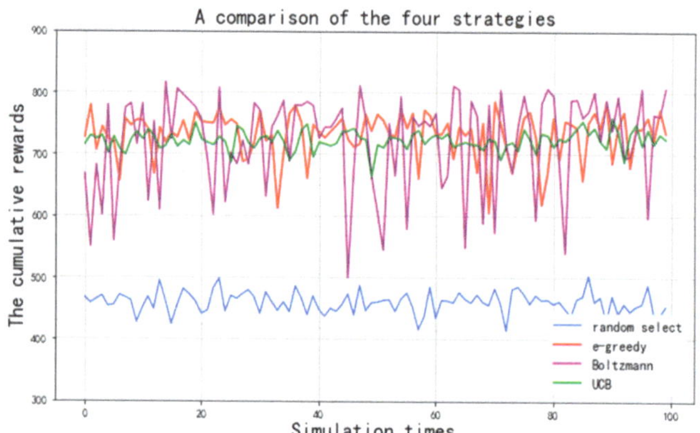

Figure 5. Comparison of four arm selection methods.

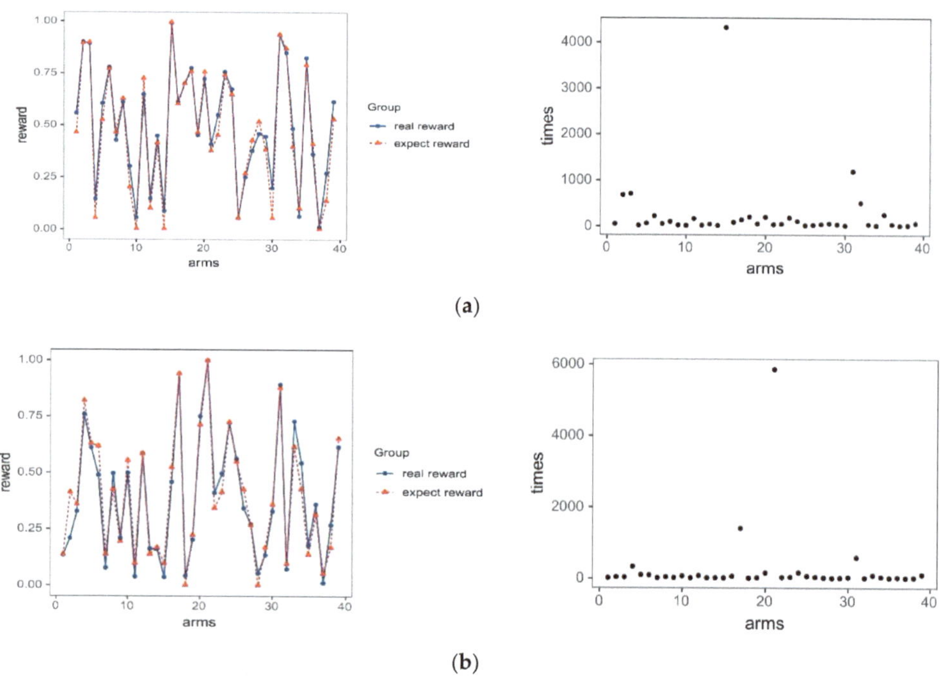

Figure 6. Left: reward prediction and right: arm selection when leakages occur in the (**a**) 15th and (**b**) 21st pipes. (On the left figure, the Y axis shows the reward for each pipe, and the X axis shows the number of pipes (i.e., the arm to be selected). The Y axis in the right figure represents the number of choices made by the agent, and the X axis represents the pipe to be selected).

4.3.3. Parametric Analysis

The explore–exploit mechanism in the algorithm is balanced by tuning the value of α. Several different mechanisms are used to identify which α value works the best. The α values are taken as 1, 0.001, 0.0001, $1/\sqrt{t}$, and 0.001/(correct-selections/10), respectively. A comparative analysis of the various α values shows how accuracy of the algorithm varies based on different α values in Figure 7.

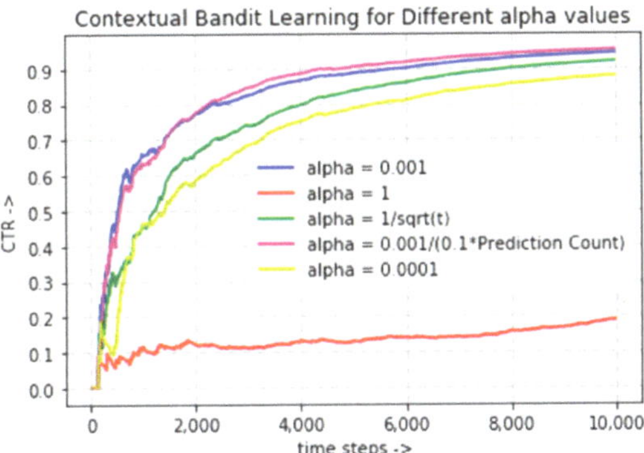

Figure 7. CTR values for different α values.

As is evident from the plots, the best CTR value is achieved with α as "0.001/(number of correct selections/10)", when the CTR value is 0.95. Subsequently, the CTR values are 0.94, 0.87, 0.91, and 0.20 for α values of 0.001, 0.0001, $1/\sqrt{t}$ and 1, respectively.

When the α value is 1, we observe that the selected counts are almost the same for all arms due to the minimal number of exploitations, thus giving it a very poor CTR value of 0.2. A significant improvement in the results can be seen when changing the value of α as a function of the square root of time step. This is mainly because the agent is regulating the degree of exploration and exploits the most out of the trained algorithm as the time passes.

A better result is obtained when using an α value of 0.001. The reason for this is that in this case we are limiting the exploration to a very small value and exploiting the most. This assures a positive outcome for the experiment.

In order to improve the CTR and achieve better results on this dataset, we assigned $\alpha = 0.001/(\text{correct-selections}/10)$ and obtained the best result so far. This increased the exploitation, particularly for the arms which gave us better results, and increased the exploration of the arms which have not been the best selections so far. This approach raised the CTR value of the LinUCB algorithm even further to 0.9508, which has been the best CTR rate of all the α values experimented with.

Moreover, it is apparent from the experimentation that the choice of α is very important as it governs the exploitation versus exploration tradeoff and can drastically improve the results, if selected wisely.

5. Conclusions

In this paper, a new leakage fault detection method based on Contextual Bandit is proposed. The entire experimental results show that the LinUCB algorithm is helpful to solve the challenge of context-independence and construct an effective pipe selection model for leakage faults.

Our method has three major advantages, including a high accuracy of 95.08%, low investment and online real-time detection capabilities. As for the low-investment problem, our method does not require additional sensors and installation of other equipment, and the current existing sensors from substations and end-users are enough to obtain data. As for the online learning and real-time detection problem, the SCADA system or IBMS system can obtain real-time data online, which can provide a software basis for rapid fault detection. At the same time, LinUCB is also an online learning algorithm. Therefore, the LinUCB algorithm just needs to collect the sensor data in real time to train an agent, which can be used to identify the right leakage pipe. However, it is different from the traditional online learning method (such as Follow the Regularized Leader (ftrl), OpenDayLight (ODL), etc.). Two main differences are as follows: (1) traditional methods try to construct a unified model for the entire scenario, while each pipe in LinUCB is a separate model. (2) Traditional online learning methods use a greedy strategy for making decisions based on the learned knowledge without exploration (but greedy strategies are often not optimal). However, on the other hand, LinUCB has a more complete exploitation and exploration mechanism, and focuses on long-term cumulative rewards, which is much more appropriate for reflecting the optimal policy.

Since DHNs are closed recurrent networks, the amount of make-up water can be an indicator to identify if a leakage occurs in the network. The delayed alarm triggering algorithm is used to trigger an alarm when a malfunction occurs and reduce the measuring errors and the interference of noise at the same time [31]. As the uptime of the DHS is much longer than the downtime, real leakage data are relatively rare. Therefore, the established model is used to simulate and obtain data for all possible leakage faults. When the leakage signal is sent, the change rate vectors from the installed sensors are input into the trained model, which can quickly output the leaking pipe label. The experimental results show that the existing number of sensors can obtain enough data to ensure the LFD model achieves an excellent detection performance, and the detection accuracy can reach 95.08%. It also shows that this method can accurately and effectively detect leaking pipes, allow operators to evaluate system performance, support troubleshooting decision mechanisms, and provide assistance in the arrangement of maintenance [32]. At the same time, we think that our method is also applicable to the leakage fault detection of air conditioning water systems.

Although our method can achieve a fairly high accuracy, it relies heavily on accurate data and suitable pre-processing. Therefore, combining the sensor fault detection method and our method can perhaps increase the robustness of FDD. In addition, based on the investigation of single agent, future work will consider using a multi-agent to detect multi-point leakage faults. Moreover, the application of reinforcement learning in fault detection and diagnosis is our research plan in the future.

Author Contributions: Conceptualization, Y.S.; Data curation, Y.S.; Formal analysis, Y.S.; Funding acquisition, J.C.; Investigation, H.W., Y.W. and Y.L.; Methodology, Y.S. and Q.F.; Project administration, J.C., H.W., Y.W. and Y.L.; Software, Y.S.; Supervision, J.C., Q.F. and Y.L.; Validation, Y.S.; Writing—original draft, Y.S.; Writing—review & editing, Q.F. All authors have read and agreed to the published version of the manuscript.

Funding: The research was funded by the National Key Research and Development Program of China, grant number 2020YFC2006602, the National Natural Science Foundation of China, grant number 62072324, 61876217, 61876121, 61772357, and the Key Research and Development Program of Jiangsu Province, grant number BE2020026.

Conflicts of Interest: The authors declare no conflict of interest.

References

1. Wang, J.; Cao, S.-J.; Yu, C.W. Development trend and challenges of sustainable urban design in the digital age. *Indoor Built Environ.* **2021**, *30*, 3–6. [CrossRef]
2. Hering, D.; Cansev, M.E.; Tamassia, E.; Xhonneux, A.; Müller, D. Temperature control of a low-temperature district heating network with Model Predictive Control and Mixed-Integer Quadratically Constrained Programming. *Energy* **2021**, *224*, 120140. [CrossRef]
3. Hussam, J.; Richard, M. Heat pipe based thermal management systems for energy-efficient data centres. *Energy* **2014**, *77*, 265–270. [CrossRef]
4. Bai, L.; Liu, H.; Yu, C.W.; Yang, Z. Optimal diameter of district heating pipe network based on the hybrid operation of distributed variable speed pumps and regulating valves. *Indoor Built Environ.* **2021**. [CrossRef]
5. Liu, G.; Zhou, X.; Yan, J.; Yan, G. A temperature and time-sharing dynamic control approach for space heating of buildings in district heating system. *Energy* **2021**, *221*, 119835. [CrossRef]
6. Li, H.; Long, E.; Zhang, Y.; Yang, H. Operation strategy of cross-season solar heat storage heating system in an alpine high-altitude area. *Indoor Built Environ.* **2020**, *29*, 1249–1259. [CrossRef]
7. Yan, K.; Chong, A.; Mo, Y. Generative adversarial network for fault detection diagnosis of chillers. *Build. Environ.* **2020**, *172*, 106698. [CrossRef]
8. Zhou, S.; O'Neill, Z.; O'Neill, C. A review of leakage detection methods for district heating networks. *Appl. Therm. Eng.* **2018**, *137*, 567–574. [CrossRef]

9. Zhao, Y.; Zhuang, X.; Min, S. A new method of leak location for the natural gas pipeline based on wavelet analysis. *Energy* **2010**, *35*, 3814–3820. [CrossRef]
10. Jia, Z.; Liang, R.; Li, H. Pipeline Leak Localization Based on FBG Hoop Strain Sensors Combined with BP Neural Network. *Appl. Sci.* **2018**, *8*, 146. [CrossRef]
11. Xue, P.; Jiang, Y.; Zhou, Z. Machine learning-based leakage fault detection for district heating networks. *Energy Build.* **2020**, *223*, 110161. [CrossRef]
12. Lei, C. Research on Leakage Fault Diagnosis of Heating Pipeline Network. Harbin Institute of Technology: Harbin, China, 2010.
13. Morteza, Z.; Mehdi, S.; Karim, S. Pipeline leakage detection and isolation: An integrated approach of statistical and wavelet feature extraction with multi-layer perceptron neural network (MLPNN). *J. Loss Prev. Process. Ind.* **2016**, *43*, 479–487. [CrossRef]
14. Berg, A.; Ahlberg, J.; Felsberg, M. Enhanced analysis of thermographic images for monitoring of district heat pipe networks. *Pattern Recognition Letters.* **2016**, *83*, 215–223. [CrossRef]
15. Wang, J.; Hou, J.; Chen, J.; Fu, Q.; Huang, G. Data mining approach for improving the optimal control of HVAC systems: An event-driven strategy. *J. Build. Eng.* **2021**, *39*, 102246. [CrossRef]
16. Martin, D.M.; Johnson, F.A. A Multiarmed Bandit Approach to Adaptive Water Quality Management. *Integr. Environ. Assess Manag.* **2020**, *16*, 841–852. [CrossRef]
17. Gittins, J.C. Bandit Processes and Dynamic Allocation Indices. *J. R. Stat. Soc. Ser. B Stat. Methodol.* **1979**, *41*, 148–164. [CrossRef]
18. Xue, P.; Jiang, Y.; Zhou, Z. *Data for: Machine Learning-Based Leakage Fault Detection for District Heating Networks*; Harbin Institute of Technology: Harbin, China, 2020. [CrossRef]
19. Savchenko, A.V.; Milov, V.R. Decision Support in Intelligent Maintenance-planning Systems Based on Contextual Multi-armed Bandit Algorithm. *Procedia Computer Science.* **2017**, *103*, 316–323. [CrossRef]
20. Sutton, R.; Barto, A. *Reinforcement Learning: An Introduction*, 2nd ed.; The MIT Press: Cambridge, MA, USA, 1992.
21. Auer, P.; Cesa-Bianchi, N.; Fischer, P. Finite-time analysis of the multiarmed bandit problem. *Mach. Learn.* **2002**, *47*, 235–256. [CrossRef]
22. Huang, K.H.; Lin, H.T. Linear Upper Confidence Bound Algorithm for Contextual Bandit Problem with Piled Rewards. In Proceedings of the Pacific-Asia Conference on Knowledge Discovery and Data Mining, Delhi, India, 11–14 May 2021; Springer: Cham, Swizterland, 2016; Volume 9652, pp. 143–155. [CrossRef]
23. BI, W.; Guo, L. Product Pricing Algorithm Based on Multi-armed Bandit. *Comput. Eng. Appl.* **2021**, *57*, 224–231.
24. Mark, A.M. Reinforcement Learning: MDP Applied to Autonomous Navigation. *Mach. Learn. Appl. Int. J.* **2017**, *4*, 1–10. [CrossRef]
25. Kim, J.; Frank, S.; Braun, J.E.; Goldwasser, D. Representing Small Commercial Building Faults in EnergyPlus, Part I: Model Development. *Buildings* **2019**, *9*, 233. [CrossRef]
26. Kim, J.; Frank, S.; Im, P.; Braun, J.E.; Goldwasser, D.; Leach, M. Representing Small Commercial Building Faults in EnergyPlus, Part II: Model Validation. *Buildings* **2019**, *9*, 239. [CrossRef]
27. Barone, G.; Buonomano, A.; Forzano, C.; Palombo, A. A novel dynamic simulation model for the thermo-economic analysis and optimisation of district heating systems. *Energy Convers. Manag.* **2020**, *220*, 113052. [CrossRef]
28. Lei, C.; Zou, P. Application of neural network in heating network leakage fault diagnosis. *J. Southeast Univ. Engl. Ed.* **2010**, *26*, 173–176.
29. Zhou, Z. *Machine Learning*; Tsinghua University Press: Beijing, China, 2016.
30. Walsh, T.J.; Szita, I.; Diuk, C. Exploring compact reinforcement-learning representations with linear regression. In Proceedings of the Twenty-Fifth Conference on Uncertainty in Artificial Intelligence, Montreal, QC, Canada, 9 May 2012.
31. Li, L.; Chu, W.; John, L.; Robert, E.S. *A Contextual-Bandit Approach to Personalized News Article Recommendation*; Association for Computing Machinery: New York, NY, USA, 2010; pp. 661–670. [CrossRef]
32. Xue, P.; Zhou, Z.; Fang, X. Fault detection and operation optimization in district heating substations based on data mining techniques. *Appl. Energy* **2017**, *205*, 926–940. [CrossRef]

MDPI AG
Grosspeteranlage 5
4052 Basel
Switzerland
Tel.: +41 61 683 77 34

Buildings Editorial Office
E-mail: buildings@mdpi.com
www.mdpi.com/journal/buildings